Lipid Mediators in the Immunology of Shock

NATO ASI Series

Advanced Science Institutes Series

A series presenting the results of activities sponsored by the NATO Science Committee, which aims at the dissemination of advanced scientific and technological knowledge, with a view to strengthening links between scientific communities.

The series is published by an international board of publishers in conjunction with the NATO Scientific Affairs Division

A	Life Sciences	Plenum Publishing Corporation
B	Physics	New York and London
C	Mathematical and Physical Sciences	D. Reidel Publishing Company Dordrecht, Boston, and Lancaster
D	Behavioral and Social Sciences	Martinus Nijhoff Publishers
E	Engineering and Materials Sciences	The Hague, Boston, Dordrecht, and Lancaster
F	Computer and Systems Sciences	Springer-Verlag
G	Ecological Sciences	Berlin, Heidelberg, New York, London,
H	Cell Biology	Paris, and Tokyo

Recent Volumes in this Series

Series A: Life Sciences

Lipid Mediators in the Immunology of Shock

Edited by

M. Paubert-Braquet

Centre de Traitement des Brules
Hopital d'Instruction des Armees Percy
Clamart, France

Associate Editors

P. Braquet

Institut Henri Beaufour
Le Plessis-Robinson, France

B. Demling

Longwood Area Trauma Center
Harvard Medical School
Boston, Massachusetts

J. R. Fletcher

St. Thomas Hospital
Nashville, Tennessee

and

M. Foegh

Georgetown University Medical Center
Washington, D.C.

Plenum Press
New York and London
Published in cooperation with NATO Scientific Affairs Division

Proceedings of a NATO Advanced Research Workshop on
Lipid Mediators in Immunology of Burn and Sepsis,
held July 20–25, 1986,
in Helsingor, Denmark

Library of Congress Cataloging in Publication Data

NATO Advanced Research Workshop on Lipid Mediators in Immunology of Burn
    and Sepsis (1986: Helsingor, Denmark)
    Lipid mediators in the immunology of shock.

    (NATO ASI series. Series A, Life sciences; v. 139)
    "Proceedings of a NATO Advanced Research Workshop on Lipid Mediators in
Immunology of Burn and Sepsis, held July 20–25, 1986, in Helsingor, Denmark"—
T.p. verso.
    "Published in cooperation with NATO Scientific Affairs Division."
    Includes bibliographies and index.
    1. Shock—Immunological aspects—Congresses. 2. Burns and scalds—Im-
munological aspects—Congresses. 3. Interleukins—Physiological effect—
Congresses. 4. Leukotrienes—Physiological effect—Congresses. I. Paubert-
Braquet, M. II. North Atlantic Treaty Organization. Scientific Affairs Division. III.
Title. IV. Series. [DNLM: 1. Burns—immunology—congresses. 2. Lipids—metab-
olism—congresses. 3. Shock—immunology—congresses. 4. Shock—physio-
pathology—congresses. QZ 140 N2791 1986]
RB150.S5N37   1986   617'.21079   87-29239
ISBN-13:978-1-4612-8245-7   e-ISBN-13:978-1-4613-0919-2
DOI: 10.1007/978-1-4613-0919-2

PREFACE

This book contains the proceedings of the ARW NATO conference on Lipid Mediators in Immunology of Burn and Sepsis held in Helsingor, Denmark, July 20-25, 1986.

This meeting brought together some of the most distinguished researchers in the fields of thermal injury, the immune system and lipid mediator biochemistry. It is well known that there is a substantial impairment of the immune response during sepsis, burn, trauma and other kinds of shock. These conditions are characterized by a massive inflammatory process which occurs during the early phase following injury. Among the various mediators released at this time are leukotrienes, thromboxane, histamine and platelet-activating factor. This latter autocoid possesses potent proinflammatory properties and together with the other mediators may account for some of the post-injury pathophysiological phenomena such as extravasation, hypotension, chemotaxis... It is of great interest to note that recently leukotrienes and platelet-activating factor have been shown to be potent mediators of the immune response. Thus, the purpose of this meeting was to bring together clinicians, immunologists and biochemists in order to examine and hopefully clarify the putative role of various lipid mediators prominent in the early stages after injury. This book is divided into the following six sections.

Section 1 provides a general overview of the physiological consequences of burn, sepsis and shock. The profound clinical and biological alterations induced by these conditions are considered. Section 2 examines the different mediators produced in response to the above pathologies with particular attention being focussed on the pharmacology of lipid mediators. The impairment of the immune response in critically ill patients is described in Section 3, while the specific mechanisms responsible for the depressed immune activity are considered in Section 4. This section includes discussions on alterations in activity of T cells, NK cells and various cytokines after shock or during other injuries such as acute myocardial infraction. Section 5 is devoted to the relationship between lipid mediators and the immune response. Accumulating evidence which suggests that platelet-activating factor and leukotrienes are important modulators of the defence system is reviewed here. Finally, in Section 6 there is a brief consideration of several new drugs which may prove to be valuable therapeutic agents in trauma, shock and related conditions.

In conclusion, the excellent contributions to this volume highlight the complex nature of this new and rapidly developing field of research. Although we are only just beginning to gain insight into the immune consequences of shock and trauma, an integrated approach to the problem such as that promoted at the NATO ARW documented here, may eventually provide (i) a better understanding of the pathophysiological events involved in thermal injury and (ii) a rationale for the development of new drugs in the treatment of shock, burn and sepsis.

The Editors

CONTENTS

I - BURN, SEPSIS AND SHOCK OVERVIEWS

2 - MEDIATORS OF BURN AND SEPSIS

3 - THE IMMUNE RESPONSE IN CRITICALLY ILL PATIENTS

4 - MECHANISMS OF IMMUNE RESPONSE IMPAIRMENT IN CRITICALLY ILL PATIENTS

5 - THE ROLE OF LIPID MEDIATORS IN THE IMMUNE RESPONSE

6 - NEW DRUGS IN SHOCK

Chapter 1

Burn, Sepsis and Shock

Overviews

BURNS AND SEPSIS : IMMUNOLOGICAL OVERVIEW

J. Guilbaud
Centre de Traitement des Grands Brûlés,
Hôpital d'Instruction des Armées Percy, 92141 Clamart

A closed world in all respects, the world of burn injured patients is poorly known. Fire has for a long time terrorized man and burns, a constant threat in our modern world, continue to make people afraid. Assessed as being 2 million per year in the United States of AMERICA and 500 000 in France, the number of burn injured patients represents every year 1 % of the population of industrialized countries. In other words, it is a subject which concerns us all.

A man or a woman's fullfilling professional, emotional and family life can quikely turn to chaos. Their hopes will disappear, all their projects will be wiped out and they will be confined to a bed, a prey to anguish and pain for weeks and sometimes months. They then realise that their skin allowed them to exist and to be autonomous, and that the destruction of this protective barrier against bacteria and temperature variations endangers them, compromises their motor functions, and may possibly affect their looks. Without it they have lost the limits of their identity and without it, they may even lose their life. Indeed, although only few pathological fields have evolved as rapidly and fortunately as that of burns, and although very severely burnt patients can be saved nowadays, the main cause of mortality is still infection.

The skin functions as an interface between all of the internal organ system of the host including the immune system and the external environment. Furthermore the skin which represents the primary target for all types of burns, is itself an important immunologic organ.
Tissular necrosis and the intense inflammatory reaction it entails lead to 2 main periods in the clinical evolution of burns :

- Firstly, a short initial period, which lasts around 48 to 72 hours and which is mainly marked by a high but transient precocious hemodynamic disequilibrium. Plasmatic exsudation and the concomitant formation of edemas indeed lead to a constant tendency for arterial pressure collapse which can be avoided only by adequate liquid ressucitation.

- Secondly a long period which ends with spontaneous or surgical re-covering of the lesions and characterised by high hypermetabolism and by a marked tendency for infection. During this period complications are not infrequent, denutrition (1) and infection occupying the forefront of the scene.

A local burn injury can indeed induce a series of effects which act to severely compromise vital host defense mechanisms (2). The presence of devitalized tissue, and the fact that the burn leads to the breakdown of a natural mechanical barrier provide an ideal environment for local infections. At the same time, the systemic failures of normal homeostatic mechanisms markedly increase the susceptibility of burn patients to severe infections (3).

According to the authors, infection represents 75 to 85 % of causes of death in severe burn injured patients.

Moreover, experimental thermal injury increases susceptibility to bacterial infections in animals which become refractory to treatment with antibiotics (4,5). Thereby, considerations of normal defense mechanisms and a review of abnormalities occuring in these patients appear to be important both in understanding the pathogenesis of sepsis and developing effective therapies to reduce mortality.

Patients with large burn injuries very rapidly reach maximal responses and have minimal ability to make further adjustments to additional stress, particularly events associated with systemic infection.

As a matter of fact, after a burn injury, profound hormonal, metabolic and biochemical disturbances occur, directly related to the size of the burns. In addition, complex immunologic alterations occur, and numerous reports have described a multitude of defects in both specific and non specific immune function following thermal injury. These alterations include impaired neutrophil function, suppressor-T-cell production, reticulo-endothelial-system (R.E.S.) depression, deficiencies in complement components, in immunoglobulins, fibronectin and other serum proteins, and the generation of circulating inhibitor factors (6,7,8).

This burn-induced immunodepression results in an enhanced susceptibility to microbial infection, in which active inhibition of the immune response plays a critical role (9,10,11).

Normal host defenses are subdivided into 2 major mechanisms, specific immune and non specific systems : (Fig. 1).

- The Specific Immune System allows a cell-mediated immunity and a humoral immunity to be present. Cell-mediated responses are effected by the T-Lymphocytes whereas the humoral immune responses depend upon the functions of the B-Lymphocytes, capable of synthesizing and secreting specific antibodies termed immunoglobulins. Under the effect of an antigenic stimulus, the T-Lymphocytes initiate a series of cellular transformations followed by mitoses which expend the pool of antigen reactive cells, the final effectors being different subpopulations of small lymphocytes which fulfil distinct functions including helper, suppressor, cytotoxic and memory activities. During these stages, T-lymphocytes elaborate lymphokines, soluble mediators which regulate the activities of other cells and act to recruit macrophages to participate in a cell-mediated immune response while

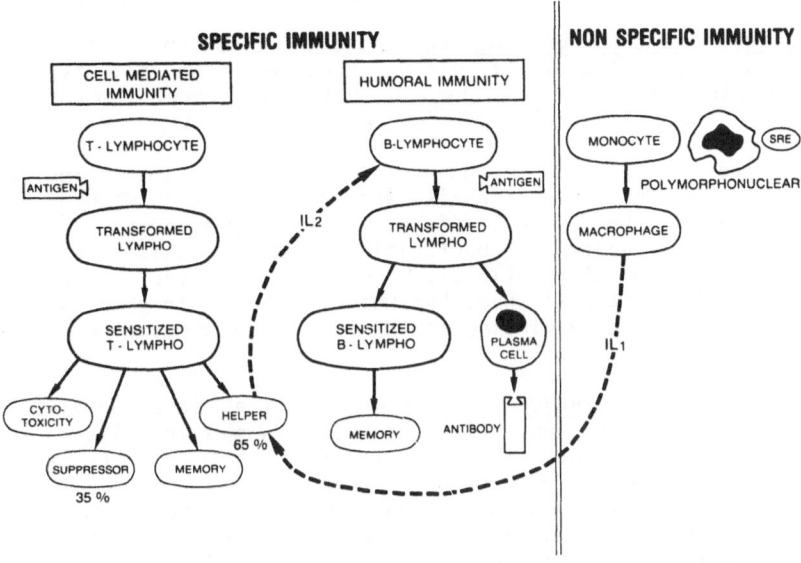

FIGURE 1

SERUM SUPPRESSIVE FACTORS

- Bacterial endotoxins
- Denatured proteins
- Immune complexes
- Histamine
- Corticosteroids
- Prostaglandins
- Leukotrienes
- PAF
- Iatrogenic ... drugs
  etc ...

FIGURE 2

helper lymphocytes act to instruct B cells to respond to antigens. The
B-lymphocyte responses parallel those of T-cells, but B-cell responses
require the instructive function of T-helper cells and macrophages.

- **Non Specific host defenses** encompass the fixed and circulating
phagocytes and a spectrum of plasma components which serve as
mediators of inflammatory reactions, including complement components,
coagulation and fibrinolytic systems, opsonins, pyrogens,
fibronectin...
    The fixed macrophages are termed the RES, and the principal
circulating phagocytes are neutrophils and monocytes which mature into
macrophages, one of the key cells of host defense mechanisms.
    The immune response to bacterial invasion consists in the
production of humoral antibodies which fix the infecting
micro-organisms, do not kill them, but which favour their phagocytosis
by neutrophils by allowing binding of the complement components on the
bacteria.

- **In Specific host defense mechanisms** a series of abnormalities
have been described in patients with extensive thermal injuries ; they
involve the T-cell system to a greater extent than the B-cell system
(12,13). The principal T-cell abnormalities (12) are a T-cell
lymphopenia with a decrease in the number of T-helper cells (OKT4) on
which the various authors do not all agree (14,15) with a significant
reduction in the T4/T8 ratio (16).
    The depression in the T-cell function (17) is marked by reduced
in vitro cytotoxicity, decreased lymphoproliferative responses to non
specific mitogens specific antigens and histoincompatible cells,
excessive T-suppressor cell activity (18,12,19,20), defective natural
killer activity and decreased lymphokine production. Data from Mannick
and Antonacci show that the ability of burn patients to produce iL2 is
significantly depressed as compared to normal controls, while burn
patients'mononuclear cells make IL1 perfectly well, and this appear to
be an important abnormality of T-lymphocyte function in burn patients
which has been detected.

    The double clinical outcome of these abnormalities of the T-cell
system is a prolonged survival of skin allografts and an inability to
elicit delayed hypersensitive skin reactions (anergic state) (21,22).
Most burn patients survive who have initial and sustained skin
reactivity. or who convert from negative to positive skin reactivity.
A very few patients survive who have initial and sustained negative or
who convert from positive to negative skin reactivity. One of the
causes of this depressed reactivity seen in patients with severe burn
injuries could be the presence of serum inhibitory factors (23,24,25).
Several factors as different as bacterial endotoxins, denatured
proteins, immune complexes, histamine, corticosteroids and especially
prostaglandins, leukotrienes and PAF have been implicated in the
depressed immune reactivity. (Fig. 2)

    For Mannick, the majority of the suppressive activity resides in
a low molecular weight polypeptide fraction of approximately 5000
Daltons. Studying the predictive value of a potentially important
inhibitory factor, the α 1 immunoregulatory globulin, Constantian has
shown that serum concentrations can be correlated with the severity of
the burn injury. For his part, Winkelstein using a T-cell colony assay
to test for serum inhibitory components (26) has shown that a high
proportion of patients with burn injuries have serum inhibitory
factors, and that there is a correlation between suppressive
activities and ultimate survival. Using the Mixed Lymphocyte Reaction
(MLR) as the assay system, serum from patients with thermal injuries

are reported to contain components, not found in normal serum, which are profoundly suppressive. Such experiments have led to the conviction that circulating factors play a major role in the lymphocyte changes observed following burn injury. Finally, very

often, a normal response can be restored in these lymphocytes by washing. It is doubtfull that serum suppressive factors can explain all the abnormalities that have been reported. But these factors are probably very important, and preliminary observations from Mannick and our Laboratory (Fig. 3) would suggest that suppressive serum from

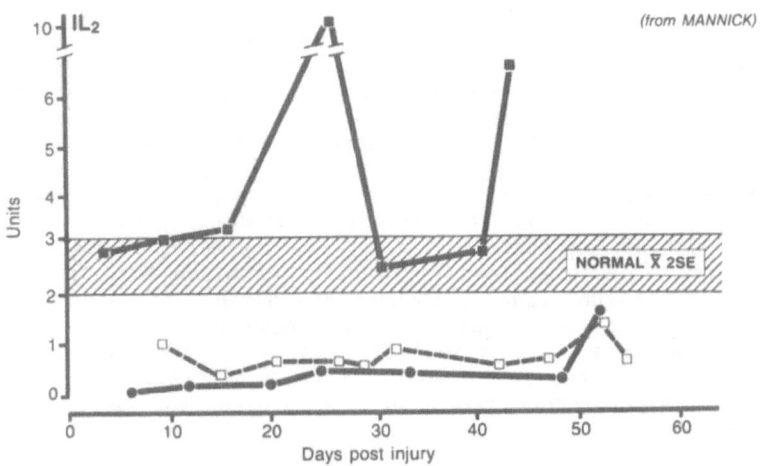

FIGURE 3.   Interleukin 2 Synthesis by Mononuclear Leukocytes for 2 Patients with Major Burn and One Patient with a little Burn.

burn patients inhibits normal lymphocytes from making IL2 in response to a phytohemagglutinin stimulus (27). An additional cause of this immunosuppression is the excessive activation of suppressor cells, both inhibitory macrophages and activated T-suppressor cells.

Finally, the potential role of metabolites of arachidonic acid as regulators of immune reactivity has been discussed by many authors, especially Ninnemann.

With regard to the B-cell system its abnormalities are less well defined. The number of circulating B-cells is not reduced by thermal injury, but in vitro B-cells from burn patients synthesize fewer antibodies than normal ones, probably due to the excessive activities of T-suppressor cells (28). The reports on the concentrations of immunoglobulins in severe burn injured patients do not always tally (29,30) : in some studies, the concentrations are decreased in a transient manner in the adult except for IgM, the rates of which remain depressed for a long time.

In addition to defects in immune reactivity, a series of abnormalities have also been described in the non specific host defense mechanisms of patients with severe burns, involving circulatory phagocytic cells, non circulating or reticulo-endo-thelial cells and serum mediators of inflammation (31).

The functions of both phagocytic cells, neutrophils and macrophages are impaired (7,32). Abnormalities in function such as neutrophil chemotaxis and chemiluminescence, degranulation and oxygen consumption, have been detected.

A very potent leukocyte activating agent appears after systemic activation of tne complement system, triggered by the thermal injury. The presence of this chemotactic peptide within the plasma results in aggregation and activation of circulating neutrophils. The sequestered agregates of these leukocytes within the pulmonary vasculature can cause direct injury of the pulmonary endothelial cells. Otherwise, neutrophil activation is attendant with an impairment of the production of superoxide anion, hydrogen peroxide and probably other toxic products.
The monocyte/macrophage is depressed in its response with a production of immunologically active factors such as plasminogen activator and Migration Inhibition Factor (M.I.F.). Furthermore, in patients with severe burns a profound loss of function of alveolar macrophages can be seen : their functional responses are diminished, as indicated by the loss of membrane depolarization responses to a variety of different agonists that stimulate them - phagocytic, chemotactic stimulation - and they lose their ability to generate superoxide anion and other oxygen products which are needed for defense against microbial organisms.

The non circulating phagocytic cells act to remove cellular debris, bacteria, denatured proteins and activated clotting factors. The activity of the RE System is markedly depressed by burn injuries as shown by Saba (33) and different authors.
The defective clearing appears to result from a deficit of a serum factor, called α-2-surface-binding glycoprotein or Fibronectine, which is capable of binding a variety of materials, thus permitting their phagocytosis primarily by Kupffer cells in the liver (34).

In severely burned patients, there is both a decrease in the concentration of Fibronectin and a progressive deterioration of Reticulo-Endothelial function, and studies in animals show a close correlation between the activity of the R.E. System and Serum Fibronectin concentrations.

In the cellular system of immunitary defenses, the cells of the epidermis also have a role to play.

The Langerhans cells, which bear FC and C3 receptors (35,36) and class II molecules on their surface, have been demonstrated functionally to exhibit antigen presenting cell potential in vivo and in vitro. Sauder et al. (37) investigating whether Langerhans cells populations were capable of IL1 production, found that epidermal cell preparations secrete a substance with IL1-like activity. But the removal of the Langerhans cells did not reduce the titer of this activity. The keratinocytes are therefore considered as a major producer of IL1-like activity associated with the skin. The skin-derived molecule termed Epidermal cell-derived Thymocyte Activating Factor (ETAF) is chemically and biologically the same as macrophage-derived IL1 (38).

With regard to antigen presenting cell (APC) function, there is probably a division of labor : the Langerhans cell represents the only Ia positive (39) cell in normal epidermis and the keratinocyte can provide a second signal with ETAF-IL1. Thus the skin is to be considered as actively involved in the immune modifications observed in burn patients. ETAF IL1 exhibits lymphocyte stimulating activity, is responsible for neutrophil release from the bone marrow stores, and can mediate a specific neutrophil degranulation so as to release lysozyme and lactoferrin. This hormone-like molecule can also stimulate fibroblasts to enhance the rate of prostaglandins and collagenase biosynthesis. Another dysfunction of a host defense mechanism which can contribute to organ failure is an abnormal activation of the complement system (40). Burn injury is associated with abnormalities of complement and complement activity (41,42,43,44). Complement components serve as important mediators of inflammatory responses, and it is now demonstrated that the complement system is activated by thermal injury and sepsis, thereby releasing biologically potent activation products (45). A number of immunologically active cells have receptors for some or all of these components (46). Both classical and alternate pathway activation can occur.

The mechanism of action could be as follows : tissue damage can activate the Hageman factor or intrinsic coagulation system and fibrinolysis. In their turn, Hageman factor activation products (and plasmin) can directly activate C1 (47) thus providing classical pathway activation. (Fig. 4) But plasmin can also directly activate

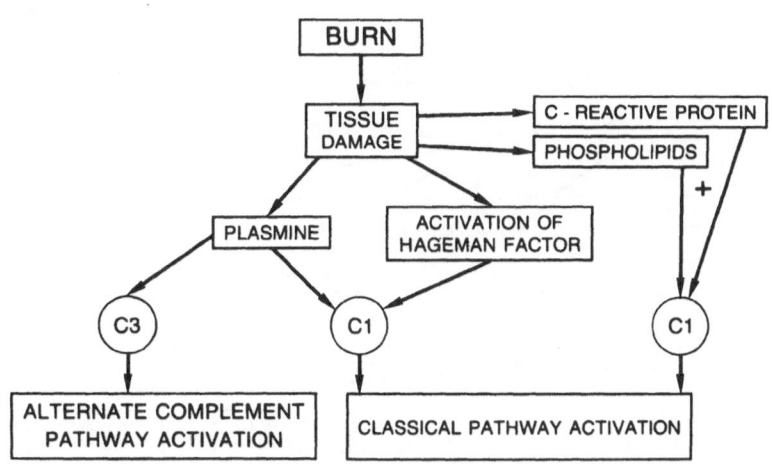

FIGURE 4

C3, thus providing alternate complement pathway activation. At the same time, with levels much higher than normally C-reactive protein bound to the phospholipids of damaged tissue can activate the classical complement pathway via C1 and may provide a continuing stimulus during burn injury. Burned animals experimentation by Burke, Donelan and Jeffrey has shown that a dramatic and immediate fall in alternate complement pathway occured, proportional to the size of the burn. 75 % of the depletion occured in the first 15 minutes and activity was nil one hour post-burn. Under these circumstances, this depletion cannot be due to a decreased synthesis, malnutrition, or to the synthesis of a de novo inhibitor, and it suggests complement activation. Evidence that the complement deficiency is due to activation include the presence of complement cleavage products and the occurence of immune complexes. Burn initiated complement cleavage products (48) can produce a series of changes of cell-mediated immune functions, some of which are beneficial while others are harmful : C3a and C5a increase vascular permeability ; C3a binds preferentially to eosinophils and basophils, and induces histamine release from basophils and mast cells. It inhibits specific antibody response.

C3b induces the release of enzymes from neutrophils, including histaminase. C5a promotes the aggregation and adhesiveness of circulating neutrophils, resulting in the formation of microemboli which are sequestered in the small vessels of the pulmonary circulation ; these microemboli can induce both pulmonary hypertension and arterial hypoxemia. (Fig. 5)

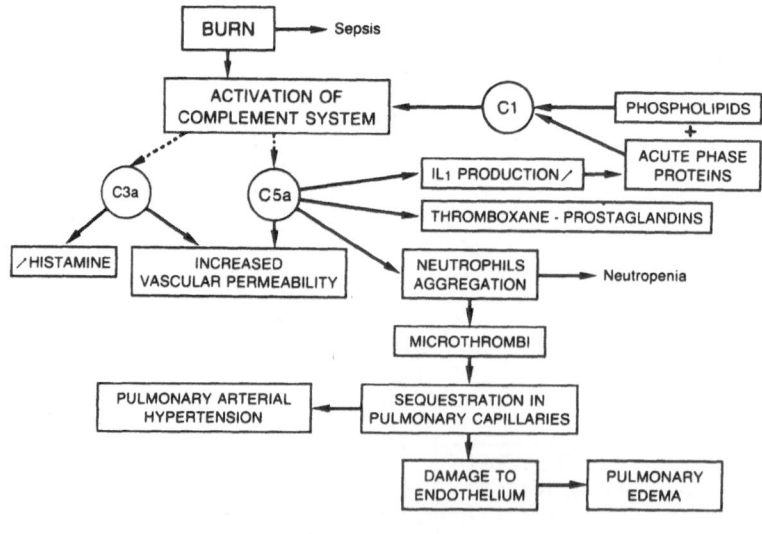

FIGURE 5

C5a induces secretion of lysosomal enzymes from macrophages and neutrophils, and induces interleukin 1 production from macrophages and the release of thromboxane and other prostaglandins ; it enhances cell-mediated and humoral immunity working at the level of the macrophage. Via macrophage production of IL1 (Fig. 6) B-cells may be stimulated to increase antibody production and T-cells may be

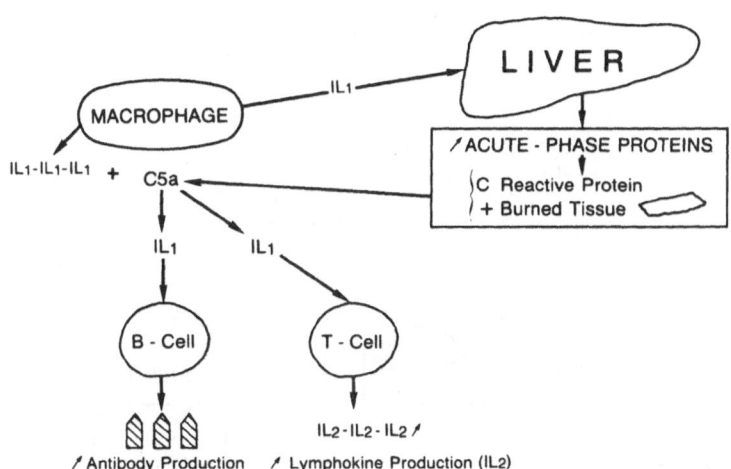

FIGURE 6. Macrophage Production of IL_1.

stimulated to increase lymphokine production, especially IL2. Via IL1 production (49), the liver may synthesize acute phase proteins. Then, C Reactive Protein could combine with burned tissues to generate more C5a which would stimulate the macrophage to make more IL1.

Thus, the complement system plays a critical role in host defence and an important role in immunoregulation (43,40,50). It is probably one of the very first links in entry into action of immunitary defences. The immune system is highly dynamic and regulated by an intricate network of interactions among different cell types as well as among cells and antibodies, antigen-antibody complexes, cells and soluble mediators.

11

These mediators play an important role in modifying the immune response in burn patients. A great number of immunologically active mediators are present following burn injury. Many are mediators of inflammation like histamine, serotonin, kinins and the products of phospholipid metabolism, especially PAF, leukotrienes and prostaglandins. Indeed, they are members of a recently recognized class of compounds which are generated from the cell membrane phospholipids. These molecules have very potent biological activites which indicate their potential as major mediators of a number of inflammatory and immunological events. Cells membranes are largely composed of lipid bilayers, of which phospholipids constitute a major component class. For a specific cell type, an activator activates enzymes which are capable of cleaving the membrane phospholipids and elaborating their constituent fatty acids.

When so released, arachidonic acid is oxidatively processed into Prostaglandins and Leukotrienes by two important enzymes, cyclooxygenase and lipoxygenase. (Fig. 7) <u>Cyclooxygenase</u> acts with

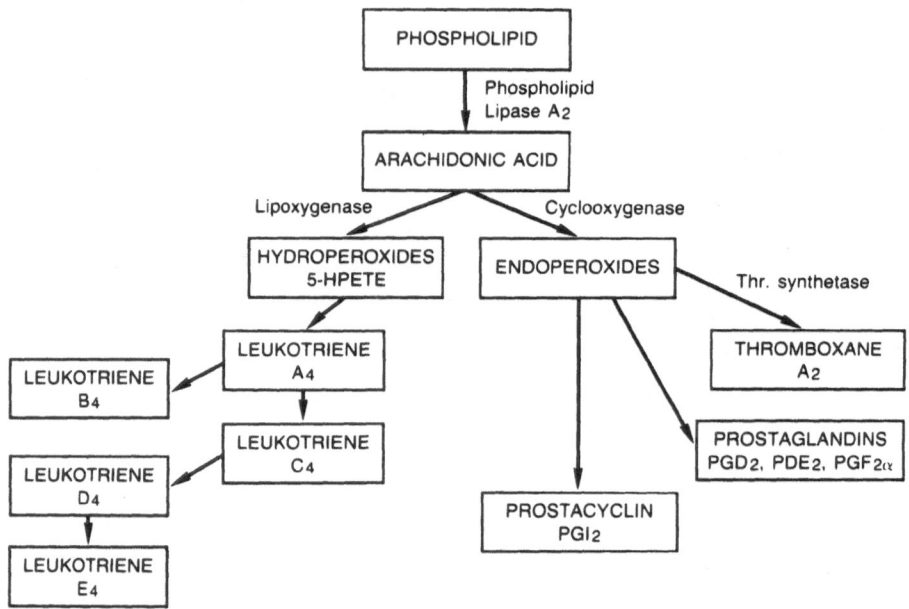

FIGURE 7

free arachidonic acid to form endoperoxides which can then be metabolized to thromboxane, prostacyclin or prostaglandins. The biosynthesis of each of these compounds requires at least one additional enzyme following the action of cyclooxygenase itself (51) and no cell type contains all of these additional enzymes : Platelets generate large quantities of thromboxane when activated (52) while endothelial cells mainly generate prostacyclin (53) and mast cells generate PGD2 (54,55).

Anggard, Arthurson and Jonsson were the first to demonstrate that thermal injury is followed by an increased local biosynthesis of prostaglandins. Mainly PGE1 was found during the first hours post-burn. Later, PGE2 (56,57,58) and PGF 2 α predominated (59). PG2 and TXB2 have been identified in burn blisters'fluid in patients (60,61,62).

After a scald injury, an increased excretion of 5β-7α-dihydroxy-11-ketotetranorprostanoïc acid, the main urinary metabolite of PGE1 and PGE2 was demonstrated in the urine of guinea-pigs (57).

In lipid extracts of scalded skin excised two hours after injury, the content of PG-like material is about 20-40 times higher than in lipid extracts of scalded skin excised simultaneously with control skin. Thus, a major part of the urinary excretion of the PGE metabolite is due to an increased local biosynthesis of PG triggered by burn injury. Prostaglandin release has been shown by a number of investigators to result directly from injury, but endotoxin release due to gut permeability changes also leads to both activation of the alternate complement pathway and to prostaglandin release.
The cyclooxygenase pathway products have systemic effects following release from burn injured tissues into the general circulation (63,64,65) and local effects.

After systemic administration, PGE compounds have been shown to be potent hypotensive agents, acting by vasodilatation and decreased peripheral resistance (66,67). They increase coronary blood flow and cause bronchial dilatation. Prostaglandins are also involved in different immunoregulatory mechanisms : they induce suppressor macrophage activity and directly activate suppressor T-cells. Prostaglandin-induced suppressor T-cells produce a low molecular weight suppressor peptide. The result is an impaired T-lymphocyte response. E Prostaglandins can inhibit cell-mediated cytolysis and IgE-mediated histamine release. PGF compounds induce effects opposite to those of PGE compounds (67).

Locally, in the microcirculation, Thromboxane possesses vasoconstrictive properties and stimulates platelet aggregation (68) Prostacyclin on the other hand, is vasodilatative and is a potent inhibitor of platelet aggregation.

The other oxidative pathway for arachidonic acid metabolism, termed the 5 lipoxygenase pathway leads to biosynthesis of leukotrienes (L.T.) through an ordered cascade of enzymes (69). Named hydroperoxides (5 HPETE), the primary products of the cascade can be metabolized to L.T. which can be divided into two groups, the LTB4 group with hydroxyl groups, and the C4, D4 and E4 group with a cysteinyl peptide chain, which together constitute the Slow Reacting Substance of Anaphylaxis. SRSA is a very potent group of compounds in causing bronchoconstriction and in increasing vasopermeability in relationship to injuries of the well-vascularized tissues. Many cells possess both cyclooxygenase and lipoxygenase, and are thus capable of enzymatic oxidation of arachidonic acid by the 2 major pathways. Thromboxane and Leukotrienes have been shown to be important mediators in early burn edema (70).

Leukotrienes generation occurs from a variety of human leukocytes and related cells. Pulmonary mast cells produce LTC4 after activation by an antigen (71,72,73,74). Alveolar macrophages produce large quantities of LTB4, even in comparison to human neutrophils (75,76). Eosinophils generate mainly LTC4 and LTB4. Eosinophils and neutrophils are capable of degrading LTC4, LTD4 and LTE4 extracellularly, while neutrophils can degrade LTB4 intracellularly to compounds that lack their biological activities. This provides an example of a reaction producing its own negative feedback system.

Injected in the human skin, LTB4 has been shown to exert a very potent chemotactic effect on neutrophils, but it can also deactivate cells so that they will no longer migrate (77). Injected intradermally, LTC4, LTD4 and LTE4 elicit a wheal, with an increased local vascular permeability. Injected intravenously to the animal, LTC4 and LTD4 induce a constrictive effect on the peripheral arterial vasculature with an increase in mean arterial blood pressure and a disastrous effect on cardiac output. With respect to the airway, LTC4 and LTD4 are significantly more potent than histamine in effecting bronchoconstriction.

At PERCY Military Hospital Burn Center, in CLAMART, it was shown for the first time (78,79,80) that lipoxygenase pathway was altered after thermal injury : as early as the first hours following burn injury (6-12 hours) there is a considerable but transient increase in LT concentrations, particularly LTB4. This peak is followed by a collapse in concentrations which remain very much below normal. This evolution is parallel with that of lymphocytary numeration and that of the OKT4/OKT8 ratio, and is contemporary with the state of anergy.

Eicosanoids thus modulate a variety of multi-faceted host responses. This class of compounds seems to have many different actions on cells, tissues and organs ; further study will therefore be necessary to better assess them. Of great interest is the last lipid mediator, PAF, synthesized by different types of cells such as platelets, basophils, eosinophils, macrophages. It originates from membrane phospholipids under the action of cytosolic hydrolases and acetyl-transferases. PAF binds to a specific binding site and triggers the activation of proteins via the phosphatidine-inositol cycle. These proteins induce the mobilization of calcium which in its turn triggers the arachidonic acid cascade. Thus PAF seems to be the most important lipidic mediator since it can be situated upstream from eicosanoids in triggering the cascade of arachidonic acid metabolites.

The intervention of PAF in shock phenomena, allergy, thrombosis and inflammation in general is a well established fact. PAF also plays a role in organ graft rejection.

An injection of PAF leads to a fall in arterial pressure (Fig. 8) with capillary leakage and an increase in hematocrit, pulmonary hypertension and bronchoconstriction, all of which are reversible signs after injection of a specific anti-PAF. A superfusion of PAF induces local formation of a thrombus which is accompanied by deendothelialization with leukocytic adhesion and migration of platelets towards the site of inflammation. Injection of immunoglobulin to an antigen sensitized animal is followed by a fall in arterial pressure with capillary extravasation. Likewise the experimental injection of endotoxin is followed by a sharp fall in arterial pressure. All these phenomena are reversible after injection of an anti-PAF. Similarly the asthma crisis induced by injection of PAF improves after injection of an anti-PAF. After activation of leukocytes of burn injured animal treated with a solution containing 1 % of an anti-PAF, a preliminary study by PERCY Burn Center in CLAMART, allowed observing a decrease in Leukotrienes and PAF production, as well as a decrease in the production of $O^{\cdot 2-}$ free radicals.

Furthermore burn injured animals under anesthesia and treated with Ringer Lactate but not receiving an anti-PAF, die in the state of shock within 24 to 36 hours following thermal injury without having regained consciousness, whereas animals treated with Ringer Lactate and a 1 % anti-PAF Solution wake up rapidly, return to a normal P.A. within one hour and eat 3 hours after accident.

mean ± S.E.M. , n = 5.

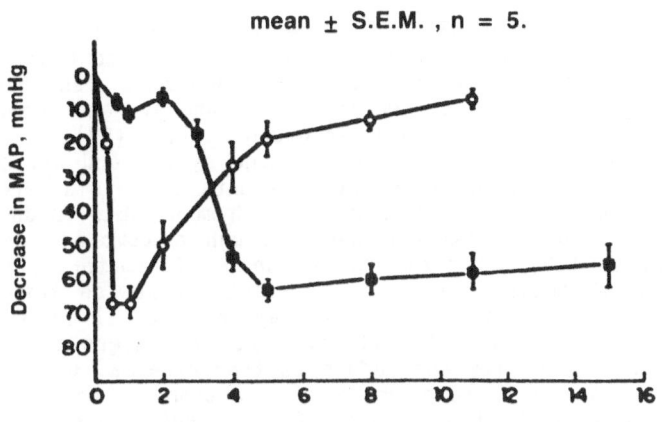

Time after infusion of PAF or Endotoxin, min.

FIGURE 8.   Decrease in mean arterial Pressure (MAP) in mmHg Following i.v. Infusion of PAF-Acether (0.50 nmole/kg) (o) or Enteriditis Endotoxin (50 mg/kg). (■).

Thus, besides the generalized physiological alterations occurring in severely burned patients, major immune dysfunction becomes clear at both the humoral and cellular levels.
Current therapy - limiting pain, rapidly excising wounds, feeding the patient and providing a warm environment, removing the toxic factors - serves to minimize stress and protects the immunitary functions in one way or another but sometimes impairs them.

Protein loss due to changes in vascular permeability and at the origin of hemodynamic instability is compensated by the adminisStration of human proteins. But at the same time, albumine binds prostaglandins and when albumin concentration is low, prostaglandins are free to exert their suppressive activity. As observed by Gelfan, the use of fresh-frozen plasma in the early phase of burn injury could do harm to the patient by adding more substrate to generate C5a and thus increasing vascular permeability and adding to shock lung. It is much more justified later as a source of alternative pathway activity for opsonization. While the early excision and graft incorporate many factors that are known to be immunosuppressive - administration of blood products, drugs, multiple anesthesia - these procedures seem to profoundly influence the immunologic depression in burn patients, and the net effect on the immune system is a positive one (81).

The effect of plasma exchange/plasmapheresis on post-burn lymphocyte suppression by the removal of circulating factors as measured by MLR demonstrate a statistically significant decrease in suppressive activity (82). Immune function and metabolic status are importantly interrelated in the burn patient. Consequently, limiting metabolic stress appears to result in beneficial immunological effects. Large amounts of nitrogen relative to caloric intake are necessary to achieve nitrogen equilibrium in patients with large burns.

Several authors (83,84,85) have shown that survival is higher in high protein groups of patients than in the control group. Malnutritional states are associated with decreased resistance to infection with an increased incidence of anergy and complement consumption (86) as well as an impaired wound healing (87).
A variety of drugs act directly on the immune system and are or have been used with varying success in burn patients : steroïds are known to decrease the release of arachidonic acid from phospholipids and to inhibit the formation of Leukotrienes through a complex series of cellular responses (88) but their use as anti-inflammatory agents is limited because of their adverse side effects. Likewise thromboxane synthetase inhibitors (imidazole, dipyridamole, etc...) result in a beneficial reduction of dermal ischemia, and scavangers inhibit the phospholipid-lipase activity and reduce post-burn edema, but they act only upon one link of a long chain of reactions ; we don't know exactly where the chain begins and if the acute phase reactants intervene at the time of injury, or wether they are continuously released for several days following injury.

In a similar respect, immunomodulators have been employed successfully, increasing survival in animals. It is certainly very early in the inflammatory process that histamin, bradykinin, complement by-products, prostaglandins, prostacyclin, cysteinyl leukotrienes and PAF act directly on the endothelial cells and cause an increased microvascular permeability and all the disorders which lead to immunological dysfunction, (Fig. 9) but the question is : what is the primum movens ?

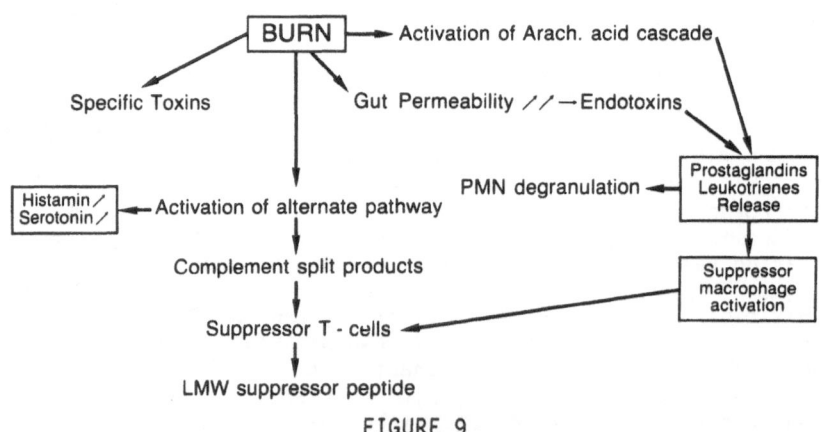

FIGURE 9

Natural products such as interleukins, leukocyte dialysates are being tested. We now need to develop methods for reversing immune defects in burn patients, such as the use of substances which may even induce cells such as macrophages and lymphocytes to produce mediators of immune competence thereby restoring immune function in immuno-depressed burn patients.

REFERENCES

1. Hiebert JM., McGough M. and Rodeheaver G., Tobiasen, J., Edgerton, MT., Edlich, RF. Surgery, 86 : 242 (1979).

2. Alexander JW., Ogle CK. and Stinnett JD. Ann. Surg. 188 : 809 (1978).

3. Alexander JW. J. Surg. Res. 8 :128 (1968).

4. McRipley RJ. and Garrison DW. Proc. Soc. Exp. Biol. Med. 115 : 336 (1964).

5. Millican RC., Evans G. and Markley K. Ann. Surg. 163 : 603 (1966).

6. Deitch EA. J. Burn care Rehab. 4 : 344 (1983).

7. Howard RJ. Surg. Clin. N.A. 59 : 199 (1979).

8. Ninnemann JL. J. Burn Care Rehab 3 : 355 (1982).

9. Bjornson AB., Altemeier WA and Bjornson HS. Ann. Surg. 188 : 93 (1978).

10. Foley FD., Grenwald KA. and Nash G. New Engl. J. Med. 282 : 652 (1970).

11. Pulaski EJ. Bahama Internat. Conf. on Burns. 1961 : 29.

12. Miller CL. and Baker CC. J. Clin. Invest. 63 : 202 (1979).

13. Warden GD and Ninnemann JL. Eds J.L. Ninnemann : 1 (1981).

14. Neilan BA., Taddeini L. and Strate RG. J.A.M.A. 238 : 493 (1977).

15. Wood, GW., Volence, FJ. and MM. Mani. following thermal injury, Clin. Exp. Immunol. 31 : 291 (1978).

16. O'Mahony JB., Palder SB., Wood J., Demling R. and Mannick JA. J. of Trauma vol. 24, 10 : 869 (1984).

17. Thangam M., Sundararaj T., Subra Manian S., Murugesan R. and Sunderarajan C. J. of Trauma vol. 24, 3 :220 (1984).

18. McIrvine AJ., O'Mahony JB., Saporoschetz I. and Mannick JA. Ann. Surg. 196 : 297 (1982).

19. Munster AM. Lancet 1 : 1329 (1976).

20. Stein M., Gamble D., Klimpel K., Herndon D. and Klimpel G. Cell. Immunol. 86 : 551 (1984).

21. Rapaport FT., Milgrome F. and Kano K. Ann. Ny Acad. Sci. 150 : 1004 (1968).

22. Wolfe JHN., Wu AV. and O'Conner NE. Arch. Surg. 117 : 1266 (1982).

23. Anwar AH. J. of Trauma. 17 : 12 (1977).

24. Ninnemann JL., Condie T. and Davis SE. J. Trauma 22 : 837 (1982).

25. Wolfe JHN, Saporoschetz I. and Young AE. Ann. Surg. 193 : 513 (1981).

26. Winkelstein A. J. of Trauma Vol. 24, 9 : S72 (1984).

27. Mannick JA. J. of Trauma vol. 24 n° 9 : S117 (1984).

28. Alexander JW., Stinnett JD. and Ogle CK. In NINNEMANN J.L. (ed) : The Immune consequences of Thermal Injury, Baltimore Williams and Wilkins, p. 21-35 (1981).

29. Alexander JW., Stinnett JD. and Ogle CK. Surgery 86 : 94 (1979).

30. Markley K., Smallman E. and Evans G. Surgery. 61 : 896 (1967).

31. Alexander JW. and Wixson D. Surg. Gynecol. Obstet. 130 : 431 (1970).

32. Masters BSS., Baxter CR. and Dobke M. Compromised leukocyte function in the burn patient. In NINNEMANN J.L. (ed), Traumatic injury. Baltimore, University Park Press, 163 (1983).

33. SABA TM. and Jaffe E. Am. J. Med. 68 : 577 (1980).

34. Mosher DF. Prog. Hemostasis Thromb. 5 : 111 (1980).

35. Berman B. and Gigli I. J. Immunol. 124 : 685 (1980).

36. Stingl G., Wolff-Schreiner EC., Pichler WJ, Oschnait F., Knapp W., Wolff K. Nature 268 : 45 (1977).

37. Sauder DN., Carter CS. and Katz SI. J. Invest. Dermatol. 79 : 34 (1982).

38. Luger TA and Oppenheim JJ. Adv. Inflam. Res. 5 : 1 (1983).

39. Rowden G., Lewis MG. and Sullivan AK. Nature 268 : 247 (1977).

40. Gelfand JA., Donelan MB., Hawiger A. and Burke JF. J. Clin. Invest. 70 : 1170 (1982).

41. Dhennin C., Pinon G. and Greco JM. J. of Trauma 18: 129 (1978).

42. Fjellstrom KE. and Arturson G. trauma. Acta Pathol. Microbiol. Scand. 59 : 257 (1963).

43. Gelfand JA. Donelan MB. and Burke JF. Ann. Surg. 198 : 58 (1983).

44. Heideman M. J. Surg. Res., 26:670 (1979).

45. Gelfand JA. J. of Trauma Vol. 24 n° 9, 118.

46. Whalin B., Perlmann H., Perlmann P., Schreiber RD. and Muller-Eberhard HJ. fragments, J. Immunol. 130 : 2831 (1983).

47. Ghebrehiwet B. Silverberg M. and Kaplan AP. J. Exp. Med. 153 : 665 (1981).

48. Sharma VK., Agarwal DS. and Satyanand   J. of Trauma 20 : 976 (1980).

49. Sipe JD., Vogel SN. and Sztein MB. Ann. N. Y. Acad. Sci.New York, 389 : 137 (1982).

50. Lowbury EJL. and Ricketts CR. J. Hygiene 55 : 266 (1957).

51. Samuelsson B., Goldyne M. and Granstrom E. Ann. Rev. Biochem 47 : 997 (1978).

52. Borgeat P. and Samuelsson B. acid. J. Biol. Chem. 254 : 2643 (1979).

53. Moncada S., Higgs GA. and Vane JR. Lancet, 1 : 18 (1977).

54. Lewis RA. J. of Trauma Vol. 24 n° 9 : 125.

55. Lewis RA., Soter NA. and Diamond PT. J. Immunol. 129 : 1627 (1982).

56. Anggard E. and Arturson G. and Jonsson CE. Acta Physiol.Scand. 80 : 46A (1970).

57. Hamberg M. and Jonsson CE. Acta Physiol. Scand. 87 : 240 (1973).

58. Jonsson CE. Acta. Univ. Ups. Suppl. 134 (1972).

59. Jonsson CE, Shimizu Y., Fredholm BB., Granstrom E. and Oliw E. Acta Physiol. Scand. 107 : 377 (1979).

60. Arturson G., Hamberg M. and Jonsson CE. Acta Physiol. Scand. 87 : 270 (1973).

61. Heggers JP., KO J., Robson MC., Heggers R. and Craft KE. Reconstr. Surg. Traumatol. 65 : 798 (1980).

62. Jonsson CE., Granstrom E. and Hamberg M. Scand. J. Plast. Reconstr. Surg., 13:45 (1979).

63. Bergstrom S., Danielsson H. and Samuelsson B. acid. Biochem. Biophys. Acta 90 : 207 (1964).

64. Hinman JW., Prostaglandins. Ann. Rev. Biochem 41 : 161 (1972).

65. Horton EW. Prostaglandins. Monographs on Endocrinology Vol VII Berlin, Springer-Verlag (1972).

66. Horton EW. and Main IH. Br. J. Pharmacol. 21 : 182 (1963).

67. Weiner R. and Kaley G. Am. J. Physiol. 217 : 563 (1969).

68. Hamberg M. and Samuelsson B. Proc. Nat. Acad. Sci. (USA) 71 : 3400 (1974).

69. Samuelsson B. Science 220 : 568 (1983).

70. Alexander F. Mathieson M., Teoh K., Huval W. and Lelcuk S. J. of Trauma, 24:709 (1984).

71. McGlashan DW., Schleimer RP. and Peters SP. J. Clin. Invest. 70 : 747 (1982).

72. Mencia-Huerta JM., Razin E. and Ringel EW. J. Immunol. 120 : 1885 (1983).

73. Razin E., Mencia-Huerta JM. and Lewis RA. Proc. Nat. Acad. Sci. USA. 79 : 4665 (1982).

74. Razin E., Mencia-Huerta JM. and Stevens RL. J. Exp. Med. 157 : 189 (1983).

75. Felds AOS., Pawlowski MA. and Cramer EB. Proc. Nat. Acad. Sci. USA. 79 : 7866 (1982).

76. Godard P., Damon M. and Michel FB. Clin. Res. 31 : 548A (1983) (abstract).

77. Goetzl EJ. and Pickett WC. J. Immunol. 125 : 178 (1980).

78. Braquet M., Ducousso R., Garay R., Carsin H. and Guilbaud J. The Lancet 8409 vol II : 976 (1984).

79. Braquet M., Ducousso R., Garay R.. Guilbaud J., Carsin H. and P. Braquet. Procedings of the 17 th Miami Winter Symposium Miami, Florida, USA, february 11-15 (1985).

80. Braquet M., Lavaud P., Dormont D., Garay R., Ducousso R., Guilbaud J., Chignard M., Borgeat P. and Braquet P. Prostaglandins29, 5: 747 (1985).

81. Stratta R., Warden G., Ninnemann J. and Saffle J. J. of Trauma vol. 26, 1 : 7 (1986).

82. Ninnemann JL., Stratta RJ. and Warden GD. Arch. Surg. 119 : 33 (1984).

83. Alexander JW., McMillan BG. and Stinnet D. Ann. Surg. 192 : 505 (1980).

84. Aulick LH., Hander EH. and Wilmore DW. J. Trauma 19 : 559 (1979).

85. Serog P., Baigts, F., Apfelbaum, M., Guilbaud, J., Chauvin, B. and Pecqueur, ML. Burns Incl. Therm. Inj. 9 : 422 (1983).

86. Alexander JW., McClellan MA., Ogle CK. and Ogle JD. Ann. Surg. 184 : 672 (1976).

87. Levinson SM. and Seifter E. Clin. Plast. Surg. 4 : 375 (1977).

88. Kuehl FA. and Egan RW. Science 210 : 978 (1980).

BURNS : CLINICAL AND BIOLOGICAL OVERVIEW

R. H. Demling

Associate Professor of Surgery, Harvard Medical School Director, Longwood Area Trauma Center, Boston, Massachusetts.

There have been major advances in the care of the burn patient over the last ten years leading to improved survival rates, in particular for the young patient with the massive burn, the improvements have been in a number of areas of burn management.

CARDIOPULMONARY RESUSCITATION

New information on the physiologic changes responsible for the dramatic shifts in bodily fluids and protein that occur after burns has led to improved resuscitation techniques. Increased microvascular permeability in burn tissue and a generalized cell-membrane defect that results in intracellular swelling have been well described.[1] The increasing importance of the interstitium in the transvascular fluid and protein flux is a topic of considerable research interest.[2,3] Changes in the interstitium in both burned and adjacent non-burned tissues, particularly in the basement membrane, now also appear to have a major role in edema formation.[4,5]

Many potent vasoactive mediators are known to be released from burn tissue. These include the vasoconstrictor and vasodilator prostaglandins, kinins, serotonin, histamine, oxygen radicals, and various lipid peroxides.[6-9] Although these factors are thought to play some part in the edema process, the use of specific inhibitors to modify the burn-edema process has been, to date, unsuccessful. Plasma-exchange transfusions have occasionally been used in the early postburn period in an attempt to remove these circulating agents.[10]

The edema that occurs in nonburned soft tissues does not appear to be due to altered protein permeability,[11] but to the severe hypoproteinemic state of a burned patient.[18,12] There also does not appear to be any altered permeability in the lungs after burns, unless severe smoke inhalation has occurred.[28] A number of experimental and clinical studies have found no early increase in water in the lungs during a controlled fluid resuscitation after a major burn.[13] Early colloid infusion is therefore feasible and, in fact, has been shown to minimize edema in the nonburned tissues and to increase blood volume better than does crystalloid.[26,14,15]

Postburn lung dysfunction remains a major cause of mortality. The incidence of smoke-inhalation injury is now known to be several times higher than previously reported; such injury is currently found in up to 20 percent of patients admitted to burn centers.[16] Routine fiberoptic bronchoscopy or xenon ventilation-perfusion scanning in patients suspected of having this injury has resulted in more frequent and earlier diagnoses.[33] Neither method, however, can predict the magnitude of damage accurately.

Our understanding of the complex chemical burn to the tracheobronchial mucosa caused by the toxic components of smoke has improved substantially.[17,18] Water-soluble gases that are found in smoke from burning plastics or rubber, such as ammonia, sulfur dioxide, and chlorine, react with water in mucous membranes, and edema. Lipid-soluble compounds such as nitrous oxide, phosgene, hydrogen chloride, and various toxic aldehydes are transported to the lower airways on carbon particles that adhere to the mucosa. All these agents damage the cell membrane directly and impair the ciliary clearance of bacteria. In addition, alveolar macrophages are activated to release potent chemotoxins, which further increase inflammation.[19]

Even with severe mucosal injury, early symptoms are absent in many cases because mucosal edema and bronchorrhea may not develop for 24 to 48 hours. Yet recent human studies that used the multiple inert-gas technique have demonstrated an early decrease in alveolar ventilation caused by the bronchospasm and edema in the airway mucosa.[20] True intrapulmonary shunting is not severe during this period, which indicates that alveolar edema is not a major component of the early disease state. Several recent clinical studies have demonstrated that lung water increases significantly only after massive inhalation injuries.[21]

Endotracheal intubation and mechanical ventilation with positive end-expiratory pressure remain the treatment of choice for severe lung injuries from exposure to smoke. Corticosteroids have been shown to be ineffective.[22,23]

INFECTIONS AND IMMUNOLOGICALLY MEDIATED RESPONSES

Organ-system failure in conjunction with sepsis remains the leading cause of death due to burns. Although infection is the primary initiator of the hyperdynamic state, it is now clear that the devitalized tissue itself can also initiate and perpetuate the mediator-induced response. This may help to explain the finding that circulating bacteria cannot be detected in more than half of the patients with burns who die from what appears to be sepsis.[24] Circulating endotoxin absorbed from the wound or from a gastrointestinal tract with an impaired mucosal barrier may have a prominent role in producing sepsis.

The lungs and the burn wound are the most frequent sites of infection and fatal infections are most often caused by highly virulent opportunistic gramnegative organisms.[25] Particularly virulent strains of Pseudomonas aeruginosa have been indentified that release exotoxin A, a factor that impairs protein synthesis both locally and systemically when absorbed. Antibiotic-resistant strains of common organisms such as methicillin-resistant Staphylcoccus aureus have been responsible for a number of epidemic infections in burn centers which result because of patient cross-contamination.

Fungal infections have also become more common, but elimination of the use of prophylactic systemic antibiotics in burn patients has decreased their incidence. Systemic antibiotics are now used only to treat documented infections, which are detected in the burn wounds by quantitative bacteriologic analyses of full-thickness biopsy specimens.[25] However, since sampling errors can affect the results of these tests, particularly if only superficial eschar is analyzed, histologic documentation of tissue invasion remains the most reliable approach.

Topical antibiotics continue to be the mainstay of wound-infection control. However, bacterial resistance to these agents has also been observed with increasing frequency as plasmids that contain antibiotic-resistant genes are transmitted to local wound bacteria. New topical agents such as chlorhexidine hydrochloride (which has never been shown to be carried on a plasmid marker) are continually being developed to overcome this problem. Sophisticated isolation equipment such as laminar air-flow units have been reported to decrease infection rates,[26,27] but equally improved survival rates have been reported with the use of more standard isolation techniques.[28]

Researchers have identified many complex alterations in both the cellular and humoral components of the immune systems of burn patients that predispose such patients to infection. Impaired phagocyte function appears to correlate most closely with the onset and degree of sepsis.[29] At least some of the phagocytic abnormalities probably result from a circulating inhibitory factor in the serum.[30] One group has reported that this factor is a polypeptide that is biochemically similar to a fragment of collagen released by the injured skin.

The lymphocyte system of burn patients is also severely altered, as indicated by an impaired response to mitogen stimulation.[31] Currently, several immunomodulators, such as thymopentin (a fragment of thymopoietin with five amino acids), that may reverse the lymphocyte and phagocyte abnormalities[32] are undergoing clinical trials.

Burn patients also have a decrease in the humoral components fibronectin (a plasma opsonin) and gamma globulin,[33] but no convincing data indicate that administration of these agents improves survival.

In both human and animal studies, early burn excision with wound closure has been reported to reverse many of the immunologic defects that occur, which emphasizes the role of the burn wound.

METABOLIC AND NUTRITIONAL ASPECTS

Advances have been made in both the understanding and treatment of the protein catabolism initiated by burn injury. The increase in metabolic rate that begins in the postresuscitation period is much greater than that seen with any other form of trauma or severe sepsis; a doubling of the normal metabolic rate of 35 to 40 kcal per square meter of body-surface area per hour occurs when burns affect more than 50 percent of body surface.[34] Associated with the hypermetabolic response are protein catabolism, ureagenesis, lipolysis, and accelerated gluconeogenesis. The increased heat production, which is accompanied by a 1 to 2°C increase in core temperature, appears to be due to a resetting of the hypothalamic temperature center. The hypothesis that a causal relationship exists between the increased evaporative heat loss from the impaired barrier of burned skin and the hypermetabolic response remains controversial.[35] However, excessive heat loss, which will occur if a burn patient is placed in a room with an average ambient temperature, will clearly exaggerate the stress response. Thus, maintaining the ambient temperature in the rooms of burn patients at or above 30°C has been found to decrease total energy expenditure, and such a practice is now standard.

The efferent mediators responsible for the metabolic manifestations appear to result partly from an excess of the counterregulatory hormones (the catecholamines, glucagon, and glucocorticoids)[36] that impair the glucose transport into tissues by insulin. Moreover, the combined infusion of glucagon, cortisol, and epinephrine and normal persons have been shown to produce many of the postburn responses, which suggests that several hormones interact to cause the metabolic changes.[37]

Afferent signals from the burn wound appear to initiate and maintain the hormonal imbalance.[38] These signals are not totally neural in origin, since spinal anesthesia or peripheral-nerve transection does not eliminate them. Circulating inflammatory mediators such as prostaglandin E_2 and interleukin-1 are now known to be involved in the perception of the afferent signal by the central nervous system. Interleukin-1 and endogenous pyrogens that are released from wound macrophages have been reported to stimulate in vitro skeletal-muscle proteolysis by means of the formation of muscle prostaglandin E_2.[39] The afferent signal can be accentuated by the brain, particularly in the presence of excessive pain or anxiety.[40]

The metabolic rate in burn patients can be decreased by administration of anesthetic agents or high-dose morphine or by early wound closure. However, the effect on outcome of lowering the metabolic rate has not been defined.[41]

The marked hepatic gluconeogenesis can be partially suppressed by infusion of exogenous glucose. Glucose must account for at least 50 percent of the total caloric intake to maximize its effect on decreasing protein breakdown, but excess glucose will result in fat production and increases in the formation of carbon dioxide. Approximately 15 to 20 percent of the calories should be proteins or amino acid equivalents, and the remainder should be given as fat, which has been found to be well used in burn patients. Recent studies have demonstrated substantial benefits, as assessed by improved immune function and increased survival, from high-protein diets (especially those perfused enterally) that are aimed at achieving a 100:1 calorie-nitrogen ratio.[42] In general, the increased attention paid to nutrition has had a major impact on burn management.[43]

WOUND MANAGEMENT

Important advances have been made in our understanding of pathophysiologic changes that occur in burn wounds and in the techniques available to close these wounds.

WOUND HEALING

Although neutrophil and platelet sequestration in the microvessels of burned tissue is evident immediately, major tissue infiltration with neutrophils and macrophages is delayed for several days.[44,45] The inflammatory cells are potent factories of vasoactive substances such as prostanoids, leukotrienes, platelet-activating factors, and complement components. When the wound mediators are released and absorbed in sufficient quantities, particularly when stimulated by local endotoxin, pulmonary dysfunction results.[46]

The wound macrophage is now known to be the cell that controls healing by the release of a number of factors.[47] Angiogenesis factor, a substance with a molecular weight of 2000 to 20,000, is secreted by hypoxic macrophages at the wound edge or outer surface.[48,49] This factor, which initiates neovascularization, appears to be a chemoattractant for mesothelial cells and vascular endothelial cells that migrate to the wound edge to form new blood vessels. The release of angiogenesis factor is accentuated by a low partial pressure. However, angiogenesis and, in turn, the formation of granulation tissue are suppressed beneath the burn eschar even when the partial pressure of oxygen there is low.

The macrophage-derived growth factor is released from macrophages below the wound surface where the tissue partial pressure of oxygen is increased. This growth factor stimulates fibroblast mitosis and subsequent fibroblast deposition of collagen fibronectin and glycosaminoglycan.[50] The rate of fibroblast proliferation and secretion depends on the availability of oxygen and therefore on the local blood flow.[51] Platelet-derived growth factor has properties of both angiogenesis factor and macrophage-derived growth factor.[52]

Infection-producing quantities of bacteria in a wound will retard the healing process not only by using the available tissue oxygen, but also by degrading the new protein being formed through release of proteases from the phagocytizing neutrophils. The best environment for wound healing therefore appears to be one in which the partial pressure of oxygen is low at the wound surface, thereby stimulating angiogenesis, and high at the subsurface, increasing the secretion of macrophage-derived growth factor and fibroblasts.[53] Thus, surface accumulation of phagocytes and bacteria, as on an open wound, should be kept to a minimum.

The rate of reepithelialization of the superficial would also appears to be controlled by wound factors, some of which are also derived from the macrophage. The gene for epidermal growth factor, one such substance, has now been cloned, so that it may eventually be available in large quantities for therapeutic purposes.[54] It is anticipated that many other growth factors will also soon be available for research and for potential therapeutic use.

EARLY WOUND CLOSURE

Since the presence of inflammatory tissue or eschar on the burn surface has been shown to produce hypermetabolism and immune deficiency that can lead to severe illness or death, many practitioners have recently returned to an aggressive surgical approach in which burn wounds are closed as early as possible.

Current surgical approaches to early wound closure vary from immediate, complete wound excision to the fascia and closure with a combination of autografts and skin substitutes within the first week[55] after the burn to sequential excision and graftings, beginning about two to four days after the burn and continuing every four to five days until wound closure.[56,57] The latter appears to be the safest and more common approach for large burns.

The complications that can arise with this aggressive surgical approach have been minimized by limiting the length of the surgical procedure, usually to two hours; the area excised to no more than 15 percent of body surface; and the blood loss to no more than 50 percent of blood volume. Excisions performed the first days after the burn, before infection has occurred, appear to result in fewer postsurgical complications than those performed more than seven days after the burn, when infection and inflammation have become severe.[57] However, maintenance of hemodynamic stability can be a major problem when surgery is performed early; i.e., prior to the development of increased blood flow to the burn area which usually occurs after day 5. Moreover, two surgical teams usually operate on an individual patient to expedite the procedure and thereby decrease blood loss.

Early surgical excision and skin-graft closure of deep, partial-thickness burns are also increasing in popularity. Burns of this depth are not only prone to infection and conversion to deeper injuries, but they also heal with considerable fibrosis and hypertrophic scarring. During the surgical procedure, burn tissue is removed in thin layers (0.025 to 0.5 cm in depth) until healthy bleeding tissue is reached, while as much viable dermis as possible is preserved to improve the functional and cosmetic result. This technique is known as tangential excision.[58]

Early debridement and wound closure that begins within three days of the burn have been reported to increase survival among children with full-thickness burns that affect more than 60 percent of the body surface to a greater extent than the approach that uses spontaneous eschar separation and delayed grafting. Although it is less convincing, some evidence now indicates that the treatment also produces a similar increase in survival among adults, although a statistically significant improvement has yet to be documented.[59] Decreased morbidity has been reported, as evidenced by a sixfold decrease in septic complications and a decrease in catabolism and in immune deficiencies, in comparison with the conventional approach that uses topical antibiotics.[60] The hospital stay is also decreased, by 20 to 30 percent, for patients with deep burns that affect 20 to 40 percent of the body surface. Although the findings are still controversial, several centers have reported that improvements in longterm function and cosmetic acceptability result from early wound closure, particularly when the patient has deep dermal burns, because it minimizes hypertrophic scarring and the need for later reconstruction.

SKIN SUBSTITUTES (ARTIFICIAL SKIN)

There has been a tremendous interest in the development of substances that restore, temporarily or permanently, the important barrier functions of the skin; i.e., functions that prevent invasive infection and water and heat loss through evaporation. There has been considerable confusion over the term "artificial skin", which has been used to describe both the temporary and permanent skin barriers. There are major differences between the properties and production of these two types of skin substitutes.

TEMPORARY SKIN SUBSTITUTES

The drying of exudate on a superficial wound with resulting scab or eschar formation has been shown to retard reepithelialization. In contrast, wounds have been shown to reepithelialize more rapidly and with less pain and inflammation when they are occluded and a thin layer of fluid from the wound maintains contact with the surface.[61] This sealed wound also creates a more favorable environment for the clearing of surface bacteria by wound-defense mechanisms. Application of topical water-soluble antibiotics controls infection in an open wound, but it also appears to increase inflammation and decrease the healing rate in comparison with the rate in a sealed wound.[62] A number of temporary skin substitutes have therefore been developed to improve healing of partial-thickness wounds, as well as to protect clean, excised wounds when they are not immediately autografted.[62] The properties required of a temporary skin substitute have been well defined.[63,64] Adherence to the wound is essential for maximum reepithelialization and minimal inflammation and fibrosis. The dressings must be permeable to water vapor and oxygen so that an anaerobic environment is not produced at the wound surface, yet they must not be permeable to bacteria. Elasticity and durability are major advantages, yet the biochemical structure of the material cannot be antigenic or toxic; otherwise, it will cause a local rejection reaction.[63,64]

There are two types of temporary skin. Biologic dressings (previously living tissue, including amniotic membranes, xenografts and homografts, and cadaver skin) have been used for a number of years, although, because of its limited availability, cadaver skin is used primarily to cover excised wounds. Recently, many synthetic skin substitutes have been developed because of the need to increase availability and shelf life over that of the biologic materials. Solid silicone polymers have been the most widely used because they are microporous and uniquely permeable to water vapor. Polyurethane and polyvinylchloride polymers are also being used. Synthetic substitutes that depend on fibrin entrapment in the porous material for adherence appear to be less successful than those in which there is a direct chemical bond with the wound. Although useful when only minor infection is present, these synthetic dressings adhere poorly to grossly contaminated wounds.

PERMANENT SKIN SUBSTITUTES

Improvements in resuscitation, cardiopulmonary and nutritional support, and infection control, combined with early wound closure that uses autografts and temporary skin substitutes, have substantially decreased the mortality and disability from severe burns. A small group of patients (representing less than 5 percent of admissions to burn units), however, have deep burns that affect more than 70 percent of the body surface and do not have enough unburned skin for wound closure. Although unburned skin can be used up to three times (and that on the scalp can be used at least six times) for autograft donations, the amount of skin still may be insufficient, particularly if patients have burns that affect more than 90 percent of the body surface. Until recently, most of these patients would not have survived long enough for skin replacement to be an issue.[8] Now, however, for a portion of this group, lack of skin is the limiting factor to survival, although functional ability will be a major problem for all these patients since covering the burn wounds with only epidermis, as is done when the same donor sites must be reused a number of times, will result in considerable scar formation.[65-69] Reused donor split grafts, with the exception of those from the scalp, have very little remaining dermis; instead, they consist mostly of an epidermal sheet that is rather easily disrupted by minimal shearing forces, which leads to continual blistering and focal sloughing. Because of these problems, permanent skin substitutes are being designed.

REFERENCES

1. Baxter CR.
   Clin Plast Surg 1 : 693-703 (1974)
2. Arturson G.
   Acta Physiol Scan (Suppl) 463 : 111-22 (1979)
3. Nanney LB.
   J Invest Dermatol 76:227-30 (1981)
4. Kramer GC., Harms BA., Bodai BI., Demling RH., Renkin EM.
   Am J Physiol 243 : 803-9 (1982)
5. Mullins RJ., Bell DR.
   Circ Res 51 : 305-13 (1982)
6. Harms BA., Bodai BI., Smith M., Gunther R., Flynn J., Demling RH.
   J Surg Res 31 : 274-280 (1981)
7. Carvajal HF., Brouhard BH., Linares HA.
   J Trauma 15 : 969-75 (1975)
8. Saez JC., Ward PH., Gunther B., Vivaldi E.
   Circ Shock 12 : 229-39 (1984)
9. Sasaki J., Cottam G., Baxter C.
   J Burn Care Rehab 4 : 251-5 (1983)
10. Warden GD., Stratta RJ., Saffle JR., Kravitz M., Ninnemann JL.
    J Trauma 23 : 945-51 (1983)
11. Harms BA., Bodai BI., Kramer GC., Demling RH.
    Microvasc Res 23 : 77-86 (1982)
12. Demling RH., Kramer G., Harms B.
    Surgery 95 : 136-44 (1984)
13. Tranbaugh RF., Lewis FR., Christensen JM., Elings VB.
    Ann Surg 192 : 479-88 (1980)
14. Goodwin CW., Dorethy J., Pruitt BA Jr.
    Ann Surg 197 : 520-31 (1983)
15. Demling RH.
    JAMA 250 : 1438-40 (1983)
16. Moylan JA.
    J Trauma 19 : 917 (1979)
17. Cahalane M., Demling RH.
    JAMA 251 : 771-3 (1984)
18. Crapo RO.
    JAMA 246 : 1694-6 (1981)
19. Loke J., Paul E., Virgulto JA., Smith GJW.
    Arch Surg 119 : 956-9 (1984)
20. Robinson NB., Hudson LD., Robertson HT., Thorning DR., Carrico CJ.,
    Heinbach DM.
    Surgery 90 : 352-63 (1981)
21. Tranbaugh RF., Elings VB., Christensen JM., Lewis FR.
    J Trauma 23 : 597-604 (1983)
22. Levine BA., Petroff PA., Slade LC., Pruitt BA Jr.
    J Trauma 18 : 188-93 (1978)
23. Marshall WG Jr., Dimick AR.
    J Trauma 23 : 102-5 (1983)
24. Teplitz C.
    Phildelphia: WB Saunders, 45-94 (1979)
25. Demling RH.
    Phildelphia: Harper & Row 1348-51 (1984)
26. Burke JF., Quinby WC., Bondoc CC., Sheehy FM., Moreno HC.
    Ann Surg 186 : 377-87 (1977)

27. Demling RH., Maly J.
    Ann NY Acad Sci 353 : 294-9 (1980)
28. Demling RH.
    J Trauma 23 : 179-84 (1983)
29. Alexander JW., Ogle CK., Stinnett JD., MacMillan BG.
    Ann Surg 188 : 809-16 (1978)
30. Wolfe JHN., Wu AVO., O'Connor NE., Saporoschetz I., Mannick JA.
    Arch Surg 117 : 1266-71 (1982)
31. Miller CL., Baker CC.
    J Clin Invest 63 : 202-10 (1979)
32. Stinnett JD., Loose LD., Miskell P., Tenney CL., Gonce SJ.,
    Alexander JW.
    Ann Surg 192 : 776-82 (1980)
33. Lanser ME., Saba TM., Scovill WA.
    Ann Surg 192 : 776-82 (1980)
34. Wilmore DW., Aulick LH.
    Surg Clin North Am 58 : 1173-87 (1978)
35. Wilmore DW., Mason AD Jr., Johnson DW., Pruitt BA Jr.
    J Appl Physiol 38 : 593-7 (1975)
36. Shamoon H., Hendler R., Sherwin RS.
    J Clin Endocrinol Metab 52 : 1235-41 (1981)
37. Bessey PQ., Watters JM., Aoki TT., Wilmore DW.
    Ann Surg 200 : 264-81 (1984)
38. Herndon D.
    J Trauma 21 : 701-7 (1981)
39. Dinarello CA.
    N Engl J Med 311 : 1413-8 (1984)
40. Taylor JW., Hander EW., Skreen R., Wilmore DW.
    J Surg Res 20 : 313-20 (1976)
41. Caldwell FT Jr., Bowser BH., Crabtree JH.
    Ann Surg 193 : 579-91 (1981)
42. Alexander JW., MacMillan BG., Stinnett JD., et al.
    Ann Surg 192 : 505-17 (1980)
43. Dominioni L., Trocki O., Mochizuki H., Fang CH., Alexander JW.
    J Burn Care Rehab 5 : 106-12 (1984)
44. Hunt TK.
    J Trauma 24 : Suppl:s 39-49 (1984)
45. Hunt TK., Sheldon G., Fuchs R.
    Burns 1 : 210-6 (1975)
46. Goetzl EJ.
    Med Clin North Am 65 : 809-28 (1981)
47. Hunt TK., Andrews WS., Halliday B., et al.
    Philadelphia: Lea & Febiger 1-18 (1981)
48. Banda MJ., Knighton DR., Hunt TK., Werb Z.
    Proc Natl Acad Sci USA 79 : 7773-7 (1982)
49. Knighton DR., Hunt TK., Scheuenstuhl H., Halliday BJ., Werb Z.,
    Banda MJ.
    Science 221 : 1283-5 (1983)
50. Martin BM., Gimbrone MA Jr., Unanue ER., Cotran RS.
    Fed Proc 40 : 335 abstract (1981)
51. Hunt TK., Pai MP.
    Surg Gynecol Obstet 135 : 561-67 (1972)
52. Knighton DR., Hunt TK., Thakral KK., Goodson WH III.
    Ann Surg 196 : 379-88 (1982)

53. Alvarez OM., Mertz PM., Eaglstein WH.
    J Surg Res 35 : 142-8 (1983)
54. Sporn MB., Roberts AB., Shull JH., Smith JM., Sodek J.
    Science 219 : 1329-31 (1983)
55. Burke JF., Quinby WC Jr., Bondoc CC.
    Surg Clin North Am 56 : 477-94 (1976)
56. Engrav LH., Heimbach DM., Reus JL., Harnar TJ., Marvin JA.
    J Trauma 23 : 1001-4 (1983)
57. Demling RH.
    J Trauma 24 : 830-4 (1984)
58. Janzekovic Z.
    J Trauma 10 : 1103-8 (1970)
59. Wolfe RA., Roi LD., Flora JD., Feller I., Cornell RG.
    JAMA 250 : 763-6 (1983)
60. Gray DT., Pine RW., Harnar TJ., Marvin JA., Engrav LH., Heimbach DM.
    Am J Surg 144 : 76-80 (1982)
61. Barnett A., Berkowitz RL., Mills R., Vistnes LM.
    Am J Surg 145 : 379-81 (1983)
62. Park GB.
    Biomater Med Devices Artif Organs 6 : 1-35 (1978)
63. Thornton JW., Taves MJ., Harney JH., et al.
    Burns 3 : 23-9 (1978)
64. Chvapil M.
    J Biomed Mater Res 16 : 245-63 (1982)
65. Zachary L., Heggers JP., Robson MC., Leach A., Ko F., Berta M.
    J Trauma 22 : 833-6 (1982)
66. Gallico GG III., O'Connor NE., Compton CC., Kehinde O., Green H.
    N Engl J Med 311 : 448-51 (1984)
67. Green H., Kehinde O., Thomas J.
    Proc Natl Acad Sci USA 76 : 5665-8 (1979)
68. Billingham RE., Reynolds J.
    Br J Plast Surg 5 : 25-36 (1952)
69. Burke JF., Yannas IV.
    Fort Sam Houston, Tex.: United States Army Institute of Surgical
    Research, 174-6 (1983)

INFECTION : CAUSE OR EFFECT OF PATHOPHYSIOLOGIC CHANGE IN BURN AND TRAUMA PATIENTS

B. A. Pruitt, Jr.
US Army Institute of Surgical Research Fort Sam Houston, TX

The survival of burn patients has increased significantly over the past 40 years as a result of improvements in both general care and burn specific treatment.[1] The use of effective topical chemotherapy to control microbial proliferation within injured tissue has significantly reduced the incidence of invasive burn wound sepsis, even in patients with extensive burn injury, and has altered the characteristics of the burn wound flora and hence the causative organisms of the invasive infections that do occur, i.e., Pseudomonas burn wound infections have become relatively rare and yeast and fungal infections relatively common.[2] Even though present day management has reduced the occurrence of burn wound infections, infection in other sites remains the most frequent cause of morbidity and mortality in successfully resuscitated patients with burns and other injuries (Table I). The incidence of infection appears to be proportional to the severity of injury, e.g.,burn size, and to reflect both systemic and local effects of injury which predispose such patients to infection, confound its diagnosis and make it difficult to differentiate the cause and effect relationships of injury and infection.[3]

LOCAL AND SYSTEMIC EFFECTS OF BURN INJURY

Exposure to thermal energy of sufficient magnitude and duration causes tissue damage of variable degree. Coagulation necrosis and cell death involve the entire thickness of the skin in third-degree injury and are associated with immediate microvascular thrombosis and permanent occlusion of the local blood supply. The zone of immediate coagulation is surrounded by a concentric zone of stasis in which cells sustain potentially reversible injury and the blood supply is impaired but amenable to restoration, provided hypovolemia is promptly corrected, wound surface desiccation is avoided, and infection is prevented. The zone of stasis is, in turn, surrounded by a zone of hyperemia characterized by vasodilatation and increased blood flow.[4]. Vascular endothelial injury is produced and permeability edema occurs in tissue subjected to temperatures over 43° C.[5] Increased vascular permeability results in transcapillary efflux of fluid into the interstitium and a decrease in circulating blood volume ensues. In the area of burn injury, precapillary resistance appears to decrease and permits transmission of near arteriolar pressure to the capillary wall as post-capillary resistance remains unchanged or even slightly elevated. Additionally, Arturson and Mellander have reported an early increase in interstitial fluid and venous effluent osmolality indicative of osmotically induced transfer of intravascular fluid into burn injured tissue[6].

Physiologically active materials are released from burned tissue as evident in various studies of burn blister fluid, burn wound lymph, and venous blood, in which increased levels of histamine, serotonin, bradykinin, prostaglandins, leukotrienes, and interleukin I, as well as activation products of the alternative complement pathway have been identified and implicated in the local hemodynamic changes, remote organ responses, and immunologic changes that occur following burn injury.[7],[8],[9],[10],[11] These materials of burn wound origin, in combination with neurohormonal responses to injury and hypovolemia, induce generalized physiologic changes that involve all organ systems and are proportional to burn size ; such changes are characterized by a biphasic pattern of immediate post-injury hypofunction and later hyperfunction in successfully resuscitated patients[1].

Locally, the injury destroys the mechanical barrier of the skin and permits microbial invasion of the protein rich avascular eschar. Unbridled microbial proliferation occurs and local vascular obstruction precludes delivery of both systemically administered antibiotics and the cellular components of the host defense system. Burn wound edema of sufficient magnitude may further compromise tissue viability by increasing intercapillary distance across which oxygen and metabolites must diffuse, and beneath areas of third-degree injury tissue pressure may increase to a level that impedes blood flow in nutrient capillaries and causes ischemia of unburned tissue. The latter situation is corrected by incision of the overlying unyielding eschar. The blood supply as related to the depth of the wound is a critical factor in local susceptibility to infection,[12] i.e., avascular full-thickness burns are more readily colonized, have a higher bacterial density, and are more often the sites of invasive wound infection than are partial-thickness burns or split-thickness skin graft donor sites that appear to be resistant to invasive wound infection except when systemic hypotension causes further cell injury. Recent studies have emphasized the importance of the extent of burn in both local and systemic susceptibility to infection. Yurt, et al. in studies of a murine model, found that a 30 percent partial-thickness burn previously resistant to surface inoculation of bacteria became susceptible to microbial invasion following infliction of an additional 30 percent full-thickness burn that remained unseeded[13] (Table II). Further studies from that laboratory have shown that even though the number of circulating neutrophils is similar following either a 30 percent or 60 percent burn, there was only half as many neutrophils in the wounds of the animals with the larger burn. Four hours after injury, in vivo testing showed that the neutrophils in the animals with the larger burns were more sensitive to infusion of zymosan activated serum.[14] These findings are consistent with indiscrete margination that compromises local wound resistance and may also play a role in the disturbance of function of remote organs such as the lung.

The microbial population on and in the wound also influences the outcome of wound care. This population is initially sparse and consists principally of gram-positive cocci ; it increases in density with time, and by the second postburn week gram-negative bacilli become predominant.[2] Topical chemotherapy retards the proliferation of bacteria but exerts little influence on the time related changes in the character of the wound flora.[15] Treatment pressures exerted by topical and systemic antimicrobials may ultimately result in a wound largely populated by yeasts and fungi. Viral infections also occur in immunocompromised burn patients, presumably due to reactivation of latent virus in the case of herpetic infections of the wound (most common in healing partial-thickness

TABLE I

CAUSES OF DEATH IN BURN PATIENTS
1983 - 1984

NUMBER OF FATAL BURNS	74
PRINCIPAL CAUSE OF DEATH	
SEPSIS	51 (69 %)
INHALATION INJURY	9 (12 %)
CARDIOVASCULAR DISEASE	8 (11 %)
ALL OTHERS	6 (8 %)

TABLE II

EFFECT OF BURN SIZE ON SUSCEPTIBILITY
TO INFECTION AND MORTALITY

EXTENT AND DEPTH OF BURN	DEPTH OF "SEEDED" BURN	NUMBER OF ANIMALS	PERCENT MORTALITY
30 % FULL-THICKNESS	FULL	12	100.0
30 % PARTIAL-THICKNESS	PARTIAL	16	12.5
30 % PARTIAL, 30 % FULL	PARTIAL	16	56.3
30 % PARTIAL, 30 % FULL	—	8	0.0

burns about the face) and to transfusion of blood products in the case of systemic cytomegalovirus infections.[2]

The occurrence of wound or other infections depends upon the balance between host defenses and microbial invasiveness. In addition to the local blood supply and the extent of burn, age influences the susceptibility to infection, with wound infections being most common in burned children, least common in young adults of 15 to 40 years, and of intermediate incidence in older patients. Pre-existing diseases may increase the risk of infection (phycomycotic infections are a particular hazard in diabetics)[16] as do foreign bodies and secondary complications such as hypovolemic shock due to any cause. Suppression of the humoral and all cellular limbs of the immune system proportional to burn size also predisposes the burn patient to both wound and systemic infections.

In addition to microbial density (invasive wound infections are rare when there are fewer than 10^5 organisms per gram of tissue even in the immunocompromised burn patient and are increasingly common as bacterial density increases above that level) enzymes such as collagenase, elastase, lipase, proteases, nucleases, and hemolysin as well as other metabolic products such as slime and vascular permeability factor produced by microorganisms exert destructive local tissue effects and influence local invasiveness and systemic virulence.[15] Endotoxin, a common product of enteric bacilli, and exotoxins produced by many bacteria also exert both local effects by activation of cellular components within the wound and systemic effects as a result of complement activation. Microbial motility appears to be important for invasion of a surface wound.[17]

The use of effective topical chemotherapy has significantly reduced the incidence of invasive burn wound infection.[18] Unfortunately, none of the three commonly used agents of verified effectiveness, i.e., mafenide acetate burn cream, 0.5 percent silver nitrate soaks, or silver sulfadiazine burn cream, sterilize the burn wound, and invasive infection still occurs in certain wounds, usually those of patients with injuries involving more than 30 percent of the body surface. The imperfect protection provided by topical agents necessitates daily examination of the entirety of the burn wound to identify signs of infection at the earliest possible time. The examination is best carried out at the time of daily wound cleansing when all dressings and topical medications have been removed, but the presence of nonviable tissue in the burn wound may obscure local signs of infection and otherwise impair evaluation. A leathery insensate eschar may compromise assessment of apin, sensitivity, tenderness, and tissue turgor. The tinctorial changes characteristic of eschar maturation as well as thermal charring may obscure infection related erythema. Increased blood flow to the area of injury commonly incréases local tissue temperature in the absence of infection, and edema due to the injury per se and to the infusion of resuscitation fluid makes local swelling an imprecise index of infection in the early postburn period.

In spite of these limitations, certain tinctorial and physical changes are characteristic of burn wound infections caused by various microorganisms. The most reliable clinical sign of invasive wound infection is rapid conversion of an area of partial-thickness injury to full-thickness necrosis.[19] Infrequently, surface desiccation of an exposed deep partial-thickness burn may cause such extension of tissue injury. The most common wound change indicative of infection is the appearance of focal dark red, brown, or black discoloration in the eschar, but focal hemorrhage secondary to local minor trauma may mimic that sign of infection. Unexpectedly rapid separation of an eschar often occurs in the presence of invasive fungal infection, but similar early separation can occur in areas of severe thermal injury where sufficient heat has been delivered to cause liquifaction of the subcutaneous fat.

The imprecision and unreliability of local signs and symptoms necessitates that other means be used to diagnose burn wound infection. A variety of surface culture techniques useful in epidemiologic monitoring and the characterization of the microbial flora of burn wounds are subject to both falsely positive and falsely negative results with such frequency as to make even quantitative culture techniques unreliable in diagnosing burn wound infection. Even quantitative cultures of burn wound biopsy specimens are subject to severe limitations, as indicated by the discordance of paired samples reported by Woolfrey, et al.[20] and a recent study at the US Army Institute of Surgical Research which found a generally good correlation between microbial densities of 10^5 organisms or less per gram of tissue with negative biopsy histology but a very poor correlation between microbial densities as high as 10^8 organisms per gram of tissue and histologic evidence of wound invasion.[21]

Histologic examination of a burn wound biopsy is the most reliable means of differentiating colonization of nonviable tissue from invasive infection of viable tissue.[3] A scalpel is used to obtain from that area of the wound showing the most prominent changes indicative of infection, a 500 mg lenticular tissue sample that must include underlying unburned subcutaneous tissue. The tissue sample is processed by either rapid section technique or a newly described frozen section technique[22] with which slides can be prepared for histologic examination within 30 minutes.[23] Identification of microorganisms (bacteria, fungi, or viruses) in viable tissue confirms the diagnosis of invasive burn wound infection. Falsely negative histologic readings can be obtained if the biopsy specimen is obtained from a non-representative area of the wound and does not include infected tissue, or as a result of an erroneous histologic interpretation. A falsely positive histologic reading may occur if the biopsy specimen does not include unburned tissue and only bacteria laden burned tissue is examined or if there is an interpretation error. Even with adequate biopsy sampling, the new frozen section method is associated with a 3.6 percent rate of falsely negative histologic readings, but such errors can be corrected by subsequent review of permanent sections. The results of the histologic examination of biopsy tissue must always be interpreted in the light of the overall condition of the patient. A histologically negative biopsy in the presence of systemic signs of sepsis should prompt a repeat biopsy and, if that too is negative, infection in a site other than the burn wound should be sought.

Treatment of invasive burn wound infection includes institution of general supportive measures, alteration of topical therapy, institution of systemic antibiotic therapy, and the subeschar infusion of a semi-synthetic penicillin as a prelude to surgical removal of the infected tissue. Timely biopsy diagnosis of invasive burn wound infection with prompt institution of the treatment program described has arrested the infection in 53 percent and been associated with survival of 26 percent of a small group of patients with histologically confirmed invasive burn wound infection.[24] The importance of early diagnosis was evident in the surviving patients, in whom the infection was identified before sufficient microvascular involvement had occurred to produce hematogenous dissemination, i.e., all blood cultures were negative.

As a result of global impairment of the immune system and other systemic effects of severe injury, infection remains the most common cause of death in hospitalized burn patients and pneumonia has supplanted burn wound infection as the most common form of spesis[25] (Table III). The reduction in invasive burn wound infection has also resulted in a marked change in the predominant form of pneumonia. The incidence of hematogenous pneumonia, an infection caused by lodgement in pulmonary capillaries of blood borne microorganisms arising in a remote source, has decreased, and the incidence of airborne or bronchopneumonia has increased.[26] An extensive burn of the chest wall and edema of the underlying tissues may attenuate auscultatory findings and compromise physical examination of the chest (burn wounds of the abdominal wall may also compromise examination of the abdomen and make diagnosis of intraabdominal sepsis difficult). Radiographic diagnosis of pneumonia, too, may be imprecise, since the roentgenographic signs of pneumonia may be mimicked by those of pulmonary edema or by early peribronchial inflammatory changes in patients with severe inhalation injury.

TABLE III

INFECTIONS IN 74 FATAL BURNS
1983 - 1984

INFECTION*	NUMBER OF PATIENTS
PNEUMONIA	43
BURN WOUND INFECTION	29
ENDOCARDITIS	9
SUPPURATIVE	
THROMBOPHELEBITIS	6
ALL OTHERS	4

* SEPTICEMIA ALSO PRESENT IN 42.

Because of the equivocal nature of roentgenographic findings, reliance has been placed on the staining characteristics and culture of endobronchial secretions to diagnose pulmonary infection, but the rapid polymicrobial colonization of the airway of patients requiring endotracheal intubation often results in the culture recovery of multiple organisms, making it difficult to determine a predominant organism and to specify appropriate antibiotic therapy. Bronchoscopically directed lavage of the involved segment of lung has been proposed as a technique to enhance the recovery of specific causative organisms and the accuracy of diagnosis of pneumonia, but a recent study at the US Army Institute of Surgical Research has shown that culture results from endoscopically directed lavage were no different from those of transtracheal aspiration.

Recovery of microorganisms from or identification of microorganisms within viable tissue constitute the definitive means of diagnosing infection. The problems surrounding such confirmation in severely injured patients have served as rationalization for the use of systemic signs of "sepsis" as an alternate means of diagnosing the presence of infection. A change in body temperature is a generally reliable but usually late sign of severe sepsis. Sudden ocurrence of hypothermia or even reversion to normothermia in a previously hyperthermic injured patient are ominous signs of systemic life-threatening gram-negative sepsis. Similarly, ileus in a patient with a previously functioning gastrointestinal tract and disorientation in a previously lucid patient may be systemic manifestations of infection.

Many of the physiologic criteria of sepsis that have been advanced, i.e., hyperthermia, tachycardia, hyperventilation, disorientation, and impaired gastrointestinal motility are also elicited by burn or mechanical injury in the absence of infection.[3] The multisystem pathophysiologic changes evoked by burn injury that may be mirrored by the response to sepsis are of greatest intensity in the early post-resuscitation phase. The magnitude of organ response diminishes in intensity with time as the burn wound heals or is closed by grafting, and organ function returns to normal as convalescence proceeds.[27] Even though infection may elicit systemic changes resembling those due to trauma and be unapparent in a patient with an extensive burn in the early postburn period, sepsis occurring later will commonly exaggerate the injury related state present at that time, and life-threatening sepsis may even cause reversion to a "primitive" early post-injury condition. It is essential to evaluate physiologic indices in patients in whom infection is suspected in terms of the level anticipated for each variable in relation to extent of burn and the time post-injury ; even so, physiologic interactions often make it difficult to identify infection specific changes.

The similarity and non-specificity of local and systemic changes due to injury and infection have focused attention on biochemical, hematologic, and hormonal assays and the quantification of the physiologic manifestations of biologically active materials either as indicators of an increased susceptibility to infection (cause) or criteria of the presence of infection (effect). It now appears as if similar changes in these assays and measurements may be induced by either injury or infection, and the interpretation of changes occurring in serial measurements and the identification of an observed change as a significant causative factor of infection is further complicated by the time required for culture confirmation of infection. Even with the use of radiometric techniques, cultures may not be positive for 24 hours and speciation of the causative organism requires an additional day. Verification of yeast, fungal, and viral infections requires even longer. A biologic change representing an effect of sepsis occurring during either the in vivo or in vitro incubation period may be erroneously assigned causative significance.

Hyperglycemia, increased circulating levels of catecholamines, corticosteroids, and glucagon, increased levels of complement activation products, increased production of arachidonic acid metabolites, and increased production of interleukin I, as well as decreased circulating levels of thyroid hormones, decreased insulin-glucagon molar ratio, decreased fibronectin levels, and deceased levels of certain coagulation factors are all components of the multisystem response to injury.[3] Initial marked leukocytosis followed by leukopenia, and secondary rebound leukocytosis, and changes in neutrophil function as well as changes in lymphocyte subpopulations also characterize the response to injury. These hematologic changes are widely believed to predispose injured patients to infection, particularly the decrease in T-lymphocyte helper-suppressor ratio,[28] but both leukocytosis and leukopenia and changes in lymphocyte subpopulations are also consequences of sepsis. The temporal relationship between observed changes and the onset of sepsis is crucial in determining any cause and effect relationship between these changes and subsequent infection.

The use of an established animal model of burn wound infection has

TABLE IV

EFFECT OF BUNR INJURY AND INFECTION ON
HELPER AND SUPPRESSOR LYMPHOCYTES

TRAITEMENT GROUP	HELPER-SUPPRESSOR RATIO (Mean value $\pm$ sd)
CONTROL	2.28 ± 0.30
BURN	2.13 ± 0.32
BURN PLUS INFECTION	1.03 ± 0.54*

*$p < .001$ vs. CONTROL
$p < .001$ vs. BURN GROUP

overcome some of the previously noted limitations of clinical studies and provided evidence that changes in lymphocyte subpopulations are often an effect, rather than a cause, of infection. Burleson, et al., in a murine model of a 30 percent burn with and without infection, have found density gradient lymphocyte fractions to be contaminated by a variable number of various non-lymphoid cells that can distort functional assays of these cell fractions. Using monoclonal antibodies and light scatter sorting technology, these investigators found that an infected burn, but not a burn without infection, caused a relative decrease in T-lymphocytes and a decrease in the ratio of helper to suppressor lymphocytes (Table IV). With gated light scatter determinations, which further reduced non-lymphoid cell contamination, the purity of the T-lymphocyte fraction increased, the helper subset fraction increased and the suppressor subset fraction decreased slightly following infection. These investigators conclude that the decrement in the helper-suppressor ratio in this model was induced by infection rather than by burn injury. The decrement in helper/suppressor ratio occurred because of a relatively greater decrease in helper than in suppressor cells.[29]

Use of the standard murine model of Pseudomonas burn wound infection has also permitted assessment of the role of microbial toxins in the pathogenesis of burn wound sepsis. Exotoxin A, produced by some Pseudomonas strains, inhibits protein synthesis and has been proposed as the cause of high mortality in mice following intraperitoneal injection of Pseudomonas organisms producing exotoxin A. In this model, protection was afforded by prechallenge immunization with toxin A toxoid. Studies employing the rat model of Pseudomonas seeding of the burn wound surface, however, demonstrate consistent invasive infection and hematogenous spread of the infection to remote organs in similarly immunized test animals, with no protection afforded by prechallenge immunization with the toxoid.[30] (Table V) These contrasting results suggest that the mouse model represents a toxicity model, rather than an infection model, and that exotoxin A is of little importance in invasive infection of a burn wound.

Animal models of burn injury have also been used to elucidate the role of vasoactive agents in the local and systemic changes that occur early post-injury. Yurt, et al., using a murine model of burn injury in which baseline histamine values in unperturbed animals were comparable to baseline values in man, has identified plasma histamine concentration elevations within one minute after thermal injury. The increase in plasma histamine concentration was proportional to the extent of surface area injured, with a single time-related plasma histamine peak observed in animals subjected to a partial-thickness burn and a biphasic elevation noted after full-thickness injury.[31] These results justify consideration of histamine as a mediator of both the local and systemic response to burn injury. In further studies, these investigators found that degranulation of mast cells by intra-peritoneal administration of polymyxin B depleted histamine stores, as confirmed by marked suppression of the post-injury increase in central venous histamine concentration, but did not alter wound edema formation[32] (Table VI). These results suggest that histamine may not be responsible for burn wound edema and that other mechanisms influence local changes following such injury. Studies by others, in various animal models, have shown that administration of inhibitors or antagonists, such as anti-histamines, histamine H2 receptor antagonists, ketoconazole, and calcium channel inhibitors only partially block edema formation, further supporting a multifactorial etiology of the local effects of burn injury.[9,33]

Studies in both injured man and animal models have revealed a variety of humoral indicators common to both injury and infection, and others that appear to be infection specific. Powanda, et al., have detected, in burned and burned infected rats, one circulating factor that absorbs light at 398 nm and two that fluoresce at 340 nm and 420 nm, respectively. The factor fluorescing at 340 nm is a response to either infection or injury. The other two factors were correlated with the extent of burn during the first 48 hours post-injury, but were subsequently related only to the presence of infection.[34] Subsequent studies by Lin and Burleson have shown that the factor fluorescing at 420 nm is a reliable indicator of infection in both the murine model and burn patients. In patients it was evident only in advanced terminal stages of sepsis. This factor showed spectral similarity to a conjugated diene and its blood concentration correlated with increased malondialdehyde concentration in plasma filtrates and tissue, particularly renal tissue, from both burned and burned infected animals[35] (Table VII). These findings suggest that lipid peroxidation may generate this circulating factor in injured and injured infected animals. These investigators, on the basis of chromatographic similarities, propose that pterin derivatives also contribute to the fluorescence at 420 nm. Burleson, more recently, has reported that the factor fluorescing at 340 nm has an apparent molecular weight of approximately 76,600 and is comprised of at least two major acid glycoprotein components with significantly different amino acid compositions.[36]

Other investigators have identified elevated plasma levels of serine proteases in the blood of critically ill and septic patients[37] and Powanda et al., have also reported that infection in burn patients is associated with significant decreases in plasma levels of haptoglobin and IgM and a significant increase in plasma levels of alpha $_1$-acid glycoprotein.[38] In both the animal and human studies reported, the humoral indicators appear to be present only in advanced or pre-terminal sepsis and do not facilitate early diagnosis of clinically unapparent sepsis.

This review permits one to draw certain conclusions and offer the following recommendations : (1) The local and systemic responses to injury

TABLE V

EFFECT OF PSEUDOMONAS EXOTOXIN A TOXOID,
ON MURINE BURN WOUND INFECTION MORTALITY

IMMUNIZATION	CHALLENGE	OBSERVED MORTALITY
NONE	LIVE PSENDOMONAS	100 %
LIVE PSEUDOMONAS STRAIN	IMMUNIZING STRAIN	0
LIVE PSEUDOMONAS STRAIN	NON-IMMUNIZING STRAIN	100 %
ALUMINUM PHOSPHATE ABSORBED TOXIN A TOXOID	LIVE PSEUDOMONAS	100 %
ALUMINUM PHOSPHATE ADJUVANT	LIVE PSEUDOMONAS	100 %

TABLE VI

EFFECT OF MAST CELL DEGRANULATION ON BURN WOUND EDEMA

TREATMENT GROUP	MAST CELLS PER VESSEL	PERCENT WATER CONTENT OF WOUND (Mean $\pm$ SEM)
SHAM BURN :		
SALINE	1.97 $\pm$ 0.16)	64.71 $\pm$ 0.24
	)**	
POLYMYXIN B	0.70 $\pm$ 0.17)	64.01 $\pm$ 0.21
30 % PARTIAL-THICKNESS BURN :		
SALINE	1.67 $\pm$ 0.22)	71.41 $\pm$ 0.46)
	)***	)N.S.
POLYMYXIN B	0.27 $\pm$ 0.05)	70.62 $\pm$ 0.51)

N.S. No significant difference

** p < 0.01)
) ONE-WAY ANOV
*** p < 0.001)

TABLE VII

RENAL TISSUE AND PLASMA SEPSIS FACTOR AND MALONDIALDEHYDE FLUORESCENT ACTIVITY

420 nm FLUORESCENT FACTOR ACTIVITY

TREATMENT GROUP	DAYS POST-INJURY	PLASMA rfu/0.1 ml	KIDNEY rfu/0.2 gm	MALONDIALDEHYDE CONCENTRATION KIDNEY OD UNITS AT 530
CONTROL	3	11.0 $\pm$ 2.5	816 $\pm$ 79.2	.093 $\pm$ .013
BURN	3	17.6 $\pm$ 6.0	1128 $\pm$ 77	0.244 $\pm$ .039
BURN-INFECTED	3	44.3 $\pm$ 24.8	1140 $\pm$ 342	0.303 $\pm$ .001

or sepsis are strikingly similar and confound the clinical diagnosis of both local and systemic infections. (2) The biochemical, hematologic, immunologic, and endocrine changes that accompany injury mimic those occurring with sepsis, making it tenuous to attribute susceptibility to infection to any of those factors. (3) The evaluation of any relationship between post-injury change and the subsequent occurrence of sepsis is best carried out in appropriate animal models in which specific temporal relationships can be defined. (4) Studies in an animal model indicate that change in the helper and suppressor lymphocyte population are an effect not a cause of infection. (5) The use of relevant animal models has permitted definition of the clinical importance of microbial factors in the pathogenesis of infection following burn injury. (6) The use of animal models is also essential in the evaluation of the physiologic significance of vasoactive mediators of the pathophysiologic response to injury. (7) Studies in animal models of burn injury suggest that local burn wound changes such as edema formation and some remote effects of injury are multifactorial in etiology. (8) Humoral factors have been identified in both animal models and patients that appear to be infection specific but the clinical importance of such factors and the usefulness of their measurement remains undefined.

REFERENCES

1. Pruitt B.A. Jr, American College of Surgeons Bulletin 70:2 (1985).
2. Pruitt B.A. Jr, Burns 11 : 79 (1984).
3. Pruitt B.A. Jr, Arch Surg 121 : 13 (1986).
4. Jackson DMcG, J Trauma 9 : 839 (1969).
5. Mustafa K.Y., Selig W.M., Burhop K.E., Minnear F.L. and Malik A.B. J Appl Physiol 60 : 1980 (1986).
6. Arturson G. and Mellander S., Acta Physiol Scand 62 : 457 (1964).
7. Shea S.M., Caulfield J.B. and Burke J.F., Microvasc Res 5 : 87 (1973).
8. Herdon D.N., Abston S. and Stein M.D., Surg Gynecol Obstet 159 : 210 (1984).
9. Alexander F., Mathieson M., Teoh K.H, Huval W.V., Lelcuk S., Valeri C.R., Shepro D. and Hechtman H.B., J Trauma 24 : 709 (1984).
10. Bjornson A.B., Bjornson H.S., J Infect Dis 153 : 1098 (1986).
11. Gelfand J.A., Donelan M., Burke J.F., Ann Surg 198 : 58 (1983).
12. Order S.E., Mason A.D. Jr, Switzer W.E., Ann Surg 161 : 502 (1965).
13. Yurt R.W., McManus A.T., Mason A.D. Jr and Pruitt B.A. Jr, Arch Surg 119 : 183 (1984).
14. Yurt R.W. and Pruitt B.A. Jr, Surgery 98 : 191 (1985).
15. Pruitt B.A. Jr, Pseudomonas aeruginosa, Huber, Berne : 55 (1979).
16. Pruitt B.A. Jr, Problems in General Surgery, 1 : 664 (1984).
17. McManus A.T., Moody E.E. and Mason A.D., Burns 6 : 235 (1980).
18. Pruitt B.A. Jr, O'Neill J.A. Jr, Moncrief J.A. and Lindberg R.B. JAMA 203 : 1054 (1968).
19. Pruitt B.A. Jr, Infection and the Surgical Patient : 113 (1982).
20. Woolfrey B.F., Fox J.M., Quall C.O., Am J Clin Pathol 75 : 532 (1981).
21. McManus A.T., Kim S.H., Mason A.D. Jr, Arch Surg, in press.
22. Kim S.H., Hubbard G.B., Worley B.L. J Burn Care Rehab 6 : 433 (1985).
23. Kim S.H., Hubbard G.B., McManus W.F., Mason A.D. Jr and Pruitt B.A. J Trauma, 25 : 1134 (1985).
24. McManus W.F., Goodwin C.W. Jr, Pruitt B.A. Jr, Arch Surg 118 : 29 1 (1983).

25. Pruitt B.A. Jr, <u>Current Problems in Surgery</u>, 16 : 52 (1979).
26. Pruitt B.A. Jr, DiVincenti S.C., Mason A.D. Jr, Foley F.D. and Flemma R.J., <u>J. Trauma</u> 10 : 519 (1970).
27. Pruitt B.A. Jr and Goodwin C.W. Jr, <u>Nutritional Support of the Seriously Ill Patient</u> : 63 (1983).
28. McIrvine A.J., O'Mahony J.B., Saporoschetz I and Mannick J.A., <u>Ann Surg</u> 196 : 297 (1982).
29. Burleson D.G., Vaughan G.K., Mason A.d., <u>Arch. Surg.</u>, in press.
30. Walker H.S., McCleod C.G. Jr, Lepla S.H. Evaluation of Pseudomonas aeruginosa toxin A in experimental rat burn wound sepsis. Annual Research Progress Report, US Army Institute of Surgical Research, Brooke Army Medical Center, For Sam Houston, TX pp 286-281, 1978.
31. Yurt R.W. and Pruitt B.A. Jr, <u>J. Appl. Physiol</u> 60 : 1782 (1986).
32. Yurt R.W., Mason A.D. Jr and Pruit B.A. Jr, <u>Surg Forum</u> 33 : 71 (1982).
33. Boykin J.V. Jr, Eriksson E., Sholley M.M. and Pittman R.N., <u>Science</u> 209 : 815 (1980).
34. Powanda M.C., Dubois J. and Villareal Y, Annual Research Project Report, US Army Institute of Surgical Research, Brooke Army Medical Center, Fort Sam Houston, TX pp 342-351, 1981.
35. Lin K-T.D., Burleson D.G. and Powanda MC, Annual Research Progress Report, US Army Institute of Surgical Research, Brooke Army Medical Center, Fort Sam Houston, TX 1984.
36. Burleson D.G., Lin K-T.D., Anders J.C., Annnual Research Progress Report, US Army Institute of Surgical Research, Fort Sam Houston, TX 1985.
37. Rund T.E., Kierulf P., Godal H.C., Aune S. and Aasen A.O., <u>Adv. Exp. Med. Biol.</u> 167 : 449 (1984).
38. Powamnda M;C., Moyer E.D., Wilmore D.W.

BIOCHEMICAL MEDIATORS IN ACUTE RESPIRATORY DISTRESS SYNDROME (ARDS) AFTER BURNING INJURY

M.E. Faymonville[1], M. Lamy[1], J. Duchateau[2], A. Adam[3],
G. Deby-Dupont[4], J. Micheels[1], D. Jacquemin[5]
[1] Dept. of Anesthesiology, University Hospital, Liège.
[2] Dept. of Immunology, University Hospital, Brussels.
[3] Lab. of Clinical Biology, Centre Hospitalier, Sainte-Ode.
[4] Lab. of Radioimmunology, University of Liège, Belgium.
[5] Dept. of Plastic Surgery, University Hospital, Liège.

SUMMARY

Seventeen burned patients, 5 of whom developed the adult respiratory distress syndrome (ARDS), were investigated. Occurrence of ARDS was correlated with the severity of burn in ARDS patients (UBS : 82 ± 27) and non ARDS patients (UBS : 36 ± 18) respectively ($p < 0.005$) and with inhalation injury.

Blood samples were collected immediately after admission, 6 - 12 h after injury, am/pm the first day and every day during a fortnight. This prospective study demonstrated abnormal C3 consumption as measured by the C3d/C3 ratio in all burned patients, not related to the presence of ARDS. A significant protease-antiprotease imbalance was found in ARDS and non ARDS patients : leukocyte elastase was increased throughout the observation period, α_2 macroglobulin drastically decreased especially in ARDS patients, α_1 proteinase inhibitor early decreased below normal levels in ARDS patients. Finally, we found a time course delayed but persisting acute phase response : C-reactive protein, haptoglobin and α_1 acid glycoprotein reached a plateau on about day 5-7.

These biochemical investigations give no evidence for a specific mediator for acute respiratory failure.

- - - - - - -

The treatment of burns has continuously been improved, based on results from experimental and clinical research. Mortality from major burns has decreased, burn shock has been nearly eliminated as a major cause of death by widespread understanding of the need for early vigorous fluid resuscitation. Advances have been made in monitoring and supporting the body's hemodynamic and metabolic response to trauma; however, respiratory complications have emerged as one of the dominants killers of individuals with major thermal injury. Pulmonary pathology, primarily as a result of inhalation injury, now accounts for 20 to 84 per cent of burn mortality (1,2,3). The presence or absence of inhalation injury may be a stronger determinant of mortality than the size of the burn injury. The lung lesions that are evident clinically in the thermally injured patient are multifactoral in etiology. Pulmonary edema, airway infection and obstruction occur from thermal injury to the skin, inhalation injury and sepsis. The lung lesions associated with pure, thermal damage to the dermis involve an increase in lung lymph flow and edema formation (4,5,6) but the exact mechanism of the microvascular permeability increase remains still unclear. The sequence of events leading to the sudden respiratory failure seen in the adult respiratory distress syndrome (ARDS) after burning injury still lacks complete characterization. Mediators responsible or implicated in the pathogenesis of the syndrome are yet to be identified. Therefore, we looked specifically at patients developing ARDS after burning injury and we decided a prospective study in order to assess differences in biochemical evolution in burned ARDS and non ARDS patients.

The present study was undertaken to determine whether or not acute burning injury with or without ARDS is accompanied by complement system activation, protease-antiprotease imbalance and if the acute phase reactant response is different in burned ARDS and non ARDS patients.

MATERIAL AND METHODS

Patients : 17 patients (34 $\pm$ 17 years, 69 $\pm$ 18 kg BW), referred to the University Hospital of Liège, were investigated. The percent body surface area (BSA) of the burn and its distribution were recorded with a Lund and Browder diagram. The average size of the burn was 31.5 $\pm$ 16 % of the body surface and the extent of the burn, assessed by the measurements of "Unit Burn Standard" (UBS) (7) was 49.5 $\pm$ 31 %. Patients received conventional acute burn care immediately after their admission to the Emergency Unit. The standard regimen that was followed included local debridement of cutaneous burns and skin graft, topical application of 1 % Silver sulphadiazine cream. Fluid resuscitation was performed according to the modified Parkland formula (8,9). When inhalation injury was suspected, a fiberoptic bronchoscopy was performed routinely on admission. Positive findings are airway inflammation or edema, mucosal necrosis, and the presence of soot in the airway.

Patients were separated into 2 groups on the basis of ARDS. Clinical assessment was performed, independently of knowledge of the biochemical data, using predetermined criteria defining the clinical onset of ARDS; ARDS was defined as being present if the patient developed non-hemodynamic pulmonary edema (radiologic whitening of both lungs with no increase in pulmonary capillary wedge pressure) with severe arterial hypoxemia requiring artificial ventilation, with a fraction of inspired oxygen greater than 0.5, a tidal volume of 12 to 15 ml/kg body weight, and a positive end expiratory pressure of at least 5 cm H_2O in order to normalize gas exchange.

Sampling procedure : Blood samples were collected from an indwelling central venous catheter at admission to the emergency unit, 6 and 12 hours after injury, twice daily the first day after injury and once the following days.

For complement factor determinations, the EDTA anticoagulated blood samples were immediately centrifuged, and plasma was stored at - 30°C until assay. Heparinized blood samples were taken for immunoreactive trypsin measurements and blood was drawn in glass tubes containing sodium citrate (0.129 M) for plasma leukocyte elastase determination. Blood was collected in glass tubes and allowed to clot, and serum was used for α_1 proteinase inhibitor (α_1PI), α_2 macroglobulin (α_2M) and acute phase reactants determinations (C-reactive protein, α_1 acid glycoprotein and haptoglobin). After centrifugation (10 min, 1,500 g), the plasma or serum was decanted with plastic pipette and stored at - 30°C.

Plasma and serum assays

Complement fraction 3 (C3) was measured by nephelometry; C3d fraction, a breakdown product of C3, was measured according to the method of Perrin and coworkers (10). Absolute levels of C3d were correlated with total C3 content of normal plasma; therefore, the ratio of C3d to total C3 concentration (C3d/C3 ratio) was calculated in order to assess true in vivo complement consumption. For more details, we refer to the paper of Duchateau and coworkers (11).

Leukocyte elastase was determined by an enzyme-linked immunoassay for human granulocyte elastase in complex with α_1 proteinase inhibitor (12).

Immunoreactive trypsin (IRT : free trypsin, trypsinogen and the complex trypsin-α_1 proteinase inhibitor) was measured by radioimmunoassay (RIA), technique using a rabbit antiserum to human cathodic trypsin, ^{125}I-labelled human cathodic trypsin, and non labelled cathodic trypsin standards (13,14). The lower limit of sensitivity was 5 µg.l^{-1} plasma. However, using this immunological technique, the complex typsin-α_2M could not be detected.

Alpha-1 proteinase inhibitor, α_2M and acute phase proteins were determined by laser nephelometry.

Statistical analysis

Means and standard deviations were calculated using a TRS80 computer. Variance analysis was used for comparison of 2 groups (ARDS and non ARDS).

RESULTS

The physical and clinical characteristics of patients are summarized in table 1.

Table 1 : Physical and clinical characteristics of patients*

	Burn Patients ARDS n=5	Burn Patients non ARDS n=12
Age, yr	35.2 + 10	34.4 + 16
Sex, M : F	1 : 4	2 : 10
Weight, kilogram	75.4 + 12	65.4 + 18
Percent BSA burn**	46.8 + 12	25.2 + 13
Percent UBS burn**	81.8 + 27	36 + 18
Inhalation injury, n (%)	3/2 (60 %)	2/1C (16.6 %)
Mortality	2/5 (40 %)	0/10

Definitions of abbreviations : BSA = body surface area
UBS = unit burn standard

* Values are expressed as mean $\pm$ SD
** p < 0.005

There were no statistically significant differences between ARDS and non ARDS patients with regard to age, sex and weight. The burned surface area (BSA) and the unit burned standard (UBS) were statistically significantly different. Inhalation injury was more frequent in the ARDS than in the non ARDS group (60 % compared to 16.6 %). After inhalation injury, ARDS appeared in 24-48 hours (3 patients); in patients without direct lung injury, ARDS occurred the 6th and 10th day after injury.

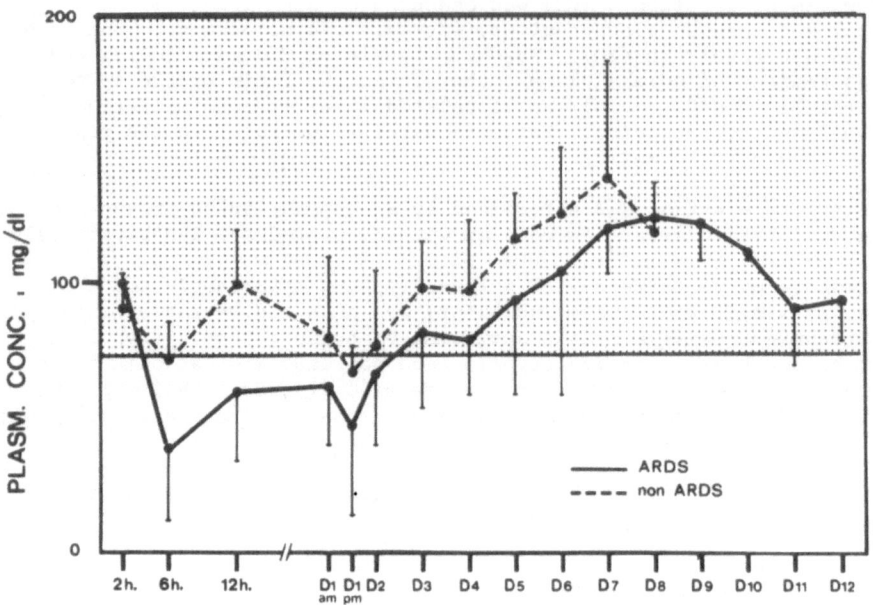

Fig. 1. Evolution of complement fraction 3 (C3) plasma values (mean ± SD) in ARDS and non ARDS patients. The normal range is represented by the dotted line area.

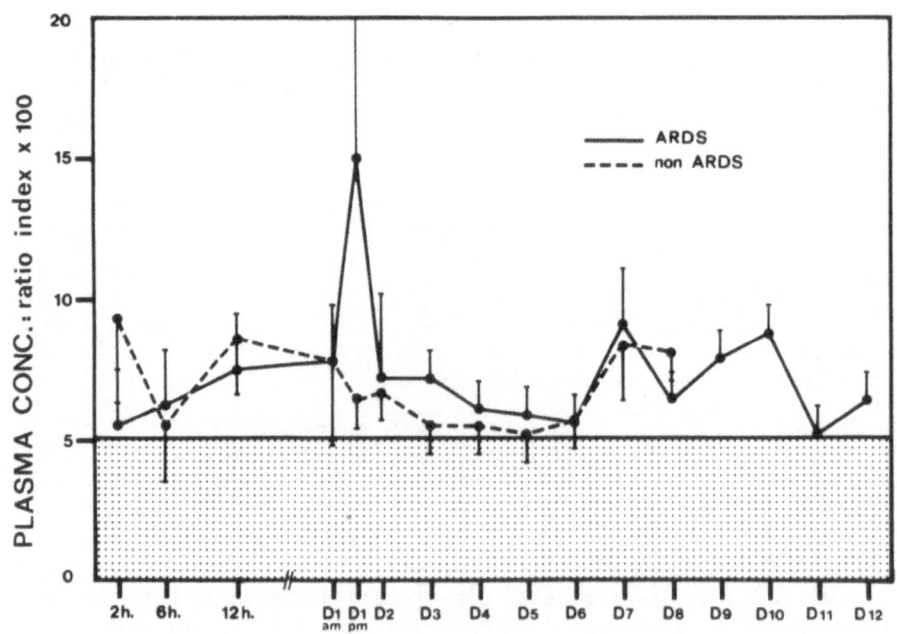

Fig. 2. Evolution of C3d/C3 ratio index in the two burned patients groups.

Evolution of complement components

In figures 1 and 2, a temporal profile of C3 and C3d/C3 ratio values is illustrated in burned ARDS and non ARDS patients. Upon admission to the Emergency Care Unit, in the two patients groups, C3 levels were in lower limit of normal values (reference value $\pm$ 2 SD). Six hours after injury, C3 levels were widely below normal range in the ARDS group, increased slowly to normal range values on day 3, reaching reference value (116 mg/dl) on day 8. In the non ARDS group, C3 levels were below reference value until day 6 but remained always in the lower limit of normal values. Throughout the observation period, the C3d/C3 ratio remained above normal limits in the two patients groups, but no statistically significant differences were found in complement activation between the two patients groups.

Immunoreactive trypsin (IRT)

Figure 3 shows plasma concentrations changes of IRT (ng/ml) in the two patients groups from day of injury to day 12, in comparison to normal values (mean $\pm$ 2 SD) measured in healthy volunteers. Normal range is represented by a shaded area. Higher IRT values were found in the ARDS group than the non ARDS group but these values remained in the upper limit of normal range and there were no statistically significant differences between the two groups.

Leukocyte elastase

As shown in figure 4, leukocyte elastase levels were highly abnormal in burned patients, even 2 h after injury. ARDS burned patients showed a more severe increase in plasma levels (6 fold increase) than non ARDS patients (2 fold), especially during the first 12 h after injury. Afterwards, the levels remained high in the two patients groups without reaching normal levels, even on day 12.

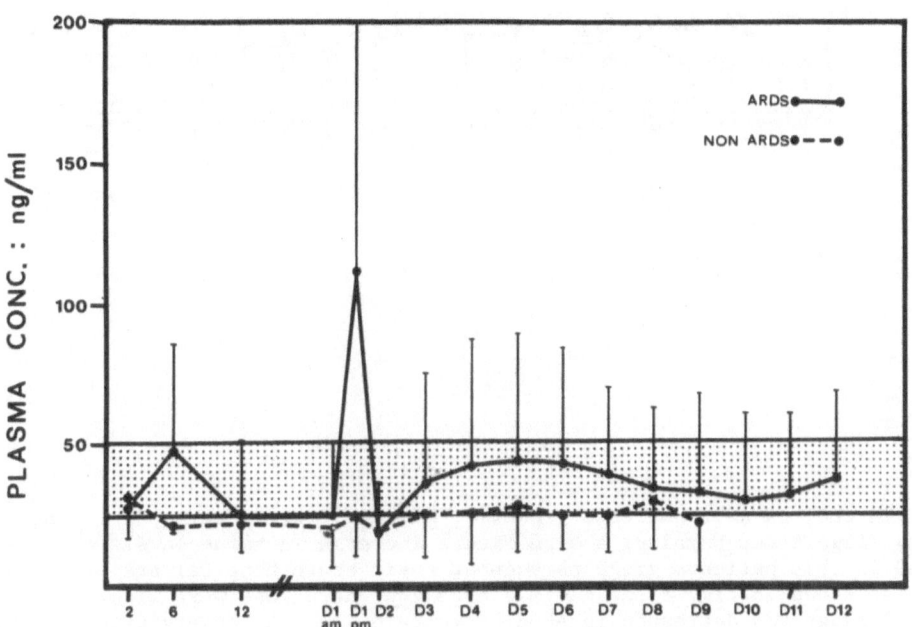

Fig. 3. Time course of plasma concentration changes of immunoreactive-trypsin in burned patients.

<u>Plasma inhibitors</u>

Alpha-1 proteinase inhibitor concentrations were decreased and out of normal range 2 and 6 hours after burning injury in ARDS patients; minimal levels were 1.53 ± 0.53 g/1. Normal range values were obtained 12 h after injury; afterwards these levels increased significantly until day 12.

BURNED PATIENTS ENUTROPHIL ELASTASE

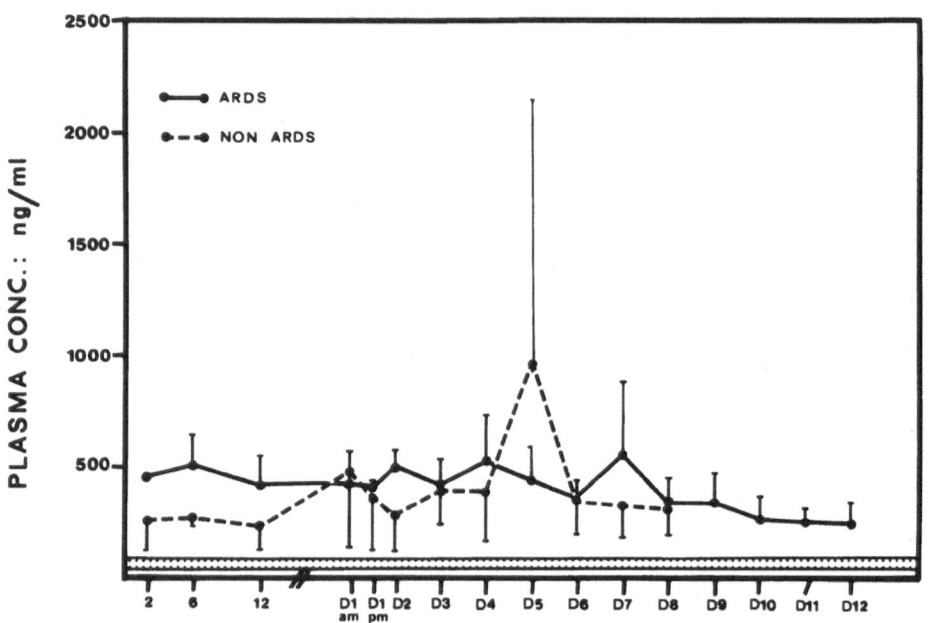

Fig. 4.   Evolution of plasma concentration of the complexe elastase α_1PI in burned patients.

In the non ARDS patients group, α_1 PI levels were decreased in the early stage after burning; a significant increase in serum levels was also noted in this patients group throughout their acute hospital stay (fig.5).

As shown in fig. 6, mean α_2M plasma concentrations were widely below normal range and decreased in ARDS patients during the observation period. These levels were significantly lower than in non ARDS patients. In non ARDS patients, α_2M levels were found to be low, but in normal range until day 1; afterwards these levels decreased also.

48

BURNED PATIENTS α_1-PROTEINASE INHIBITOR

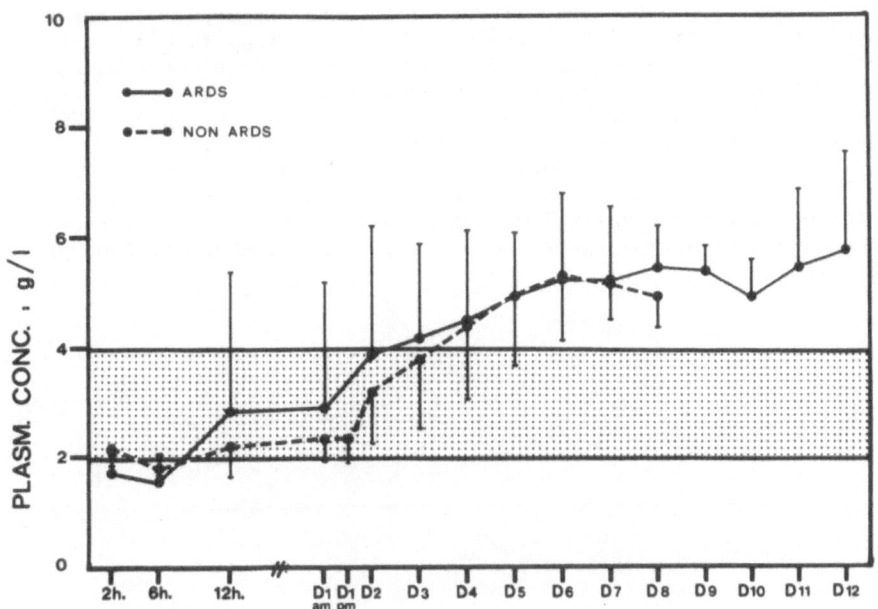

Fig. 5.   Evolution of α_1PI plasma concentration (mean $\pm$ SD) in the two
           patients groups.

BURNED PATIENTS   ALPHA$_2$ - MACROGLOBULIN

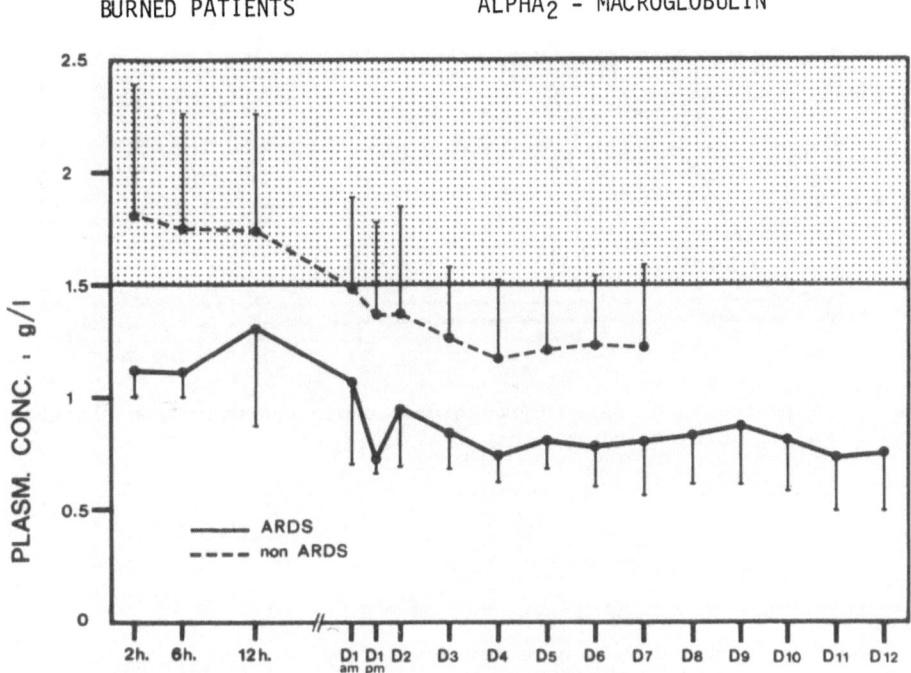

Fig. 6.   Evolution of α_2M plasma concentration (mean $\pm$ SD) in burned
           patients.

Acute phase proteins

The proteins were found to increase after burning injury. C-reactive protein showed the most rapid rise (6-12 h after injury). In the two patients groups, high plasma levels reached a plateau from day 5 to day 7; afterwards these levels decreased in non ARDS patients but remained high throughout the observation period in ARDS patients. CRP values were higher the first day after injury in ARDS patients than in non ARDS patients, but not statistically significantly (fig. 7).

Time course of serum alpha-1 acid glycoprotein concentrations changes after burning injury is shown in fig. 8. Serum concentrations were low, but in normal range, especially in ARDS patients; on day 4-5, these levels increased out of normal range, respectively in non ARDS and ARDS patients (fig. 8).

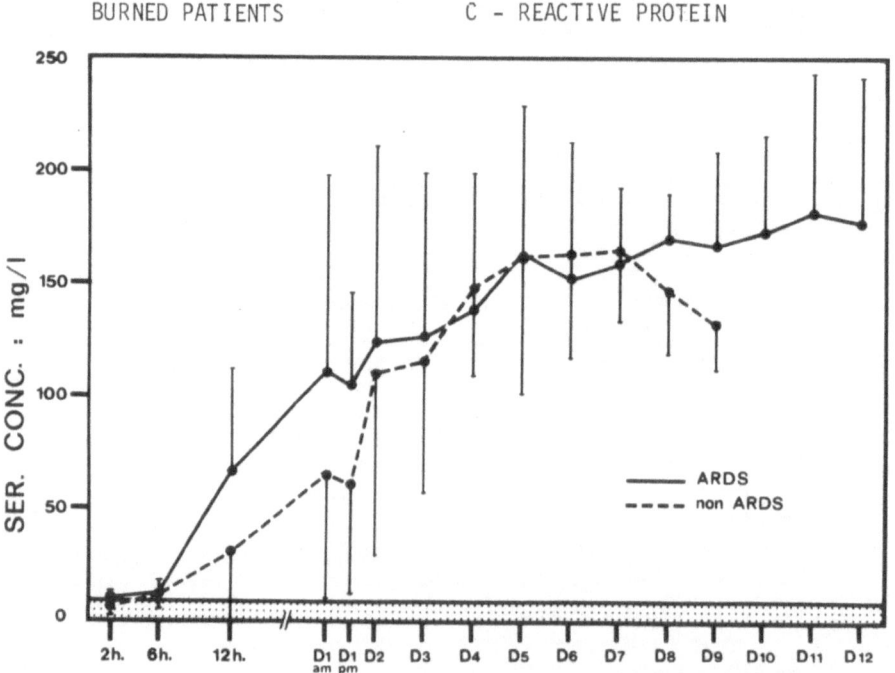

Fig. 7. Time course of serum C-reactive protein concentration changes after burning injury.

Serum haptoglobin levels were low during four days after burning injury in ARDS patients; a slight increase in serum concentrations was noted on day 7 with a constant increase in these levels throughout the observation period. Higher serum levels were observed in non ARDS patients, but essentially the same pattern of changes was observed in the two groups (fig. 9).

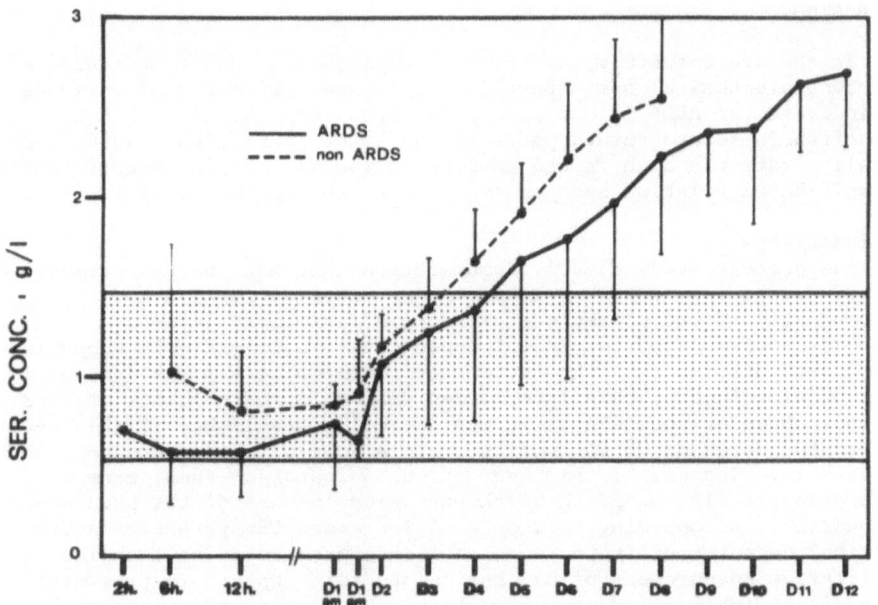

Fig. 8. Time course of serum α_1 acid glycoprotein concentration changes after burning injury.

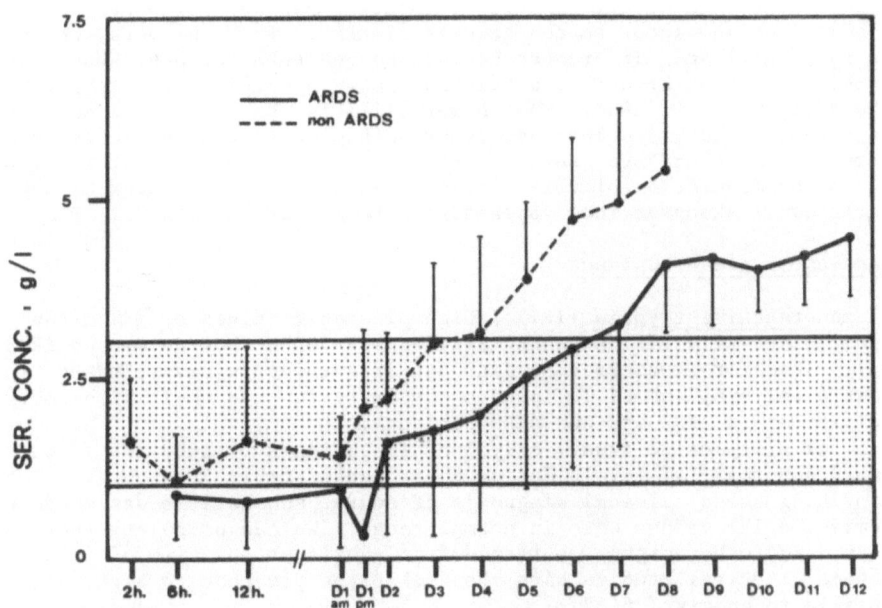

Fig. 9. Time course of serum haptoglobin concentration changes after burning injury.

In the present study, we compare the evolution and time course of different biochemical parameters in burned ARDS and non ARDS patients. The incidence of ARDS in our population group was 29.4 %. ARDS occurred more often in severe burned patients and after inhalation injury. The mortality rate was high in the ARDS group (40 %). These results confirm the well-known relation between severe ARDS and mortality (15).

Complement system

The present study clearly demonstrated that most burned patients at risk of developing ARDS show in vivo complement activation and this independently of the presence or absence of pulmonary lesions. Immunologic disorders, complement dependent or not, are well known in burned patients (16,17). We found that total C3 levels are lowered in acute burning injury; this information was confirmed by other studies (18,19). However, until now, it was not well established if this decrease was due to increased C3 catabolism or to a loss of this protein by the burned skin. The use of this test as an indicator of complement consumption is also hampered by the transient nature of the phenomenon. In addition, the opposing influence of increased C3 synthesis (acute phase reaction) do not facilitate to cut off this problem. The technique for quantitative measurement of C3d has gained firm support in providing an index of in vivo complement consumption. We also found an abnormal increase of C3d, a breakdown product of C3; this C3d assay detected more easily abnormal consumption. A combination of the two tests, expressed as the C3d/C3 ratio, first described by Duchateau and coworkers (11) give certainly a better information of C3 catabolism than each parameter alone; but it is also restricted by vascular permeability changes. If vascular permeability increases, the ratio C3d to C3 is underestimated. This may partially explain the variation observed during the acute hospitalisation stay of our patients and the similarity between the curves of ARDS and non ARDS patients. There is certainly an in vivo complement activation, by heat denatured plasma and tissue proteins (20,21), but actually it is not possible to quantify this response.

Complement-induced granulocyte aggregation has been proposed as a major contributing factor to the genesis of ARDS. From our data, which showed no significant differences between burned ARDS and non ARDS patients, a question arised of a relation between complement activation and the development of ARDS. The answer will be, that in the genesis of this syndrome, probably other factors or pathophysiologic sequences than complement activation have been implicated (22,23). The role of the lung itself in the modulation of this activity, as well as the status of the patient's own leukocytes, now appears as potentially important.

Protease-antiprotease balance

Immunoreactive trypsin (IRT) : High plasmatic values of IRT often appeared in severely ill patients, particularly when they developed ARDS (24,25). Therefore, we tried to find out if burning injury released pancreatic enzymes, and if IRT plasma concentration and their time course were different in ARDS and non ARDS patients.

In our prospective study, only 3 of 17 patients (2 ARDS and 1 non ARDS) had abnormal IRT levels during their acute stay. This observation is surprising as no clinical diagnosis of acute pancreatitis was suspected and admission IRT values were in normal range. We can postulate that the pancreas, like other organs, suffered from anoxia at the time of injury with shock. Several studies have shown that the pancreas is highly susceptible to anoxia (26,27).

From this study, we can conclude that pancreatic disorders are not responsible for deterioration of lung function but accompany it in some cases.

Leukocyte elastase : increased plasma levels of this most prominent of the neutral proteinases are found as early as two hours after burn injury. Levels were higher in the ARDS group, especially the day of injury, but also in the non ARDS group, highly abnormal levels were observed throughout the acute hospitalization stay.
The increased extracellular concentrations of leukocyte elastase may thus be explained by an enhanced granulocyte turnover and/or activity. Until now, it can not be retained as a specific marker of ARDS.

Plasma contains several proteinase inhibitors that serve to confine and terminate proteolytic events.

Alpha-2 macroglobulin (α_2M), a large proteinase inhibitor, is normally present in high concentrations and is probably a final common pathway for degradation of proteinases. Our observation that plasma levels of α_2M were low after burn injury suggests a consumption and clearance of this inhibitor; but extravasation in blister fluids at the burning side and in lymph can not be excluded and may partially explain why ARDS patients with higher UBS have lower values than non ARDS patients. However, when looking for α_2M evolution according to the severity of burns, we didn't find differences at 2 and 6 h after injury (28). Therefore, we suggest a consumption of this protein by the formation of a complex with proteolytic enzymes in the intersitial fluid early after injury. The decrease of α_2M during the observation period may be further due to an increased rate of metabolism of this protein. This may be due to a further protease release.

Alpha-1 proteinase inhibitor (α_1PI), a broad-spectrum inhibitor, is thought to be especially important in controlling neutrophil elastase in the extracellular space, and may be important in permitting extracellular proteolysis. Early after injury (2 - 6 h), we found abnormally low levels of α_1PI with minimal concentrations 6 h after burn. This could probably be related to a consumption of this inhibitor and/or an extravasation and loss in blister fluids and lymph and alveolar space, as shown in lung lavage after diverse injuries (29,30). Afterwards, α_1PI increased slowly, even in the presence of high elastase levels and behaved as an acute phase reactant.
The present study clearly demonstrated that most patients, after burning injury and especially when ARDS is associated, presented a protease release with severe decrease in proteinase inhibitor.

The acute phase response
In this study, we investigated the acute phase proteins together with their concentrations and the time course of their response in ARDS and non ARDS burned patients. CRP has been found in increased quantities 12 h after injury, in ARDS and non ARDS patients. There were no statistically significant differences between the two groups; CRP levels were higher (not statistically significantly), at the beginning, in ARDS than in non ARDS, but a plateau was reached in the two groups between day 5 and day 7. Independently of its short half life time, CRP concentration remained high; this may be due to the often encountered sepsis of burned patients.

Serum levels of alpha-1 acid glycoprotein increased to abnormal levels earlier in the non ARDS group than in the ARDS group (day 4 and 5 respectively). Time course of mean plasma concentration seemed to be delayed in the ARDS group. These levels increased constantly throughout the observation period, in relation with surinfection problems encountered in burned patients.

Haptoglobin :serum levels of this protein were lower in ARDS burned patients than in non ARDS patients, and the increase in serum levels out of normal range appeared later. One could propose different hypothesis : extravasation of this protein, hemodilution, but most probably an increased haptoglobin metabolism due to the rapid removal from the circulation of haptoglobin-hemoglobin complexes (31). In fact, burning injuries are often complicated by intravascular hemolysis, especially in the acute stage after burning injury and in patients with high UBS. In our ARDS patients group, UBS was significantly greater than in the non ARDS group, explaining the differences in intensity and time course of haptoglobin serum concentrations between the two groups.

When looking for acute phase proteins in these groups of patients, we found no statistically significant differences between the two groups; however in ARDS patients the magnitude of the acute phase response appeared to be less important, delayed in time and persisted longer compared with other trauma (32,33). This may be due to an increased vascular permeability and protein extravasation in more severe burned patients, an increased catabolism and elimination from the circulation, or to a metabolic depression with discrepancies in rate of synthesis.

CONCLUSION

In conclusion, this prospective biochemical study in burned ARDS and non ARDS patients shows in the two groups an in vivo complement activation, a significant protease-antiprotease imbalance, a time course delayed but persisting "acute phase response", but give no evidence for a specific mediator for acute respiratory failure. However, there seemed to be more quantitative severe changes in patients developing ARDS. This may be due to either a more severe reactivity of some patients to trauma, or, even to a preexistent deficit in their protection system (e.g. antiproteases), or a combination of the two. There seemed to be a close interrelationship between all these systems but no system has been pointed out as a reliable predictor of ARDS. However, especially in severe burned patients, a modification in the vascular permeability and extravasation of proteins in the extravascular space and blister fluids may hamper interpretation of biochemical results.

REFERENCES

1. Philips A.W., Cope O., Ann. Surg. 155: 1 (1962)
2. Di Vincenti F.C., Pruitt B.A.. J. Trauma 11: 109 (1971)
3. Silverstein P., Dressler D.P. Ann. Surg. 171: 124 (1970)
4. Demling R.H. J. Burn Care Rehab. 138 (1982)
5. Rapaport F.T., Nemirovsky M.S., Bachvaroff M.R. et al.
    Ann. Surg. 177: 472 (1972)
6. Traber D.L., Bohs C.T., Carvajal H.F. et al. Surg. Gynecol.
    Obstet. 148: 753 (1979)
7. Sachs A., Watson T. Lancet 1: 718 (1969)
8. Artz C.P., Moncrief J.A., Pruitt B.A.Jr. W.B. Saunders
    Company Philadelphia pp. 169 (1979)
9. Munster A.M.. Surgery 87: 29 (1980)
10. Perrin L.H., Lambert P.H., Miescher P.A. J. Clin. Invest.
    56: 165 (1975)
11. Duchateau J., Haas M., Schreyen H. et al. Am. Rev. Respir.
    Dis. 130: 1058 (1984)

12. Neumann S., Henrich N., Gunzer G., Lang H. In: Proteases Potential role in health and disease. Hörl WH, Heidland A, eds. Plenum Press, New York (1984)
13. Malvano R., Marchisio M., Massaglia A. et al. Scand. J. Gastroenterol. 15(suppl.62): 3 (1980)
14. Geokas M.C., Largman C., Brodrick S.W. et al. Am. J. Physiol. 236: E77 (1979)
15. Zapol W., Snider M.T., Hill J.D. et al. JAMA 242(20): 2193 (1979)
16. Ninnemann J.L. Williams and Wilkins Company, Baltimore, Maryland (1981)
17. Ninneman J.L., Condie J.T., Davis S.E., Crockett R.A. Journal of Trauma 22: 837 (1982)
18. Farrel M.F., Day N.K., Tsakraklides V. et al. Surgery 73: 697 (1973)
19. Turinsky J., Loose J.D., Saba T.M. Burns 6: 114 (1979)
20. Heideman M. J. Surg. Res. 26: 670 (1979)
21. Heideman M., Gelin L.E. Burns 5: 245 (1979)
22. Rinaldo J.E., Rogers R.M. N. Engl. J. Med. 306: 900 (1982)
23. Brigham K.L. Clin. Chest. Med. 3: 9 (1982)
24. Deby-Dupont G., Haas M., Pincemail J. et al. Intensive Care Med. 10: 7 (1984)
25. Nicod P.H., Leuenberger C., Seydoux F. et al. Am. Rev. Respir. Dis. 131: 696 (1985)
26. Warshaw A.L., O'Hara P.S. Ann. Surg. 188: 197 (1978)
27. Jones R., Garcia J., Mergner W. et al. Arch. Pathol. 99: 634 (1975)
28. Faymonville M.E., Micheels J., Adam A. et al. Burns, in press (1986)
29. Fowler A.A., Walchak S., Gielas P.C. et al. Chest. 81(5): 50S (1982)
30. Janoff A. Chest 83(5): 545S (1983)
31. Putman F.W. in "The Plasma Proteins", Putman F.W. Ed. Vol. II, Academic Press New York (1975)
32. Faymonville M.E., Adam A., Malengreaux P. et al. Marker Proteins in Inflammation, Vol. 2, Arnaud, Bienvenu, Laurent, Walter de Gruyter et co. Eds, Berlin, New York (1984)
33. Lamy M., Faymonville M.E., Adam A. et al. in: Proceedings of the International Workshop "Innovation in Intensive Care - Shock and ARDS" Cortina, March 15-21 (1986)

"SEPSIS PARAMETERS" IN PATIENTS WITH LETHAL BURN

U. Schöffel[1], T. Lenz[1], G. Ruf[1], Ch. Mittermayer[2] and M. Lausen[1]
[1] Chirurgische Universitätsklinik Freiburg, FRG.
[2] Abteilung Pathologie der RWTH Aachen, FRG.

The overwhelming inflammatory response during the development of a septic state causes a pronounced activation of cellular and plasmatic systems. This response can be recorded by determining the levels of parameters which were shown to represent the activation state of the single systems as well as the inhibitory capacity of the plasma. Such a "grading" might be of value in monitoring the clinical course and the therapeutic management. Additionally, the question as to whether a septic complication may occur can be answered before pathological manifestations are to be observed clinically. In order to record the inflammatory response in a variety of septic states with different underlaying diseases (i.e. peritonitis, postoperative sepsis, major trauma) we extended our studies to a series of patients with lethal burn.

The time-sequence pattern of the levels of inflammatory parameters were measured in eight patients with lethal burn injury at intervals from 12 to 24 hours. This population consisted of five men and three women with a mean age of 45 ($\pm$ 18) years. The studies were begun at the time of admission if the burned surface area was estimated to be more than 60%. They were continued until death occurred. The parameter-levels were correlated with routinely obtained laboratory data, clinical variables, and possible therapeutic influences (i.e. fluid management, fresh frozen plasma-therapy etc.). Fibrinopeptida A (FPA), the complement split products C3a, C3d, and C3c, the elastase-a_1 proteinase inhibitor-complex (Ea_1PI), fibronectin, prekallikrein, plasminogen, the coagulation factor XII (F XII), fibrinogen, and the proteinase inhibitors antithrombin III (antigen: ATIIIa and functional activity: AT III f), a_2 macroglobulin (a_2MG), and a_2 antiplasmin were determined by use of commercially available assay systems with the exception of C3d which was measured by rocket-immunoelectrophoresis. Endotoxin plasma levels were determined by a quantitative spectrophotometric method, utilizing a chromogenic peptide substrate and limulus lysate (Kabi-Diagnostica). We performed internal standardization to avoid the influence of individual variations in the "detoxifying capacity" of plasma[1,2].

The mean values of the individual parameters, determined in blood samples taken at the time of admission to our hospital, are shown in figure 1. The time intervals between the burn injury and the time of blood sampling ranged from two to ten hours with a mean of five hours. The burned area (second and third degree wounds) was estimated to include 65% to 85% of the body-surface (mean 75%). In figure 1, the investigated parameters are shown in a radial presentation with their scales depending on the normal range, which is indicated for each parameter by the ring which surrounds the center. The indented figure within the inner circle results from the connection of the actual zeros which touch the normal range in case of FPA and C3a. The connection of the values obtained for each parameter results in a star-like figure, which is characteristic for the actual inflammatory state. Values obtained from control samples were always shown to be located within the ring. At the time of admission the mean values for the patients with lethal burn show a high spike for FPA as a sign for fibrinogen proteolysis resulting from the activation of the coagulation cascade. A slight elevation of C3a and of the elastase-a_1PI-complex indicates the activation of the complement system and of leukocytes. Reduced levels were found for some protease-inhibitors, for fibronectin, prekallikrein, and plasminogen.

In order to compare these results with high-activation states of other causes, the mean values of patients with preseptic lethal peritonitis (n=8) and with lethal trauma (n=6) are also indicated in figure 1 The three groups were comparable with respect to the time interval between the blood sampling and the traumatic event. While there were no significant differences between the trauma group and the peritonitis group, the signs of complement activation and of leukocytic release reaction were much more pronounced in the former two than in the patient group with lethal burn. On the other hand, the reduction of some of the serum proteins, especially of proteinase-inhibitors, was less obvious. This finding was paralleled by a slight reduction in the total serum protein content (5.6 g%) in the burn group, while the protein levels of both of the other groups stayed within the normal range. Concerning the intravenous fluid administration during the emergency treatment before the admission, there was a mean infused volume of 1.6 and 1.2 liters of mainly cristalloid solutions to be registered in the burn and in the trauma group respectively. Patients with lethal peritonitis never received more than 500 ml of cristalloids before the first blood sampling.

The time sequela of the activation pattern in an individual patient with lethal burn is shown in figure 2. This 44 year-old male patient with a third degree burn of 80% of his body-surface area, survived for six days after the injury and developed "sepsis" at the third day. Patients were considered septic if at least five of the following conditions occurred: temperature more than 39 °C; acute deterioration of mental status; oliguria with an urine output of less than 30 ml/hr for at least two consecutive hours; respiratory failure; mandatory catecholamine treatment to sustain an adequate systolic blood pressure; bacteremia; leukocytosis of more than 15.000 cells/mm^3; thrombocytopenia of less than 80.000 cells/mm^3. Using these criteria only, in the two patients of the burn group who lived for more than for four days, "sepsis" was to be diagnosed. Figure 2 shows that the early FPA elevation slightly declined at day 2 and 4, reaching, however, its highest level at day 6. The C3 split products C3a and C3d rose continuously to excessive values at day 6, as did the elastase a_1PI complex. A slight fibrinogen elevation during the clinical course was accompanied by a rising of the proteinase-inhibitor levels which was constant but for a_2 antiplasmin, yet did not reach the normal range with the exception of a_2 macroglobulin. Markedly lowered fibronectin levels, subnormal values for factor XII and plasminogen and a constant decline of the prekallikrein level further characterized the clinical course. At day 6, eight hours prior to death, hemodialysis was started.

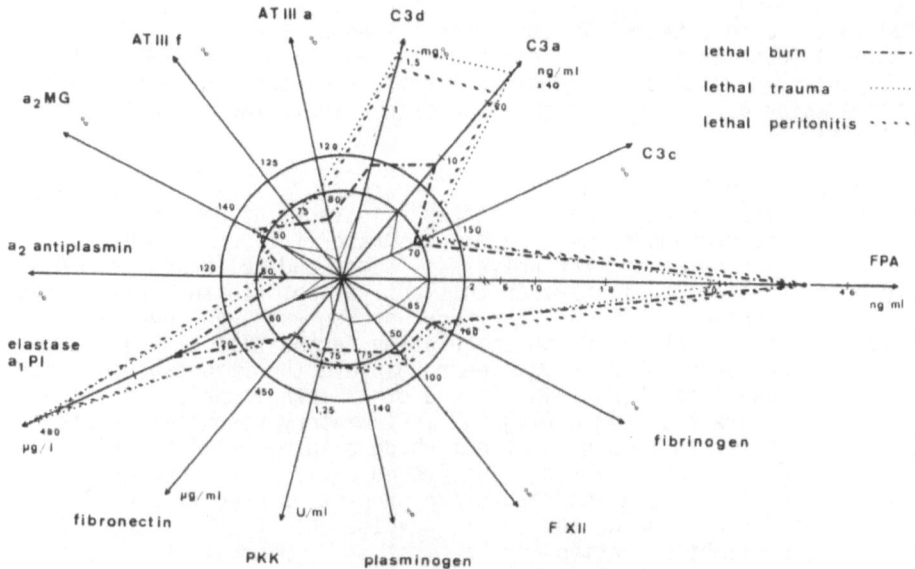

Figure 1: mean values of three patient groups at the time of admission.
See text for further explanations and abbreviations.

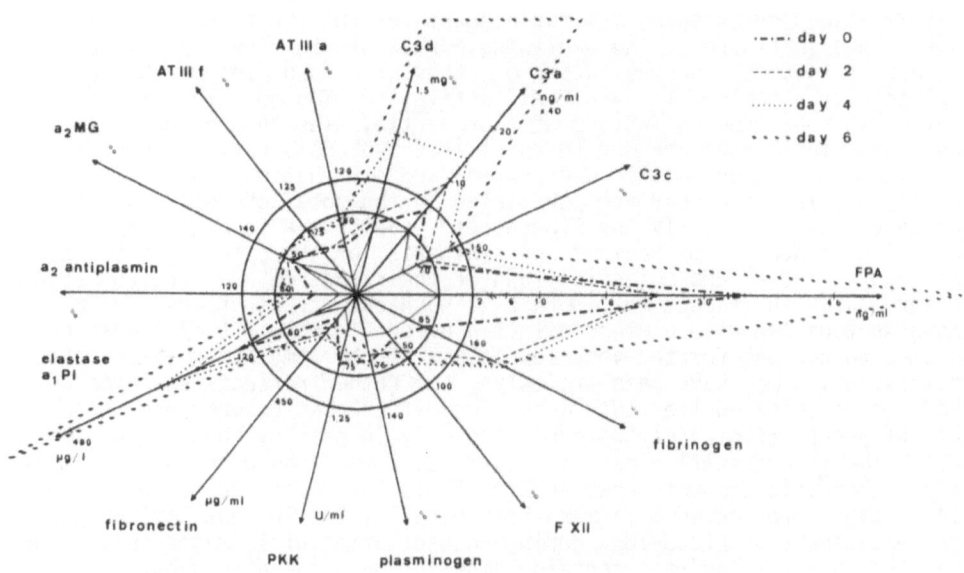

Figure 2: values of an individual patient with lethal burn reflecting
the development of the overwhelming inflammatory response
until he died at the end of the sixth day.

There were no major differences between the patients in the burn group as to the input/output balances, the administration of fresh frozen plasma or the need of chatecholamines. Mechanical ventilation had to be applied in six cases, hemodialysis in three, and partial necrosectomy was performed in four cases. In none of these patients an endotoxin level could be observed which differed significantly from the controls, nore were there major differences in the endotoxin detoxifying capacity of their plasma.

Several parameters of the inflammatory response were recently shown to be of prognostic value in septic states[3-7]. Every inflammatory response leads to an activation of the major plasmatic (i.e. complement-, kinin-forming-, coagulation/fibrinolysis-systems) and cellular systems. Certain parameters are indicative of this activation: Fibrinopeptida A is a marker of the thrombin-induced fibrinogen-proteolysis and thus, indirectly, of the activation of the coagulation factor XII[6]. The early elevation of FPA noted in our study seems to prove the importance of that particular pathway after burn injuries and other lethal challenges. The role of the complement system during the inflammatory response is rather complex: Apart from its cytotoxic effect which also may lead to the so-called "innocent bystander lysis" of unaltered cells of the own organism, its chemotactic, opsonizing and vasoactive properties make it one of the most important factors during the septic development. Figure 1 shows that there is only a slight elevation of the first split product of the complement factor 3, C3a with C3d in the normal range, and a low level of C3c which reflects the total amount of C3. During the later course (figure 2), however, an increasing activation of the complement cascade is to be observed within the intravascular space. The activation of leukocytes can be recorded by the complex of leukocytic elastase and its main inhibitor a_1 proteinase inhibitor. After burn injury local vasodilatation may be more pronounced than leukocyte activation in the first period, while in the later course toxic products and metabolites as well as physiologic stimuli like, for example, C5a may account for the steep increase of Ea_1PI shown in figure 2. The declining PKK levels in figure 2 - despite a tendency towards renormalization of other serum proteins-may be best explained by a consumption via F XII activation. The question as to whether the observed variations of fibronectin, plasminogen, or of the proteinase inhibitors are due to an altered synthesis rate, a specific consumption or an unspecific degradation and elimination, must remain open. Within certain range they are paralleled by the total protein content of the serum, but especially the fibronectin levels have to be interpreted carefully, since the mechanisms of synthesis, of the exchange between free and surface-bound molecules, and of their control are far from being clear[5,8]. With respect to figure 2, a wash-out effect of most serum proteins together with a minor dilution effect (hematocrit 42%) must be assumed to account for the very low levels at the time of admission. Fluid administration may have been excessive, but rather reflects the high demand due to an extensive fluid loss. The differences between the three patient groups (fig.1) may be explained only in part by these mechanisms. Again, the interpretation may be hazardous. First, the wash-out effect has to be taken into account. However,high fluid loss also occurs in peritonitis, which is reflected by a mean hematocrit of 54% in that patient group. The replacement of blood loss during resuscitation after major injury, on the other hand, leads to a certain hemodilution. Regarding substances of high molecular weight like a_2MG or fibrinogen, which might be less influenced by minor vascular leakage, one must keep in mind, that they react as acute phase proteins, even if a_2MG levels were reported to rest relatively stable during the early response[9]. All these influences are difficult to evaluate. Thus correction or correlation factors can hardly be used, and statistical analysis may be doubtful under these circumstances.

(The differences between the burn group and both of the other patient groups would be considered to be significant with $p < 0.005$ for C3a and Elastase a_1PI (< 0.01 for C3d) using the Wilcoxon U Test).

Concerning the results presented in both of the figures, and with the adequate cautiousness, some conclusions can be drawn: Contrary to lethal peritonitis and lethal trauma, the intravascular appearance of complement split products and leukocyte release products seems to be less pronounced in the early phase after lethal burn injury, while the fibrinogen split product FPA reaches similar levels in all three groups. Later, during the development of a septic state, the response becomes uniform, reflecting similar pathways which may account for the irreversibility of the state. In this context, the use of parameters which are not influenced by a hampered synthesis-rate may be of value in monitoring the overwhelming inflammatory response in sepsis.

REFERENCES

1. B. Urbaschek, B. Ditter, K.-P. Becker, and R. Urbaschek, Protective effects and role of endotoxin in experimental septicemia, Circ. Shock 14:209-222 (1984).
2. U. Schöffel, J. Shiga, and Ch. Mittermayer, The proliferation -inhibiting effect of endotoxin on human endothelial cells in culture and its possible implication in states of shock, Circ. Shock 9:499-508 (1982).
3. A. O. Aasen, N. Smith-Erichsen, and E. Amundsen, Studies on pathological plasma proteolysis in patients with septicemia, Scand. J. clin. Lab. Invest. 45, Supp.178:37-45 (1985).
4. J. K. S. Nuytinck, R. J. A. Goris, H. Redl, G. Schlag, and P. J. J. van Munster, Posttraumatic complications and inflammatory mediators, Arch. Surg. 121:886-890 (1986).
5. P. S. Richards and T. M. Saba, Effect of endotoxin on fibronectin and Kupffer cell activity, Hepatology 5:32-37 (1985).
6. J. Fareed, R. L. Bick, G. Sqillaci, J. M. Walenga, and E. W. Bermes jr., Molecular markers of hemostatic disorders: implications in the diagnosis and therapeutic management of thrombotic and bleeding disorders. Clin. Chemistry 29:1641-1657 (1983).
7. F. D. Moore jr., C. Davis, M. Rodrck, J. A. Mannick, and D. T. Fearon, Neutrophil activation in thermal injury as assessed by increased expression of complement receptors, N Engl. J. Med 314:948-953 (1986).
8. J. Grossman, T. Pohlman, F. Koerner, and D. Mosher, Plasma fibronectin concentration in animal models of sepsis and endotoxemia, J. Surg. Res. 34:145-150 (1983).
9. G. Sganga, J. H. Siegel, G. Brown, B. Coleman, Wiles III C. E., H. Belzberg, S. Wedel, and R. Placko, Reprioritization of hepatic plasma protein release in trauma and sepsis, Arch. Surg. 120:187-198 (1985).

COMPUTER ASSISTED 10 MHZ SONOGRAPHY OF BURN DEPTH IN MAN - A QUANTITATIVE GUIDE FOR THE SURGICAL STRATEGY OF SEPSIS PREVENTION

J. A. Bauer[1], P. Breitenberger[1] and St. Grommek[2]

[1] Chirurgische Klinik Innenstadt und Chirurgische Poliklinik der Universität München, Nußbaumstr. 20, D-8000 München 2.
[2] Isomag München, Linprun-Str. 49 D-8000 München 2.

Aim of study

Characterisation of the thermic injury is a pre-condition for the surgical treatment of burn patients. The depth of the injury, the third dimension of the burns is of considerable clinical importance because necrotic tissue must be excised and the exposed areas covered. In practice the experienced surgeon relies on clinical inspection when assessing the depth of the burn.

Method

In an animal experiment a standardised scald wound was induced in rats and pigs and 10 MHz ultrasonics was used to measure the depth of the heat lesions. The sonograms can be regarded as a histological slice of the tissue layers affected by the thermic lesions. Changes in the thickness of the cutis can be measured and correlated with the severity of the burn. The depth of the lesion can be accurately determined within hours after it has been induced (1).

Results

When this method of measurement was applied to humans, the following problems were solved:

1. Maintenance of sterilitiy when doing the measurements on the patient.

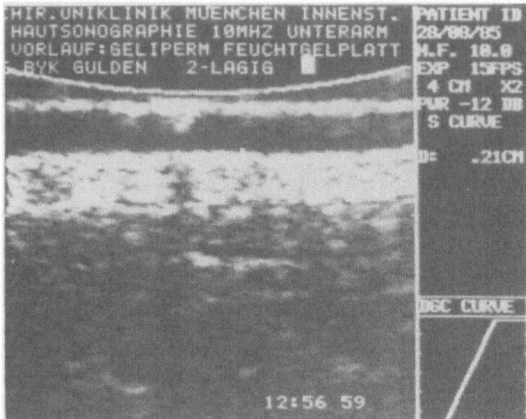

Fig.1. Geliperm, a sterile polyacryleamid-gel developed for treatment of burns (2) used as sonogel

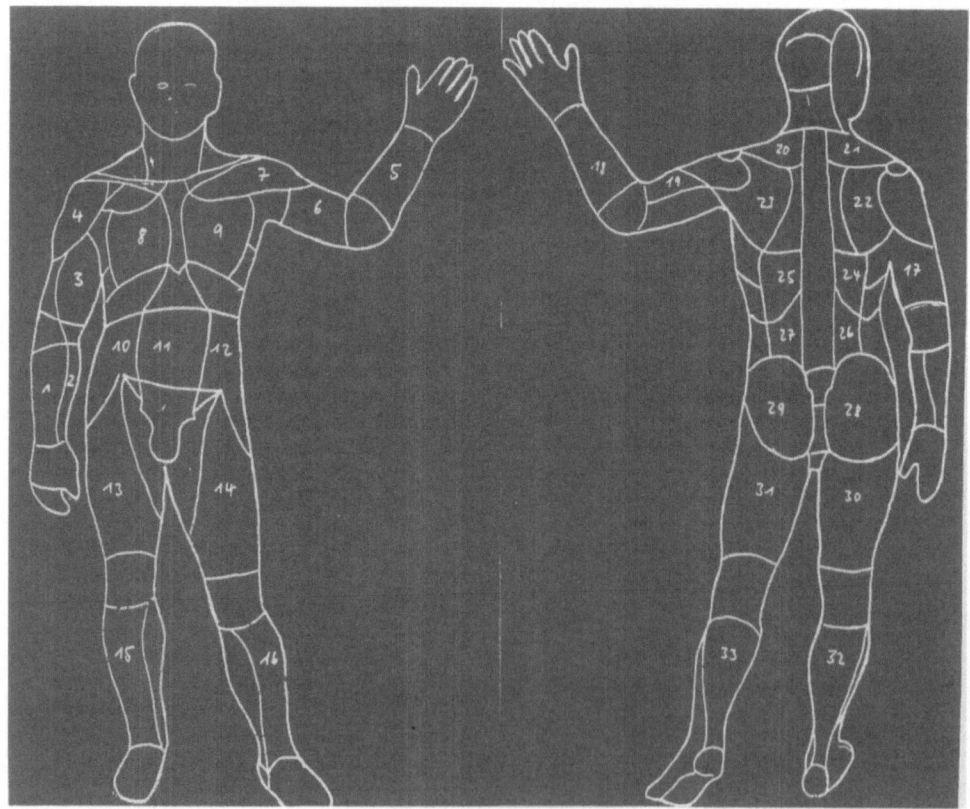

Fig.2 Anatomical areas

2. Normal values were established by measuring the skin thickness of test subjects. Age, sex and anatomical site were taken into consideration(3,4).

AGEGROUPS (A)

AREA	1-10	10-20	20-40	40-50	50-60	60-80
1	14	22	20	20	20	18
2	12	17	20	19	17	14
3	12	16	20	18	17	16
4	14	18	24	20	21	21
5	13	18	20	19	18	17
6	13	18	20	17	19	13
7	15	19	23	20	20	20
8	13	18	21	21	20	19
9	13	19	21	20	21	18
10	13	18	21	20	20	19
11	14	18	22	21	20	19
12	14	19	21	19	21	19
13	14	16	20	19	19	17
14	15	17	22	20	20	17
15	16	18	20	22	19	18
16	15	18	20	20	21	18
17	15	21	22	20	20	18
18	14	21	21	21	19	16
19	13	21	22	21	20	16
20	17	23	28	26	27	23
21	17	24	29	25	28	24
22	17	23	27	25	24	30
23	17	24	33	24	23	29
24	16	22	22	23	24	25
25	15	23	28	23	23	25
26	14	21	22	23	22	24
27	14	21	23	23	21	24
28	15	20	21	21	22	23
29	14	20	22	21	21	23
30	14	21	20	19	20	20
31	13	20	21	19	19	19
32	14	20	21	20	19	18
33	15	20	21	20	19	16

Fig. 3.

Values for the anatomical areas (skin thickness in mmx 10^{-1})

Remarks

1. SD of areas with light (UV) exposure: 0,025 - 0,058 cm
   SD of areas without ligth (UV) exposure: 0,006 - 0,025 cm
2. n=8 for every age groups, 4 males and 4 females, no sex
   - linked differences.
3. In elder persons (60-80 a) posterior areas of the chest show an increase of skin thickness. The other areas return to the thickness-values of the age group 10-20 a.
4. From 10 to 60 years dermal thickness is only area-dependent.

3. Development of soft-ware which draws on data gathered on the test persons and uses the thickness measurements of the intact areas of skin to calculate the thickness of the skin before it was scalded.

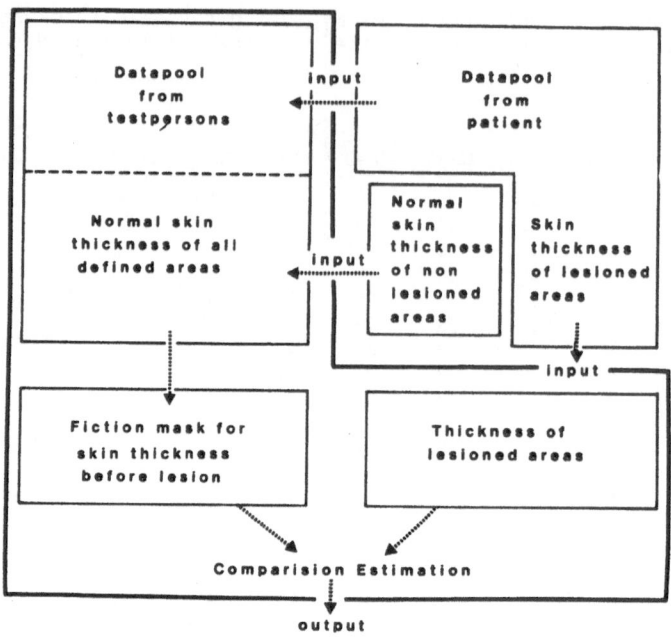

Fig. 4: Software for quantified diagnosis of depth and percentage of burns

Conclusion

The estimated thickness of the skin before scalding is compared with the
actual thickness of unscalded areas. This data feedback increases the
accuracy of the estimations as the number of patients examined increases.
Data on scalds and burns are gathered in the same way, but are kept separate and are compared. This makes further animal experiments unnecessary

References

1. Bauer J., Lehn N., Sauer Th.
Use of a real-time scanner with a 10 MHz transducer head for quantitative
ultrasound assessment of depth and area of skin scald injuries in rats,
pigs, and humans and of therapeutic of etofenamate on scald injuries in
rats.
In: Langenbecks Arch.Chir.Suppl.Chir.Forum (ed.: H.J.Streicher, M.
Schwaiger). Springer (Berlin, Heidelberg, New York) p. 45 (1986)
2. Spector M., Weissgerber P., Reese N., Harmon S.L.
A polyacrylamide agar hydrogel material for the treatment of burns, 8th
congress of society for biomaterials, Orlando, Florida USA (April 1982)
3. Nichter L.S., Bryant C.A., Edlich R.F.
Efficacy of burned surface area estimates calculated from charts - the
need for a computer based model.
J. Trauma, 25:477 (1985)
4. Feldman K.W.
Help needed on hot water burns.
Pediatrics, 71:145 (1983)

PHARMACOLOGICAL CONTROL OF HEMORRHAGIC SHOCK : DOES HYPERTONIC RESUSCITATION QUALIFY ?

M. Rocha e Silva
Divisao de experimentaçao, Instituto do Coraçao, Facultade de
Medicina, Universidade de Sao Paulo, Caixa Postal 11450
Sao Paulo, O5499, SP, Brasil

The critical function of the mammalian circulatory system is the handling of respiratory gases, required at variable levels, over the many conditions to which each individual is normally exposed during his lifetime. This problem is handled through variable blood flow. An average healthy human adult at rest produces a cardiac output of 5 l/min, but activity will change this substantially. A sedate, untrained person can raise output to 10 l/min, trained people are capable of 20 or 25 l/min, while exceptionally gifted and well conditioned athletes produce outputs in the 30 l/min range. In contrast, total blood volume (normal value 5 liters) is a relatively fixed parameter, which should not vary by more than 10% around the mean. In this sense, hemorrhage may be regarded as an unsual problem: how may large blood volume losses be reconciled with fixed (basal) cardiac output. Light blood loss is of course a very simple and trivial matter: mammals encounter it frequently during their lives and cope easily: human blood donors, regular or occasional, give 10% of their total blood volume and simply walk back to normal activity. Severe blood loss is quite a different matter: it is a fairly rare, probably a once-in-a-lifetime occurrence, and the problem it poses, and the ways in which the natural response is dealt with in a civilized therapeutical environment is the object of this discussion.

THE CONTROL OF BLOOD FLOW AND BLOOD VOLUME

Blood flow and blood volume are both controlled through the same basic mechanism, namely variation of the vascular radius. But these two controls are in fact independent and specialized functions. Flow is controlled by radius variations in the resistive network, as a function of R^4, while blood volume is controlled in the capacitive network, as a function of R^2. It is obvious that radius variations in resistive vessels also affect vascular capacity (and hence segmental blood volume), while radius variations in capacitive vessels affect vascular resistance and flow. Specialization derives from the fact that resistive vessels concentrate a large fraction of total resistance but contain only a small fraction of total volume/capacity, the inverse being true for capacitive vessels. Just how much vascular resistance is concentrated in resistive vessels, how much capacity in capacitive vessels is a disputed point: different numerical estimates vary, but most people would agree that resistive vessels (small and terminal arteries, arterioles, capillaries, venules) are effective flow controllers because they account for approximately 80% of the total resting systemic vascular resistance (SVR), but only contain about 10% of the total blood volume/vascular capacity (V-C); capacitive vessels (veins) derive their effectiveness from the fact that they contain some 60% of V-C, but only account for 5% of SVR. If this (or any other appropriate) set of distributional values is adopted, SVR, cardiac output (CO), and V-C variations may be independently analysed in resistive or capacitive vessels as functions of R variations. For simplicity, blood flow will be assumed as laminar, blood vessels as perfect cylinders of constant length and variable radius, and the pressure drop (ΔP) constant. Under these conditions, $SVR = K_1 * R^{-4}$, $CO = K_2 * SVR^{-1}$, and $VOLUME = CAPACITY = K_3 * R^2$.

Resistive vessel analysis may start from resting SVR (set equal to 1 arbitrary unit, AU):

$$SVR = \underset{\text{(resistive vessels)}}{0.8} + \underset{\text{(other vessels)}}{0.2} = 1 \text{ AU.}$$

If the entire set of "resistive" radii were to vary in uniform proportion, while "other" radii remained fixed, one could write:

$$SVR = (0.8 * r^{-4} + 0.2) \text{ AU,}$$

"r" being the (uniform) relative radius change throughout the resistive system.

The relation for cardiac output (resting CO = 5 l/min) could be written

$$CO = [5 * (0.8 * r^{-4} + 0.2)^{-1}] \text{ l/min,}$$

while V-C analysis for resistive vessels would follow the relation (resting V-C = 5 liters):

$$V-C = [5 * (0.1 * r^2 + 0.9)] \text{ liters.}$$

Capacitive vessel analysis follows the same principles, appropriate distributional values being of course substituted; thus resting SVR may be written:

$$SVR = \underset{\text{(capacitive vessels)}}{0.05} + \underset{\text{(other vessels)}}{0.95} = 1 \text{ AU.}$$

Again, if "r'" is the (uniform) relative "capacitive" radius variation throughout the capacitive system, while "other" radii remain constant:

$$SVR = (0.05 * r'^{-4} + 0.95) \text{ AU,}$$

$$CO = [5 * (0.05 * r'^{-4} + 0.95)^{-1}] \text{ l/min,}$$

$$V-C = [5 * (0.6 * r'^2 + 0.4)] \text{ liters.}$$

Fig 1 displays blood flow (CO) and V-C as functions of resistive (Fig 1A) and capacitive (Fig 1B) radius variations, within the 0.5<r<2 range (i.e. for 50% reductions up to 100% increases in radii). Within this range, variations of the radii of resistive vessels command very large flow variations, with negligible V-C changes. Towards the upper end of the range, the slope of the flow-radius function (which may of course be regarded as the gain of the flow control system) decreases, because the resistance of resistive vessels tends to zero, which means that SVR tends to the resistance of "other" vessels, as a minimum. The gain of this flow control function is highest within the 4-12 l/min output range, with a maximum near the 7 l/min mark. It is very important to note at this point that resistive vessels are not normally controlled by uniform "r" variation, but the concept is still applicable: "r"

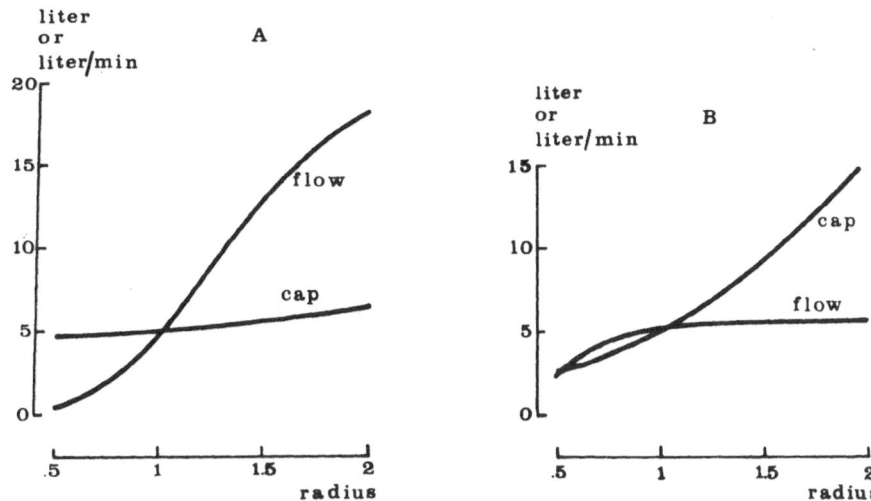

Fig. 1. Cardiac output and vascular capacity/blood volume as a
function of variations of the radii of resistive (A) or
capacitive (B) vessels.

(defined as a uniform relative radius variation) may equally stand for the
weighted mean of regionally patterned changes. Capacitive radii variations, in
contrast, produce very large V-C changes (actually larger than real life at
the upper end of the range) with negligible effects on flow. Little is known
about the exact patterns of regional capacitive control, but it may be
acceptable, at present, to suppose that it occurs more uniformly throughout
the system than does resistive control. The point is however far from settled.

The effects of exercise on vascular control (Fig 2) are a fine example of
efficient vascular adjustment. Large CO levels may be obtained with small V-C
increases within the resistive segment (Fig 2A). The blood volume required to
meet this increased V/C comes of course from the capacitive side, at the cost
of insignificant resistive change (Fig 2B). It may however be noted that at
extremely high levels of CO, this model predicts a transfer of nearly 1.5
liters of blood into the resistive side; such a blood "loss" would certainly
meet most people's definition of moderate to severe hemorrhage: it is probably
the physiological basis for a number of unconfirmed reports that high performance athletes receive transfusions of previously collected own blood, immediately before important sporting events. Conversely, it has been recently
demonstrated that "athletically trained" rats are significantly less susceptible to the effects of hemorrhage than untrained controls (1).

When blood loss of this, or of larger, magnitude does in fact occur, a
very different corrective adjustment takes place (Fig 3), largely as a
consequence of baroreceptor and volume receptor reflex mechanims: both
resistive and capacitive vessels constrict (it may incidentally be noted that
during exercise, baroreceptor adjustment is strongly inhibited). Resistive
vasoconstriction (Fig 3A), which completely fails to correct the vascular
capacitive imbalance, restricts flow to specific territories, namely enteric,
renal, muscular and cutaneous. If hemorrhage is light, reductions are
correspondingly slight, with little or no damage to the restricted organs or
systems. But severe blood loss entails flow reductions which may in time
become critical. Capacitive constriction (Fig 3B), on the other hand, should
theoretically be able to handle blood losses as large as 40% of total blood
volume, but in practice this simply does not occur. It has already been

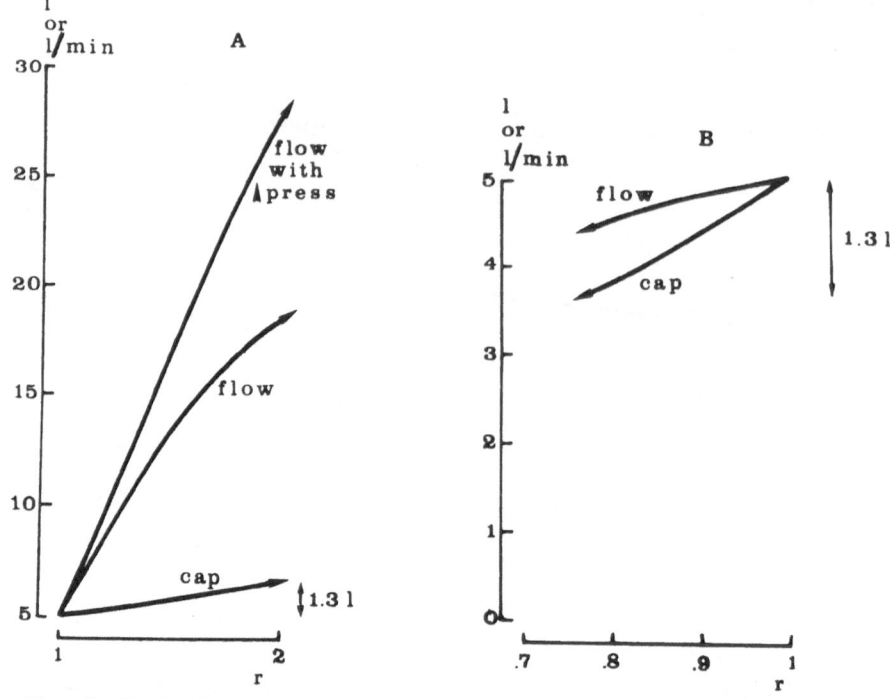

Fig. 2. Resistive (A) and capacitive (B) radii adjustments in
exercise. Note that in this condition, output is further
enhanced by arterial pressure increase.

noted that light hemorrhage is a fairly trivial encounter in the life of a
mammal; effective recovery from the condition is loaded with adaptive
advantage; the slight reductions of resistive and capacitive vessel radii
which normally occur correct volume imbalance without serious flow

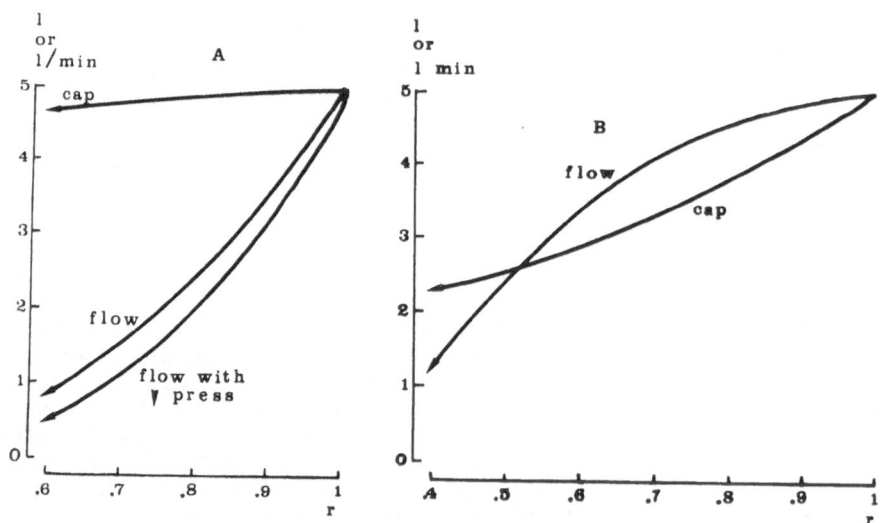

Fig. 3. Resistive (A) and capacitive (B) radii adjustments in
hemorrhage. Pressure drops enhance flow restrictions.

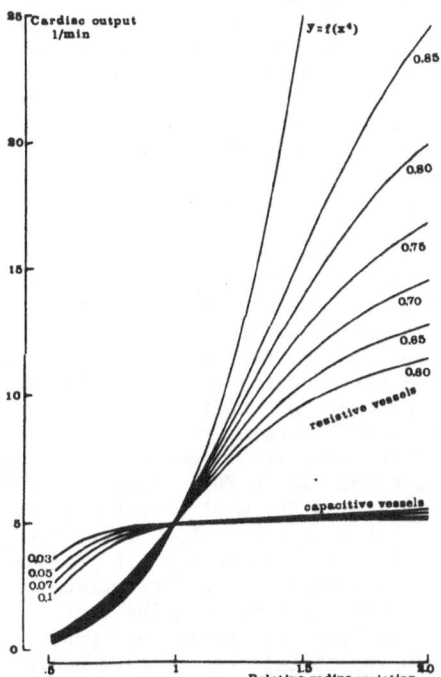

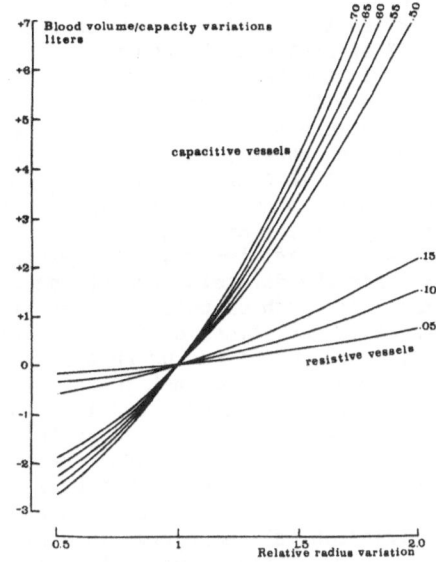

Fig. 4. Parametric variation of "a"
in the flow/radius function.

Fig. 5. Parametric variation of "b"
in the V-C/radius function.

restrictions. In contrast, severe blood loss, in the wild, is a condition
associated with virtually no probability of survival: more likely than not, a
larger, faster, stronger predator has caught up, which turns preservation of
vital organ funtion into a meaningless issue, as far as evolutionary adaptive
drive is concerned. The intense enteric and renal constriction, which are only
quantitatively different from the response which occurs in light hemorrhage,
will possibly lead to ischemia and necrosis, but this will have little impact
on survival chances. It may therefore be said that the natural response to
severe hemorrhage is probably inapropriate, because there is no adaptive drive
for anything better, even though the machinery for selective venoconstriction
and ultimate recovery may be there.

It is of course possible to alter the mathematical analysis, by taking
different distributional values for the concentration of resistance, or of
capacity in any given segment of the circulation. The general equations are:

$$SVR = (a*r^{-4} + 1 - a) \; AU$$

$$CO = [5*(a*r^{-4} + 1 - a)^{-1}] \; l/min$$

$$V-C = [5*(b*r^2 + 1 - b)] \; liters$$

where "r" is the relative (uniform or weighted mean) radius variation of the
segment under analysis, "a" is the fraction of total resting SVR, and "b" the
fraction of total resting V-C concentrated in the segment. In Figs. 1 to 3 a
particular set of values for "a" and "b" were adopted but the general approach
to the problem is displayed here. Fig. 4 shows flow x radius functions when
"a" is varied parametrically in the upper and lower extremes. The upper range

71

of "a" (60 to 85% concentration of resistance) covers possible values for resistive vessels. The mere inspection of these curves makes it clear that real life "a" must be at least 70% (i.e., that resistive vessels must concentrate at least 70% of total resting SVR), since lower values produce maximal cardiac outputs which fall short of values actually observed during exercise. Fig. 4 also shows that high values of "a" produce a family of curves which are all similar below resting CO. The low range of "a" values (3, 5, 7 or 10% of resting SVR) apply to capacitive vessels. No matter which of the 4 curves is adopted, capacitive vessels may dilate to accomodate any volume increase without affecting flow. But for radii < 1, capacitive vessels act more and more as resistive vessels and here there is an important difference between the curves: the larger the value of "a", the bigger the resistive effect. But in fact, the 5% value for "b" is by far the most likely: resistive distribution along the vascular bed is very simply measured, because it is directly proportional to the respective segmental pressure decrement. In contrast, capacitive segmental distribution is not so easy: the best estimates come from the determination of geometrical parameters (radius, length and number) for each vascular segment within organs and systems. Fig. 5 displays parametric variations of "b" (possible values for capacitive concentration at rest) in the V-C x radius function: the upper range (50 to 70%) applies to the capacity of veins. Geometric estimates suggest 60% as the most probable value, but here the margin of error is considerable. All curves show that large volumes may be accomodated with smaller than 50% increases in radius. Venoconstriction, on the other hand, reduces capacity: a 30% reduction in venous radii would reduce V-C by 1.5 to 2.5 liters, depending on the value of "b". The lower range of "b" values (5 to 15%) applies to resistive vessels: constriction is ineffective in terms of reducing total vascular capacity, irrespective of the value of "b"; in contrast, resistive dilation is strongly dependent on "b". If it were to equal 15%, maximal cardiac output would require a transfer of 2 liters of blood into the resistive system, which practically rules out such a high value for the "b" parameter.

CLINICAL TREATMENT OF HEMORRHAGE

The treatment of severe blood loss may be considered in terms of fluid replacement and pharmacological control. Fluid repalcement is, in theory, the ideal procedure, but in practice there are quite a number of pitfalls. A simple practical rule is that replacement should roughly equal estimated losses if blood, plasma, or artificial colloids are used, but should exceed losses by 3:1 if crystalloid solutions are employed. Since one only deals with large losses (light hemorrhage does not require treatment) replacement volumes are always necessarily large. Blood is an expensive commodity, its supply is limited, rapid initiation of treatment is hindered by typing and cross-matching procedures, and field use is virtually out of question. Moreover, stocked blood deteriorates rapidly, and the risks of contamination and infection transmission are ever present. Last, but certainly not least, some people express religious objections to blood transfusion. Many but obviously not all of these problems beset the use of plasma, whereas artificial colloids pose few of these problems, but, when used in large volumes have been implicated in undesirable side effects, such as allergic/anaphylactic reactions, and disruption of the blood clotting machinery, amongst others. Crystalloid solutions, which pose none of these problems, have their own limitations: because they exhert no oncotic pressure across the endothelial wall, they distribute within the larger extracellular space, and hence must be administered in larger volumes. Overhidration problems, which must be solved later, are frequently encountered (2). When all is weighed, fluid replacement is a relatively simple, safe, not overexpensive procedure when blood loss occurs inside a hospital or to a patient rapidly admitted to hospital. It becomes a serious problem in field conditions, or when transfer time is prolonged, and may be described as formidable in mass accidents, or in military situations.

The basic deficiency of conventional pharmacological control of severe hemorrhage is that it is incapable of completely correcting the condition. Its basic aim is to counteract resistive vasoconstriction, while maintaining or enhancing capacitive venoconstriction. The standard pharmacological control of hemorrhage goes through alfa-adrenergic antagonists, beta-adrenergic agonists, as well as dopamine (a selective splanchnic and renal precapillary dilator). Nitroprusside, an efficient overall precapillary dilator is also widely employed. Combinations of these agents are effective tools in vascular adjustment, but no known drug is capable of exherting all the proper effects, and selective venoconstrictors are simply non-existent.

The second part of this paper will be devoted to the analysis of recent developments in the field of hypertonic resuscitation of hemorrhagic shock, which at first sight does not appear to be a pharmacological procedure. But we shall endeavour to demonstrate that very hypertonic sodium salt solutions exhert actions which have little to do with their mere physical properties as hypertonic fluids.

HYPERTONIC RESUSCITATION

Hypertonic resuscitation is by no means a novelty. The earliest recorded use is by Penfield (3), who used NaCl at a concentration of 1.8% (600 mOsm/l) for the treatment acute hemorrhage. Hypertonic solutions have been used experimentally or clinically because they induce increased myocardial contractility (4), widespread pre-capillary dilation (5), and expanded plasma volume, through an osmotic fluid shift into the vascular compartment, all of which are obviously beneficial to hypovolemic subjects. Hypertonic sodium chloride and bicarbonate were used in concentrations of up to 1800 mOsm/l (6) glucose up to 2500 mOsm/l (7), but positive effects were always described as transient and requiring conventional fluid replacement not long after the initial hypertonic therapy.

In 1980 however, we described results (8) showing that acutely bled dogs (shed blood volume: 42 ml/kg, equivalent to approximately 50% of total blood volume) could be permanently resuscitated within 45 min of the start of bleeding by a single intravenous bolus of very hypertonic NaCl (concentration: 7.5%, equivalent to 2400 mOsm/l; volume: 10% of total shed blood, equivalent to an average 4.2 ml/kg). It should be stressed that after hypertonic resuscitation in this manner, blood was not reinfused, nor was any other treatment, or intravenous fluid replacement given to sustain the response: the animal itself restored its blood volume, at first by plasma expansion, later by red cell formation. The overall survival rate for this procedure in dogs, over a 5 year span, was higher than 95%. The basic pattern of the cardiovascular response to intravenous hypertonic NaCl includes the restoration of mean arterial pressure, cardiac output, acid-base equilibrium and mean circulatory filling pressure to stable, near control levels, in spite of the very transient nature of plasma volume expansion (8, 9). Regional flows are diversely affected: muscular resistance vessels constrict, while renal, mesenteric, portal and coronary vessels all dilate in response to hypertonic NaCl injections (10). It has also beeen demonstrated that hypertonic resuscitation induces a prolonged increase of myocardial contractility (11) and a reduction of intracranial pressure, with normal cerebral blood flow (12) These data are summarized in Table 1.

We have also demonstrated (13) that a pulmonary vagal reflex is essential for long-term survival: injections given upstream of the pulmonary microcirculation (intravenous, into the right cardiac chambers or into the pulmonary artery) produced typical survival responses with stable, high

TABLE 1. A SUMMARY OF DESCRIBED EFFECTS OF HYPERTONIC RESUSCITATION
FROM SEVERE HEMORRHAGIC SHOCK WITH 7.5% NaCl SOLUTIONS.

SPECIES	DESCRIBED EFFECTS	REFERENCE
dog	stable recovery of arterial pressure stable recovery of cardiac output stable recovery of acid-base equilibrium increased mesenteric blood flow high (> 95%) long-term survival	8
sheep	transient recovery of arterial pressure	14
dog	increased myocardial contractility	11
dog	reduced intracranial pressure normal cerebral blood flow	12
dog	increased renal blood flow increased urine flow	15
dog	recovery of mean circulatory filling pressure	9
dog	muscular vasocontriction renal vasodilation	10

cardiac output, arterial pressure and acid-base equilibrium; in contrast, post-pulmonary injections resulted in near zero survival rates, with only transient recovery of arterial pressure, output and acid-base equilibrium. We also found that high survival rates, after pre-pulmonary injections, were only observed in the presence of normal neural conduction through the cervical vagi: blockage by a local anesthetic, or by cooling, at the time of hypertonic resuscitation prevented survival. These results have been elegantly confirmed by Younes et al (16): in chronically prepared dogs, with unilaterally denervated lungs, they showed that hypertonic resuscitation was effective when injected into sham-denervated, or into innervated lungs, but not so when given directly to the denervated lung. It has been demonstrated that the restoration of mean circulatory filling pressure (9), and of muscular vasoconstriction (Rocha e Silva et al, 1986) are also dependent upon intact vagal conduction.

TABLE 2. THE EFFECTS OF DIFFERENT HYPERTONIC (2400 mOsm/l)
SOLUTIONS ON LONG TERM SURVIVAL AND MEAN CIRCULATORY
FILLING PRESSURE IN SEVERELY BLED DOGS.

SOLUTE	SURVIVAL RATES % (SURV/TOTAL)	MEAN CIRCULATORY FILLING PRESSURE
Na chloride	98% (165/168)	restored
Na acetate	72% (13/18)	restored
Na bicarb.	61% (11/18)	restored
Lithium Cl	5% (1/18)	not restored
Tris Cl	22% (4/18)	not restored
Urea	33% (6/18)	not restored
Mannitol	11% (2/18)	not restored
Glucose	5 (1/18)	not restored

Other hypertonic solutions at 2400 mOsm/l have also been tested: other Na salts, such as acetate and bicarbonate, at 2400 mOsm/l, induce high survival rates (71% and 62%, respectively), with high arterial pressure, cardiac output, mean circulatory pressure and a recovery of acid-base equilibrium (17, 18, 19); in contrast low levels of survival (5% to 33%) were noted following lithium chloride, tris chloride, urea, glucose or mannitol, all at 2400 mOsm/l (20). Table 2 summarizes these data. It may be noted that the 3 sodium salts, which induce high survival rates restore mean circulatory filling pressure, while other chlorides and non-electrolytes, which do not restore mean filling pressure, are associated with very low survival rates. It thus appears that venoconstriction and long term survival, both of which has been shown to depend on intact vagal connections are dependent on the presence of sodium ions in the hypertonic solution.

The duration of shock appears to be critical. Mermel and Boyle (15) have found that, in the dog, a longer lasting shock condition (75 min) only responds transiently to hypertonic resuscitation with NaCl. Similar results have been reported by Nakayama et al (14) in the sheep, after a 2 hour period of shock. In both conditions the total volume of removed blood was higher than 50% of estimated or measured initial blood volume. No long term survival rates were however referred by either group. Experiments under course in our laboratory (Rocha e Silva et al, unpublished) confirm the observations of Mermel and Boyle (15), and additionally show that long-term survival rates are drastically reduced by the extension of the period of hypovolemia. The combination of hyperosmotic and hyperoncotic solutions (NaCl 7.2% + dextran 7%) has also been tested: in unanesthetized sheep submitted to shock lasting for 2 hours at 40 mm Hg, hypertonic NaCl-dextran produced significantly higher levels of plasma volume and cardiac output, versus hypertonic NaCl, or dextran, over a 2 hour post-resuscitation observation period (17).

In terms of eventual clinical applications, these data open the prospect of small volume resuscitation for severe blood loss. A number of clinical tests has been reported where 2400 mOsm/l NaCl solutions have been employed (21, 22, 23), but it is not known whether man can be resuscitated by a single bolus injection of hypertonic NaCl, as sole treatment. It goes without saying that it would be entirely unethical to actively try to answer such a question. Moreover the search for an ideal small volume therapeutic agent is still on. Exclusively on the grounds of survival rates, sodium chloride must be a part of any such solution, but acetate may be useful to increase output and correct metabolic acidosis. The addition of dextran may prove a useful therapeutic tool, although it may not be exempt from undesirable collateral efects (24). The likelyhood of such side effects is however small because of the very small volumes to be used. Limited clinical trials with hypertonic NaCl are presently in progress in the São Paulo University Medical School Hospital and with hypertonic NaCl-dextran at The University of California, Davis, Medical School Hospital.

In summary, it has been shown that 2400 mOsm/l NaCl is effective in the resuscitation of severe hemorrhage, not only because it produces fluid shift into the vascular compartment, but also because it has important pharmacological actions: it dilates renal and mesenteric vascular territories, reversing the trend to ischemic necrosis; but perhaps more importantly, it induces selective muscular resistive constriction and overall venoconstriction; muscular vasoconstriction shifts blood flow into the dilated territories, without materially affecting a territory which is perforce at rest and notoriously capable of the highest degree of anaerobiosis; venoconstriction actively reduces vascular capacity to match reduced blood volume and helps create a hyperdynamic circulatory condition, which safely carries the animals over a critical period of time till they become capable of restoring, first plasma volume, later blood volume to normality. It is still unknown whether human hemorrhagic hypotension will respond entirely in the manner described for dogs, but it is certain that the early beneficial effects of hypertonic resuscitation are observable. It may thus be suggested that the passage of highly concentrated sodium salts through the lungs will ellicit a neural reflex which triggers intense active venoconstriction. If this be the case, even the relatively large mesenteric and renal dilations, which are simultaneously observed, could be accounted for by the better adjustment of capacity to reduced circulating volume.

REFERENCES

1. Bond, R.F., Armstrong, R.B. and Johnson III, G. (1986) Circ Shock 19: 257.
2. Stein, L., Beraud, J.J., Morisete, M., da Luz, P.L., Weil M.H. and Shubin, H.. (1975) Circulation. 52: 483.
3. Penfield, W.G. (1919) Am J Physiol 48: 121.
4. Wildenthal, K., Mierzwiak D.S. and Mitchell, J.H. (1969) Am J Physiol 216: 898.
5. Gazitua, S., Scott, J.B., Swindall, B. and Haddy, F.J. (1971) Am J Physiol 220: 384.
6. Baue, A.E., Tragus, E.T. and Parkins, W.M. (1967) Am J Physiol 212: 54.
7. McNamara, J.J., Molot, M.D., Dunn R.A. and Stremple, J.F. (1972) Ann Surg 176: 176.
8. Velasco, I.T., Pontieri, V., Rocha e Silva, M. and Lopes, O.U. (1980) Am J Physiol 239: H664.
9. Lopes, O.U., Velasco, I.T., Guertzenstein, P.G., Rocha e Silva, M. and Pontieri, P. (1986) Hypertension 8 (Suppl I): I195.
10. Rocha e Silva, M., Negraes, G.A., Soares, A.M., Pontieri, V. and Loppnow, L. (1986) Circ. Shock 19: 165.
11. Kien, N., Reitan, J, White D. and Wu, C. (1985) Fed Proc 44: 1353.
12. Prough, D.S., Carson Johnson, J., Poole, G.V., Jr, Stulken, E.H., Johnston, W. E.and Royster, R. (1985) Crit Care Med 13: 407.
13. Lopes O.U., Pontieri, V., Rocha e Silva, M. and Velasco, I.T. (1981) Am J Physiol 241: H883.
14. Nakayama, S., Sibley, L., Gunther, R.A., Holcroft, J.W. and Kramer, G.C. (1984) Circ Shock 13: 149.
15. Mermel, G.W. and Boyle, W.A. (1986) Fed Proc 45: 880.
16. Younes, R.N., Aun, F., Tomida, R.M. and Birolini, D. (1985) Surgery 98: 900.
17. Jeffrey-Smith, G., Kramer, G.C., Perron, P., Nakayama, S.I. Gunther, R.A. and Holcroft, J.W. (1985) J. Surg Res 39: 517.
18. Rocha e Silva, M., Velasco, I.T., Oliveira, M.A. and Nogueira da Silva, R.I. (1986) Circ Shock 18: 342.
19. Saragoça, M.A., Mulinari, R.A., Bessa, A.M.A., Draibe, S.A. and Ramos, O.L.. (1986) Circ Shock 18: 339.
20. Velasco, I.T., Rocha e Silva, M., Oliveira M.A. and Negraes, G.A. (1986) Circ Shock 18: 345.
21. De Felippe, J., Jr., Timoner, J., Velasco, I.T., Lopes, O.U.and Rocha e Silva, M. (1980) Lancet ii: 1002.
22. Auler, J.O.C., Jr, Pereira, M.H.C., Rocha e Silva, M., Jatene, A.D. and Pileggi, P. (1985) Circ Shock 16: 94.
23. Auler, J.O.C. Jr, Pascual, J.M., Pereira, M.H.C., Amaral, R.G. and Rocha e Silva, M. (1986) Circ Shock 18: 339.
24. Houston, M.C., Thompson, W.L. and Robertson, D. (1984) Arch Int Med 144: 1433.

Chapter 2

Mediators of
Burn and Sepsis

OXYGEN, FREE RADICALS, SHOCK AND TRAUMA

C. Deby[1], M. Lamy[2], P. Braquet[3], R. Goutier[1]

[1] Laboratoire de Biochimie et Radiobiologie, Université de Liège, Institut de Chimie, B6, Sart-Tilman 4000 Liège I, Belgium
[2] Service d'Anesthésiologie, Université de Liège, Hôpital de Bavière, Bd de la Constitution, 66 - 4020 Liège, Belgium
[3] IHB, 17 avenue Descartes, 92350 Le Plessis Robinson, France

FREE RADICALS (1)

In atoms and molecules, the electrons generate, by rotation, a magnetic field represented by a vector, the spin. Generally, all the electrons are paired, opposing their spins, the resulting magnetic field being null. In molecules, electrons of neighbouring atoms are paired, forming covalent bonds, which can be broken by a sufficient amount of energy; after this rupture, two electrons become unpaired and the implicated atoms or molecules acquire a magnetic field (Fig. 1). The energy can be furnished by radiations (UV, ionizing radiations) and by chemical or enzymatic reactions. Such atoms or molecules presenting an odd number of electrons, and thus a magnetic field, are called <u>free radicals</u>.

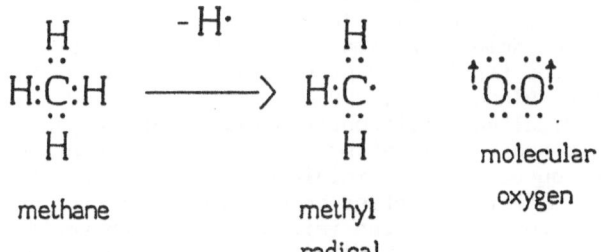

methane   methyl radical   molecular oxygen

Fig. 1. Right : Formation of a free radical by rupture of a covalent bond (by deshydro-genation, in this example). The arrows symbolize the spins of unpaired electrons. Left : The biradical structure of oxygen, at its fundamental state.

Free radicals are generally considered as very strong reagents, reacting with the most stable organic molecules. Generally, a newly formed radical reacts instantaneously with the surrounding molecules, taking out an hydrogen atom, with its electron, in order to pair its unpaired electron (Fig. 2). A new covalence is broken, and another free radical is generated. This is the initiation of a chain of reactions, which can be rapidly terminated, by the collision of two radicals, which dimerize. But, in aerobic conditions, oxygen can intervene, changing completely the above scheme.

$$\cdot OH + R:H \longrightarrow H_2O + R\cdot$$

$$R\cdot + X:R \longrightarrow R:H + X\cdot$$

$$X\cdot + Z:H \longrightarrow Z\cdot + X:H$$

....

$$Z\cdot + Z\cdot \longrightarrow Z:Z \text{ (dimer.)}$$

Fig. 2. Chain of radicalar reactions
initiated by hydroxyle radical $\cdot OH$, and dimerization,
terminating the chain process.

OXYGEN AND PEROXIDATION (2,3)

Oxygen is a biradical, characterized by the presence of two unpaired electrons (Fig. 1). Surprisingly, the quantum mechanics demonstrate that such biradical cannot react with non radicalar molecules, which constitute the great majority of biological molecules (4).

But, oxygen reacts very fastly with free radicals, forming peroxy radicals. This phenomenon threatens particularly polyunsaturated fatty acids (PUFA), constituents of biological membranes. Autoxidation of PUFA is initiated by the break of the covalence binding of hydrogen and carbon atoms, near double bonds (Fig. 3). The resulting free radical is relatively stabilized by electron delocalization (resonance). This stabilization sufficiently delays the terminating process of dimerization to allow the arrival of an oxygen molecule (4). Oxygen reacts with radicalar PUFA ($R\cdot$) forming a peroxy radical $ROO\cdot$ (lipoperoxidation). Hydroxyle radical $\cdot OH$ is classically implicated as initiating agent (5). It is a small radical, soluble as well in aqueous and in lipidic mediums, and is extremely reactive, forming water, after extracting an hydrogen atom from surrounding molecules. However other agents can initiate the lipoperoxidation, particularly perferryl cation ($Fe^{4+} - O_2^=)^{++}$ (6). Peroxy radical $ROO\cdot$ extracts an hydrogen atom from a new PUFA molecule, RH, forming a new lipidic radical $R\cdot$, and becoming itself a hydrolipoperoxide, ROOH. The peroxy radical $ROO\cdot$ extracts an hydrogen atom from a new unsaturated fatty acid, which, in its turn, becomes radicalar, $ROO\cdot$ becoming ROOH, an hydrolipoperoxide.

Hydrolipoperoxides are not reactive by themself, but, meeting ferrous ion, Fe^{2+}, they are reduced into alkoxy radicals $RO\cdot$, which are strong reagents, able to initiate another loop of peroxidation, and replacing easily hydroxyle radical or perferryl cation. Thus, in the presence of oxygen, and thanks to the presence of iron, which is an universal component

of the living matter, the lipoperoxidation processes become rapidly
extensive, in aerobic conditions, and are dreadful agents of membrane
destruction (7).

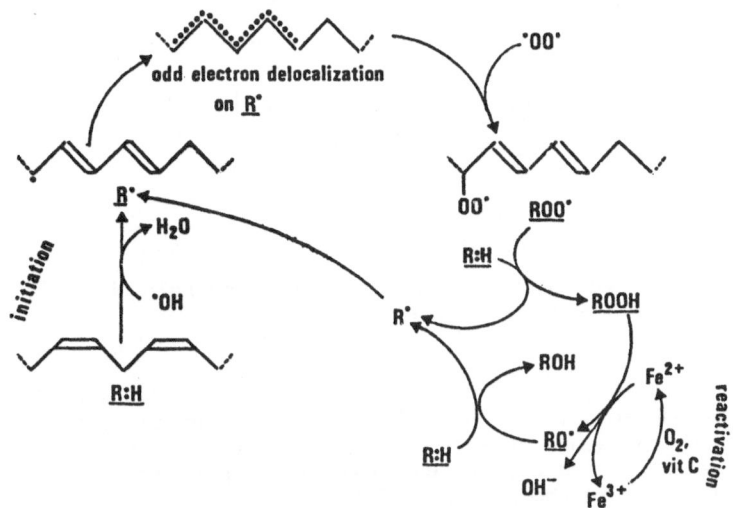

Fig. 3. Autoxidation cycle of unsaturated molecule and
iron-induced reactivation mechanism. R:H : unsaturated fatty acid. R· : radicalar fatty acid,
formed after H· removing from R:H. ·OO· : oxygen
molecule. ROO· : peroxy radical. ROOH : hydrolipoperoxide. RO· : alkoxy radical. ·OH : hydroxyle radical. OH⁻ : hydroxyle anion. Superoxide
anion or ascorbate can regenerate Fe^{2+}.

BIOCHEMICAL SOURCES OF FREE RADICALS

Dangerous radicals arise from oxygen metabolism (5,7). Oxygen in an
universal electron acceptor, in aerobic organisms. In mitochondria,
normally, at the end of the respiratory chain, oxygen is completely and
instantaneously reduced into water :

$$O_2 + 4 e^- + 4 H^+ \quad ----\blacktriangleright \quad 2 H_2O$$

But, in abnormal conditions, such as during metabolic troubles following hypoxia or anoxia, the mitochondrial electron transport through the
cytochromes chain is stopped, and the oxygen molecules go up to the level
of ubiquinone where they are reduced electron by electron (8). This univalent reduction of O_2 generates free radicals, and particularly, hydroxyle
radical. The first step is the formation of superoxide anion, O_2^-, by an
enzymatic mechanism associated with ubiquinone. Then, O_2^- dismutates into
hydrogen peroxide, H_2O_2. The coexistence between O_2^- and H_2O_2, in the
presence of ferrous iron, initiates a Haber-Weiss cycle, producing hydroxyle
radicals (Fig. 4).

The mechanism of free radical production, at the level of mitochondria, during reoxygenation succeeding to severe hypoxia, is evoked on
fig. 5. Haber-Weiss cycles can also be produced by the activity of several enzymes. Xanthine-deshydrogenase, which intervenes in the last steps

$$\text{'O-O'} + e^- \xrightarrow[\text{enz.}]{} (\text{:O-O'})^- \quad \text{or} \quad O_2^{\overline{\cdot}}$$

$$2\,O_2^{\overline{\cdot}} + 2\,H^+ \longrightarrow H_2O_2 + O_2 \quad (\text{dismutation})$$

$$H_2O_2 + Fe^{2+} \longrightarrow \text{'OH} + OH^- + Fe^{3+}$$

$$Fe^{3+} + O_2^{\overline{\cdot}} \longrightarrow Fe^{2+} + O_2$$

Haber-Weiss cycle

Fig. 4. Univalent reduction of oxygen, and role of Fe^{++}
in the production of 'OH.

of purine catabolism, can be converted into xanthine-oxidase, in pathological situations inducing severe calcium unbalance particularly in ischemia
(9,10). In the presence of xanthine-oxidase, purine catabolism produces
superoxide anion, which dismutates into hydrogen peroxide, generating free
radicals such 'OH, in the presence of Fe^{2+}.

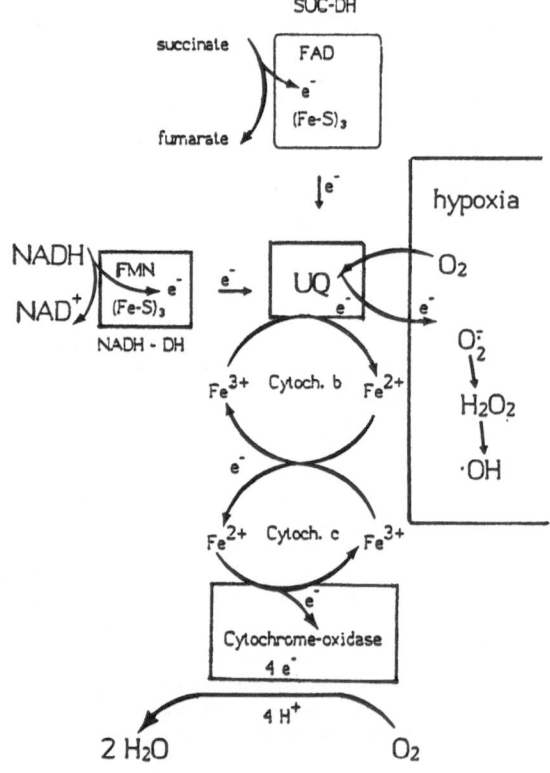

Fig. 5. Mechanisms of production of activated
oxygen species by the mitochondrial respiratory chain, after hypoxia. NADH-DH :
NADH-deshydrogenase. SUC-DH : succinodes-
hydrogenase. UQ : ubiquinone.

Other sources of lipid radicals are the activities of lipoxygenases and of cyclooxygenase. During prostaglandin synthesis, the conversion of endoperoxide H_2 is accompanied by the generation of free radicals (12). The mixed oxidasic function of the liver can transform several molecules into free radicals : it is the case for carbon tetrachloride and certain antibiotics such as doxorubicin (13,14).

ROLE OF IRON IN RADICALAR REACTIONS

Because of its electronic structure, fundamental oxygen cannot react directly with non radical molecules (4). Monoradicals can play the role of mediators, between oxygen and non radical molecules, as in lipoperoxidations, for instance. But this role is mainly supported by transition metals, and particularly by iron. Transition metals are characterized by the property to be linked to certain molecules, called ligands, by coordination links. A well-known example is the coordination of iron by adenosine diphosphate (ADP). The particular properties of transition metals are explained by the presence, in their atoms, of five uncompletely filled d orbitals (Fig. 6).

During the coordinative process (15),3 d orbitals are redistributed in two sublayers, separated by an energy level depending on the nature of the ligand (splitting). In the case of ferrous iron, which exhibits six d electrons, a low splitting energy corresponds to a wide redistribution of electrons. There are four unpaired electrons in the complexed iron atom. On the contrary, some other ligands, giving a higher splitting energy, inhibit the redistribution of electrons to the two orbitals of higher energy : all the six electrons of ferrous iron remain paired on the three orbitals of lower energy. This phenomenon is observed when iron is complexed with chelators such as DETAPAC and deferroxamine; these ligands inhibit oxidative activities of ferrous ion.

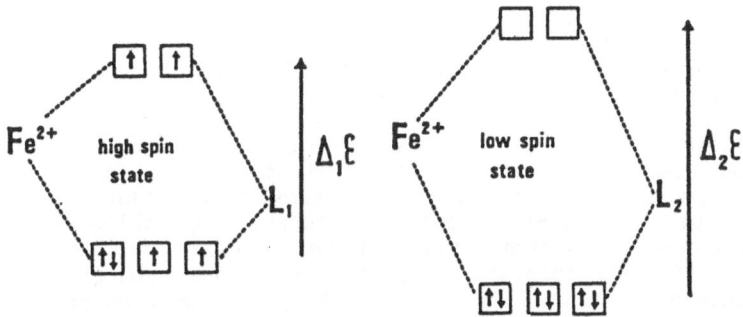

Fig. 6. The two modes of coordination for Fe^{2+}. In the high spin state, the low splitting energy $\Delta \mathcal{E}_1$ permits the occupation of five d orbitals : there are 4 unpaired electrons. In the low-spin state, the high energy splitting $\Delta \mathcal{E}_2$ impairs the jump of the electrons of low energy to the orbitals of higher energy : the 6 electrons are paired. L_1 and L_2 : ligands

Ferrous complexes, which exhibit a high number of unpaired electrons, are particularly efficient to promote the combination of oxygen with organic molecules. ADP complexed iron plays this role. The presence of conveniently complexed iron is as efficient to induce lipoperoxidation

as hydroxyle radical. Inversely, iron chelation protects against lipoperoxidation (16). Intracellular iron is combined to ferritin and is maintained at the state of Fe^{3+} (17). Physiopathological conditions, such as
ischemia or hypoxia, release iron in cytoplasm, perhaps after reduction
into Fe^{2+} by O_2^- (11). Then, Fe^{2+} can be coordinated into oxidative complexes, not only by ADP, but also by particular sequences of cytoplasmic
proteins.

GENERATION OF RADICALS BY ACTIVATED LEUKOCYTES

Polymorphonuclear leukocytes (PMNLs) and macrophages are potent generators of oxidant species. Activation of complement produces C_{3a} and C_{5a},
which are activators of polymorphonuclear leukocytes and trigger the phenomenon called <u>respiratory burst</u> in these cells (18,19).

The O_2 consumption increases dramatically, amiboid movements are
observed, and a degranulation process occurs, which consists in a release
of the lysosomal content out ot the cell. A microsomal enzyme, NADPH-
oxidase, reduces intensively O_2 into superoxide anion, which dismutes into
H_2O_2; H_2O_2 is released out of the cell, together with O_2^- (20) and in the
presence of Fe^{2+}, regenerated by reductants such as ascorbate, can lead to
the formation of $^{\cdot}OH$, and to peroxidation to living membranes (20,21).
One of the lysosomal enzymes released out of PMNLs is myeloperoxidase, which
PMNLs synthesizes hypochlorous acid, HClO. HClO, reacting with H_2O_2,
produces singlet oxygen, a highly reactive oxidant species, and reacting
with several amines, produces N-amine chlorides, highly oxidant species (22).

PHYSIOPATHOLOGICAL GENERATION OF FREE RADICALS

A first pathological situation generating activated oxygen species is
observed in tissues during refeeding oxygen, after hypoxia and anoxia, and
results from the univalent reduction of oxygen at the level of ubiquinone,
in mitochondria (Fig. 5).

Purine catabolism can generate O_2^- and subsequent reactions of fig. 4,
in hypoxic conditions (10).

Pathological consequences can arise from an excess of defense response
against immunological stimuli. Massive PMNLs activation results in aggregation of these cells, which can be immobilized in capillary beds, and
particularly in the lung (19). There, an active cooperation is established
between proteases and oxidant species, released by the activated leukocytes
in the blood. Oxidant species, and particularly amine chlorides, destroy
plasma antiproteases (23). Therefore, leukocytes proteases are free to
attack the capillary walls. An amplification phenomenon occurs, by action
of superoxide on plasma factors, creating leukotaxic factors, by stimulation
of platelet by H_2O_2, releasing not only thromboxane, but leukotaxic lipoxygenase products. This scheme can explain some aspects of the development
of adult respiratory distress syndrome (ARDS), which is often observed
after severe trauma and sepsis (24).

Cooperativity between proteases and oxidant agents, produced together
and released by PMNLs, was suggested by different authors. It was demonstrated that high levels of both enzymatically and immunologically active
trypsin were measured in the plasma of patients suffering from severe
sepsis and polytrauma, and developing ARDS (25). Trypsin activity was
improved by the fall of antiproteasic activity, induced by PMNLs oxidative
factors.

ROLE OF ACTIVATED OXYGEN SPECIES IN SHOCK AND TRAUMA : A SYNOPSIS

Shock induces deep modifications of membrane permeability, and modify the normal calcium transport. Xanthine-deshydrogenase is then activated and becomes xanthine oxidase, generating activated oxygen species (O_2^- , $\cdot OH$, lipoperoxides, alkoxy radicals). Severe tissular hypoxia is often the consequence of shock : during reoxygenation, the disturbed mitochondrial respiratory chain reduces univalently oxygen, with the cascade leading to free radicals and lipoperoxidation, as in the case of xanthine-oxidase.

Severe trauma are often complicated by shock and by sepsis. Sepsis, particularly abdominal sepsis, stimulates PMNLs, via the alternative pathway of complement activation. Activated PMNLs are generators of multiple oxidant agents : hydroxyl radicals, HClO, amine chloride, singlet oxygen, etc... .

It appears thus, from these above considerations, that multiple causes are at the origin of activated oxygen metabolism. The best proofs of this activation are given by the demonstration of a fall of the antioxidant potential, measured, either by a global method (25) or by the determination of a determined antioxidant, such as tocopherol. Preliminary research on the evolution of plasma tocopherol in ARDS showed a fall of vitamin E in severe cases (26). Studies continued on this subject confirmed that the fall of plasma tocopherol seems to be proportional to the gravity of the case.

In shock and sepsis, there are thus many reasons to invoke the presence of deleterious oxygen species, injuring cellular membranes by peroxidation of polyunsaturated fatty acids. But, up to now, no direct, unequirocal proof have been furnished, of the in vivo existence of lipoperoxidative processes.

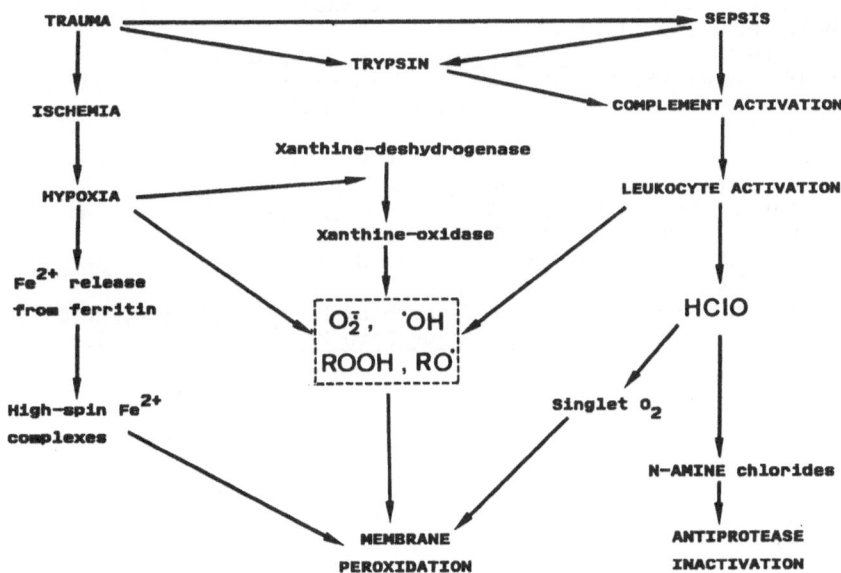

Fig. 7. Overview of the physiological and physiopathological mechanisms which cooperate for membrane peroxidation, by generation of free radicals and activated oxygen species.

REFERENCES

1. Pryor WA. in : <u>Introduction to free radical chemistry</u>. Prentice-Hall, Inc., Englewood Cliffs (New-Jersey) (1966).
2. Hill HAO. in : <u>Oxygen free radicals and tissue damages</u>. Ciba Found. Symp. 65. Excerpta Medica, p 5 (1979).
3. Buege JA, and Aust SD. <u>Methods Enzymol</u>, 51 : 302 (1978).
4. Hamilton GA. in : <u>Molecular mechanisms of oxygen activation</u> (ed : O. Hayashi). Academic Press (New-York) p 405 (1974).
5. Del Maestro RF, Thaw HH, Bjork J, and Arfors KE. <u>Acta Physiol Scand</u> , suppl. 492 : 43 (1980).
6. Sugioka K, Nakano H, Tero-Kubota S, and Ikegani Y. <u>Biochem Biophys Acta</u>, 753 : 411 (1983).
7. Porter NA. <u>Methods Enzymol</u>, 105 : 273 (1984).
8. Boveris A, Turrens JF. in : <u>Chemical and biochemical aspects of super-oxide and superoxide dismutase</u> (eds : J.V. Bannister and H.A.O Hill). Elsevier/North-Holland (New-York) p 84 (1980).
9. McCord JM. <u>New Engl J Med</u>, 312 : 159 (1985).
10. Roy RS, and McCord JM. <u>Fed Proc</u>, 41 : 767 (1982).
11. White BC, Krause GS, Aust SD, and Eyster GE. <u>Ann Emerg Med</u>, 14 : 804 (1985).
12. Deby C, Pincemail J, Hans P, Braquet P, Lion Y, Deby-Dupont G, and Goutier R. in : <u>Cerebral Ischemia</u> (eds : E. Bes, P. Braquet, R. Paoletti and B.K. Sjesjö). Excerpta Medica (Amsterdam, New York, Oxford) p 249 (1984).
13. McCay PB, Noguchi T, Fong KL, Lai EK, and Poyer JL. in : <u>Free radicals in biology</u>, vol. IV. Academic Press (New-York) p 155 (1980).
14. Doroshow JH, Locker GY, and Myers CE. <u>J Clin Invest</u>, 65 : 128 (1980).
15. Henrici-Olivé G, and Olivé S, in : <u>Coordination and catalysis</u>. Verlag Chemie (Weinheim, New-York) (1977).
16. Aust SD, White BC. <u>Adv Free Radical Biol Med</u>, 1 : 1 (1985).
17. Crichton RR. in : Ciba Found. Symp. 65. <u>Excerpta Medica</u>, p 57 (1979).
18. Badwey JA, and Karnovski ML. <u>Annu Rev Biochem</u>, 49 : 695 (1980).
19. Hammerschmidt DE, Harris PD, Wayland H, Craddock PR, and Jacob HS. <u>Am J Pathol</u>, 102 : 146 (1981).
20. Root RK, and Metcalf JA. <u>J Clin Invest</u>, 60 : 1266 (1977).
21. Babior BM, Kipnes RS, Curnutte JT. <u>J Clin Invest</u>, 52 : 741 (1973).
22. Grisham MB, Jefferson MM, Melton DF, and Thomas EL. <u>J Biol Chem</u>, 259 : 10404 (1984).
23. Weiss SJ, Lampert MB, and Test ST. <u>Science</u>, 222 : 625 (1983).
24. Freeman BA, and Tanswell AK. <u>Adv Free Radical Biol Med</u>, 1 : 133 (1985).
25. Deby-Dupont G, Hass M, Pincemail J, Braun M, Lamy M, Deby C, and Franchimont P. <u>Intensive Care Med</u>, 10 : 7 (1984).
26. Deby C, Deby-Dupont G, Hans P, Pincemail J, and Neuray J. <u>Experientia</u>, 39 : 1113 (1983).
27. Bertrand Y, Artoisenet A, Allard B, Barbier B, de Meulder A, Raymaert M, Mathieu P, and Dumont E. <u>Intensive Care Med</u>, 11 : 65 (1985).

ROLE OF FREE RADICAL LIPID PEROXIDATION IN BURN AND ENDOTOXIN SHOCK

T. Yoshikawa, N. Yoshida, H. Miyagawa, T. Takemura, T. Tanigawa, M. Murakami and M. Kondo
First Department of Medicine, Kyoto Prefectural University of Medicine, Kamigyo-ku, Kyoto 602, Japan

INTRODUCTION

Enhanced free radical generation leading to lipid peroxidation has been claimed to be associated with various disorders (Bulkey, 1983). There is considerable indirect evidence supporting a role for oxygen radicals in circulatory shock. Crowell et al. (1969) and Cunningham and Keaveny (1978) reported that allopurinol, a competitive inhibitor of xanthine oxidase, substantially increased the survival rate of dogs subjected to hemorrhagic shock. Severe burns result in both local and systemic hemodynamic changes (Gilmore and Handford, 1956; Wolfe and Miller, 1976; Adams et al., 1981), as well as in endocrine (Turinsky et al., 1977), neuroendocrine (Cova and Glaviano, 1968; Goodall and Moncrief, 1965), and metabolic (Robinson and Miller, 1981; Wolfe et al., 1977) alterations. Acute endotoxin shock is rapidly produced with Escherichia coli endotoxin (Balis et al., 1978).

It was recently suggested that in every kind of shock many coexisting mechanisms may lead to the overproduction of oxygen free radicals and that a pathogenetic role of such radicals may by considered highly probable (Novelli, 1983). To explore this hypothesis, we examined the effects of superoxide dismutase (SOD) and catalase which are enzymatic oxygen radical scavengers, on burn and endotoxin shock in rats.

MATERIALS AND METHODS

Female Wistar rats weighing 190–220 g, from Keari Co., Ltd., Osaka were housed for at least 7 days in our animal quarters prior to the experiments. The animals were not fed for 24 h, but were allowed free access to water. Endotoxin (Escherichia coli 055: B5 lipopolysaccharide B; Difco Lab., Detroit, Mich) was dissolved in pyrogen-free physiological saline before every experiment. Burn shock was produced by immersing half of the animal's body into 80°C water for 10 s. Endotoxin shock was induced by a single injection of 100 mg/kg of endotoxin, diluted in 1.0 ml of physiological saline, into the femoral vein using a syringe.

To examine the effects of SOD and catalase on burn shock, the rats were injected with SOD from bovine blood (3200 U/mg protein; Sigma Chemical Co., St. Louis, Mo) at 50 mg/kg or catalase from bovine liver (40,000 Sigma U/ mg protein; Sigma Chemical Co., St. Louis, Mo) at 2.0 mg/kg, intravenously 30 s before the burn stress. To examine the effects of these enzymes on endotoxin shock, the animals were injected with SOD from bovine blood (3050

U/mg protein; Sigma Chemical Co., St. Louis, Mo) at 50 mg/kg or catalase from bovine liver (40,000 Sigma U/mg protein; Sigma Chemical Co., St. Louis, Mo) at 1.0 mg/kg, subcutaneously 12 and 1 h before the injection of endotoxin (100 mg/kg).

The severity of shock was determined with systolic blood pressure and serum lysosomal enzymes, such as acid phosphatase and β-glucuronidase.

Systolic blood pressure and heart rate count were measured by the method of Bunag (1973) using a sphygmomanometer PS-100 (Riken Kaihatsu Co., Tokyo). Serum acid phosphatase activity was assayed according to the method of Andersch and Szczypinski (1947) using p-nitrophenylphosphate as the substrate. Serum β-glucuronidase activity was measured with Sigma reagents (Sigma Chemical Co., St. Louis, Mo), based essentially on the method of Fishman et al., (1967), using phenolphthalein glucuronic acid as the substrate. Thiobarbituric acid (TBA) reactive substances in serum were determined by the method of Yagi (1978). The concentration of TBA reactive substances was expressed in terms of malondialdehyde using tetramethoxy-propane as a standard.

Results are expressed as mean value ± standard deviation (SD).

RESULTS

Burn Shock

Systolic blood pressure of rats before the immersion into 80°C water was 121 ± 8 mmHg (Mean ± SD, n=7). The blood pressure was significantly reduced to undetectable value (under 50 mmHg) 10, 60, 120 and 180 min after the burn. Immediately after the immersion into 80°C water (10 min), acid phosphatase and β-glucuronidase activities were significantly increased (Fig. 1) and serum TBA reactants were significantly increased (Fig. 2). All rats were killed at 3 h after the burn shock and changes in following parameters of rats treated with SOD or catalase were examined.

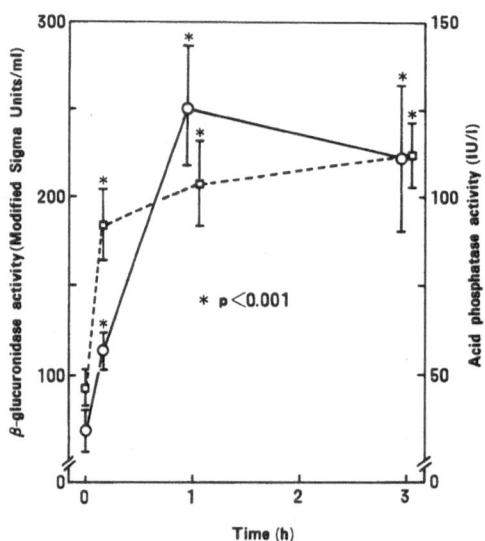

Fig. 1. Changes in serum lysosomal enzymes before and after burn stress. Results are expressed as mean value ± SD of 7 rats.

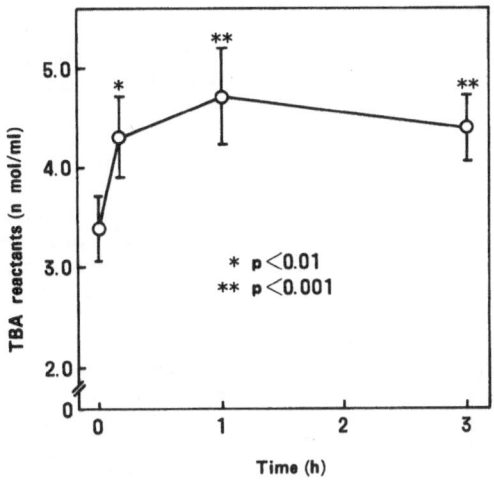

Fig. 2. Changes in serum TBA reactants before and after burn stress. Results are expressed as mean value ± SD of 7 rats.

The changes in serum acid phosphatase and β-glucuronidase activities

The increase in serum acid phosphatase activity after the burn stress was significantly inhibited by the treatment with SOD and catalase, but could not be by SOD or catalase alone (Fig. 3). The increase in β-glucuronidase activity was not reduced by these treatment.

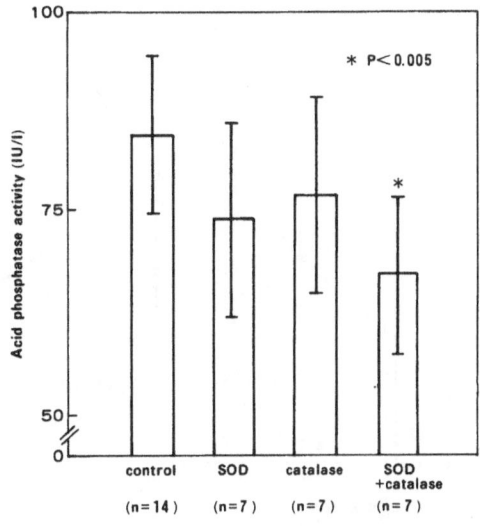

Fig. 3. Effects of SOD or/and catalase on serum acid phosphatase activity 3 h after the burn stress. Control rats were injected with 1.0 ml of saline.

The changes in TBA reactive substances

The increase in serum TBA reactants 3 h after the burn stress was not significantly reduced by the treatment with SOD or/and catalase.

Endotoxin Shock

Immediately after the injection of endotoxin, systolic blood pressure was reduced and heart rate was increased (Fig. 4). Acid phosphatase and β-glucuronidase activities, and serum TBA reactants were significantly increased (Fig. 5, 6).

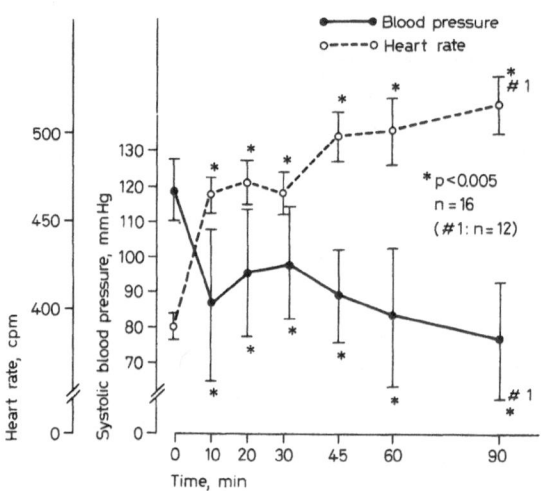

Fig. 4. Changes in systolic blood pressure and heart rate of rats injected with endotoxin (100 mg/kg).

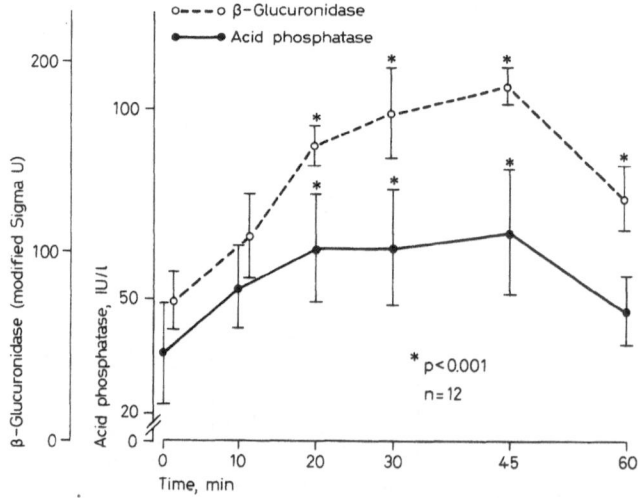

Fig. 5. Changes in serum acid phosphatase and β-glucuronidase activities after the injection of endotoxin (100 mg/kg).

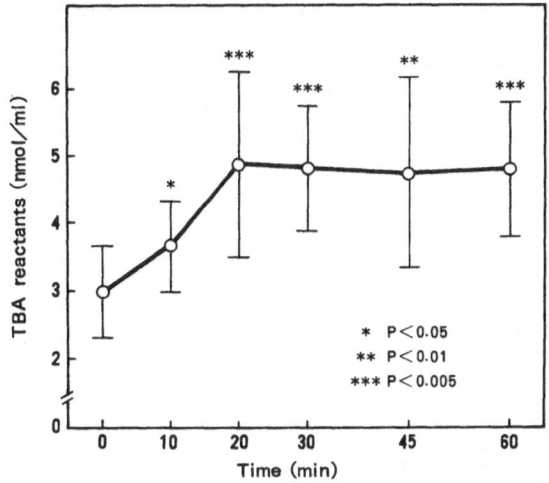

Fig. 6. Changes in serum TBA reactants after the injection of endotoxin (100 mg/kg).

These changes were most remarkable at 45 min after the injection of endotoxin. All rats were killed at 45 min after the endotoxin injection, and the changes in following parameters of rats treated with SOD or catalase were examined.

The changes in systolic blood pressure

The blood pressure was significantly reduced 45 min after the injection of endotoxin. The reduction was significantly inhibited by the treatment with SOD or catalase (Fig. 7).

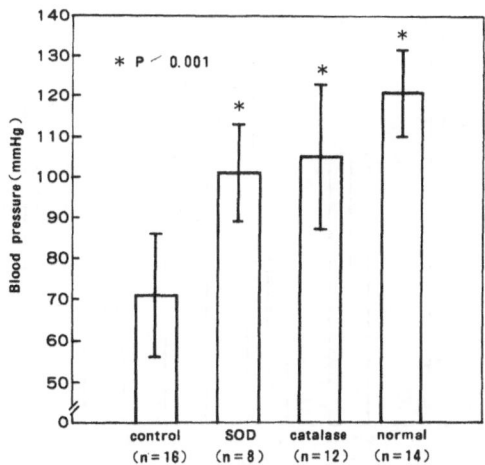

Fig. 7. Effects of SOD or catalase on endotoxin shock. SOD or catalase was injected sc 12 and 1 h before the injection of endotoxin (100 mg/kg).

The increase in acid phosphatase activity after endotoxin shock was significantly inhibited by the administration of SOD or catalase (Fig. 8). The increase in β-glucuronidase activity was inhibited by the treatment of SOD, but could not be by catalase (Fig. 9).

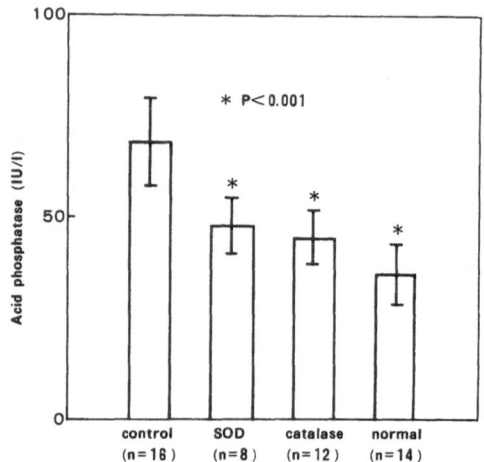

Fig. 8. Effect of SOD or catalase on serum acid phosphatase activity 45 min after the endotoxin injection.

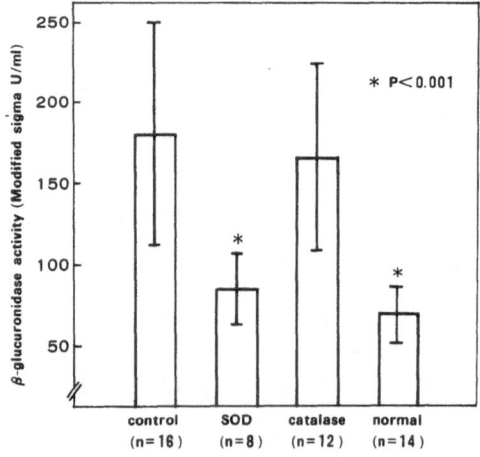

Fig. 9. Effect of SOD or catalase on serum β-glucuronidase activity 45 min after the endotoxin injection.

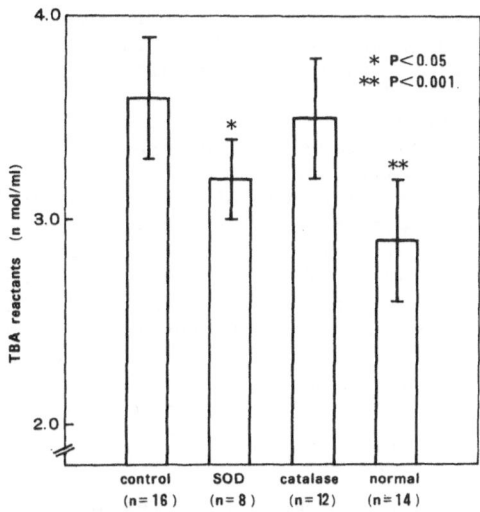

Fig. 10. Effect of SOD or catalase on serum TBA reactants 45 min after the endotoxin injection.

The changes in TBA reactive substances

Serum TBA reactants were significantly increased 45 min after the injection of endotoxin. The increase was inhibited by the administration of SOD but not by catalase (Fig. 10).

DISCUSSION

After injection of endotoxin (100 mg/kg) or immersion in 80°C water, the blood pressure was significantly decreased, and serum lysosomal enzyme levels were increased. These findings indicate that these treated rats were in shock state. Serum TBA reactive substances were significantly increased after the shock. The aggravation of endotoxin shock was prevented by the treatment with SOD or catalase. However, only a slight prevention was observed in burn shock.

The implication of oxygen radicals and free radical lipid peroxidation in the pathogenesis of shock was reported in different shock models; endotoxin shock (Yoshikawa et al., 1986; Maiarino et al., 1986; Kunimoto et al., 1986), burn shock (Saez et al., 1984; Till et al., 1986), mesenteric artery occlusion shock (Novelli et al., 1986), and hypovolemic-traumatic shock (Schlag and Redl, 1986).

In shock, an insufficient defence system and excessive free radical generation may result in severe cell damage. These produced oxygen radicals may initiate lipid peroxidation and thus directly damage the cell membrane, which contains a large amount of polyunsaturated fatty acids. Furthermore, lipid peroxides produced by peroxidation of lipid injure several organs. These processes enhance vascular permeability, histamine secretion, platelet aggregation, and generation of chemotactic factors.

The source of oxygen-derived free radicals produced in shock states seems to be different in several shock models. Oxygen radicals are generated from activated granulocytes, from xanthinoxidase activity, and by damage to mitochondrial oxidative capacity.

The present data confirm the previous report that lipid peroxidation occurs during endotoxemia (Yoshikawa et al., 1983). Furthermore, SOD and catalase can prevent the aggravation of endotoxin shock suggesting the involvement of the oxygen radicals. On the other hand, SOD and catalase did not show a strong preventive effect against burn shock. Dimethyl sulfoxide, a hydroxyl radical scavenger, can prevent the aggravation of burn shock (unpublished data).

The origin and the chemical nature of lipid peroxidation products appearing in serum of rats with burn and endotoxin shock are not yet known. However, from all this rather indirect evidence we can conclude that a pathological mechanism involving the oxygen-derived free radicals is indeed operative in burn and endotoxin shock. Therefore, scavengers of free radicals may become an important therapeutic agent in the future.

REFERENCES

Adams, H.R., Baxter, C.R., Parker, J.L., and Senning, R. Circ. Shock 8: 613 (1981).
Andersch, M.A., and Szczypinski, A.J. Am. J. Clin. Path. 17: 571 (1947).
Balis, J.U., Rappaport, E.S., Gerber, L., Fareed, J., Buddingh, F., and Messmore, H.L. Lab. Invest. 38: 511 (1978).
Bulkley, G.B. Surgery 94: 407 (1983).
Bunag, R. J. Appl. Physiol. 34: 279 (1973).
Cova, R., and Glaviano, V.V. Proc. Soc. Exp. Biol. Med. 128: 642 .(1968).
Crowell. J.W., Jones, C.E., nad Smith, E.E. Am. J. Physiol. 216: 749 (1969).
Cunningham, S.K., and Keaveny, T.V. Eur. Surg. Res. 10: 305 (1978).
Fishman, W.H., Kato, K., Antiss, C.L., and Green, S. Clin. Chim. Acta 15: 435 (1967).
Gilmore, J.P., and Handford, S.W. J. Appl. Physiol., 8: 393 (1956).
Goodall, McC., and Moncrift, J.A. Ann. Surg. 162: 893 (1965).
Kunimoto, F., Morita, T., and Ogawa, R. Oxygen free radicals in shock, G.P. Novelli, and F. Ursini, eds., Karger Basel (1986).
Maiorino, M., Bordi, L., Consales, G., Ursini, F., and Novelli, G.P., Oxygen free radicals in shock, G.P. Novelli, and F. Ursini, eds., Karger, Basel (1986).
Novelli, G.P., and De Guadio, A.R., Shock research, D.H. Lewis, and U. Haglund, eds., Elsevier Science Publishers, Amsterdam (1983).
Novelli, G.P., Livi, P., Ghinassi, M.L., Brunelleschi, S., and Fantozzi, R. Oxygen free radicals in shock, G.P. Novelli and F. Ursini, eds., Karger Basel (1986).
Novelli, G.P., Livi, P., Ghinassi, M.L., Lisi, L., Brunelleschi, S., and Fantozzi, R. Oxygen free radicals in shock, G.P. Novelli and F. Ursini, eds., Karger Basel (1986).
Robinson, K.M., and Miller, H.I. Circ. Shock 8: 1 (1981).
Saez, J.C., Ward, P.H., Gunther, B., and Vivaldi, E. Circ. Shock 12: 229 (1984).
Schlag, G., and Redl, H. Oxygen free radicals in shock, G.P. Novelli and F.Ursini, eds., Karger, Basel (1986).
Till, G.O., Hatheri, J.R., and Ward, P.A. Oxygen free radicals in shock, G.P. Novelli, and F. Ursini eds., Karger, Basel (1986).
Turinsky, J., Saba, T.M., Scovilli, W.A., and Shesnut, T. J. Trauma 17: 344 (1977).
Wolfe, R.R., and Miller, H.I. Am. J. Physiol. 231: 892 (1976).

Wolfe, R.R., Elahi, D., Spitzer, J.J., and Miller, H.I. Surg. Gynecol.
    Obstet. 144: 359 (1977).
Yagi, K. Biochem. Med. 15: 212 (1976).
Yoshikawa, T., Murakami, M., Furukawa, Y., Kato, H., Takemura, S., and
    Kondo, M. Thromb. Haemostas. 49: 214 (1983).
Yoshikawa, T., Seto, O., Itani, K., Kakimi, Y., Sigino, S., and Kondo,
    M. Oxygen free radicals in shock, G.P. Novelli, and F. Ursini,
    eds., Karger Basel (1986).

THE SIGNIFICANCE OF PENTANE MEASUREMENTS IN MAN

C. Deby[1], J. Pincemail[1], Y. Bertrand[2], M. Lismonde[3]

[1] Laboratoire de Biochimie et de Radiobiologie, Université de Liège, Institut de Chimie, B6, Sart-Tilman 4000 Liège I, Belgium
[2] Clinique Sainte-Elizabeth, Bloc opératoire, 5000 Namur, Belgium
[3] Service d'Anesthésiologie, Hôpital de Bavière, Université de Liège, 66 boulevard de la Constitution, 4020 Liège, Belgium

OVERVIEW OF THE METHODS OF LIPOPEROXIDATION DETERMINATION

Lipoperoxidation of polyunsaturated fatty acids, constituents of cell membranes, is the main feature running out of in vitro experiments on oxygen toxicity (1). Theoretical considerations let us suppose that, in in vivo conditions, lipid autoxidation can occur, explaining some aspects of oxygen toxicity, particularly at the pulmonary level (2). However, until now, no absolutely safe methods have been elaborated, to unequivocally demonstrate the in vivo lipoperoxides production. Theoretically, a safe method would determine a lipoperoxide derivating product which is not further catabolized and which arises exclusively from lipoperoxidative pathway. Different procedures were proposed, each one based on the determination of a particular step, during hydroperoxide decomposition. The steps are the following :
a) Peroxidation of polyunsaturated fatty acids, after diene conjugation (3);
b) Spontaneous or catalytically accelerated breakdown of hydroperoxides leading to malonaldehyde (MDA) and to other fragments, which become finally hydrocarbons, such as ethane and pentane, or pentanol (4,5).

It is not often possible to directly detect the presence in vivo of hydroperoxide ROOH function, because its instability, particularly high in biological conditions. However methods such as sensitive iodometry (6) or the use of a chromogen (7,8) can reveal the presence of hydroperoxide in biological samples (7), abnormally poor in antilipoperoxidant agents. Preference was given to the dosage of decomposition products arising from ROOH, in biological sampling.
Malonaldehyde (MDA) determination is the most popular procedure to estimate in vitro and in vivo lipoperoxidation processes, by means of thiobarbituric acid. But it cannot be considered as a safe method because MDA can arise from other pathways than lipoperoxidation, and particularly, from thromboxane biosynthesis (9). On the other hand, MDA itself enters in the formation of lipofuscin, reacting with aminoacids to give Schiff bases (10). Another argument against MDA technique is that the lipoperoxide decomposition does not lead automatically to MDA (11), and thus, that "MDA production is only a small proportion of the total PUFA loss" (12).

Conjugated dienes measurements by measuring optical absorbance at 233 nanometers was another technique proposed for lipoperoxidation estimation (13). Dienes appear at the onset of peroxidation (3), and seem to be linked to several steps of peroxide decomposition. However we have observed that conjugated dienes decreased in contact with blood plasma. Detection of pentane in exhaled gases was proposed as a specific mean to evidence lipoperoxidation processes in living animal and in man (5). This method is not invasive, and it was supposed that pentane could only arise from lipoperoxidation and that it could not be metabolized. However the possible metabolism of pentane by hepatic metabolism constitutes a risk recently evoked (14).

DECOMPOSITION OF LIPOPEROXIDES INTO HYDROCARBONS

This phenomenon was predicted by Frankel et al. (1), in 1961, and experimentally confirmed by subsequent works. Autoxidation of methyl linoleate produced, according to Horvat et al. (3), methane, ethane, propane, butane, and pentane, this latter representing 90% of total emission of hydrocarbons. The improvement of GLC techniques permitted to demonstrate that ethane and pentane were the mean hydrocarbons resulting from lipoperoxide decomposition (5).

Frankel et al. (1) has proposed that hydrocarbon formation resulted from the splitting of the fatty chain on the alkyl side of the carbon chain supporting the -OOH function. A demonstration of this theory was afforded by Evans et al. (6), showing that linoleate hydroperoxide decomposition formed pentane , but that another pathway was also possible, forming non volatile pentanal. Further research showed that ethane arised from $\omega 3$ polyunsaturated fatty acid (linolenic acid family), while pentane resulted from $\omega 6$ family of polyunsaturated acids, such as linoleic, and arachidonic acid (7,8). The role of iron in hydrolipoperoxide decomposition clearly appeared (8) :

$$Fe^{2+} + ROOH \longrightarrow RO^{\cdot} + Fe^{3+} + OH^-$$

EXHALED GASES AS INDEX OF IN VIVO LIPID PEROXIDATION

Already in 1974, Riely et al. (11) proposed the ethane measurement as an index of peroxidation in tissues homogenates. However Dillard et al. (12) prefered the determination of pentane in expirated air, during in vivo lipid peroxidation experiments on animal, in place of ethane measurement, considering the quantitative dominance of $\omega 6$ family in most animal tissues. They estimated, for instance, that in the study of tocopherol effect in protection of rodents against in vivo lipoperoxidation, pentane determination afforded best results than ethane. However feeding a fat source high in $\omega 3$ fatty acids, such as cod liver oil, enhanced greatly the ethane production in rats, relatively to those fed with lard ($\omega 6$ rich), during carbon tetrachloride-induced in vivo peroxidation (14).

THE PROBLEM OF HYDROCARBON METABOLISM

Some years ago, it has been observed that liver microsomes metabolized hydrocarbons into corresponding alcohols (15).

Frank et al. (16) exposed rats to an atmosphere containing ethane, propane, butane and pentane, and observed a diminution in the hydrocarbon concentration, proportional to the chain length. The phenomenon was inhibited by administration of three inhibitors of hepatic hydroxylation :

98

tetrahydrofuran, dithiocarb or carbon monoxide. However, metyrapone, recognized as a potent inhibitor of cytochrome P450 system, was without effect. Müller et al. (17) observed that n-pentane disappeared from the chamber containing perfused livers, although they were submitted to Fe^{2+}, a treatment which induces massive peroxidation, and concluded that this hydrocarbon was metabolized by hepatic tissue. Recently, Lawrence and Cohen (18,18b) found that ethane measurement was a better index of lipid peroxidation than pentane measurement, because ethane was practically not metabolized. Other authors gave also a preference for ethane, for in vivo lipoperoxidation estimations (19,20). According to Lawrence et al. (21), ethane measurements would furnish better results for the study of the role of vitamin E on lipoperoxidation inhibition, in the rat, than pentane determination, proposed by Tappel (22). Nevertheless, recent attempts to evidence lipoperoxidation processes in human were performed measuring pentane (23,24), following the first works of Dillard et al., in 1978 (25). Wade and Rij (26), studying ethane and pentane exhalation in volunteers, established a mathematical formula, in order to calculate the correction of the measured pentane values, taking in account the rapid metabolism of pentane, and its solubility in tissues. In accord with Tappel (13) and Dillard (12), these authors estimated that pentane would provide a more reliable measure of in vivo lipoperoxidations, because ω6 fatty acids are the predominant polyunsaturated fatty acids in the membranes.

We have observed (27) that pentane emission from autoxidized linoleate, occuring in sealed vials, was significantly lowered in the presence of hepatic microsomes from phenobarbital-treated rats. As it appears on fig. 1, this phenomenon was not observed with boiled microsomes. Metyrapone, an inhibitor of the cytochrome P450 system, restored, at 80% of the control values, the amount of pentane emitted by linoleate peroxides, in the presence of microsomes. Cytochrome P450 seems thus to intervene in the metabolism of pentane, as earlier suggested (15).

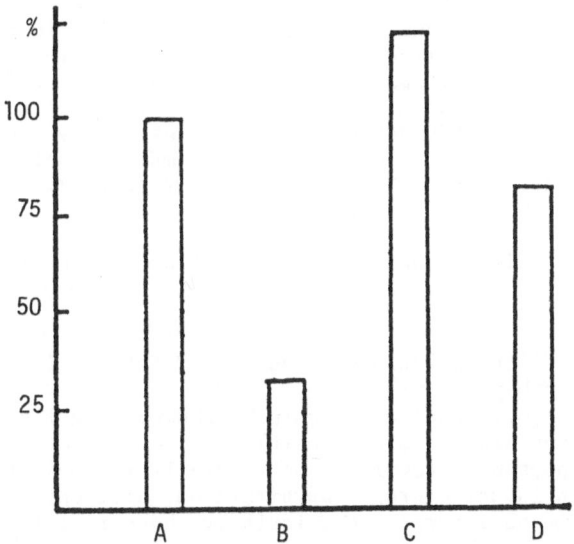

Fig. 1. Emission of pentane from autoxidized linoleate peroxide emulsion, in percent of A
(controls). A : emulsion alone. B : emulsion + hepatic microsomes from phenobarbital-treated rats. C : emulsion + boiled
microsomes. D : emulsion + microsomes +
metyrapone 10^{-3} .

An important cause of artefact must be underlined, in experiments measuring pentane emission from microsomal suspensions : it is the retention of pentane by lipoproteins suspensions, at room temperature. Pentane is known to be soluble in fat, in a greater measure than ethane (26). However after heating microsomal suspensions, at temperature higher than 37° C (the boiling point of pentane), pentane is entirely desengaged from the aqueous phase.

RELIABILITY OF IN VIVO LIPOPEROXIDATION ESTIMATION BY PENTANE MEASUREMENTS

A first doubt on the safety of this method was emitted relatively to the experiments on rodents. It was suggested that the major source of pentane, in the rat, during experiments with vitamin E-deficient diet particularly, was the intestinal bacteria. This hypothesis was sustained by the fact that antibiotics depleted pentane exhalation in rats (28).

Another reason of uncertainty was the great variability of results, not only from one publication to another, but also in a same group of experiments. For example, Dillard et al. (25) found, in 5 normal volunteers, amounts of pentane varying from approximately 70 pmoles to 1300 pmoles in expired air, in 2 minutes. If one estimates that 20 liters air are approximately exhaled in 2 minutes, these data would become 2.8 to 52 pmoles per liter. Morita et al. (24) demonstrated that the basal pentane emission varied ten fold in expired air, among individuals. Wade and Rij (26), using a rebreathing technique in human experiments, found 40 pmoles/liter of expired pentane, at the onset of the experiment, raising to 90 pmoles/ liter after 1 hour. These authors calculated, using a coefficient of pentane metabolism, that the normal pentane concentration was 120 ± 50 pmoles/liter, in expired air.

We have used, for pentane exhalation determination in humans, a new procedure to trap the expired hydrocarbons, using charcoal cartridges at room temperature. Charcoal was extracted with carbon disulfide, and sample of 5 μl CS_2 were injected on a Porapak T column, in a GLC Barber-Colman 2000, equipped with a modified FID detector, presenting a very low electronic noise. As Dillard et al. (25), we found, on 22 normal male volunteers, a wide distribution of pentane values, in expired air (Fig. 2), raising from 5 pmoles/liter to 140 pmoles/liter.

These discrepancies in pentane concentrations can be explained by several causes :

1. Feeding habits, with variations of $\omega 6$ and $\omega 3$ fatty acids intake, from a subject to another.
2. Variations in peroxidized fatty chain breakdown, leading not only to hydrocarbons, but also to aldehyde (6).
3. Variations in pentane metabolism, particularly in subjects receiving cytochrome P450 inducing drugs (barbiturates, anticonvulsivants, etc).
4. Variations in metabolic rate, particularly those in relation with the muscular activity , which can deeply enhance pentane exhalation (until to 800%) as shown by Dillard (25).

Arguments 3. and 4. seem to be the most valuable.

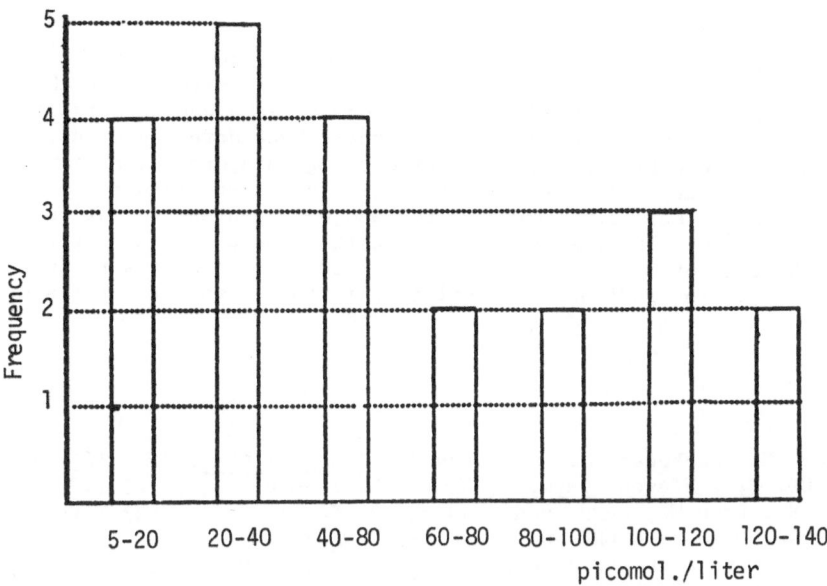

Fig. 2. Distribution of pentane values among 22 normal volunteers.

OPPORTUNITY OF HYDROCARBON MEASUREMENTS

It is not possible to obtain normal values with a better precision for expirated pentane, imprecisions caused either by metabolic rate variations (let us remember the difficulties to measure basal metabolism by oxygen consumption), either by cytochrome P450 inducers; it is however possible to obtain significant results during experiments where a subject is his own control, and when pentane variations considerably raise up the initial value. The results of Dillard et al. (25) obtained by this methodology, concerning the role of vitamin E in prevention of pentane emission during strong exercise, were very significant. Repeating the experiments of Dillard, and using our own method of pentane trapping, we also found that, in man, exercise produced a rise of more than twenty fold of expirated pentane, and that a treatment with vitamin E during 3 weeks reduced this increase to 6 fold only.

Exhaled hydrocarbons determination is the sole method applicable for in vivo lipoperoxidation study, and is thus very important for human research. It is a method which is not more or less safer than the other techniques designed to measure the decomposition products of lipoperoxides. We think that these methods are complementary, and that it is preferable, when it is possible, to perform diene, malonaldehyde and pentane measurements together. These measurements constitute the first approach of the problem of lipoperoxidation evaluation. The second approach, which completes the results obtained with the first, is generally poorly known. It consists to quantify the total antilipoperoxidant and antiradicalar activities of tissues (29), in order to determine the diminution in the intracellular defenses to oxidants.

CONCLUSIONS

At present, in vivo lipoperoxidation phenomenons can be detected and estimated by analysis of products arising from the lipoperoxide-breakdown. Pentane measurements in the breathed air, proposed as a non invasive method, and estimated to be safer than diene or malonaldehyde determinations, is yet submitted to several risks of inaccuracy, regarding to the hepatic metabolism of pentane, particularly. However significant results can be obtained when a subject is his own control and when pentane variations are sufficiently marked. The results obtained with these different determinations must be completed by a poorly known method which consists to quantify the variations of antiradicalar or antilipoperoxidant contents of tissues or plasma.

REFERENCES

1. Frankel EN, Nowakowska J, and Evans CD. J Am Oil Chem, 38 : 161 (1961).
2. Evans CD. Proc Flavon. Chem. Symp. Campbell Soup Co., p 123 (1961).
3. Horvat RJ, Lane WG, Ng H, and Sheperd AD. Nature, 203 : 523 (1964).
4. Smouse TH, Mokherjee BD, and Chang SS. Chem Ind, 29 : 1301 (1965).
5. Scholz RG, and Ptak LR. J Am Oil Chem Soc, 34 : 596 (1966).
6. Evans CD, List GR, Dolev A, McConnell DG, and Hoffmann RL. Lipids, 2 : 432 (1967).
7. Donovan DH, and Menzel DB. Experientia, 34 : 775 (1978).
8. Dumelin EE, and Tappel AL. Lipids, 12 : 894 (1977).
9. Evans CD, List GR, Hoffmann RL, and Moser HA. J Am Oil Chem Soc, 46 : 501 (1969).
10. Jarvie PK, Lee GD, Erickson DR, and Burkus EA. J Am Chem Oil Chem Soc, 48 : 121 (1971).
11. Riely C, Cohen G, and Lieberman M. Science, 183 : 208 (1974).
12. Dillard CJ, Dumelin EE, and Tappel AL. Lipids, 12 : 109 (1977).
13. Tappel AL, and Dillard CJ. Fed Proc, 40 : 174 (1981).
14. Hafeman DG, and Hoekstra WG. J Nutr, 107 : 656 (1977).
15. Frommer U, Ullrich V, Staundinger HJ. Hoppe-Seyler's Z Physiol Chem, 351 : 903 (1970).
16. Frank H, Hintze T, Bimboes D, and Remmer H. Toxicol Appl Pharmacol, 56 : 337 (1980).
17. Müller A, Graf P, Wendel A, and Sies H. FEBS Lett, 126 : 241 (1981).
18. Lawrence GD, and Cohen G. Anal Biochem, 122 : 283 (1982).
18b. Lawrence GD, and Cohen G. Methods Enzymol, 105 : 305 (1984).
19. Filser JG, Bolt HM, Muliawan H, and Kappur H. Arch Toxicol, 52 : 135 (1983).
20. Müller A, and Sies H. Methods Enzymol, 105 : 311 (1984).
21. Lawrence G, Cohen G, Machlin LJ. Ann NY Acad Sci, 393 : 227 (1982).
22. Tappel AL. Free Radical Biol, 4 : 1 (1980).
23. Thomas MJ, Hinson TR, Adair NE, Hiner L, McLee BD. in : IV Int. Conf. Superoxide and Superoxide Anion. Roma, September 1985. Abstract book p 98.
24. Morita S, Snider MT, and Inada Y. Anesthesiology, 64 : 730 (1986).
25. Dillard CJ, Litov RE, Savin WM, Dumelin EE, and Tappel AL. Appl Physiol Respir Environ Exercise Physiol, 45 : 927 (1978).
26. Wade CR, and Van Rij AM. Anal Biochem, 150 : 1 (1985).
27. Deby C, Pincemail J, Bertrand Y, Lismonde M, Lamy M, and Goutier R. in : 2e Réunion Groupe de Contact "Oxygen Metabolism". Liège, June 6, 1986.
28. Gelmont D, Stein RA, and Mead JF. Biochem Biophys Res Commun, 102 : 932 (1981).
29. Deby C, Deby-Dupont G, Hans P, Pincemail J, and Neuray J. Experientia, 39 : 1113 (1983).

DETECTION OF PENTANE AS A MEASUREMENT OF LIPID PEROXIDATION IN HUMANS USING GAS CHROMATOGRAPHY WITH A PHOTOIONIZATION DETECTOR

K.F. Heim[1], U.-M. Makila[1], R. Leveson[2], G.S. Ledley[1], G. Thomas[1], C. Rackley and P.W. Ramwell[1]

[1] Georgetown University Medical Center, Departments of Physiology and Biophysics, and Medicine 3900 Reservoir Road, N.W. Washington, D.C. 20007
[2] Photovac, Inc. Thornhill, Ontario, Canada

Introduction

Lipid peroxidation (LP), which involves the oxidation of polyunsaturated fatty acids, has been implicated in several pathological conditions, including cancer (1) and atherosclerosis (2-4), and in the ageing process (5,6). Many of the products of LP can react with oxygen to form epoxyhydroperoxides, ketohydroperoxides, and cyclic and bicyclic peroxides (7). These secondary products can in turn decompose to monohydroperoxides, aldehydes such as malonyldialdehyde (MDA), and volatile hydrocarbons (HCs)(7).

Markers of Lipid Peroxidation

Investigators have been developing methods to measure LP _in vitro_ in order to confirm the presence of this process in altered physiological states. The most commonly used technique is the detection of MDA via its reaction with thiobarbituric acid (8). Other methods include the use of fluorescence spectroscopy for direct measurement of MDA (9), ultraviolet spectroscopy to detect diene conjugation in damaged fatty acids (10), and chemiluminescence to detect the presence of free radicals in tissue samples (11). However, none of these techniques can be carried out _in vivo_ and most of them require lengthy procedures or yield non-quantitative results.

Recently the volatile HCs ethane (12) and pentane (13) have been used to measure LP. Ethane is formed from the peroxidation of w-3 fatty acids, such as linolenic acid, while pentane is a product of the peroxidation of w-6 fatty acids, such as linoleic acid and arachidonic acid (13). Detection of these gases appears to be a sensitive, specific, efficient, inexpensive, easy method of measuring LP _in vivo_.

Measurement of Ethane and Pentane

Volatile HCs in expired air have been detected using gas chromatography (GC). The type of detector attached to the GC apparatus is of great importance to the sensitivity of the results. The flame ionization detector (FID) is the most widely used type in the analysis of HCs due to its sensitivity and versatility in detecting a wide range of organic compounds. However, it requires that breath samples be concentrated using an adsorbent such as activated charcoal prior to injection into the column of GC (14,15).

The photoionization detector (PID) has recently been used to study HCs and appears to be much more sensitive than the FID for measurement of ethane and pentane (16). There is no need to concentrate the sample before injection.

Previous Studies of Exhaled Alkanes

Many authors have reported on the use of GC to detect changes in HC levels in the breath of animals under changing conditions. In rats, several agents suspected of inducing LP have been found to cause an increase in the production of one or more HCs. Pentane levels were elevated in expired air after administration of iron-dextran (17), polyunsaturated fatty acids (18), and carbon tetrachloride (19,20). Exhaustive exercise also caused a moderate increase in pentane (21).

Some compounds have the ability to protect lipids from being oxidized. One study showed that the presence of vitamin E in the diet decreased levels of expired pentane in rats (22). In monkeys, vitamin E reduced the production of HCs in the presence of ozone, a known LP inducer (23). Other substances have also shown antioxidant properties in rats, including selinium, methionine (24), and mannitol (25).

A few investigators have extended their work to include human subjects. As was found in rats, exercise led to an increase in expired pentane while intake of vitamin E prevented that increase (26). In newborn infants, administration of a lipid emulsion led to an increased exhalation of ethane and pentane (27). These results show that the method of measuring LP in vivo by the detection of volatile HCs in the breath can be applied in human studies.

Materials and Methods

1) Gas Chromatography

The gas chromatographic system consisted of an SE-30 column (5% on Chromosorb G) and a photoionization detector, model 10A10 (Photovac, Inc., Thornhill, Ontario Canada). The PID system is based upon the emission of photons from an ultraviolet lamp containing a low pressure gas. Emission occurs when the gas is excited by a radio frequency source. Compounds in the passing sample will be ionized, provided their ionization potentials are below the energy of the emitting photons (28-30). In the case of pentane, the lamp output must be above 10.35 eV. Any ionized compound will cause a response depending on its concentration, while unionized substances will pass undetected.

The carrier gas used was zero-grade compressed air (Roberts Oxygen Co., Washington, D.C.). This was passed through a hydrocarbon trap (Applied Science, State College, Pa.) and then into the column at a flow rat of 10 ml/min. The entire PID analysis was done at room temperature (22 +/ -3°C).

2) Calibration and analysis of pentane

Liquid pentane (Fisher Scientific Co., Fairlawn, N.J.) was used to prepare pentane standards for calibration. The mass (m) of pentane required to obtain desired concentrations was determined by the formula $m = 4.1 \times 10^{-8}$ M x V x C, where M = molecular mass of pentane (72.15), V = volume in liters, and C = concentration desired in ppm. In order to express the volume of pentane in ml, m was divided by the density of pentane (0.62 g/ml). The calculated amount of liquid pentane was pipetted into a 1 liter glass gas sampling vessel with teflon stopcocks and a teflon-coated low-bleed septum (Supelco, Inc., Bellefonte, Pa). The pentane was allowed to evaporate within the nitrogen-containing vessel and a 1 ml sample of the gas was then diluted in another 1 liter vessel filled with HC-free air to yield the required concentration in ppbs. Using this dilution technique, five different standards were prepared with concentrations of 1048, 838, 629, 419, and 210 ppbs of pentane. Standard samples peak in cm was measured from a strip-chart recorder (Packard Co., Downsgrove, Il.). This was repeated eight times with each of the standards. The accuracy of the calibration was evaluated by using two different concentrations of pentane gas standards (Matheson Gas Products, Baltimore, Md.), 1.2 and 6.6 ppm. Between 0.05 - 1.0 ml of both standards were injected into the GC.

A single-channel integrator (3390 A, Hewlett Packard, Pa.) was used for computing and reporting the results on the basis of peak heights. The retention time of pentane was between 0.43 - 0.65 min.

Samples were injected into the GC via a 1 ml glass tuberculin syringe (Bectom-Dickenson, Rutherford, N.J.). All measurements were done in triplicate. In order to convert the results from ppb to pmol/ml, a factor of 0.041 (from the equation P1 x V1 / T1 = P2 x V2/T2) was used.

3. Sample Collection

Informed consent was obtained from healthy volunteers (n=26; 14 females, 12 males) aged 34 +/-16 years. The subjects breathed room air for one minute through a 3-way Triple-J valve (W.E. Collins, Inc., Braintree, Ma.) and expired air was collected in a 22 liter 5-layer Tedlar gas sampling bag (Calibrated Instruments, Inc. Ardsley, N.Y.). Tedlar was chosen due to its inertness and impermeability to the HCs of interest. The sampling bag was carefully flushed three times with nitrogen (Roberts Oxygen Co., Washington, D.C.) between samples in order to avoid was evaluated by analyzing a sample from the nitrogen filled bag prior to each use. In addition, ambient air samples were measured daily to determine the ambient pentane concentration.

Using the same method, breath samples were collected from hospital patients undergoing renal transplantation and cardiac catheterization.

Results

The results of the pentane calibration procedure showed that the standard curve is linear up to 1050 ppb (43 pmol/ml) with a detection limit of 10 ppb (0.41 pmol/ml).

The concentration of pentane in the ambient air of the laboratory was 0.69 + 0.05 pmol/ml (mean + SE; n=35). The pentane concentration of the nitrogen-filled bag was 0.38 + 0.03 pmol/ml (m=45). When analysis of variance was used to test the differences between triplicate measurement from the same sample, the F value was 0.017 (NS). The coefficient of variation (CV) was 5%. When known concentrations of pentane were injected into the GC, the recovery was 92%.

The mean concentration of pentane in the minute volume of expired air of the healthy human subjects was 11.45 + 1.1 pmol/ml ranging from 3.06 - 21.5 pmol/ml. This concentration correlated with the surface area in m^2 (r=0.759; p 0.001, n=26; fig. 1). There were no significant differences in the concentration of pentane between females and males.

Figure 2 shows levels of pentane in the expired air of patients following renal transplantation. The increases demonstrated in patients A,B and D correlate with periods of transplant rejection.

In the case of patients undergoing cardiac catheterization, samples were taken both before and after the procedure was performed. Although there appeared to be a trend toward an increase in expired pentane immediately after catheterization was completed, this increase was not found to be statistically significant (data not shown).

Discussion

The biological effects of LP products and their roles in several disease states have attracted much attention in the past. They have been implicated in cancer, atherosclerosis, diabetes, reperfusion injury, and in toxicity induced by chemicals such as ethanol, carbon tetrachloride, alloxan, and certain pesticides. This has led to the development of several analytical techniques to measure LP products, both in vitro in in vivo (9-13). Most of these methods are non-specific and time-consuming. Although the measurement of MDA by its reaction with thiobarbituric acid (TBA) has been popular, the TBA reaction is not specific for MDA since many other LP products give positive reactions. Also, MDA may not be a useful marker for in vivo LP due to its rapid metabolism (31).

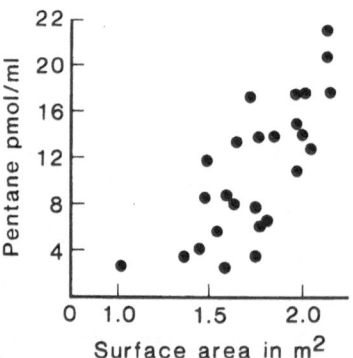

Figure 1. Plot of pentane levels in expired air vs. body surface area of human volunteers (r=0.759, n=26).

Lately, much attention has been given to the measurement of HCs in the expired air of experimental animals as an index of _in vivo_ LP. We have used a gas chromatographic atechnique with a photoionization detector for the measurement of pentane in human expired air. Unlike previous GC techniques which are cumbersome, time-consuming, and required extensive sample concentration, our method is very simple, sensitive, rapid, and does not require any tedious procedures for sample concentration. In addition, the entire separation and measurement can be performed at room temperature.

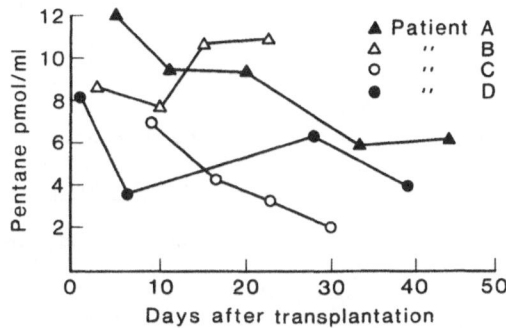

Figure 2. Pentane levels in four patients having undergone renal transplantation. In patients A,B and D the time of increase in pentane levels coincides with signs of transplant rejection. Each point is the mean of three readings from the same patient.

Our pentane values in humans (3-21.5 pmol/ml) are higher than those reported by others. This discrepancy may be due to difference in sample preparation and treatment, or inadequate sample concentration in previous methods. Also, the PID system is more sensitive than the earlier FID systems, as the detection limit of our PID was (0.41 pmol/ml). It also allows measurement to be done directly and permits immediate comparison with a known standard.

The variation seen in the values of pentane among individuals may be due to factors such as exercise status, diet, smoking, and metabolic rate. Very little information is available regarding the metabolism of pentane in humans but it has been shown that pentane is metabolized in rats by liver alcohol dehydrogenase (31).

Our studies also show that a positive correlation exists between pentane levels and body surface area, suggesting that the expired pentane may reflect whole body pentane, and hence, whole body LP. However, the source of pentane in expired air is still unknown.

We have used this technique to measure pentane levels in hospital patients, as well as healthy individuals. The equipment is easily transported to the patient's bedside. Our results show that there is a decrease in pentane levels as the patient recovers from renal transplantation, while an increase is observed during rejection. This is an exciting observation and needs further evaluation. As for patients undergoing cardiac catheterization, an increase in pentane was seen following the procedure as compared to measurements taken before the catheterization, but the difference was not significant.

In conclusion, we feel that measurement of expired pentane using GC-PID is a sensitive, rapid procedure which can be used clinically to detect LP. It is particularly attractive due to its non-invasive nature, portability, and ease of operation.

Acknowledgements

We are grateful for the initial work of Dr. Julie Du and for grants supported from the National Institutes of Health (HL 32319 & HL 34974).

References

1. Ames B.N. Science 221: 1256 (1983)
2. Valles J., Aznar J., Santos M.T. and Fernandez M.A. Thromb. Res. 27: 585 (1982)
3. Rao P.S., Cohen M.V. and Mueller H.S. J. Mol. Cell. Cardiol. 15: 713 (1983)
4. Loeper J.., Goy J. and Emerit J. Bull. Acad. Nat. Med. 168: 91 (1984)
5. Sagai M. and Ichinose T. Life Sci. 27: 731 (1980)
6. Harman D. Proc. Natl. Acad. Sci. 78: 7124 (1981)
7. Frankel E.N. JAOCS 61: 1908 (1984)
8. Nair V. and Turner G.A. Lipids 19: 804 (1984)
9. Lunec J. and Dormandy T.L. Clin. Sci. Mol. Med. 56: 53 (1979)
10. Diluzio N.R. Food Chem. 20:486 (1972)
11. Namedov T.G., Konev V.V. and Popov G.A. Biofizika. 18: 643 (1973)
12. Riely C.A., Cohen G. and Lieverman M. Science 183: 208 (1974)
13. Dumelin E.E. and Tappel A.L. Lipids 12: 894 (1977)
14. Lawrence G.D. and Cohen G. Anal Biochem. 122: 238 (1982)
15. Lawrence G.D. and Cohen G. Methods Enzymol. 105: 305 (1984)
16. Driscoll J.N., Ford J., Jaramillo L.F. and Gruber E.T. J. Chromatogr. 158: 171 (1978)
17. Dillard C.J., Downey J.E. and Tappel A.L. Lipids 19: 127 (1984)
18. Dillard C.J., Litov R.E. and Tappel A.L. Lipids 13: 396 (1978)
19. Koster U., Albrecht D. and Kappus H. Toxicol. Appl. Pharmacol. 41: 639 (1977)
20. Lindstrom T.D. and Anders M.W. Biochem. Pharmacol. 27: 563 (1978)
21. Gee D.L. and Tappel A.L. Life Sci. 28: 2425 (1981)
22. Dillard C.J., Dumelin E.E. and Tappel A.L. Lipids 12: 109 (1977)
23. Dumelin E.E., Dillard C.J. and Tappel A.L. Environ Res. 15: 38 (1978)
24. Hafeman D.G. and Hoekstra W.G. J. Nutr. 107: 656 (1977)
25. Dillard C.J., Kunert K.J. and Tappel A.L. Arch. Biochem. Biophys. 216: 204 (1982)
26. Dillard C.J., Litov R.E., Savin W.M. et al. J. Appl. Physiol.: Respirat. Environ Exercise Physiol. 45: 927 (1978)
27. Wispe J.R., Bell E.F. and Roberts R.J. Pediatr. Res. 19: 374 (1985)
28. Lovelock J.E. Nature 188: 401 (1960)
29. Locke D.C. and Meloan C.E. Anal. Chem. 37: 389 (1965)
30. Freedman A.N. J. Chromatogr. 190: 263 (1980)
31. Frank H., Hintze T., Dimboes D. and Remmer H. Toxicol. Appl. Pharmacol. 56: 337 (1980)

EICOSANOIDS AND THE PATHOPHYSIOLOGY
OF SEPSIS AND TRAUMA

J. R. Fletcher, B. C. Chernow, H. R. Alexander and M. P. Fink,
Vanderbilt University, Saint Thomas Hospital, Nashville, Tennessee
Respiratory Surgical Intensive Care Unit, Massachusets General
Hospital, Boston, Massachusetts,
Medical Corps, U.S. Navy, FPO Seattle, Washingthon,
University of Massachusets, Worcester, Massachusetts.

INTRODUCTION

Circulatory shock from sepsis continues as a major clinical problem
in the world. No new treatment modalities other than volume resuscitation
have emerged as effective agents since the early concepts of humoral
mechanisms were formulated in the early part of this century. The discovery of Northover and Subramanian in 1962 (1) that cyclooxygenase
inhibitors attenuated the hemodynamic events and prolonged the survival
in endotoxemia in dogs has stimulated a substantial interest in the area
of the role of eicosanoids in endotoxin shock, bacteremic shock, the
sepsis syndrome and septic shock (2-22).

The purpose of this chapter is to review the human studies of
sepsis and trauma that suggest the eicosanoids may be implicated in the
pathophysiology of these entities. Several clinical problems in sepsis
and trauma are well recognized: hemodynamic alterations, inflammation
and capillary permeability, immunosuppression, metabolic abnormalities
and multiple organ dysfunction. All of these areas cannot be reviewed in
detail, however there are enough reported studies in some areas to
formulate interrelationships.

HUMAN SEPSIS

The first studies that suggested the eicosanoids may be active in
human sepsis was reported by Ramwell et al. in 1975 [23]. Three
severely ill septic patients demonstrated elevated mixed venous $PGF_{2\alpha}$
values. Baracos et al. [24] reported that protein degradation by human
leukocytic pyrogen may be mediated in part by PGE_2. Utilizing an in
vitro muscle preparation, these investigators demonstrated that protein
degradation was accelerated when incubated with human leukocytic pyrogen
as compared with normal plasma. Interestingly, PGE_2 levels correlated
directly with protein degradation. Indomethacin (a cyclooxygenase
inhibitor) prevented the proteolysis and eicosanoid production. Plasma
thromboxane values in human septic shock were first reported by Reines et
al. [25]. Elevated thromboxane values were clearly present in septic
shock and these authors correlated the levels to nonsurvivors. They did
not report the relationship of these eicosanoids to other pathophysio-

logical events. A study of five septic patients and the eicosanoids was reported by Parratt and colleagues [26] in 1982. They attempted to correlate plasma $PGF_{2\alpha}$ or $iTXB_2$ to various hemodynamic or pulmonary parameters. Thromboxane values were greater than $PGF_{2\alpha}$ values, however these eicosanoids correlated only with alveolar-arterial oxygen differences. Oettinger et al. investigated the hemodynamic and respiratory parameters and their relationship to endogenous $PGF_{2\alpha}$ values in nine patients with severe sepsis [27]. Healthy volunteers and nonseptic intensive care patients were utilized as control subjects. All patients in sepsis were hyperdynamic. Endogenous arterial $PGF_{2\alpha}$ concentrations were significantly increased in the hyperdynamic state of septic shock compared to the preshock and recovery periods of time in these patients. In addition, these $PGF_{2\alpha}$ values were significantly increased in the hyperdynamic state of septic shock compared to the preshock and recovery periods of time in these patients. In addition, these $PGF_{2\alpha}$ values were significantly increased in the hyperdynamic state of septic shock compared to the preshock and recovery periods of time in these patients. In addition, these $PGF_{2\alpha}$ values were significantly greater than those of the healthy volunteers and the nonseptic intensive care patients. Of particular interest was the finding that the lungs of nonseptic intensive care patients cleared $PGF_{2\alpha}$ from 233 ± 17 to 48 ± 31 pq/ml; whereas in the septic patient the arterial concentrations of 1252 ± 119 pq/ml significantly exceeded the mixed venous values of 824 ± 89 pq/ml. These authors imply that a) the endogenous $PGF_{2\alpha}$ values were related to the hyperdynamic state of severe sepsis; b) that the lung in sepsis has a net production of eicosanoids and c) that the endogenous $PGF_{2\alpha}$ did not correlate individually with respiratory parameters or pulmonary vascular resistance data.

In a slightly different approach than pure sepsis, fifteen patients with thermal injuries were studied to determine whether or not changes in thromboxane and prostacyclin values might account for some of the observed systemic changes in burned patients [28]. TxB_2 and $6\text{-keto-}PGF_{1\alpha}$ values were determined once or twice per week in each patient. Early after burn injury TxB_2 values were significantly elevated, but gradually decreased within the ensuing three weeks. Of particular interest in their study were the nine patients who developed sepsis. All the septic episodes occurred 14 days post burn. Plasma TxB_2 values were correlated with sepsis and were extremely high (3.47 pm/ml); a 115 fold increase over control values. Patients who were septic after 14 days post injury consistently maintained elevated thromboxane B_2 levels, similar to those observed seven to 14 days post burn. The increase in plasma thromboxane B_2 coincided with the development of sepsis; the nonseptic patient did not have a similar increase in thromboxane B_2 even up to 53 days after injury. The authors suggest that the eicosanoids contribute to post-thermal injury systemic responses, both in the acute phase as well as during sepsis.

The next studies reported utilized pharmacologic agents that inhibit specific arachidonic acid metabolites. These two studies utilized Dazoxiben, a selective thromboxane synthetase inhibitor, in the adult respiratory distress syndrome. The reason for including these studies is that in both of them septic patients were included. Leeman et al. investigated the hemodynamic and gasometric effects of Dazoxiben in seven patients who developed adult respiratory distress syndrome [29]. The patients were studied for 120 minutes after a single intravenous bolus dose of Dazoxiben 1.5 mg/kg/BW. The clinical data indicated that the etiologies of the ARDS were varied. Two patients had septic shock. One patient had hemorrhagic shock, and one patient each had cardiopulmonary bypass injury, pulmonary contusion and an inhalation injury. The study was performed within the first 48 hours after recognition of ARDS as defined by an arterial oxygen pressure (PaO_2) less than 60 mmHg with a concentration of oxygen in the inspired gas (F_iO_2) equal to 0.40 or more, associated with bilateral diffuse pulmonary infiltration on the chest radiograph in the absence of left ventricular failure, pulmonary infection or chronic pulmonary disease. Patients who had received corticosteroids or nonsteroidal inflammatory drugs were excluded. Each patient was hemodynamically stable. Four of the seven patients survived, and three patients died from sepsis at least ten days after the end of the study. Hemodynamic parameters included heart rate, cardiac index, mean systemic arterial pressure, systemic vascular resistance, mean pulmonary arterial pressure, pulmonary vascular resistance and right ventricular stroke work index. Gasometric determinations included PaO_2, $PaCO_2$, PVO_2 and venous admixture. These parameters were measured before, 30, 60 and 120 minutes after Dazoxiben. Their results show that hemodynamic parameters were not affected following Dazoxiben; however, PaO_2 rose progressively from 83 $\pm$ 13 to 101 $\pm$ 13 mmHg and venous admixture decreased slightly from 34 to 29 $\pm$ 4, two hours after the injection of Dazoxiben, but none of these changes were statistically significant. There was no change in coagulation screening tests observed during the study or the following day. The authors conclude that thromboxane A_2 is <u>not</u> an important mediator in pulmonary hypertension occurring during human ARDS, at least once the syndrome has been recognized. Further, they stated that it is unlikely that Dazoxiben will provide a breakthrough in the management of patients with established ARDS, regardless of the etiology. The second study with Dazoxiben is somewhat similar, but the question asked is different from that of the first study. Reines et al. investigated the role that Dazoxiben has in human sepsis and adult respiratory distress syndrome [30]. They asked the question: Will Dazoxiben, a thromboxane inhibitor, be safe and efficacious in patients with ARDS? Their patients included five control patients and five Dazoxiben treated patients. Following initial evaluation, the patients were randomized to either saline or Dazoxiben groups. In the therapy group, Dazoxiben, 100 mg was given I.V. every four hours. There were five patients in each group. Study parameters were evaluated four hours before Dazoxiben and every four hours thereafter for a period of 72 hours. Excluded were patients with major cardiac injuries and patients with significant pre-existing renal or pulmonary problems. Cardiovascular parameters were evaluated in these patients. The results were similar to the previously mentioned study. There were no differences in the cardiovascular parameters or the pulmonary parameters. The coagulation changes were similar in both groups. The authors concluded that Dazoxiben appears safe; it does lower the immunoreactive thromboxane B_2 levels in these patients. The patients did represent a variety of etiologic causes for their ARDS. The authors questioned whether or not there was a role for thromboxane B_2 in these patients with ARDS; if there were a significant role, it probably occurs early in the course of ARDS. Further, they raised the question of whether or not immunoreactive thromboxane A_2 has a role in sepsis.

To further elucidate the potential role of the eicosanoids in patients in septic shock, Halushka et al. reported elevated plasma 6-keto-$PGF_{1\alpha}$ in patients in septic shock [31]. Fourteen patients admitted to the intensive care units with a diagnosis of septic shock were studied. Criteria for inclusion in this study were fever, abnormal white cell count, hypotension without pressor drugs and positive intraperitoneal or blood cultures. Catecholamines, antibiotics and steroids were given when clinically indicated. No patient in this study received aspirin or any other nonsteroidal anti-inflammatory drug. Treatment was not influenced by participation in this study. Three patients not in sepsis or in shock who had a central venous catheter served as controls. Six-keto-$PGF_{1\alpha}$ values were obtained as soon as possible after the diagnosis of sepsis.

Hemodynamic data were recorded for each patient at the same time that blood was sampled. Some of the patients were receiving pressor agents when hemodynamic data were recorded. The patients were then grouped in categories of survivors and nonsurvivors. The levels of 6-keto-PGF$_{1\alpha}$ were significantly higher in the patients who died from septic shock compared to those who survived and compared to the control group. Interestingly, these results were similar to those previously reported by the same group in which plasma thromboxane values were the highest levels in patients who died from septic shock. Unlike the previous studies in which plasma thromboxane levels were not elevated in the survivors of septic shock, in the present study plasma 6-keto-PGF$_{1\alpha}$ levels were also significantly elevated in survivors. The authors stated it is unknown whether or not the elevated levels of 6-keto-PGF$_{1\alpha}$ were due to agonal events, the stress of shock, inadequate tissue perfusion or the pathogenesis of human septic shock.

Another pharmacological approach in attempting to attenuate the effects of eicosanoid system on septic patients is presented in the following study. Twenty-five patients who met criteria for the sepsis syndrome and had not received salicylates were entered into the study reported by Reines et al. [32]. Twenty-one did not receive any salicylates and four received a single dose of 600 milligrams of aspirin by rectal suppository. Criteria for the sepsis syndrome included: 1) temperature of greater than 39° C; 2) white blood cell count greater than 12,000 or 20% in immature forms; and 3) a known source of infection with positive cultures or positive blood cultures or gross pus in a close space. Cardiovascular criteria included: 1) systemic hypotension (less than 85 mmHg); 2) systemic vascular resistance less than 800 dynes/centimeters/seconds^{-5}, or 3) unexplained systemic metabolic acidosis. Blood samples for immunoreactive thromboxane B$_2$ were obtained from a central venous or pulmonary artery catheter as soon as possible after diagnosis of sepsis. These samples were obtained one to four hours after a single dose of methylprednisolone was given to nine patients. The doses of methylprednisolone given intravenously varied from one to two grams per patient (15-30 milligrams/kilogram). Blood samples for immunoreactive thromboxane B$_2$ were obtained prior to and four hours after aspirin administration in four patients. A single dose of 650 milligrams of aspirin by suppository yielded a salicylate level of 28 mg/ml in the four patients tested. Of the 25 patients studied, nine received steroid treated patients, four received aspirin and 12 received neither. Plasma iTxB$_2$ values were significantly increased in nonsurvivors compared with patients who survived. There was no difference in plasma iTxB$_2$ levels between patients who received steroids and those who received neither steroids nor aspirin. There was no significant difference in prothrombin time, partial thromboplastin time and platelet counts between the patients who received steroids and those who did not. Intrapulmonary shunting, as estimated by the PaO$_2$/FiO$_2$ ratio, likewise was not different in the steroids compared to the nonsteroid group. Plasma iTxB$_2$ levels decreased in all four patients receiving aspirin from a level of 444 $\pm$ 159 pq/ml to 174 $\pm$ 53 pq/ml. The authors imply that plasma iTxB$_2$ values were utilized in the present study as a marker of the ability of steroids to inhibit arachidonic acid metabolism in septic patients. It was clear from their study that plasma iTxB$_2$ level was elevated in patients dying from sepsis. The use of glucocorticoids did not reduce plasma iTxB$_2$. It was their conclusion that glucocorticoids did not appear to significantly affect the conversion of arachidonic acid to thromboxane in septic patients nor did they seem to alter the other parameters which were evaluated.

The next series of patients were reported by Slotman et al. from the Rhode Island Hospital [33]. These authors evaluated the possible role of thromboxane and prostacyclin and clinical acute respiratory failure. Of particular importance in this study were the patients who had ARDS from severe sepsis. Of the 67 consecutive patients at risk for ARDS who were studied prospectively, 12 out of 21 with severe sepsis developed ARDS. The difference between this study and the previous studies is that these patients were studied prospectively. They evaluated thromboxane, prostacyclin, platelet and white blood cell counts for up to five days. The criteria for severe sepsis included the following: a) clinical suspicion of an infection of the urinary tract, perineum, skin or soft tissue, or female genital tract; b) hyperpyrexia (temperature greater than 101° F) or hypopyrexia (temperature less than 96° F); c) tachycardia of greater than 90 beats per minute; d) tachypnea (greater than 20 breaths per minute) accompanied by respiratory alkalosis in the absence of artificial respiratory maintenance; e) evidence of inadequate tissue perfusion exemplified by either altered mental status or arterial PO_2 of less than 75 mmHg without overt pulmonary disease as a cause or elevated plasma lactic acid levels and an urinary output of less than 30 cc's for one hour. Patients entered into this study were followed clinically for seven days. They were observed continuously for clinical signs and symptoms of ARDS. Plasma levels of $iTxB_2$ and 6-keto-$PGF_{1\alpha}$ were measured by RIA daily for five days. The authors demonstrated that plasma $iTxB_2$ and prostacyclin were significantly related to ARDS. The incidence of ARDS was significantly increased with a peak plasma thromboxane greater than 70 pq/ml. The authors utilized nonsteroidal inflammatory agents in eight patients and ketoconazole in two patients given within 48 hours of the study. The NSAID and Ketoconazole decreased the thromboxane levels, but produced no significant difference in plasma prostacyclin or in the incidence of ARDS. Exogenously administered corticosteroids were given to six patients without head injury. There were no significant differences between the patients who received these steroids and those who did not with regards to plasma eicosanoid levels or the incidence of ARDS. These authors confirm the previous work by Reines et al.: the concentration of thromboxane measured in all critically ill patients were similar to those Reines reported in patients surviving septic shock and in nonseptic control patients. The values of thromboxane however, were different. Data reported by Slotman revealed no significant difference in thromboxane concentrations between survivors and patients who died of their illness and/or injury regardless of the presence of absence of ARDS. Similarly, no association was found between thromboxane greater than 70 pq/ml and mortality. The authors believe that this discrepancy may be related to the timing of thromboxane sampling. In this study, thromboxane and prostacyclin measurements were obtained early in the hospital course of these patients and were associated with subsequent development of acute respiratory failure. The overall mortality was significant (48%) in patients with ARDS. Most thromboxane measurements in this study were performed before the patients were moribund. An unrelated, but interesting observation made by these investigators was that when patients with severe sepsis were followed prospectively there was no associated increase in the incidence of ARDS when patients with head injuries were excluded. It was their conclusion that thromboxane B_2 and prostacyclin appear to be involved in the pathophysiology of clinical acute respiratory failure. Many of these cases were patients with severe sepsis. Their final conclusions were that thromboxane and prostacyclin appear to be involved in the pathophysiology of ARDS; there were no significant differences in patients who died whether or not ARDS was present with regards to thromboxane and prostacyclin; there was no relationship between platelet counts, white blood cell counts, ARDS, thromboxane and prostacyclin; steroids did not alter the ARDS or the eicosanoids; patients with head injuries had an increase in immunoreactive thromboxane B_2 levels and finally, it was their opinion that thromboxane B_2 or prostacyclin levels could not easily be utilized in the early or late detection of ARDS in patients.

There has been increasing interest the role that the eicosanoids have in the adult respiratory distress syndrome. The next two studies, however, take a different approach. The studies attempt to utilize prostaglandin E_1 in the treatment of patients who have ARDS from sepsis or a variety of other causes. These investigations are of particular interest because most studies have reported that the eicosanoids might be involved in the pathophysiology of severe sepsis as well as ARDS. The first study is reported by Tokioka et al. and published in 1985 [34]. Ten patients, six men and four women, with a mean age of 63 years were studied. The etiology of the ARDS in these patients was severe infection; six had peritonitis (4 bacterial, 2 candidiasis) and four had bacterial pneumonia. None of them had evidence of pre-existing lung disease. Prostaglandin E_1 was administered intravenously to ten patients in order to investigate its hemodynamic effects. All patients were intubated and were ventilated by volume limited ventilators. Inspired oxygen concentration varied between 40 and 70% and was held constant for each patient during the study period. To maintain arterial pressure and urine output, seven of the ten patients received 3-5 ug/kg/min of dopamine throughout the study. Six to twelve hours after admission, prostaglandin E_1 was infused via a central venous catheter at 0.025 ug/kg/min for 30 minutes. Mean systemic arterial pressure, mean pulmonary arterial pressure, pulmonary capillary wedge pressure and right atrial pressure were measured with standard pressure transducers at the end of expiration. Cardiac output was determined by thermal dilution technique and a number of calculated variables were also accomplished: cardiac index, systemic vascular resistance, pulmonary vascular resistance, left ventricular stroke work index, right ventricular stroke work index, oxygen delivery, oxygen consumption, intrapulmonary shunt fraction and oxygen extraction ratio.

All patients had severe respiratory dysfunction. Before the administration of PGE_1, the shunt fraction was $36 \pm 3\%$ and mean pulmonary arterial pressure was 25 ± 1 torr. Pulmonary vascular resistance was elevated although systemic vascular resistance was normal. Prostaglandin E_1 infusions decreased mean pulmonary arterial pressure, mean arterial pressure and systemic vascular resistance while increasing the cardiac index. There was minimal change in the shunt fractions or the other hemodynamic parameters in the pulmonary circuit. Tokioka et al. concluded that PGE_1 increases the ventilation perfusion uneveness present in these patients; PGE_1 slightly improves pulmonary hemodynamic parameters and tissue oxygenation. The long term effects of these particular studies are undetermined.

Another very interesting study utilizing PGE_1 in the treatment of ARDS was reported by Holcroft et al. in 1986 [35]. Patients were considered for entry into the study if they had ARDS that required mechanical ventilation with an inspired oxygen concentration of at least 0.40; with end respiratory pressure of at least 5 centimeters of water and if they were not responding to conventional therapy. Patients were excluded if they were at risk for intracranial bleeding from a head injury or if they were hemodynamically unstable. A seven day infusion of prostaglandin E_1 was evaluated in a prospective, randomized placebo controlled, double-blinded trial on surgical patients with ARDS. The infusion of PGE_1 was given through a central venous catheter or through the right atrial port of a pulmonary arterial catheter. The dose was begun at 5 ng/kg/min and gradually increased over three hours to give the desired dose of 30 ng/kg/min. The mean systemic arterial pressure was not allowed to drop more than 20% from the baseline during the titration period. The infusion was given for seven days or until the patient was extubated, whichever came sooner. All pressures, both systemic and pulmonary, were measured with fluid filled catheters. There were 21

patients randomized to the PGE$_1$ group and 20 patients entered into the placebo group. In both groups, there were a variety of predisposing problems, but each group contained a number of trauma-related patients and a number of sepsis-related patients. The patients in both groups were treated similarly during the course of the PGE$_1$ infusions. None of the patients in either group received nonsteroidal anti-inflammatory drugs. The infusion of PGE$_1$ improved the pulmonary function. Two of the 21 PGE patients died with severe pulmonary failure compared with nine of the twenty placebo patients (p = 0.01 by Fisher's Exact Test). P$_a$O$_2$/FiO$_2$ indices in the PGE$_1$ increased from a value of 149 $\pm$ 40 at the time of entry to 245 $\pm$ 95 at the time of extubation or death. The P$_a$O$_2$/FiO$_2$ indices in the placebo group did not improve significantly.
In the six patients who were free of severe organ failure at the time of entry, the P$_a$O$_2$/FiO$_2$ indices increased from 185 $\pm$ 2 to 303 $\pm$ 55 at extubation. The indices of ten placebo patients initially free of severe organ failure showed no improvement at extubation or death. Survival at 30 days at the end of infusion, the predetermined endpoint, was significantly better in the patients given PGE (p = 0.03) with 15 of the 21 PGE patients (71%) alive at this time compared with seven of the twenty placebo patients (35%). Overall survival of PGE patients was not significantly better than the survival of the placebo groups. Overall survival in patients initially free of severe organ failure, however, was significantly better in the PGE$_1$ patients. Of the six PGE$_1$ patients free of severe organ failure at the time of entry, all survived to leave the hospital. Of the ten placebo patients initially free of severe organ failure, four survived. Six of the PGE$_1$ and nine of the placebo patients had uncontained peritonitis at death. The data from this study suggests clearly that this drug was safe in patients who were critically ill and that there was some improvement overall both in the ability of the lung to function, and perhaps, protecting the organs from injury by the use of the PGE$_1$. Further, the results indicate that PGE$_1$ may be a promising agent for the treatment of ARDS and perhaps, even patients with sepsis. The drug was used for seven days and did show some overall improvement. The authors feel that additional studies should be done to confirm their work and evaluate its effectiveness in a relatively smaller well-defined group of patients.

In summary, the published works of the human sepsis studies related to the eicosanoids to date imply that the eicosanoids are clearly activated in patients with sepsis. Previous studies done in animals with endotoxin do not appear to be similar to those which occur in human sepsis. Laboratory animals studies with sepsis, however, do show similarities between the animal studies and human states. From the aforementioned studies, it is uncertain what the relationship is between the plasma eicosanoid concentrations and pathophysiological events in human sepsis. Further, there is some question about the role the eicosanoids have in both pulmonary and systemic hemodynamic events, as well as in tissue oxygenation in these states. One of the most difficult aspects of these studies is that the patients represent a variety of states of disease, there is no single etiologic factor and many other therapeutic modalities have been utilized in the management of these patients. It is exceedingly difficult, if not impossible, to study these patients in a dose-reponse manner. While it appears that the eicosanoids are involved in the metabolic, hemodynamic and pulmonary derangements in patients with sepsis, their specific relationships to any of these events must await more sophisticated, well defined, clinical studies utilizing specific inhibitors to help elucidate their undetermined role in the patient with sepsis syndrome. Clearly there is sufficient data in humans to merit continued research in these areas.

The possible role of the leukotrienes in sepsis is still undetermined. The leukotrienes possess a diverse inflammatory action. Leukotriene B_4 is a potent leukocyte chemotaxin and leukocyte aggregating agent. Other leukotriene products such as LTC_4, LTD_4 and LTE_4 produce increased vascular permeability, bronchorestriction and a variety of cardiovascular responses. There has been some experimental work in animals that suggests that the leukotrienes may be involved in pulmonary hypertension and hypoxemia following endotoxin. A number of additional experimental studies have been done in smaller animals, but it is too early in the development of this area to make any prognostic statements with regards to the importance of their involvement, even in animal studies. This particular area is made even more difficult because of the lack of sensitive methods of measurement of the leukotrienes in body fluids. In a review published in 1985 of the role of the eicosanoids as mediators of ischemia in shock, one author implied that perhaps, the leukotrienes might be mediators of tissue ischemia and shock [36].

An interesting study in hyperdynamic sepsis in dogs was reported by Fink et al. in 1984 [14]. Their hypothesis was that the eicosanoids might play a role in the hyperdynamic septic state. An intra-abdominal sepsis canine model was utilized. Once animals developed the hyperdynamic state, two chemically different nonsteroidal, anti-inflammatory drugs were administered to determine the effect of these on the hemodynamic parameters in this model. The animals had increased cardiac output and heart rate with decreased systemic arterial pressure and systemic vascular resistance. Within 60 minutes of the intravenous injection of either indomethacin (2 mg/kg) or ibuprofen (25 mg/kg), the cardiac output, heart rate and pulmonary arterial pressure decreased, whilst the systemic vascular resistance increased significantly. These studies were compared to control animals in which a laparotomy alone was performed. This is a particularly well done study of sepsis and is important since the pathophysiology of this hyperdynamic state is poorly understood. The previous study reported by Oettinger in human studies supports the concept that the arachidonic acid metabolites or eicosanoids are involved with the hyperdynamic septic state. The study by Fink et al. would imply that even dissimilar types of cyclooxygenase inhibitors would improve the hyperdynamic state if it were ever considered of value to be utilized in humans. Additional studies, of course, need to be done to determine the specific inter-relationships of the eicosanoids to the hemodynamic changes that are occurring. A more recent study in humans reported the use of prostaglandin E_1 in right heart failure and pulmonary artery hypertension following mitral valve replacement [37]. Patients undergoing mitral valve replacement, particularly those with severe pulmonary hypertension and/or congestive heart failure, may develop life threatening right heart failure in the immediate post cardiopulmonary bypass period. These authors studied the effects of high dose prostaglandin E_1, 130–150 ng/kg/min, a potent pulmonary vasodilator, in combination with massive infusions of norepinephrine into the left atrium in five consecutive patients with refractory pulmonary arterial hypertension after mitral valve replacement. The pharmacologic approach utilized by these authors takes advantage of the pulmonary vasodilating effects of prostaglandin E_1 while offsetting associated systemic vasodilatation and resulting hypotension. All of these patients had rapid pulmonary vasodilator responses followed by marked improvement in right ventricular function. All patients survived the operation, and none had right ventricular infarction or chronic right heart failure postoperatively. The study demonstrates a marked beneficial action of PGE_1 in the desperate clinical setting of right failure and pulmonary hypertension after mitral

valve replacement and suggests that an approach to patients with right
heart failure and pulmonary artery hypertension might be worth considering
in a similar group of patients that have systemic sepsis. These types of
studies provide a clinical basis for perhaps a new direction in the
medical treatment of some existing conditions.

Another area of interest that is emerging is related to the possible
role of the eicosanoids in patients with graded surgical trauma. The
first preliminary paper was reported by Alexander et al. in 1984 [38].
These investigators profiled changes in urinary $iTxB_2$ during graded
surgical trauma in selective patients. It was their hypothesis that
$iTxB_2$ would reflect the relative degree of surgical (tissue) trauma.

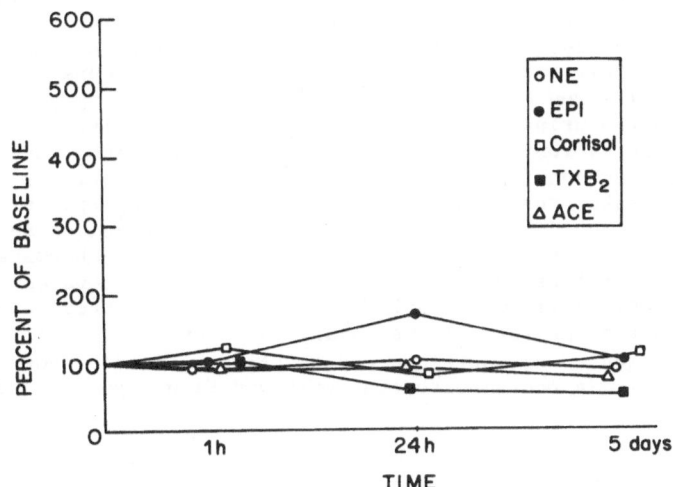

Fig. 1. Hormonal response in graded surgical trauma - I minimal tissue
injury. NE = norepinephrine: EPI = epinephrine: TxB_2 =
Thromboxane: ACE = angiotensin converting enzyme.

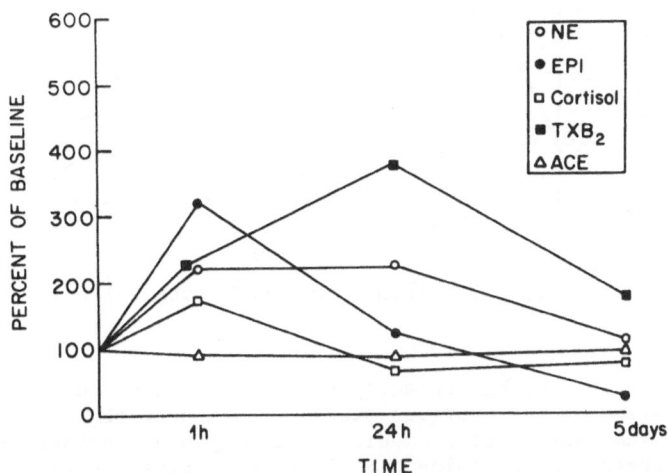

Fig. 2. Hormonal response in surgical trauma - II moderate injury.

The data as previously reported was incompletely analyzed at that time. The following data represents the complete analysis of that particular study. All surgical admissions for elective herniorrhaphy, cholecystectomy and colectomy were evaluated for suitability. The subjects were adult males that had received no prostaglandin inhibitors, had no intercurrent diseases and no perioperative complications. Group I patients had minimal tissue injury and consisted of thirteen patients who underwent elective herniorrhaphy. Group II patients (n = 9) had cholecystectomy and were considered as moderate tissue injury. Group III patients with colectomy (n = 13) were considered as severe injury. Table 1 indicates the urinary iTxB$_2$ levels in graded surgical trauma. The design was that urines were collected preoperatively, immediately postoperatively and on postop day one for the measurement of iTxB$_2$ by radioimmunoassay. Students T test was utilized on paired samples and unpaired groups with a p value of .05 considered significant. Of particular interest was that even with elective inguinal herniorrhapies, there was a significant increase in the iTxB$_2$ level in the immediate postoperative period. By postop day one, these values had decreased, but were not completely back to baseline. In the other groups of patients, cholecystectomies and colectomies, there was a increasingly significant difference in the urinary thromboxane level compared with the baseline levels in those patients as well as compared to the iTxB$_2$ response from inguinal herniorrhaphy. In both the cholecystectomy patients and the colectomy patients, the urinary concentration of iTxB$_2$ was still elevated on post-op day one compared to the baseline values for each particular group. It was our conclusion that the iTxB$_2$ hormonal response to minimal tissue injury was not impressive compared to those of moderate and severe tissue injuries. These data suggested that iTxB$_2$ in the urine might be reflect tissue injury in elective surgical patients. for hormonal agents were obtained preoperatively and one hour postoperatively, 24 hours postoperatively and five days postoperatively.

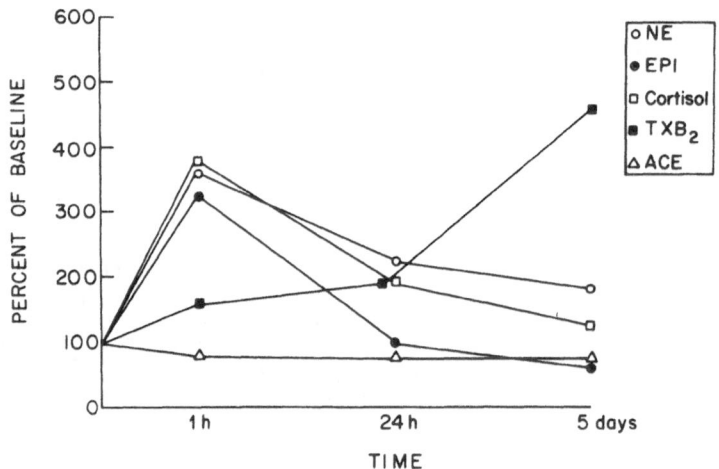

Fig. 3. Hormonal response in severe surgical trauma - III.

A follow-up study has recently been completed in our laboratory that is an extension of the previous study by Alexander et al. The concept of this study was that hormones mediate the body's homeostatic response to surgical stress. Eicosanoids and other vasoactive substances would reflect the hormonal response to tissue injury. Venous blood samples

Serum for angiotensin converting enzyme, T_4, T_3, free T_4 and T_3 were also collected. Plasma for cortisol, $iTxB_2$, norepinephrine and epinephrine were also determined. Hemodynamic parameters include heart rate, blood pressure and mean arterial pressure. The patient population included minimal tissue injury (inguinal hernia being the prototype), moderate tissue (cholecystectomy being the prototype) and severe tissue injury (subtotal colectomy being the prototype). Included in Tables II, III and IV are the profiles of patients included in this particular study. In Figure I, the hormonal response to graded surgical trauma in Group I is presented. In minimal surgical trauma, hormonal responses were mildly altered. In Figure II, with exception of angiotensin converting enzyme, the response to the tissue injury resolves within five days. The $iTxB_2$ response, however, continues to increase for 24 hours and then returns towards the baseline by the five day time period. In Figure III, the hormonal response to severe surgical trauma is presented. Again, one notes that there is a greater response at one hour following completion of surgery especially with norepinephrine, epinephrine and cortisol. Within 24 hours, these reduced to basal levels with the exception of $iTxB_2$. $iTxB_2$ values increases again at five days. Baseline values for these studies are in Table V. Dr. Chernow et al. concluded that the hormonal response to minimal tissue injury is negligible as evidenced by the data in the Group I patients. There are multiple patterns of hormonal responses in humans subjected to surgical trauma. The cortisol and catecholamine responses are similar as would be expected from the literature. Plasma $iTxB_2$ increases with increasing tissue injury as shown by Groups II and III. The exact relationship of hormonal responses to tissue injury or other parameters remains speculative. Additional studies specifically designed to examine a well defined population are indicated.

In summary of these interesting studies, the most recent papers suggest that there are new therapeutic opportunities for the use of prostaglandin E_1 and perhaps other eicosanoids as yet undetermined. Further, there is a possibility that the eicosanoid system is a very sensitive system to cellular injury, and that these particular chemicals may evolve to be an important indicator of cellular injury. At the present time, any concrete conclusions about the importance of these substances in tissue injury would be premature.

TABLE I

URINARY THROMBOXANE IN GRADED SURGICAL TRAUMA

ng/ml, mean $\pm$ SEM

	PRE-OP	POST-OP	POD 1
Herniorrhaphy (n=13)	0.48+.06	0.94+.19*	0.78+.19
Cholecystectomy (n=9)	1.75+.36	6.77+1.63*+	2.01+.70+
Colectomy (n=13)	1.08+.27	3.75+.83*+	1.47+.30+

*p < 0.02, significant from the baseline
+ < 0.005, significant from hernia

TABLE II

PROFILE OF PATIENTS

Group I

PATIENT #	AGE	SEX	OPERATIVE TIME	OPERATION DONE
1	20	F	35 mins	Laparoscopy
2	54	M	87 "	Inguinal hernia repair
3	47	M	86 "	Inguinal hernia repair
4	51	M	119 "	Inguinal hernia repair
5	26	M	100 "	Excision osteochondroma
6	27	M	68 "	Inguinal hernia repair
7	46	M	83 "	Inguinal hernia repair
8	28	F	42 "	Laparoscopy
9	25	F	26 "	Laparoscopy
10	19	M	69 "	Inguinal hernia repair

n=10   34.3±4.3   7M/3F   71.5±9.4

TABLE III

PROFILE OF PATIENTS

Group II

PATIENT #	AGE	SEX	OPERATIVE TIME	OPERATION DONE
11	57	M	290 mins	Cholecystectomy*
12	68	M	96 "	Cholecystectomy
13	32	F	136 "	Cholecystectomy
14	40	F	120 "	Vaginal hysterectomy
15	36	F	87 "	Cholecystectomy
16	18	F	125 "	Appendectomy
17	20	F	115 "	Celiotomy
18	45	F	170 "	TAHBSO **
19	42	F	83 "	Cholecystectomy
20	58	F	123 "	Cholecystectomy
21	43	M	92 "	Cholecystectomy
22	68	M	90 "	Cholecystectomy

n=12   43.9±4.8   4M/8F   127.3±16.5

* = common bile duct exploration was also performed

** = total abdominal hysterectomy, bilateral salpingo-oophorectomy

TABLE IV

PROFILE OF PATIENTS

Group III

PATIENT #	AGE	SEX	OPERATIVE TIME	OPERATION DONE
23	49	M	287 mins	Gastrectomy
24	23	M	350 "	Diverting Ileostomy
25	62	M	255 "	Aorto-bifemoral bypass
26	55	F	143 "	Hemi-colectomy
27	64	F	254 "	Colectomy
28	74	F	160 "	Colectomy
29	58	F	160 "	Hemi-colectomy
30	58	M	87 "	Colostomy/gastrostomy
31	31	M	438 "	Choledochojejunostomy

n=9 52.7±5.4 5M/4F 237.1±37.2

TABLE V

BASELINE VALUES FOR HORMONAL RESPONSE IN
GRADED SURGICAL INJURY

NE	300±40 pg/ml
Epi	32±7.3 pg/ml
Cortisol	16±23 pg/ml
TxB_2	177±80 pg/m.
ACE	12.2±1.5 u/ml

REFERENCES

1. B. J. Northover and G. Subramanian, Analgesic-Antipyretic Drugs as Antagonists of Endotoxin Shock in Dogs, J. Path. Bact. 83:463 (1982).
2. F. L. Anderson, W. Jubiz, A. C. Fralios, T. J. Tsagaris, and H. Kuida, Plasma Prostaglandin Levels During Endotoxin Shock in Dogs, Circulation 45:2 (1972).
3. J. R. Culp, E. G. Erdos, L. B. Hinshaw, and D. D. Holmes, Effects of Anti-inflammatory Drugs in Shock Caused by Injection of Living E. Coli Cells, Proc. Soc. Exp. Biol. Med. 137:219 (1971).
4. J. R. Fletcher, P. W. Ramwell, and C. M. Herman, Prostaglandins and the Hemodynamic Course of Endotoxin Shock, J. Surg. Res. 20:589 (1976).
5. C. V. Greenway and V. S. Murthy, Mesenteric Vasoconstriction after Endotoxin Administration in Cats Pretreated with Aspirin, Br. J. Pharmac. 43:259 (1971).
6. R. C. Hall, R. L. Hodge, R. Irvine, F. Katic, and J. J. Middleton, The Effect of Aspirin on the Response to Endotoxin, Austral. J. Exp. Biol. Med. Sci. 50:589 (1972).
7. L. B. Hinshaw, L. A. Solomon, E. G. Erdos, D. A. Reines, and B. J. Gunter, Effects of Acetylsalicylic Acid on the Canine Response to Endotoxin, J. Pharmac. Exp. Ther. 157:667 (1967).
8. J. R. Parratt and R. M. Sturgess, The Effect of Indomethacin in the Cardiovascular and Metabolic Responses to E. Coli Endotoxin in the Cat, Br. J. Pharmac. 50:177 (1974).
9. J. R. Parratt and R. M. Sturgess, Evidence that Prostaglandin Release Mediates Pulmonary Vasoconstriction Induced by E. Coli Endotoxin, J. Physiol. Land. 246:79 (1975a)
10. J. R. Parratt and R. M. Sturgess, The Protective Effect of Sodium Meclofenamate in Experimental Endotoxin Shock, Br. J. Pharmac. 53:466P (1975b).
11. J. R. Parratt and R. M. Sturgess, The Effects of Repeated Administration of Sodium Meclofenamate, An Inhibitor of Prostaglandin Synthetase, in Feline Endotoxin Shock, Circulat. Shock 2:301 (1975c).
12. J. R. Parratt and R. M. Sturgess, The Effect of a New Anti-inflammatory Drug, Flurbioprofen, on the Respiratory, Haemodynamic and Metabolic Responses to E. Coli Endotoxin Shock in the Cat, Br. J. Pharmac. 58:547 (1976).
13. J. R. Fletcher, The Role of Prostaglanins in Sepsis, Scand. J. Infect. Dis. Suppl. 31:55 (1982).
14. M. P. Fink, T. J. MacVittie, and L. C. Casey, Inhibition of Prostaglandin Synthesis Restores Normal Hemodynamics in Canine Hyperdynamic Sepsis, Ann. Surg. 200:619 (1984).
15. R. R. Butler, W. C. Wise, P. V. Halushka, and J. A. Cook, Gentamicin and Indomethacin in the Treatment of Septic Shock: Effects on Prostacyclin and Thromboxane A_2 Production, J. Pharmac. Exp. Ther. 225:94 (1983).
16. S. L. Kunkel, S. W. Chensue, and S. H. Phan, Prostaglandins as Endogenous Mediators of Interleukin 1 Production, J. Immuno. 136:186 (1986).
17. M. P. Fink, W. M. Gardiner, R. Roethel, and J. R. Fletcher, Plasma Levels of 6-Keto-PGF$_{1\alpha}$ but Not TxB$_2$ Increase in Rats with Peritonitis Due to Cecal Ligation, Circulat. Shock 16:297 (1985).
18. G. J. Slotman, J. V. Quinn, K. W. Burchard, and D. S. Gann, Thromboxane, Prostacyclin, and the Hemodynamic Effects of Graded Bacteremic Shock, Circulat. Shock 16:397 (1985).
19. G. J. Slotman, J. V. Quinn, K. W. Burchard, and D. S. Gann, Thromboxane Interaction with Cardiopulmonary Dysfunction in Graded Bacterial Sepsis, J. Trauma 24:803 (1984).
20. W. C. Wise, P. V. Halushka, R. G. Knapp, and J. A. Cook, Ibuprofen, Methylprednisolone, and Gentamicin as Cojoint Therapy in Septic Shock, Circulat. Shock 17:59 (1985).

21. L. C. Casey, J. R. Fletcher, M. I. Zmudka, and P. W. Ramwell, The Role of Thromboxane in Primate Endotoxin Shock, J. Surg. Res. 39:140 (1985).

22. R. H. Carmona, T. C. Tsao, and D. D. Trunkey, The Role of Prostacyclin and Thromboxane in Sepsis and Septic Shock, Arch. Surg. 119:189 (1984).

23. P. W. Ramwell, J. R. Fletcher, W. F. Flamenbaum, The Arachidonic Acid-Prostaglandin System in Endotoxemia, 6th Intl. Cong. of Pharmac., Helsinki, Clini. Pharmac. 5:175 (1975).

24. V. Baracos, H. P. Rodemann, C. A. Dinarello, and A. L. Goldberg, Stimulation of Muscle Protein Degradation and Prostaglaindin E_2 Release by Leukocytic Pyrogen (Interleukin 1), New Eng. J. Med. 308:553 (1983).

25. H. D. Reines, J. A. Cook, P. A. Halushka, W. C. Wise and W. Rambo, Plasma Thromboxane Concentrations in Human Septic Shock, Lancet 2:174 (1982).

26. J. R. Parratt, S. J. Coker, B. Hughes, A. MacDonald, I. Ledingham, I. Rodger, and I. Zeitln, The Possible Role of Prostaglandins and Thromboxanes in the Pulmonary Consequences of Experimental Endotoxin Shock and Clinical Sepsis in: "The Role of Chemical Mediators in the Pathophysiology of Acute Illness and Injury," Rita McConn, ed., Raven Press, New York (1982).

27. W. K. E. Oettinger, G. O. Walter, U. M. Jensen, A. Beyer, and A. Peskar, Endogenous Prostaglandin $F_{2\alpha}$ in the Hyperdynamic State of Severe Sepsis in Man, Br. J. Surg. 70:237 (1983).

28. D. N. Herndon, S. Abston, and M. D. Stein, Increased Thromboxane B_2 Levels in the Plasma of Burned and Septic Burned Patients, Surg. Gyn. & Ob. 159:210 (1984).

29. M. Leeman, J. M. Boeynaems, J. P. Degaute, J. L. Vincent, and R. J. Kahn, Administration of Dazoxiben, A Selective Thromboxane Inhibitor, in the Adult Respiratory Distress Syndrome, Chest 87:726 (1985).

30. H. D. Reines, P. V. Halushka, L. S. Olanoff, and P. S. Hunt, Dazoxiben in Human Sepsis and Adult Respiratory Distress Syndrome, Clin. Pharm. & Therap. 37:391 (1986).

31. P. V. Halushka, H. D. Reines, S. E. Barrow, I. A. Blair, C. T.Dollery, W. Rambo, J. A. Cook, and W. C. Wise, Elevated Plasma 6-Keto-Prostaglandin $F_{1\alpha}$ in Patients in Septic Shock, Crit. Care Med. 13:451 (1985).

32. H. D. Reines, P. V. Halushka, J. A. Cook, and C. B. Loadholt, Lack of Effect of Glucocorticoids upon Plasma Thromboxane in Patients in a State of Shock, Surg. Gyn. & Ob. 160:320 (1985).

33. G. S. Slotman, K. W. Burchard, and D. S. Gann, Thromboxane and Prostacyclin in Clinical Acute Respiratory Failure, J. Surg. Res. 39:1 (1985).

34. H. Tokioka, O. Kobayashi, Y. Ohta, T. Wakabayashi, and F. Kosaka, The Acute Effects of Prostaglandin E_1 on the Pulmonary Circulation and Oxygen Delivery in Patients with the Adult Respiratory Distress Syndrome, Inten. Care Med. 11:61 (1985).

35. J. W. Holcroft, M. J. Vassar, and C. J. Weber, Prostaglandin E_1 and Survival in Patients with the Adult Respiratory Distress Syndrome, Ann. Surg. 203:371 (1986?).

36. A. M. Lefer, Eicosanoids as Mediators of Ischemia and Shock, 68th Annal Meeting of the Federation of American Societies for Experimental Biology, St. Louis, 44:275 (1985).

37. M. N. D'Ambra, P. J. LaRaia, D. M. Philbin, W. D. Watkins, A. D. Hilgenberg, and M. J. Buckley, Prostaglandin E_1 A New Therapy for Refractory Right Heart Failure and Pulmonary Hypertension after Mitral Valve Replacement, J. Thorac. Cardiovasc. Surg. 89:567 (1985).

38. H. R. Alexander, W. R. Thompson, P. W. Ramwell, and J. R. Fletcher, Urinary Thromboxane (TxA_2) Reflects Dose-Response Tissue Injury in Humans, Curr. Surg. 42:18 (1985).

LIPID METABOLISM DURING STARVATION AND SEPSIS IN RELATION TO FATTY ACID PROFILE IN LIVER AND α-LINOLENIC- AND γ-LINOLENIC ACID-ENRICHED DIETS

C. Larsson-Backström, J. Paprocki, L. Lindmark and L. Svensson
Departments of Pharmacology and Nutrition, KabiVitrum AB.
S-112 87 Stockholm, Sweden

INTRODUCTION

Starvation and sepsis induce metabolic changes of particular impor-tance for the liver function. During sepsis, a condition characterized by starvation during the first, most severe period, the starvation induced ketonemia is markedly reduced (Flatt and Blackburn, 1974). Ketogenesis is assumed to be reduced by an inhibitory effect of malonyl-CoA (McGarry et al., 1978). Decreased responses to this inhibition have been reported in starvation (McGarry and Foster, 1979) and by long-chain fatty-acyl-CoA (Mills et al., 1983) in isolated mitochondria. The role of essential fatty acids (EFA) in liver lipids in these conditions is uncertain.

Aims of the study

This study was undertaken to evaluate: 1) the changes induced by starvation and sepsis, on fatty acid profile in liver and some metabolic parameters which mainly concern the energy substrates during starvation and sepsis; 2) the influence of dietary α-linolenic acid (18:3w3, ALA) and γ-linolenic acid (18:3w6, GLA) on these changes; and 3) the possible roles of EFA in the liver production of ketone bodies.

METHODOLOGY

Rats (90 g) were fed three diets (14% fat) for three weeks: controls (C; ~ 7% 18:3w3 and ~ 50% linoleic acid, 18:2w6), ALA (~ 20% 18:3w3 and ~ 40% 18:2w6) and GLA (~ 20% 18:3w6 and ~ 40% 18:2w6). The rats were divided into two experimental groups; fasted and septic-fasted. Peritonitis and septic shock were induced by i.p. E.coli (approximate LD_{50} dose).

Blood samples were taken from fed rats (separate rats) and from fasted and septic-fasted rats (survivors) at 24 hrs for determinations of β-hydroxybutyrate (BHB), free fatty acids (FFA) and glucose by enzymatic methods; prekallikrein (PKK; chromogenic peptide substrate assay) and platelet count (PC). Liver samples were extracted and analyzed for trigly-cerides (TGL) (enzymatically) and fatty acid profile (gas chromatography) in neutral lipids (NL) and phospholipids (PL).

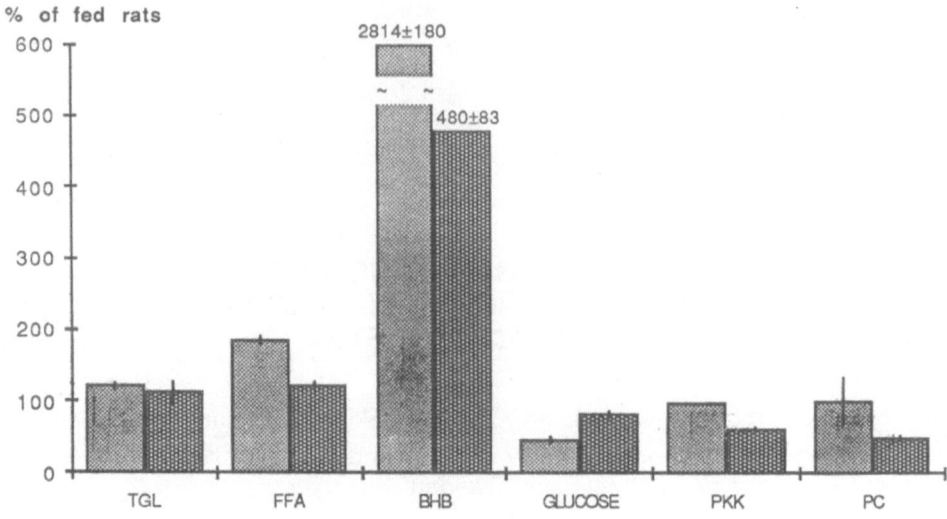

Fig. 1. Changes in metabolic parameters (TGL, FFA, BHB, Glucose), PKK and PC in fasted (■) and septic-fasted (▨) rats, in % of fed rats, on control diet at 24 hrs. ($\bar{X}\pm$SEM: n ⩾ 5).

Unsaturation index (UI) was calculated from the number of double bonds multiplied by the percentage of fatty acids.

RESULTS AND DISCUSSION

Figure 1 shows that fasting induces an increase in free fatty acids (FFA) and β-hydroxybutyrate (BHB) which is lower during sepsis in the control group. These changes and those in glucose, prekallikrein (PKK) and platelet count (PC) shown in Fig. 1 are in accordance with earlier reports and demonstrate the relevance of the model.

The results presented in Table 1 show that, during fasting, linoleic acid (18:2w6) and α-linolenic acid (18:3w3) are reduced, whereas arachidonic acid (20:4w6) and docosahexaenoic acid (22:6w3) are accumulated in neutral lipids (NL), with a resulting increment in the unsaturation index. These changes are reduced during sepsis (Table 1). Arachidonic acid (20:4w6) in phospholipids remains unchanged during fasting, whereas

Table 1. Fatty acid profile (weight %) and unsaturation index (UI) in liver neutral lipids (NL) and phospholipids (PL) from fed, fasted and septic-fasted rats on control diet at 24 hrs. ($\bar{X}\pm$ SEM; n=5). (Linoleic acid 18:2w6, ɣ-linolenic acid 18:3w6, arachidonic acid 20:4w6, α-linolenic acid 18:3w3, eicosapentaenoic acid 20:5w3 and docosahexaenoic acid 22:6w3).

CONTROLS		18:2w6	18:3w6	20:4w6	18:3w3	20:5w3	22:6w3	UI
FED	NL	40.1±1.1	0.72±0.07	3.5±0.1	3.0 ±0.1	0.45±0.04	1.1±0.1	145
	PL	16.7±0.3	0.12±0.01	24.9±0.3	0.21±0.01	0.19±0.01	7.0±0.2	195
FASTED	NL	32.8±0.5	0.75±0.08	14.5±1.2	2.4 ±0.1	0.61±0.10	3.1±0.1	179
	PL	12.9±0.4	0.10±0.01	24.4±0.5	0.14±0.01	0.08±0.01	10.3±0.7	200
SEPTIC-	NL	34.4±1.1	0.25±0.02	10.8±0.8	2.6 ±0.1	0.43±0.02	2.4±0.1	162
FASTED	PL	17.5±0.3	0.08±0.01	21.0±0.3	0.15±0.01	0.22±0.02	7.5±0.5	180

Table 2. Fatty acid profile (weight %) and unsaturation index (UI) in liver neutral lipids (NL) and phospholipids (PL) from fed, fasted and septic-fasted rats on ALA diet, at 24 hrs. ($\overline{X}\pm$SEM; n=5).

ALA		18:2w6	18:3w6	20:4w6	18:3w3	20:5w3	22:6w3	UI
FED	NL	35.9±1.4	0.42±0.03	1.9±0.2	10.8 ±0.9	1.3 ±0.2	1.3±0.2	157
	PL	17.4±0.9	0.11±0.01	22.0±0.6	0.60±0.07	1.19±0.07	7.2±0.4	197
FASTED	NL	34.7±0.6	0.37±0.02	5.4±0.4	12.5 ±0.3	1.4 ±0.1	2.1±0.1	173
	PL	14.7±0.5	0.06±0.01	25.4±0.3	0.70±0.05	0.56±0.05	8.1±0.5	200
SEPTIC-	NL	34.1±1.6	0.27±0.05	3.9±0.6	12.4 ±0.6	1.4 ±0.2	2.1±0.3	167
FASTED	PL	15.0±0.6	0.05±0.01	25.6±0.4	0.68±0.06	0.68±0.06	8.2±0.5	203

Table 3. Fatty acid profile (weight %) and unsaturation index (UI) in liver neutral lipids (NL) and phospholipids (PL) from fed, fasted and septic-fasted rats on GLA diet, at 24 hrs. ($\overline{X}\pm$SEM; n=5, n=2 for GLA, septic-fasted rats).

GLA		18:2w6	18:3w6	20:4w6	18:3w3	20:5w3	22:6w3	UI
FED	NL	24.9±1.0	4.1 ±0.3	9.6±1.0	0.42±0.07	0.06±0.01	0.31±0.03	151
	PL	9.9±0.6	0.95±0.05	30.8±0.4	0.04±0.01	0.03±0.01	3.3 ±0.1	203
FASTED	NL	25.6±0.4	4.3 ±0.3	15.2±1.7	0.49±0.03	0.12±0.01	0.88±0.15	169
	PL	8.9±0.5	0.64±0.03	29.9±0.3	0.03±0.01	0.05±0.02	5.7 ±0.3	205
SEPTIC-	NL	25.9±2.0	3.7 ±0.6	14.0±2.7	0.46±0.04	0.19±0.06	0.82±0.12	166
FASTED	PL	10.1±0.7	0.64±0.05	29.6±1.0	0.03±0	0	5.5 ±0.2	203

it is reduced in sepsis (Table 1). This decrease is presumably due to a sepsis-induced metabolism of arachidonic acid. Linoleic acid (18:2w6) and eicosapentaenoic acid (20:5w3) in phospholipids are reduced during fasting but are unaltered in sepsis (Table 1). These results indicate the existence of different pools for arachidonic, linoleic and eicosapentaenoic acids in neutral lipids and phospholipids, with specific regulatory mechanisms for each of the fatty acids.

Dietary ALA and GLA increase the incorporation of w3-fatty acids and w6-fatty acids, respectively, in liver neutral lipids and phospholipids, with a resulting increase in the unsaturation index (Tables 2 and 3). The changes in fatty acid profile shown for the control group during fasting and sepsis (Table 1) are reduced following dietary ALA and GLA (Tables 2 and 3). Linoleic acid (18:2w6) in fed rats is lower in neutral lipids and phospholipids following dietary GLA, as compared to dietary ALA, though the ALA and GLA diets contain similar amounts of linoleic acid (Tables 2 and 3). This, and the decreased levels of arachidonic acid in neutral lipids from fed, fasted and septic-fasted rats on ALA diet (Table 2), indicate inhibition by high amounts of dietary α-linolenic acid and its metabolites on linoleic acid metabolism, as shown earlier (Blond et al., 1978). The metabolism of arachidonic acid is also inhibited by dietary ALA and its metabolites (Clup et al., 1979), which may help to explain the unchanged level of arachidonic acid in phospholipids during sepsis following ALA diet, as compared to the control diet (Table 2).

The degree of unsaturation (UI) following dietary ALA and GLA, was lower during fasting and higher during sepsis, compared to control diet (Tables 2 and 3). Dietary ALA and GLA reduced liver triglycerides (LTG) and enhanced free fatty acids (FFA) in blood, both in fasted and in

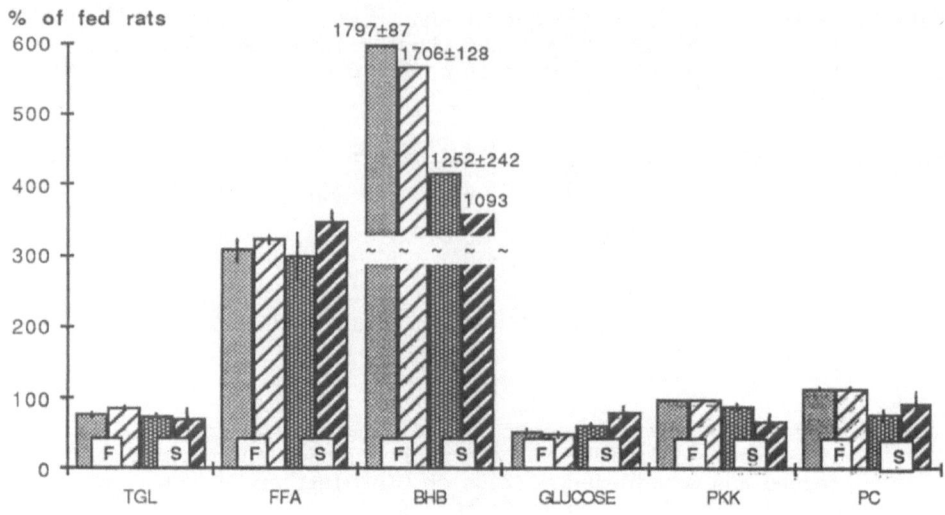

Fig. 2. Changes in metabolic parameters (TGL, FFA, BHB, Glucose), PKK and PC in fasted (F) and septic-fasted (S) rats, in % of fed rats, on ALA diet (fasted, ▨ septic-fasted) and GLA diet (▨ fasted, ▨ septic-fasted), at 24 hrs. (X̄±SEM; n ⩾ 5; n=2 for GLA, septic-fasted rats).

septic-fasted rats (Fig. 2). The level of β-hydroxybutyrate (BHB) is lower during fasting but higher during sepsis, compared to the control group.

These results taken together with those from the control group, with the fasting-induced increase in unsaturation index and β-hydroxybutyrate, which is reduced during sepsis, indicate a relationship between keto-genesis and the degree of unsaturation of liver lipids.

Conclusions — This study shows that: 1) Separate pools of the individual EFA:s exist in both phospholipids and neutral lipids in liver; 2) sepsis and starvation show different changes in the EFA-profile; 3) arachidonic acid decreases, while linoleic and eicosapentaenoic acids remain unchanged during sepsis; 4) the starvation induced transfer of EFA:s into neutral liver lipids is reduced during sepsis; 5) the septic condition, with its inhibitory influence on ketone body level, was alle-viated and slightly normalized by dietary supplementation with α-linolenic acid; 6) this is presumably a beneficial effect which seems to be due partly to an increase in the unsaturation index.

REFERENCES

J.P. Blond, J.P. Poisson and P. Lemerchal. Arch. Int. Physiol. Biochim. 86:741 (1978).
B.R. Clup, B.G.Tifus and W.E.M. Lands. Prostaglandin Med. 3:269 (1979).
J.P. Flatt and G.L. Blackburn. Am. J. Clin. Nutr. 27:175 (1974).
J.D. McGarry, Y. Takabayashi and D.W. Foster. J.Biol. Chem. 253:8294 (1978).
J.D. McGarry and D.W. Foster. J.Biol. Chem. 254:8163 (1979).
S.E. Mills, D.W. Foster and J.D. McGarry. Biochem. J. 214:83 (1983).

PATHOPHYSIOLOGY OF RENAL FAILURE IN SHOCK : ROLE OF LIPID MEDIATORS

G.E. Plante, R.L. Hébert, C. Lamoureux, P. Braquet, P. Sirois

Department of Physiology, University of Sherbrooke, Sherbrooke, Quebec, Canada

IHB, 17 avenue Descartes, 92350 Le Plessis Robinson, France

GENERAL CHARACTERISTICS OF SHOCK

Shock encountered in a variety of etiological conditions represents a generalized disease of microcirculation and involves most capillary networks. As a matter of fact, it could be considered as a disease of the endothelial layer, and perhaps of the sub-endothelial tissues of blood vessels (1). Endothelial cells are capable of synthesizing a number of vasoactive compounds which may dilate or constrict the media, as well as triggering white blood cells and platelets from the circulation, also capable of releasing vasoactive substances (2). Under most etiological conditions, shock is preceded by loss of vascular tone in most resistance vessels, by increased capillary permeability to circulatory macromolecules such as proteins, and by excessive loss of plasma volume into the interstitial space. Consequently, blood pressure collapses and perfusion of vital organs is compromised, leading to further deterioration of body fluid compartments homeostasis (1). The excess fluid and plasma proteins within the intersitial compartment where oxygen and substrates supply as well as removal of toxic end-products of cellular metabolism are dependent upon the physico-chemical characteristics responsible for their diffusion in opposite directions in this compartment of extracellular volume, probably represents the first deleterious step leading to impairment of function in vital organs during shock, and the first pathophysiological event responsible for the irreversible nature of certain shock syndromes (3, 4). Additional morbid consequences result from the hypoperfusion of specific vital organs, such as the heart which may reduce significantly its mechanical function (5), the lung which may alter the metabolism of certain vasoactive compounds released during shock (6), the gastrointestinal tract which may produce and/or release vasoactive compounds in response to ischemic injury (7), and finally the kidney which is the natural source of a variety of vasopressive peptides and numerous vasoactive metabolites of arachidonic acid, such as prostaglandins and leukotrienes (8, 9).

SHOCK AND THE KIDNEY

The kidney is almost always involved during shock and has even been incriminated as the major cause of death in several shock patients, especially before the advent of hemodialysis. This organ is mostly susceptible to anoxic injury, and because of the peculiar characteristics of its cellular metabolism, as well as of the specific architecture of the nephron segments, cellular necrosis not infrequently occurs when renal perfusion pressure is reduced, as it occurs in shock (10). The reduction of renal blood flow results in decreased glomerular filtration rate which leads to accumulation of metabolic end-products normally eliminated in the urine, such as urea, creatinine, uric acid, and probably of several other toxic substances. Since the kidney is also responsible for the inactivation and excretion of a variety of vasoactive subtances produced in situ, or generated in extrarenal tissues (11), it is likely that the impairment of such important functions might contribute to the maintenance of shock and perhaps, under peculiar circumstances, to its generation (12)! Precise identification of the mediators responsible for the renal response to hypoperfusion is therefore of critical importance, not only for understanding the pathophysiological events leading to shock, but also for the development of therapeutic tools to prevent or reverse renal failure during shock (13).

PAF-ACETHER: A TOOL TO STUDY SHOCK

PAF-Acether is a glycerophospholipid with potent vasodilatory properties which mimics, when injected intravenously to experimental animals, most of the systemic and regional hemodynamic features of shock (13, 14). This compound in fact is 1,000 times more potent than bradykinin, and 10,000 times more potent than histamine in producing plasma exudation in the cutaneous microcirculation of the rat and guinea pig (15). Its sites of production in vivo include almost the entire vascular endothelium, some blood cells such as the platelets, the polymorphonuclear and macrophage cells, as well as the kidney (16). In the latter organ, both glomerular cells as well as the interstitial cells contained in the medulla were shown to possess the enzymatic machinery to produce PAF (17). It is of interest that the synthesis of PAF occurs in parallel with the release of arachidonic acid, which gives rise to prostaglandins and leukotrienes via the cyclo-oxygenase and the lipoxygenase pathways, respectively (18). It is likely therefore that PAF-Acether, prostaglandins and leukotrienes, all produced within the cell membrane and released in the extracellular environment or in the cytosol, interact one with the other to amplify or neutralize physiological phenomena initiated by any of these lipid mediators (16, 19, 20).

PAF-ACETHER AND THE KIDNEY OF SHOCK

We utilized PAF-Acether in vivo to examine some of the physiological and pathophysiological effects of this compound on the dog kidney, trying to dissociate the potent systemic influences of this glycerophospholipid from its eventual specific effects on the kidney. Small doses of PAF-Acether were used in anesthetized Mongrel dogs to produce transient and reversible hemodynamic disturbances which resemble incipiens shock. In some experiments, intrarenal infusion of PAF-Acether was achieved to further examine the effects of this substance on renal function in a model where the peripheral consequences of PAF were minimized. Finally, a series of antagonists were used in an attempt to identify the role of eventual mediators in the physiological expression of PAF-Acether. All these procedures and methods were previously described in papers from our laboratory (21, 22).

When PAF-Acether is administered into the femoral vein of anesthetized dogs, in doses of 0.78ug given in bolus, hematocrit rises progressively from a mean control value of 43 ± 1 to 56 ± 1% 10 minutes after the end of injection (Table 1). This 30% rise in hematocrit is secondary to plasma leakage from the intravascular to the interstitial fluid spaces. As a consequence, blood pressure measured in the femoral artery is reduced by more than half, at the peak of PAF-Acether action: systolic pressure drops from 125 ± 3mmHg during the control periods to 60 ± 5mmHg 10 minutes after PAF-Acether injection. During the following 50-minute recovery period, blood pressure slowly returns to normalcy as systolic values rise to 72 ± 3 and 120 ± 4mmHg. Renal hemodynamics is markedly influenced by this reduction in systemic pressure. Renal plasma flow measured by the para-amino-hippurate clearance, decreases from a mean control value of 116 ± 5 to 59 ± 4ml/min 10 minutes after PAF injection. This parameter slowly returns to normal values during the five 10-minute clearance periods taken during recovery. Similarly, glomerular filtration rate, measured by the inulin clearance, decreases from 42 ± 2 to 26 ± 3ml/min at the peak of PAF-Acether action. During recovery, glomerular filtration rate increases to 31 ± 2 and 43 ± 5ml/min. Finally, absolute urinary excretion of sodium is reduced from an average control value of 69 ± 8uEq/min, to a minimal value of 25 ± 6uEq/min 10 minutes after the end of PAF injection. This parameter rapidly recovers to 72 ± 5uEq/min during the first clearance period after PAF. Thereafter, a rebound natriuresis reaching 117 ± 6uEq/min is observed. In summary, the net effect of systemic intravenous injection of a relatively small dose of PAF-Acether is characterized by massive plasma losses towards the interstitial compartment accompanied by arterial hypotension and reduction of renal perfusion pressure. The renal consequences thus appear to result from the systemic effect of PAF injection, and are characterized by a reduced renal plasma flow and glomerular filtration, as well as by a drop in the net urinary excretion of sodium. All these abnormalities are reversible within the hour that follows PAF injection, with the exception of natriuresis which recovers above control values as described (23).

Table 1: Effects of intrafemoral PAF on systemic and renal hemodynamics.

	Control	PAF	Recovery				
Ht %*	43±1	56±1	52±2	48±1	45±1	42±2	43±2
BP mmHg	125±3	60±5	72±3	84±4	90±3	110±5	120±4
RPF ml/min	116±5	59±4	68±6	81±5	92±6	101±5	105±4
GFR ml/min	42±2	26±3	31±2	35±4	34±2	38±4	43±5
$U_{Na}V$ uEq/min	69±8	25±6	72±5	94±8	116±6	118±5	117±6

* Ht: hematocrit; BP: blood pressure; RPF: renal plasma flow; GFR: glomerular filtration rate; $U_{Na}V$: urinary sodium excretion.

PAF-ACETHER HAS SPECIFIC AND DIRECT EFFECTS ON THE KIDNEY

In order to dissociate the renal from the systemic effects of PAF-Acether, additionnal experiments were performed again in anesthetized Mongrel dogs in which the left renal artery was canulated to allow the infusion of small doses of PAF-Acether directly into the renal microcirculation. In these experiments, PAF-Acether was given in doses of 0.15 and 0.30ug/kg, administered in single bolus. The right contralateral kidney was used as control. In this series of experiments, no systemic effects were recorded, in particular there was no change in peripheral blood hematocrit as well as blood pressure (Table 2). Therefore, one can reasonably exclude any systemic effect of PAF-Acether doses injected in the left renal artery. Of interest, during these experiments, renal hemodynamics was affected by PAF-Acether despite the maintenance of systemic blood pressure and hematocrit. Left kidney plama flow was reduced from 108 ± 5 to 62 ± 4 ml/min 10 minutes after the end of PAF injection. This parameter returned slowly to control values in the five 10 minute recovery clearance periods. Note that the renal plasma flow measured in the contralateral kidney remained essentially unchanged, suggesting the absence of significant recirculation of PAF-Acether into the general circulation. Similarly, glomerular filtration rate measured in the left kidney decreased from 32 ± 2 to 12 ± 1 ml/min from control periods to PAF injection. This parameter recovered gradually to normal values during the following 60 minutes period. Filtration rate measured in the contralateral kidney remained essentially unchanged, suggesting again that minimal amounts of PAF-Acether, if any, reached the control right kidney. Finally, urinary sodium excretion from the left kidney decreased from 140 ± 6 to 68 ± 7 uEq/min, a fall similar to that seen in the previous group of experiments. A rapid recovery of this parameter was also observed, urinary excretion rising to 98 ± 6 and 198 ± 12 uEq/min during the five consecutive 10-minute clearance periods that followed the end of the PAF-Acether injection. Contralateral natriuresis remained relatively stable throughout these experiments. In summary, the results obtained in this series of experiments indicate that PAF-Acether has a direct influence on renal hemodynamics and sodium excretion, and that systemic physical or hormonal influences are not required for the expression of the physiological effects of PAF-Acether on the kidney (23).

Table 2:   Effects of intrafemoral PAF on systemic and renal hemodynamics.

	Control	PAF	Recovery				
Ht %*	42±2	45±1	44±2	43±1	42±2	43±1	44±1
BP mmHg	120±3	118±2	121±3	117±4	120±2	120±2	116±4
RPF ml/min** L	108±5	62±4	76±5	79±4	86±5	98±4	101±6
R	112±5	109±4	107±5	102±4	110±6	108±6	105±8
GFR ml/min L	32±2	12±1	16±2	18±2	25±3	28±4	30±2
R	28±2	30±2	29±3	32±1	30±4	29±2	30±2
$U_{Na}V$ uEq/min L	140±6	68±7	98±6	156±12	160±8	186±9	198±12
R	168±11	159±12	161±8	157±10	168±9	164±7	170±9

*   Abbreviations as in Table 1.
** L and R represent left and right kidneys.

CONTRIBUTIONS OF PROSTAGLANDINS AND ANGIOTENSIN II TO THE RENAL EFFECTS OF PAF-ACETHER:

The contribution of prostaglandins to the physiological expression of PAF-Acether on systemic and renal hemodynamics was examined using indomethacin as an inhibitor of cyclooxygenase: this drug was used in a dose of 0.05mg/kg/min. After 60 minutes of equilibration under indomethacin, control clearance periods were obtained and PAF-Acether was administered in the femoral vein of anesthetized Mongrel dogs in a dose of 0.78ug/kg, again given in a single bolus. Five consecutive 10-minute clearance periods were obtained during recovery. Peripheral blood hematocrit remained essentially unchanged during PAF injection in dogs treated with indomethacin: control and experimental values obtained after PAF were 41 ± 1 and 40 ± 2%, respectively (Table 3). Similarly, pre-treatment with indomethacin fully protected systemic blood pressure from the effect of PAF injection: systolic pressure averaged 107 ± 7 and 108 ± 6mmHg during control and following PAF injection, respectively. However, this treatment failed to block the renal hemodynamic effects of PAF-Acether. As in the previous groups, renal plasma flow decreased from 55 ± 4 to 24 ± 3ml/min from control conditions to PAF injection. Glomerular filtration also decreased from mean control values of 34 ± 3 to 14 ± 2ml/min following PAF injection. The pattern of recovery of renal hemodynamics was relatively rapid for renal plasma flow which reached control values at the end of the first recovery period, that is 20 minutes after the end of PAF injection. Recovery of glomerular filtration was more gradual and was not complete, since 60 minutes after the end of PAF injection, this parameter was still 20% below control values. Finally, indomethacin treatment also failed to prevent the marked reduction in urinary sodium excretion, from 82 ± 10 to 22 ± 7uEq/min which occured during PAF injection. The pattern of recovery of sodium excretion was also gradual and incomplete, as noticed for glomerular filtration rate: there was no rebound natriuresis as seen in the two previous series of experiments during the recovery period after PAF injection. In summary, this series of experiments indicate that vasodilatory prostaglandins, presumably PGE_2, are responsible for the peripheral effects of PAF-Acether, whereas inhibition of prostaglandins synthesis with indomethacin failed to alter the renal response to PAF-Acether. However, it is of interest that indomethacin treated dogs had a pattern of recovery for glomerular filtration and urinary sodium excretion following PAF injection that differ from that seen in previous groups: this finding suggests that prostaglandins might be involved in part of the renal response to PAF-Acether.

Table 3: Effects of intrafemoral PAF during indomethacin treatment.

	Control	PAF	Recovery				
Ht %*	41±1	40±2	41±1	42±1	41±2	40±1	42±1
BP mmHg	107±7	108±6	112±6	110±5	108±6	107±5	109±6
RPF ml/min	55±4	24±3	56±5	57±6	56±6	57±7	55±6
GFR ml/min	34±3	14±2	27±2	27±2	27±1	28±1	28±2
$U_{Na}V$ uEq/min	82±10	22±7	55±6	68±8	72±7	76±6	75±6

* Abbreviations as in Table 1.

The important reduction of renal plasma flow and glomerular filtration rate which occured during and following PAF-Acether injection, while peripheral arterial blood pressure and presumably renal perfusion pressure were maintained normal in indomethacin treated dogs, suggests the intervention of vasoconstrictor substances: vasopressor peptides such as angiotensin II and/or vasoconstrictor metabolites of arachidonic acid represent potential candidates to explain the observed renal hemodynamic changes. The eventual role of vasoconstrictor prostaglandins would appear remote, however, since indomethacin blockade of prostaglandins synthesis failed to protect the renal hemodynamic consequences of PAF-Acether injection. Therefore, the role of angiotensin II in mediating this effect of PAF was examined in another series of experiments. In these animals, prepared in the same manner as previous groups, indomethacin was administered throughout experiments. The renal response to PAF injection, administered in the femoral vein at a dose of 0.78ug/kg was again examined after control periods. Thereafter, saralasin, an antagonist of angiotensin II receptors, was infused at a dose of 100ug/kg. Three clearance periods were obtained during indomethacin plus saralasin, and the PAF was administered. The same number of recovery periods were obtained as in the previous groups.

As expected, PAF injection reduced markedly renal plasma flow, glomerular filtration rate and urinary sodium excretion during indomethacin (Table 4). However, during indomethacin and saralasin, PAF injection failed to alter renal plasma flow, which remained at 105 ± 8 and 96 ± 7ml/min, before and during PAF, respectively, whereas glomerular filtration averaged 52 ± 3 and 51 ± 4ml/min, during the same periods. Of interest, urinary sodium excretion was reduced from 132 ± 12 to 70 ± 8uEq/min from control conditions to PAF injection, despite the fact that filtered sodium remained essentially unchanged due to the protective effect of saralasin on renal hemodynamics, in particular on glomerular filtration rate. In summary, one can conclude from these observations that the intervention of angiotensin II is required for the expression of PAF-Acether on renal hemodynamics, since saralasin entirely protects renal plasma flow and glomerular filtration from the effect of the glycerophospholipid. However, the effect of PAF-Acether on urinary sodium excretion appears to be independent of renal hemodynamics, since angiotensin II blockade failed to prevent the marked reduction in urinary sodium excretion which occured during PAF (13, 23).

Table 4: Effects of intrafemoral PAF during saralasin blockade.

	Control[**]	PAF	SAR	PAF	Recovery
Ht %[*]	42±2	40±3	41±2	41±2	40±3
BP mmHg	110±6	107±4	112±5	111±4	109±4
RPF ml/min	105±8	58±7	105±8	96±7	128±9
GFR ml/min	45±3	28±3	52±3	51±4	55±3
$U_{Na}V$ uEq/min	155±9	21±6	132±12	70±8	118±7

[*]   Abbreviations as in Table 1.
[**]   Indomethacin is administered throughout experiments.

Since filtration fraction, which determines peritubular oncotic pressure, was not affected during PAF injection in dogs treated with indomethacin and saralasin, it is unlikely that the reduction in urinary sodium excretion was related to an increased passive movement of sodium and water from the tubular lumen to the peritubular capillary: the forces which determine such passive movements were not altered from control conditions to PAF injection in this series of experiments. Therefore, one has to assume that the 50% reduction of urinary sodium excretion was due to increased active tubular reabsorption of this ion and/or to some impairment of the transtubular permeability. These mechanisms were not directly assessed in the present study but find some support in the recent literature (24, 25).

EFFECT OF BN-52021, A PAF RECEPTOR ANTAGONIST:

A natural antagonist of PAF receptors (BN-52021) has recently been extracted from Ginko Biloba. This compound which is entirely different structuraly from PAF-Acether, has been repeatedly tested over the past years on several in vitro and in vivo models, and found to antagonise most physiological effects of PAF-Acether (26). We utilized this compound in our dog model. The effect of PAF-Acether administered into the femoral vein at a dose of 0.78ug/kg, was examined at four different and increasing doses of BN-52021: 1.0, 2.5, 5.0 and 25.0ug. The PAF antagonist was administered 30 minutes before each injection of PAF-Acether. Complete recovery of hemodynamic parameters was allowed before each additional test injection. The expected reduction in the systemic blood pressure following PAF was abolished in a dose-related manner: the

5.0 and 25.0 doses of BN-52021 completely abolished the hypotensive action of PAF-Acether (Figure 1). The lower dose of the PAF receptor antagonist was even more evident on renal hemodynamics: the 1.0ug dose almost completely abolished the effect of PAF-Acether on renal plasma flow and glomerular filtration: these parameters averaged 74 $\pm$ 4 and 72 $\pm$ 4ml/min, and 42 $\pm$ 1 and 40 $\pm$ 1ml/min during control periods and PAF injection, respectively. Similar findings were observed for urinary sodium excretion. The 2.5ug dose of BN-52021 abolished the anti-natriuretic effect of PAF-Acether. In summary, blockade of PAF-Acether receptors with BN-52021 inhibits all the physiological effects of PAF on systemic hemodynamics, as well as on renal plasma flow and glomerular filtration rate. BN-52021 is the only antagonist that prevented the reduction in urinary sodium excretion (23).

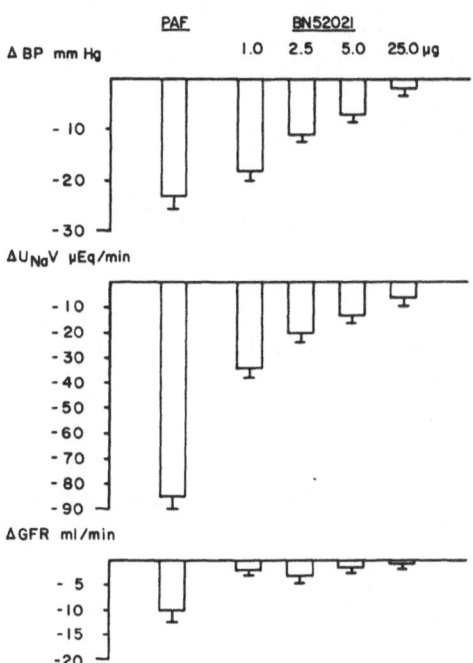

Figure 1

Effects of PAF before and during increasing doses of BN52021. Values $\pm$ SEM represent changes from control, for blood pressure (BP), urinary sodium ($U_{Na}V$) and glomerular filtration (GFR).

REFERENCES

1.   PARKER, M.M., PARILLO, J.E.. Septic Shock. Hemodynamics and pathogenesis. JAMA 250:3324, (1983).

2.   FURCHGOTT, R.F.. The role of endothelium in the responses of vascular smooth muscle to drugs. Am. Rev. Pharmacol. Toxicol. 24:175, (1984).

3.   COMPER, W.D., LAURENT, T.C.. Physiological function of connective tissue polysaccharides. Physiol. Rev. 58:255, (1978).

4.   AUKLAND, K., NICOLAYSSEN, C.. Interstitial fluid volume: local regulatory mechanisms. Physiol. Rev. 61:556, (1981).

5.   FEURSTEIN, G., BOYD, L.M., EZRA, D., GOLDSTEIN, R.E.. Effect of platelet activating factor on coronary circulation of domestic pig. Am. J. Physiol. 246:H466,(1984).

6.   LOEGERING, D.J.. Humoral factor depletion and reticulo-endothelial depression during hemorrhagic shock. Am. J. Physiol. 232:H282, (1977).

7.   GREENBERG, S., McGOWAN, C., GLENN, T.M.. Pulmonary vascular smooth muscle function in porcine splanchnic arterial occlusion shock. Am. J. Physiol. 241:H34, (1981).

8.   MUIRHEAD, E.E., BYERS, L.W., DESIDERIO, D.M., BROOKS, B., BROSIUS, W.M.. Antihypertensive lipids for the kidney: alkylether analogs of phosphatidylcholine. Fed. Proc. 40:2285, (1981).

9.   SRAER, J., RIGAUD, M., BENS, M., RABINOVITCH, H., ARDAILLOU, R.. Metabolism of arachidonic acid via the lipoxygenase pathway in human and murine glomeruli. J. Biol. Chem. 258:4325, (1983).

10. BREZIS, M., ROSEN, S., SILVA, P., EPSTEIN, F.H.. Renal ischemia: a new perspective. Kidney Int. 26:375, (1984).

11. CARONE, F.A., PETERSON, D.R.. Hydrolysis and transport of small peptides by the proximal tubule. Am. J. Physiol. 238:F151, (1980).

12. CAMUSSI, G.. Potential role of platelet-activating factor in renal pathophysiology. Kidney Int. 29:469, (1986).

13. PLANTE, G.E., HEBERT, R.L., BRAQUET, P., SIROIS, P.. Effects of PAF-Acether on renal function and its inhibitions with a new antagonist (BN-52021). New horizons in platelet activating factor research. C.M. Winslow, M.L. Lee, ed. (1986).

14. BESSIN, P., BONNET, J., APFFER, D., SOULARD, C., DESGROUX, L., PELAS, I., BENVENISTE, J.. Acute circulatory collapse caused by platelet-activating factor (PAF-Acether) in dogs. Eur. J. Pharmacol. 86:403, (1983).

15. HWANG, S.B., LI, C.L., LAM, M.H., SHEN, T.Y.. Characterization of cutaneous vascular permeability induced by platelet-activating factor in guinea pigs and rats and its inhibition by a platelet-activating factor antagonist. Lab. Invest. 52:617, (1985).

16. SCHLONDORFF, D., NEUWIRTH, R.. Platelet-activating factor and the kidney. Am. J. Physiol. 251:F1, (1986).

17. PIROTSKY, E., BIDAULT, J., BURTIN, M., GUBLER, M.C., BENVENISTE, J.. Release of platelet-activating factor, slow-reacting substance and vasoactive species from isolated rat kidney. Kidney Int. 25:404, (1984).

18. CHILTON, F.H., ELLIS, J.M., OLSON, S.C., WYKLE, R.L.. 1-O-Alkyl-2:sn-glycero-3-phosphocholine. A common source of platelet-activating factor and arachidonate in human polymorphonuclear leukocytes. J. Biol. Chem. 259:12014, (1984).

19. HEBERT, R.L., LAMOUREUX, C., SIROIS, P., BRAQUET, P., PLANTE, G.E.. Potentiating effects of leukotriene B4 and prostaglandin E2 on urinary sodium excretion by the dog kidney. Prostaglandins, Leukotrienes and Med. 18:69, (1985).

20. McGIVERN, D.V., BASRAN, G.S.. Synergism between platelet-activating factor (PAF-Acether) and prostaglandin E2 in man. _Eur. J. Pharmacol._ 102:183, (1984).

21. JOBIN, J., NAWAR, T., CARON, C., PLANTE, G.E.. Effect of acetazolamide on renal bicarbonate excretion in volume expanded dogs. _Am. J. Physiol._ 232:F484, (1977).

22. PLANTE, G.E., ERIAN, R., PETITCLERC, C.. Renal excretion of levamisole. _J. Pharmacol. Exp. Therap._ 216:617, (1981).

23. HEBERT, R.L., SIROIS, P., BRAQUET, P., PLANTE, G.E.. Hemodynamic effects of PAF-Acether and the dog kidney. _Prostaglandins_, Leukotrienes and Med. in press.

24. BAUM, M., BERRY, C.A.. Evidence of neutral transcellular NaCl transport and neutral basolateral chloride exit in the rabbit proximal convoluted tubule. _J. Clin. Invest._ 74:205, (1984).

25. PLANTE, G.E., HEBERT, R.L., FRANCO, N., ST-PIERRE, S.. Vasoactive intestinal peptide affects the renal transport of ions. _Regulatory Peptides_ 54:136, (1985).

26. BRAQUET, P.. BN-52021 and related compounds: a new series of highly specific PAF-Acether receptor antagonists. _Proc. Int. Congr. Immunopharmacol._, Florence, May 1985.

MULTIPLE ORGAN FAILURE, LIVER FAILURE AND POLYUNSATURATED FAT METABOLISM

F.B. Cerra, M.D., M. West, M.D., P. Alden, M.D., G. Keller, M.D., R.L. Simmons, M.D.
Department of Surgery, University of Minnesota, Minneapolis Minnesota 55455, USA

ABSTRACT

Multiple organ failure continues to be the pathway of death after burns, trauma and sepsis. This clinical syndrome represents the transition from a hypermetabolic response to injury that has associated respiratory dysfunction, to a setting of clinical organ failures and death. Risk factors include: perfusion deficits, persistent foci of dead or injured tissue, an uncontrolled focus of infection, the presence of the respiratory distress syndrome, persistent hypermetabolism, and preexisting fibrotic liver disease. Once in the organ failure syndrome, most treatment modalities become progressively ineffective, including: ventilation, antibiotics, nutrition, and surgery. The best treatment remains prevention with rapid control of the source and restoration of oxygen transport. To date, no single "magic bullet" has been shown to exist either experimentally or clinically.

The response to injury involves alterations in physiology and in the metabolism of carbohydrate, fat and amino acids. These changes seem to reflect the modulation of the end-organs by the mediator systems activated in response to the stress stimulus. The transition from hypermetabolism to organ failure appears to reflect the clinical appearance of liver failure. The peripheral tissue, however, appears to continue to function well until death becomes eminent. It is currently hypothesized that this liver failure represents a state of regulatory dysfunction induced by the activated hepatic macrophage, the Kuppfer cell. This same process may also influence metabolic failure in other organs as well. The activation of these macrophages is hypothesized to represent the final stage of a series of continuous stimulating events, eg. hypoxia, endotoxin, bacteria, and gut translocated toxins. The precise monokine(s) responsible are not yet completely characterized. Treatment consists of the modalities outlined above and the employment of aggressive metabolic support.

An injury is usually a local event that initiates a local redness, swelling, pain, heat, and a loss of function. With severe insults or when local host defenses fail, the response becomes systemic. The systemic response is recognized as the clinical syndrome of fever, anorexia, sympathomimetic signs, malaise altered mentation, and leukocytosis with a left-shift on differential count. This response is often associated with a hyperdynamic physiologic state and hypermetabolism. Stimuli such as infection or injured tissue, and significant perfusion deficits seen in the clinical settings of burns, polytrauma, pancreatitis, and ruptured aneurysms, can also induce the same systemic response. In the case of infection, the type of invading micro-organisms does not seem to influence the character of the host response.[1,2,3,4,5,6,7]

Once the systemic metabolic response to injury has been activated, a typical time-course usually results (Fig. 1). The initial events are summed-up with the expression, "ebb" phase[7] or "shock" phase. The ebb phase immediately precedes the flow phase which tends to peak on day 3 post injury and abate spontaneously by day 7-10 post injury. The degree of this initial response has considerable biologic variablility. Using several clinically useful sign-posts, a categorization can be defined that quantitates the degree of response (Table I). When a defined level is reached (level 2), hypermetabolism is said to be present.

When the response does not abate, a complication has usually occurred. Typically, water retention and myocardial infarctions have their maximum incidence on day 3 postinjury; liver failure tends to occur on day postoperative 3-5 in patients with cirrhosis. If the cause can be found and corrected, recovery usually occurs. After several such episodes, or when the cause cannot be controlled, a phase of persistent hypermetabolism is entered and the transition to organ failure may occur (Fig. 1). Usually the respiratory distress syndrome

TABLE I

METABOLIC CRITERIA FOR STRESS STRATIFICATION

Stress Level	Clinical Prototype	Urinary Nitrogen Gm/Day	Oxygen Consumed ml/M^2	Blood Glucose mg/dl	Plasma Lactate mM/L	Glucagon Insulin Ratio
0	starvation	<5	90+10	100+20	100+50	2+.5
1	elective surgery	5-10	130+10	150+25	1200+200	2.5+8
2	trauma	10-15	150+20	150+25	1200+200	3+.7
3	sepsis	>15	180+20	250+50	2500+500	8+1.5

A continuum exists within which ranges of values or patterns of response can be identified. An individual patient may not coincide with the clinical prototypes, e.g., all septic patients do not exhibit level 3 metabolism. The correlation between clinical setting and degree of metabolic stress is r = 0.6 to 0.7.

is seen first. Organ failure is usually heralded by a rising bilirubin followed by a rising creatinine. There are several recognized risk factors for this transition: severe, unrecognized, persistent perfusion deficits; an uncontrolled or uncontrollable infection source; the presence of respiratory distress syndrome; and preexisting cirrhosis or hepatitis. Organ failure can be divided into early and late phases (Table II). The transition usually signifies a higher mortality risk; as the organ failures worsen the mortality risk increases. There comes a point, frequently hard to define, at which all known treatment modalities seem to become supportive instead of therapeutic, including surgery, ventilation, antibiotics and nutrition[1,2,3,6,8,19].

The transition to the clinical organ failure syndrome apppear to coincide with the onset of clinical hepatic failure. Both clinical and metabolic evidence supports this point of view. At this time jaundice, biliary tract dilatation, colestasis, and biliary sludge become prominent clinical features. The R/Q tends to run over 0.9 and frequently exceeds 1.00 with an associated increase in hepatic lipogenesis; triglyceride intolerence occurs with a reduced ability to clear exogenous triglycerides of long chain fatty acids. The amino acid profile is similar to that seen in the liver failure of cirrhosis; the hepatic redox potential falls as does hepatic amino acid clearance and protein synthesis; ureagenesis increases. When fully developed, the liver failure is almost uniformly fatal. It remains the most common cause of death in surgical ICU patients and continues to consume a major share of intensive care resources.[2,6,14,19]

METABOLISM

The alterations in physiology and metabolism run a spectrum as the hypermetabolism transcends into organ failure and then into the terminal form of organ failure. The phase of hypermetabolism represents a normal systemic response to the mediator systems. With the transition to organ failure, the systemic response begins to

TABLE II

EARLY AND LATE ORGAN FAILURE

	Early	Late
Mentation	light coma	deep coma
Respiratory distress syndrome	present	advanced
Bilirubin	3-4 mg/dl	>8 mg/dl and rising
Creatinine (Cr)	2-3 mg/dl	>3 mg/dl and rising
BUN/Cr ratio	normal	↓ (off nutrition)
Muscle mass	±	autocannibalism
Lactate	1.5 mM/L	>2.0 mM/L
BOHB/AcAc* ratio	normal	increased
Phenylalanine	<80 M/L	>80 M/L
Triglyceride (12 hr fasting)	<250 mg/dl	>250 mg/dl
O_2CI	>160 ml/M^2	>160 ml/M^2
$V\dot{C}O_2$	<5 ml/kg	>5 ml/M^2
N-balance (on nutrition)	equilibrium or positive	>-5 gm/day

*BOHB/AcAc = hydroxybutyrate/acetoacetate

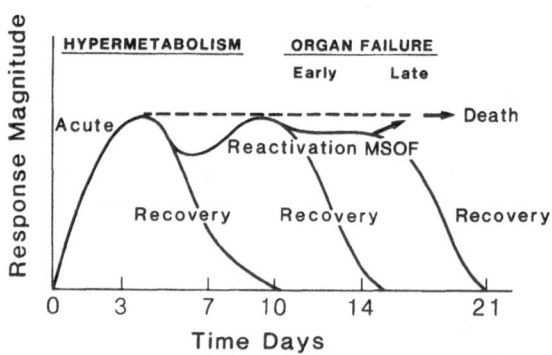

Figure 1. The presence of dead or injured tissue, an oxygen debt or invading microorganisms can activate the metabolic response to injury. The response magnitude depends on how much stimulation has occurred; the uncomplicated form usually peaks on Day 3 and abates by Day 7-10 postinjury. Hypermetabolism-organ failure represent a metabolic response spectrum. Clinically, the hypermetabolism phase has lung dysfunction and is primarily a peripheral phenomenon. With the onset of organ failure, clinically overt liver failure seems to be present. The risk factors for this transition include: A severe or persistent oxygen debt; an uncontrolled focus of infection or dead/injured tissue; and a perfusion insult followed by infection.

deteriorate, with the eventual death of the host. The metabolism is much different from that of starvation. A detailed discussion is beyond the scope of this paper; a summary of principles will be presented here.

In hypermetabolism the energy expenditure is more than twice that seen in starvation (Table III). This is reflected in the increased oxygen consumption, reduced peripheral resistence, and increased cardiac output. The combustion ration runs 0.8-0.85[2,14] with about 1/3 each of the energy expenditure deriving from glucose, fat, and amino acids. The oxidative utilization of pyruvate is reduced, presumably reflecting the reduced activity of pyruvate dehydrogenase. Lactate production is increased with a proportionate increase in pyruvate, so that there is little excess lactate. Alanine production is increased, oxygen consumption is high and there is use of alternate substrates in the Krebs cycle. These observations reflect an altered type of aerobic metabolism which utilizes substrates other than glucose, namely fat and amino acids[2,8,16]. Gluconeogenesis is increased and much less responsive to the administration of exogenous glucose[17]. Ketosis is reduced; fat oxidation increased; lipolysis increased; and lipogensis decreased. Total body protein synthesis is reduced with increased catabolism. Administration of exogenous amino acids can increase the rate of protein synthesis; matching but not reducing the catabolic rate. This autocannibalism is thought to reflect the response to mediator(s), perhaps a monokine or peptide. Thus the demand for amino acids is increased for energy production and protein synthesis; and mobilization from the mobile pools, as in skeletal muscle, occurs in response to these mediators. The result is a rapid loss of muscle protein and redistribution of the nitrogen to the organs with increased work, eg, liver, heart, kidney. Hepatic amino acid clearance is increased and is reflected in an increased rate of hepatic protein synthesis[2,8,10,11,12,13,15,18,20,21].

With the transition to organ failure, most of the processes previously mentioned are augmented. Energy production still seems adequate; the peripheral metabolism still seems to work well. However,

TABLE III

COMPARISON OF STARVATION AND SEPSIS

	Starvation	Sepsis
Energy Expenditure	↓	++
Mediator Activation	+	+++
R/Q	0.7	0.8-0.85
Fuel	glucose/fat	mixed
Gluconeogenesis	+	+++
Protein Synthesis	↓	↓↓
Catabolism	---	+++
AA Oxidation	±	+++
Ureagenesis	±	++
Ketosis	+++	+
Responsiveness to Exogenous Substrate	+++	+
Rate of Malnutrition Development	+	+++

AA = amino acid

hepatic redox potential begins to fall; hepatic lipid production increases; and triglyceride clearance and hepatic protein synthesis begin to fail. This latter event can usually still be supported with exogenous amino acids. If not, the outcome is poor. As the terminal state approaches, glucose production is reduced; amino acid extraction fails and both hepatic and total body protein synthesis decrease while catabolism increases; spontaneous hypertriglyceridemia occurs; lactate and pyruvate rise rapidly; ureagenesis is unrestricted; energy production fails; and the clinical findings of liver, renal and cardiac failure occur.[2,8,19]

The organ failure syndrome induces changes in the polyunsaturated fatty acid (PUFA) profile of the phospholipid fraction of plasma. The profile becomes that of essential fatty acid deficiency with excess levels of other PUFA (Table IV). Exogenous fat emulsion, triglycerides of the long chain fatty acids, do not appear able to correct this profile and exaggerate certain aspects of it. Of note, are the observed reductions in arachidonic acid with increased conversion to 22:5w6. Changes in these lipid fractions may reflect alterations in the cell membrane structural components, and thus may be involved in the organ failure process. Whether this picture is cause or effect, is yet unknown.

THE HEPATIC FAILURE

Liver dysfunction, then, becomes a primary focus in the organ failure process. Research has concentrated on the regulation of hepatocyte function. Bacteria and toxins circulating in the blood are removed from the circulation by Kupffer cells. In the process of removal, Kupffer cells may become activated with release of numerous mediator substances. We hypothesize that this progressive activation

TABLE IV

ALTERATIONS IN PLASMA POLYUNSATURATED FATTY ACID PROFILES

INDUCED BY HYPERMETABOLISM-ORGAN FAILURE

FATTY ACID	CONTROL	WITH LIPID	NO LIPID
N	33	10	5
16:0 Palmitic	26.7 ± .4	30 ± 1	30.4 ± 1.1
18:1w9 Oleic	8.8 ± .3	8 ± .6	15.7 ± 3.6
18:2w6 Linoleic	22.9 ± .6	13.9 ± 2.1	9.4 ± 1.3
18:3w3 Linolenic	.2 ± .03	.02 ± .02	.02 ± .02
20:3w9	.15 ± .01	.25 ± .09	1.1 ± .4
20:4w6 Arachidonic	11 ± .5	6.6 ± .7	8.8 ± 1
22:5w6	42.3 ± .6	31.7 ± 20.8	27.7 ± 1.9

Control = nonstressed, well-nourished   No Lipid = 7 days parenteral nutrition without fat
Lipid = 7 days parenteral nutrition with 500 ml 10% fat given qd

Lipid and nonlipid group were all patients in the surgical ICU and had the hypermetabolism-organ failure complex

144

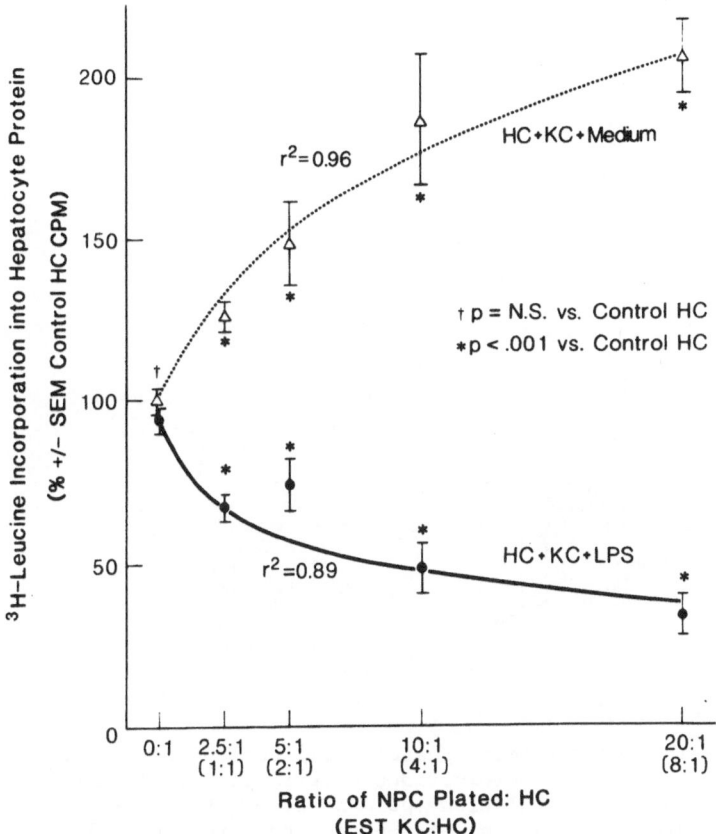

Figure 2. The hepatocyte alone in culture has a baseline level protein synthesis. When Kupffer cells are added, hepatocyte protein synthesis increases. This synthesis can be inhibited by ischemia, endotoxin and killed bacteria. The process requires Kupffer cells and seems to be mediated by a peptide produced by the Kupffer cells in response to the stimuli.

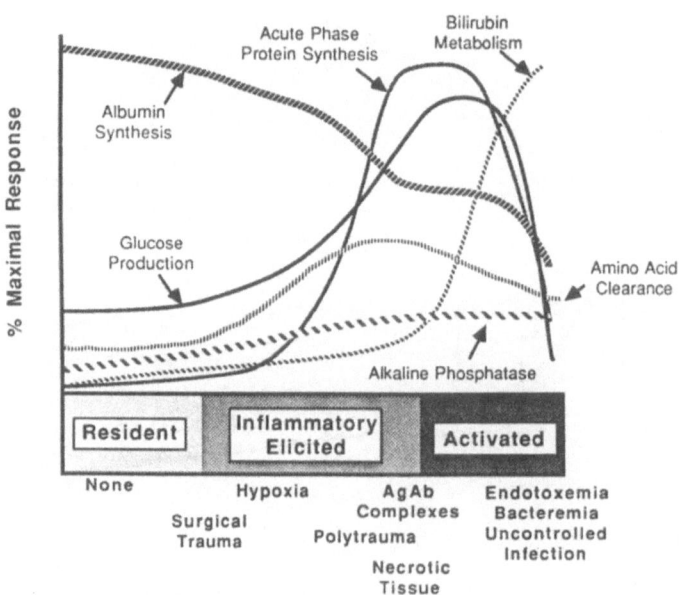

Figure 3. The Kupffer (macrophage) cells become increasingly activated by different kinds and combinations of stimuli. It is hypothesized that this repeated stimulation, perhaps from gut bactera and/or toxins, eventually overactivates the macrophage and a response with survival value (increased hepatic protein synthesis) becomes one detrimental to survival (decreased heptic protein synthesis.)

makes the liver particularly susceptible to Kupffer cell-mediated dysfunction during MSOF.[23,24,25,26,27,28]

To understand the mechanisms responsible for the alterations of liver function, an in vitro model system was used. The model uses enzymatically digested liver to retrieve hepatocytes and Kupffer cells. Protein synthesis is used as a functional parameter because it is easily measured in vitro using radioactive precursor incorporation, and because it is a highly sensitive parameter of integrated hepatocellular function. Activated macrophages/Kupffer cells induce a biphasic response in protein synthesis with an early induction followed by a depression in the synthesis of protein.[23,24]

Lipopolysaccharide (LPS) and killed E. coli (KEC) had little or no direct effect on liver cell function. When hepatocytes were cocultured with macrophages or Kupffer cells and LPS or KEC, hepatocyte protein synthesis decreased markedly (Fig. 2). These results were not reproduced if lymphocytes were used instead of macrophages. In spite of the marked alteration in cocultured hepatocyte function, hepatocyte viability or microscopic morphology did not change.[23,24]

Dexamethasone had no effect on the protein synthesis of hepatocytes cultured alone but did block the decrease in protein synthesis after addition of LPS or KEC. It did not affect the Kupffer cell-mediator increase in protein synthesis in the absence of these mediators.[26]

The effect of decreased oxygen concentrations on cocultured hepatocyte protein synthesis revealed that oxygen tensions as low as 18 torr had little effect on hepatocyte protein synthesis with or without LPS. With the coculture system, depressed hepatocyte protein synthesis was seen with hypoxia in the absence of LPS. When LPS was then added, an exaggerated depression of protein synthesis occurred. These results suggest that exposure to transient hypoxia may activate Kupffer cells or macrophages, and that LPS can further amplify these effects.

Addition of inhibitors of prostaglandin synthetase such as indomethacin had no effect. Similarly, the 5-lipooxygenase inhibitor nor-dihydroguariatic acid did not alter the coculture response. Addition of the end products of arachidonic acid metabolism, such as prostaglandin E_1, E_2, or leukotriene B_4, did not affect hepatocyte protein synthesis.[27]

Another possible group of mediators released by activated macrophages are the toxic oxygen species, including hydrogen peroxide, superoxide anion, and hydroxyl radical. Hepatic protein synthesis is decreased by the addition of hydrogen peroxide and exacerbated by depletion of intracellular oxygen radical scavengers such as reduced glutathione. However, catalase and superoxide dismutase, which degrade hydrogen peroxide or superoxide anion, respectively, did not prevent the Kupffer cell-mediated decrease in hepatocyte protein synthesis in coculture after addition of LPS or KEC.[24]

Activated macrophages or Kupffer cells release peptides that may be potential mediators of altered liver function. Among these are interleukin 1, plasminogen activator, and tumor necrosis factor/cachectin. Interleukin 1 can stimulate synthesis of several acute-phase proteins from hepatocytes, while decreasing the synthesis of albumin and transferrin.

In this model system, conditioned medium from crude nonparenchymal cells, purified Kupffer cells, or macrophages, triggered with LPS will result in decreased protein synthesis when added to hepatocytes cultured alone. The activity is heat-labile, with maximal inhibitory activity between 15 and 30 k daltons. Conditioned medium from coculture or macrophages/Kupffer cells generated in the absence of LPS has no effect. When the supernatant medium is assayed for lymphocyte-activating factor activity, it is found in high levels. However, addition of IL-1 to hepatocytes alone does not result in decreased protein synthesis as seen in coculture.[28] Interestingly, the time course of decreased protein synthesis is similar when conditioned medium is added, compared to that in coculture; that is, a four- to eight-hour lag occurs before decreased protein synthesis is detectable, with a maximal decrease seen after 16 to 24 hours. This observation suggests that the lag is not a result of mediator synthesis, but rather the time required for the hepatocyte to respond to the mediator.

A growing body of evidence suggests that leukocytic mediators are pathogenic in the mechanism of altered organ function during MSOF. In the liver, the normal anatomic relationship of Kupffer cells and hepatocytes suggests an obvious potential for modulating cell-cell interactions. The Kupffer cells within the liver normally function to clear the circulation of these potentially toxic materials; in the process, they may become activated, releasing numerous mediators (Fig. 3). In vitro, Kupffer cells can then mediate decreases in hepatocyte protein synthesis. They seem to do so via a heat-labile, soluble substance with a molecular weight of 15,000 to 30,000 (10). Whether this mediator(s) is responsible for the liver dysfunction seen clinically remains to be clarified.

CLINICAL CORRELATIONS

Given the current concepts of the organ failure syndrome, the clinical correlations fall into three general categories: prevention, mediator cell modulation, and hepatocyte support.

Prevention remains the best mode of treatment for the organ failure syndrome. Control of the source is the first principle. Whenever possible, the cause needs to be stopped so that continued stimulation of the mediator systems does not occur. The restoration and maintainance of oxygen transport, resuscitation, continues to be a mainstay of treatment. Recent data strongly suggests that in many of the clinical settings in which organ failure occurs, the invasive monitoring regimens that focus on flow and oxygen consumption have an associated reduction in organ failures and mortality.[9] Metabolic support has become an important part of the treatment regimen. It has been difficult to show that this modality has had a direct effect on the disease process itself or on mortality. Nevertheless, when applied in a setting of good surgery and critical care, there has been a reduction in organ failure mortality.[3] Presumably this cofactor effect is a result of patients not dying of malnutrition or by "running out of gas".

Mediator cell control is a new concept that is developing in the therapeutic armamentarium. The major aim would be to prevent excessive macrophage activation. Some activity seems to be essential for survival; excess activity may lead to dysmodulation of the target-cell. The modalities mentioned in prevention would also seem influential here. In addition, particular attention to gut support may be important in this regard. Animal work indicates that early gut nutrition may prevent the process from starting. Gut sterilization techniques may also play a role. The use of antibodies against core lipid components could also be influential. Once activated, little current insight exists into influencing mediator cell activity in the organ failure setting. Perhaps drug intervention or hormone manipulation will be of value. It has become increasingly apparent that a single "magic bullet" probably does not exits. Combinations of drugs that are precisely timed will probably. be necessary.

In this latter setting, the focus shifts to the hepatocyte. Its integrity and function need aggressive support. Currently, this takes the form of meticulous attention to oxygen transport, preventing substrate-limited metabolism, and modulation of the Kuppfer cell. Considerable research needs to yet be done before the details of these regulatory processes are completely understood and rational treatment regimens designed and tested.

REFERENCES

1. Carrico CJ, Meakins J, Marshall J, Fry D, Maier R: Multiple Organ Failure Syndrome. Arch Surg 121:196-208,(1986).
2. Cerra FB: Hypermetabolism, Organ Failure and Metabolic Support. Surgery, in press.
3. Madoff RD, Sharpe SM, Fath JJ, Simmons RL, Cerra FB: Prolonged Surgical Intensive Care. Arch Surg, 120:698-702,(1985).
4. Wiles JB, Cerra FB, Siegel JH Border JR: The Systemic Septic Response: Does the Organism Matter? Crit Care Med 8:55-60,(1980).
5. Deutschman C, Simmons RL, Cerra FB: The Systemic Response to Cytomegalovirus: Further Evidence for a Host Dependent Response. Arch Surg, in press.
6. Tilney N, Bailey G, Morgan A: Sequential System Failure After Rupture of Abdominal Aortic Aneurysms. Ann Surg 118:117-122, (1973).
7. Cuthbertson D, Tilstone W: Metabolism during the Post-Injury Period. Adv Clin Chem 12:1-55,(1977).
8. Cerra FB, Siegel JH, Colman B, Border J, McMenamy RH: Autocannibalism, A Failure of Exogenous Nutritional Support. Ann Surg 192:570-574,(1980).
9. Shoemaker W: Hemodynamic and Oxygen Transport Patterns in Septic Shock: Physiologic Mechanisms and Therapeutic Implications. (In) Prospectives in Sepsis and Septic Shock, Sibbald W, Sprung, C (ed), SCCM, Fullerton Ca,(1985).
10. Alberti KG, Batstone GF, Foster K, et al: Relative Role of Various Hormones in Mediating the Metabolic Response to Injury. JPEN 4:141,(1980).
11. Bessey P, Waters J, Soki T, Wilmore D: Combined Hormone Infusion Simulates the Metabolic Response to Injury. Ann Surg 200:264-281,(1984).
12. Clowes G, George B, Villes: Muscle Proteolysis Induced by a Circulating Peptide in Patients with Sepsis and Trauma. N Eng J Med 308:545-552,(1983).
13. Lefer A: Eicosanoids as Mediators of Ischemia and Shock. Fed Proc 14:275-280,(1985).
14. Giovannini I, Boldrini G, Castagnato M, Namio G, Pittiniti, Castiolini G: Respiratory Quotient and Patterns of Substrate Utilization in Human Sepsis and Trauma. JPEN 7:226-230,(1983).
15. Clowes GHA, O'Donnell TF, Blackburn G: Energy Metabolism and Proteolysis in Traumatized and Septic Man. Surg Clin North Amer 56:1169-1172,(1976).
16. Elwyn DH, Kinney JM, Juvanandum M: Influence of Increasing Carbohydrate Intake on Glucose Kinetics in Injured Patients. Ann Surg 190:117-127,(1979).
17. Long C, Kinney J, Geiger J: Nonsuppressability of Gluconeogenesis by in Septic Patients. Metabolism 25:193,(1976).
18. Birkhan R, Long C, Fitkin DL, Geiger J, Blakemore W: A Comparison of the Effects of Skeletal Trauma and Surgery on the Ketosis of Starvation in Man. J Trauma 21:513-519,(1981).
19. Cerra FB, Siegel JH, Border J, Coleman B: The Hepatic Failure of Sepsis: Cellular vs Substrate. Surgery 86:409-422,(1979).

20. Birkhan R, Long C, Fitkin D, Geiger J, Blakemore W: Effects of Major Skeletal Trauma on Whole Body Protein Turnover in Man Measured by 1, 14C Leucine. Surgery 88:294-297,(1980).

21. Long C, Birkhan R, Geiger J: Contribution of Skeletal Muscle Protein in Elevated Rates of Whole-Body Protein Catabolism in Trauma Patients. Am J Clin Nut 34:1087,(1981).

22. Nordenstrom J, Jeevanandam M, Elwyn, D, Kinney J: Increasing Glucose Intake During Total Parenteral Nutrition Increases Norepinephrine Excretion in Trauma and Sepsis. Clin Physiol 1:525-534, 1981.

23. West MA, Keller G, Hyland B, Cerra F, Simmons R: Hepatocyte Function in Sepsis: Kupffer Cells Mediate a Biphasic Protein Synthesis Response in Hepatocytes After Endotoxin and Killed E. coli. Surg 98:388-395, 1985.

24. Keller G, Barke R, Harty J, Humphrey E, Simmons R: Decreased Hepatic Glutathione Levels in Septic Shock; Predisposition of Hepatocytes to Oxidative Stress. Arch Surg 120:941-945, 1985.

25. Morris A, Henry W, Shearer J, Caldwell M: Macrophage Interaction With Skeletal Muscle: A Potential Role of Macrophages in Determining the Energy State of Healing Wounds. 25:746-751, 1985.

26. West MA, Keller FA, Hyland B, Cerra FB, Simmons RL: Kupffer Cell Modulation of Hepatocellular Function in Multiple Systems Organ Failure. J Leukocyte Biol 36(3):436, 1984.

27. Keller GA, West MA, Cerra FB, Simmons RL: Macrophage-mediated Modulation of Hepatic Function in Multiple-systems Failure. J Surg Res 39:555-563.

28. Keller GA, West MA, Wilkes A, Cerra FB, Simmons RL: Modulation of Hepatocyte Protein Synthesis by Endotoxin Activated Kupffer Cells II. Mediation by Soluble Transfer Factors. Ann Surg 201:429-435, 1985.

LEUKOTRIENES IN ENDOTOXIN SHOCK

W. Hagmann, C. Denzlinger, S. Rapp and D. Keppler
Biochemisches Institut, University of Freiburg.
D-7800 Freiburg, West Germany

INTRODUCTION

Actions of endotoxin (lipopolysaccharide, LPS) in vivo have long been suggested to be mediated by arachidonate metabolites (1,2). The important role of arachidonate-derived metabolites including leukotrienes, prostaglandins and thromboxane in experimental LPS shock has been deduced from the LPS resistance of essential fatty acid-deficient rats (3,4), from the altered arachidonate metabolism in LPS-tolerant rats (5), as well as from pharmacological evidence (1,2,6,7). Recent studies showed that the LPS-resistant C3H/HeJ mouse strain, whose macrophages are defective in prostaglandin (8,9) and leukotriene synthesis (10), can be made highly LPS-sensitive by transfer of pure macrophages from LPS-sensitive C3H/HeN mice (11). These data stress the target cell role of macrophages in the in vivo action of LPS (8,11).

Our early studies in a small-dose LPS shock model in mice (6,7) pointed to the leukotrienes, in particular to the cysteinyl leukotrienes LTC_4, LTD_4, and LTE_4, as key mediators in LPS shock. Cysteinyl leukotrienes are known to cause in minute amounts inflammatory and anaphylactic reactions (12), myocardial depression and shock-like reactions (13,14), and extensive plasma extravasation (15) in sensitive animal species. Several of the pathophysiologic symptoms mediated by cysteinyl leukotrienes are also observed in LPS shock (16). These studies are thus consistent with a mediator function of cysteinyl leukotrienes in lethal LPS action.

PREVENTION OF LETHAL ENDOTOXIN ACTION BY INHIBITORS OF LEUKOTRIENE
SYNTHESIS OR ACTION

Our pharmacological studies employed a small-dose LPS shock model in
which mice (NMRI and C57BL/6, 12 to 14 weeks of age) were sensitized
against LPS (17) (from S. abortus equi, 1-4 µg/kg, or from S. minnesota,
300 µg/kg, i.v. or i.p.) by treatment with the inhibitors of
hepatocellular RNA synthesis D-galactosamine (3.5 mmol/kg, i.p.) (18) or
α-amanitin (0.2 µmol/kg, i.p.) (19). All animals died from LPS-induced
shock within 5-12 hours under these conditions, but neither D-
galactosamine nor α-amanitin were lethal themselves within this time
period. Complete protection in this LPS shock model was achieved (6,7) by
dexamethasone, which indirectly inhibits arachidonate release via
lipocortin (20), by the dual 5-lipoxygenase and cyclooxygenase inhibitor
BW 755C (21), by diethylcarbamazine, an inhibitor of leukotriene
synthesis (22), and by FPL 55712, a selective receptor antagonist for
cysteinyl leukotrienes (23). Furthermore, LY 171883, a recently developed
LTD_4/LTE_4 receptor antagonist (24) protected 60% of NMRI mice against
lethal LPS action when injected in 3 doses (60 µmol/kg each, i.v.) at 0,
2, and 4 hours. In contrast, cyclooxygenase inhibitors like indomethacin
and indoprofen (25) could not or only partially prevent LPS lethality.

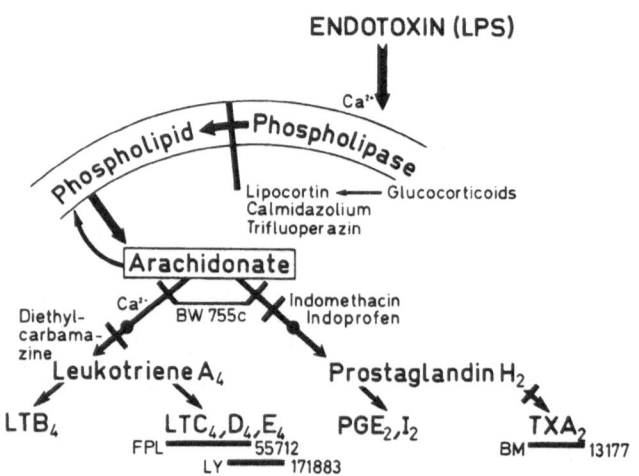

Figure 1. Pharmacological interference after endotoxin stimulation of
the arachidonate cascade. The sites of action of several drugs
interfering with eicosanoid synthesis or action (26) are indicated.
Application details were described earlier (6,7) or are given in the
text.

On account of this inhibitor profile and the efficacy of the receptor antagonists (fig.1) we argue for an important role of the cysteinyl leukotrienes in the lethal action of LPS under our experimental conditions.

GENERATION OF CYSTEINYL LEUKOTRIENES IN VIVO IN ENDOTOXIN SHOCK

According to the results from tracer studies with $[^3H]LTC_4$ (27,28) or $[^3H]LTD_4$ (7) we measured the systemic production of cysteinyl leukotrienes by analyzing bile, since these mediators are rapidly eliminated from the circulating blood with about 60% of the injected dose being recovered in bile within 30 min (fig.2). The major endogenous metabolite of cysteinyl leukotrienes that can be analyzed in rat bile by sequential high-performance liquid chromatography and radioimmunoassay

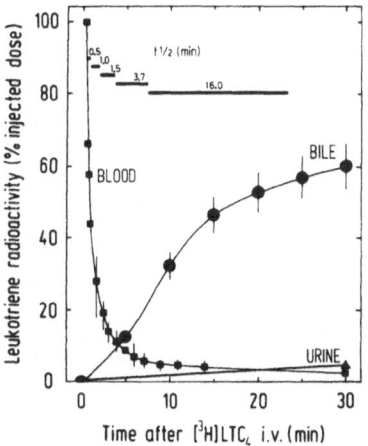

Figure 2. Elimination of $[^3H]LTC_4$ and its metabolites from blood into bile and urine of rats. Bile and small amounts of blood were collected continuously; urine was sampled from the urinary bladder within 30 min after i.v. $[^3H]LTC_4$ injection. Data give percent of injected tritium circulating in blood or accumulated in bile and urine. Horizontal bars indicate the half-life times (min) of ^3H-leukotrienes circulating in blood (27,28).

was found to be N-acetyl LTE$_4$ (fig.3) (29,30). Administration of LPS (from S. minnesota R595, 15 mg/kg, i.v.) into rats caused a rapid although transient generation of cysteinyl leukotrienes reflected by the increase in the N-acetyl LTE$_4$ production rate (7) and in its biliary concentration in anesthesized or unanesthesized rats (fig.4) (7,29,30). The measured quantities of cysteinyl leukotrienes generated in rats after LPS injection may well suffice to evoke known phenomena associated with endotoxin shock such as tissue edema and circulatory and respiratory dysfunction.

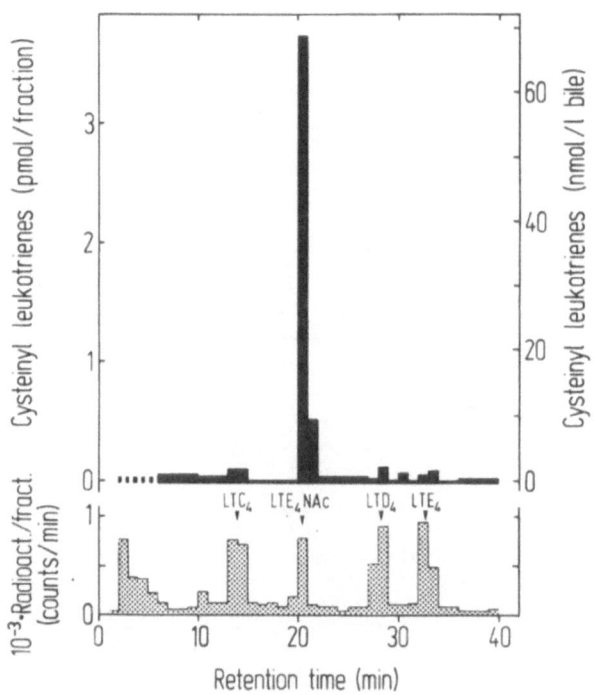

Figure 3.   Pattern of endogenous cysteinyl leukotrienes in bile of rats after endotoxin. Bile was sampled 0.5-1 hour after LPS (from S. minn. R595, 10 mg/kg, i.v.) and prepared for HPLC and RIA as described (27). Upper panel: HPLC separation giving concentrations of cysteinyl leukotrienes determined in the HPLC fractions by RIA and calculated according to their respective cross-reactivities (7,29). Lower panel: HPLC separation of internal ^{3}H-leukotriene standards.

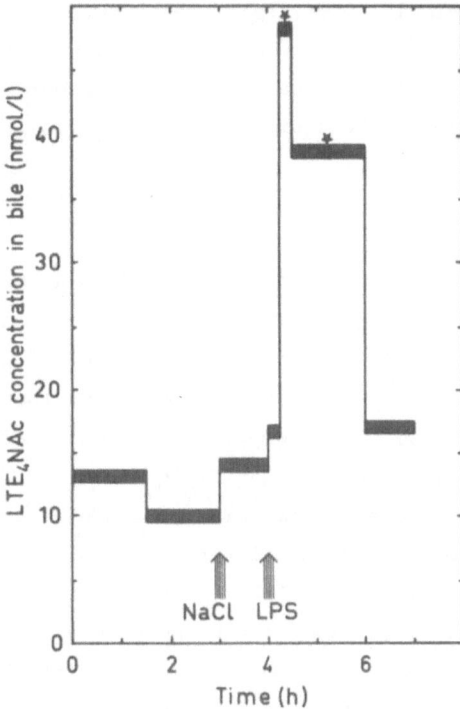

Figure 4. Endogenous cysteinyl leukotriene production induced by endotoxin in vivo. Rats were injected i.v. with saline (NaCl) as control and with LPS (from S. minn. R595, 15 mg/kg). N-Acetyl LTE_4 as indicator metabolite for leukotriene generation was measured in bile as described previously (27-30). Asterisks indicate significant elevation by $P < 0.02$.

INHIBITION OF THE HEPATOBILIARY ELIMINATION OF CYSTEINYL LEUKOTRIENES BY LPS

LPS and Lipid A, the endotoxic principle of the LPS molecule (31), severely impaired even at low doses the hepatobiliary clearance of LTC_4, LTD_4, LTE_4 and their metabolites (7). The biliary appearance of $[^3H]LTC_4$ metabolites decreased to less than 10% of control within 4 hours after LPS (5 mg/kg) (32). The concomitant LPS-induced reduction of bile flow and of tracer $[^{14}C]$taurocholate elimination, however, was much less pronounced than the severe inhibition of cysteinyl leukotriene elimination into bile (7). The latter effect resulted in elevated concentrations of cysteinyl leukotrienes in blood and liver (7,28). Lethal LPS action may therefore depend on both pathophysiological events, namely enhanced endogenous production and simultaneously impaired deactivation of cysteinyl leukotrienes, both possibly potentiating the effects of these shock-inducing mediators in vivo. It is of interest that bile duct ligation, which renders this predominant way of cysteinyl leukotriene elimination impossible (28), sensitizes rats against the lethal action of LPS (33).

MEDIATOR NETWORK IN ENDOTOXIN SHOCK: ROLE OF LEUKOTRIENES

A key role of leukotrienes, particularly of cysteinyl leukotrienes, in the lethal action of LPS is suggested from the pharmacological evidence summarized above. LPS, however, is known to elicit the release of a variety of mediators from target cells especially the mononuclear phagocytes. The LPS target cells release, in addition to cysteinyl leukotrienes (10), other lipid mediators such as several prostanoids (2,34) and platelet-activating factor (35) which in turn can induce leukotriene synthesis (36). The shock-inducing effects of platelet-activating factor, however, can be fully antagonized by inhibitors of leukotriene synthesis or by a cysteinyl leukotriene receptor antagonist (37). Additional mediators like tumor necrosis factor (38) and members of the interleukin-1 family (39,40) participate in the detrimental (38,41) or beneficial actions of LPS. Beneficial and compensatory processes after LPS injection comprise in particular the enhanced synthesis of acute-phase proteins in hepatocytes (39) resulting in a LPS-resistant condition. D-Galactosamine and α-amanitin, which suppress the hepatocellular synthesis of acute-phase proteins (39), raise the sensitivity of different animal species to LPS many-thousand-fold (17,19) possibly by this interference with acute-phase protein biosynthesis.

The mediator profile in LPS shock may vary in different species and under different experimental protocols. Additional studies in human sepsis are needed to define the role of cysteinyl leukotrienes in a way analogous to our studies in mice and rats.

REFERENCES

1.   J.R. Parratt in: Pathophysiology of endotoxin (ed.: L.B. Hinshaw). Elsevier Science Publishers (Amsterdam) p.203 (1985)
2.   J.T. Flynn in: Pathophysiology of endotoxin (ed.: L.B. Hinshaw). Elsevier Science Publishers (Amsterdam) p.237 (1985)

3. J.A. Cook, W.C. Wise, and Callihan, C.S.
    Circ Shock 6: 333 (1979)

4. J.A. Cook, P.V. Halushka, and Wise, W.C.
    Circ Shock 9: 605 (1982)

5. W.C. Wise, J.A. Cook, and Halushka, P.V.
    Advances Shock Res 10: 131 (1983)

6. W. Hagmann, and Keppler, D. Naturwiss, 69: 594 (1982)

7. W. Hagmann, C. Denzlinger, and Keppler, D.
    Circ Shock 14: 223 (1984)

8. L.M. Wahl, D.L. Rosenstreich, L.M. Glode, A.L. Sandberg, and
    Mergenhagen, S.E. *Infect Immun* 23: 8 (1979)

9. E.Th. Rietschel, U. Schade, M. Jensen, H.W. Wollenweber,
    O. Lüderitz, and Greisman, S.G. *Scand J Infect Dis*
    Suppl, 31:8 (1982)

10. T. Lüderitz, A. Roth, U. Schade, and Rietschel, E.Th.
    EOS - *J Immunol Immunopharmacol* Suppl, 6: 204 (1986)

11. M.A. Freudenberg, D. Keppler, and Galanos, C.
    Infect Immun 51: 891 (1986)

12. B. Samuelsson. *Science* 220: 568 (1983)

13. G. Feuerstein. *Prostaglandins* 27: 781 (1984)

14. A.M. Lefer. *Biochem Pharmacol* 35: 123 (1986)

15. X.Y. Hua, S.E. Dahlen, J.M. Lundberg, S. Hammarström, and
    Hedqvist, P. *Naunyn-Schmiedeberg's Arch Pharmacol* 330: 136 (1985)

16. L.B. Hinshaw in: Pathophysiology of shock, anoxia, and ischemia
    (eds.:R.A. Cowley, B.F. Trump). Williams & Wilkins
    (Baltimore) p.219 (1982)

17. C. Galaros, M.A. Freudenberg, and Reutter, W.
    Proc Natl Acad Sci USA 76: 5939 (1979)

18. D. Keppler, J. Pausch, and Decker, K. J Biol Chem, 249: 211 (1974)

19. H.W. Seyberth, H. Schmidt-Gayk, and Hackenthal, E.
    Toxicon 10: 491 (1972)

20. M. Di Rosa, R.J. Flower, F. Hirata, L. Parente,
    and Russo-Marie, F. *Prostaglandins* 28: 441 (1984)

21. R.W. Randall, K.E. Eakins, G.A. Higgs, J.A. Salmon, and
    Tateson, J.E. *Agent Action* 10: 553 (1980)

22. W.R. Mathews, and Murphy,R.C. Biochem Pharmacol, 31: 2129 (1982)

23. J. Augstein, J.B. Farmer, T.B. Lee, P. Sheard, and Tattersall,M.L.
    Nature, *New Biology* 245: 215 (1973)

24. J.H. Fleisch, L.E. Rinkema, K.D. Haisch, D. Swanson-Bean,
    T. Goodson, P.P.K. Ho, and Marshall, W.S.
    J Pharmacol Exp Therapeut 233: 148 (1985)

25. R. Ceserani, M. Colombo, and Mandelli, V.
    Prostaglandins Med 2: 337 (1979)

26. M.K. Bach in: The Leukotrienes, Chemistry and Biology (eds.: L.W. Chakrin, D.M. Bailey). Academic Press (Orlando) p.163 (1984)

27. C. Denzlinger, S. Rapp, W. Hagmann, and Keppler, D.
    Science 230: 330 (1985)

28. D. Keppler, W. Hagmann, S. Rapp, C. Denzlinger, and Koch, H.K.
    Hepatology 5: 883 (1985)

29. W. Hagmann, C. Denzlinger, and Keppler, D. FEBS Lett 180: 309 (1985)

30. W. Hagmann, C. Denzlinger, S. Rapp, G. Weckbecker, and Keppler, D.
    Prostaglandins 31: 239 (1986)

31. K. Tanamoto, U. Zähringer, G.R. McKenzie, C. Galanos, E.Th. Rietschel, O. Lüderitz, S. Kusumoto, and Shiba, T.
    Infect Immun 44: 421 (1984)

32. D. Keppler, W. Hagmann, and Rapp, S. Rev Infect Dis (1986) in press

33. E.N. Wardle, and Wright, N.A. Br Med J 4: 472 (1970)

34. A.M. Lefer in: Handbook of Shock and Trauma, Vol 1: Basic Science (eds.: B.M. Altura, A.M. Lefer, W. Schumer). Raven Press (New York) p. 355 (1983)

35. R. Roubin, and Benveniste, J.
    Comp Immun Microbiol Infect Dis 8: 109 (1985)

36. N.F. Voelkel, J.T. Reeves, P.M. Henson, and Murphy, R.C.
    Science 218: 286 (1982)

37. J.M. Young, P.J. Maloney, S.N. Jubb, and Clark, J.S.
    Prostaglandins 30: 545 (1985)

38. B. Beutler, I.W. Milsark, and Cerami, A.C.
    Science 229: 869 (1985)

39. A. Koj in: Pathophysiology of Plasma Protein Metabolism (ed.: G. Mariani). McMillan (London) p. 221 (1984)

40. A. Koj in: The acute-phase response to injury and infection. The roles of interleukin-1 and other mediators (eds.: A.H. Gordon, A. Koj). Elsevier Science Publishers (Amsterdam) p. 173 (1985)

41. B. Beutler, N. Krochin, I.W. Milsark, C. Luedke, and Cerami, A.
    Science 232: 977 (1986)

Supported by grants from the Deutsche Forschungsgemeinschaft through SFB 154, Freiburg

TIME-DEPENDENT APPEARANCE OF LTB₄ IN HUMAN BURN BLISTERS

J. A. Bauer and Th. Strasser

Chirurgische Klinik Innenstadt und Chirurgische Poliklinik der Universität München, Nußbaumstr. 20, D-8000 München 2
Medizinische Klinik Innenstadt der Universität München
Ziemssenstr. 1, D-8000 München 2

AIM OF STUDY

Burn blisters are containing LT. The level of content is difficult to be explained : immediate or delayed cold water treatment (CWT) and time interval (few minutes, hours or days) modify the measurable content of LT.

MATERIALS AND METHODS

The blister fluid from burns on the trunk and limbs of 37 outpatients who had just arrived in our hospital were analysed. 6 patients were excluded (the fluid were frozen only at -30°C). In 8 cases of "early comers" (time interval between accident and removal of blister fluid t<100 min.) fluid was removed before CWT ; in 10 cases of "early comers" fluid was removed after CWT. In all cases of "late comers" (t>100 min.) CWT was not applied. The fluid was removed from the blister with a needle under sterile conditions, frozen at -70°C and examined for LTB₄ (HPLC and RIA) and cysteinyl leukotrienes. None of the samples of fluid contained cells.

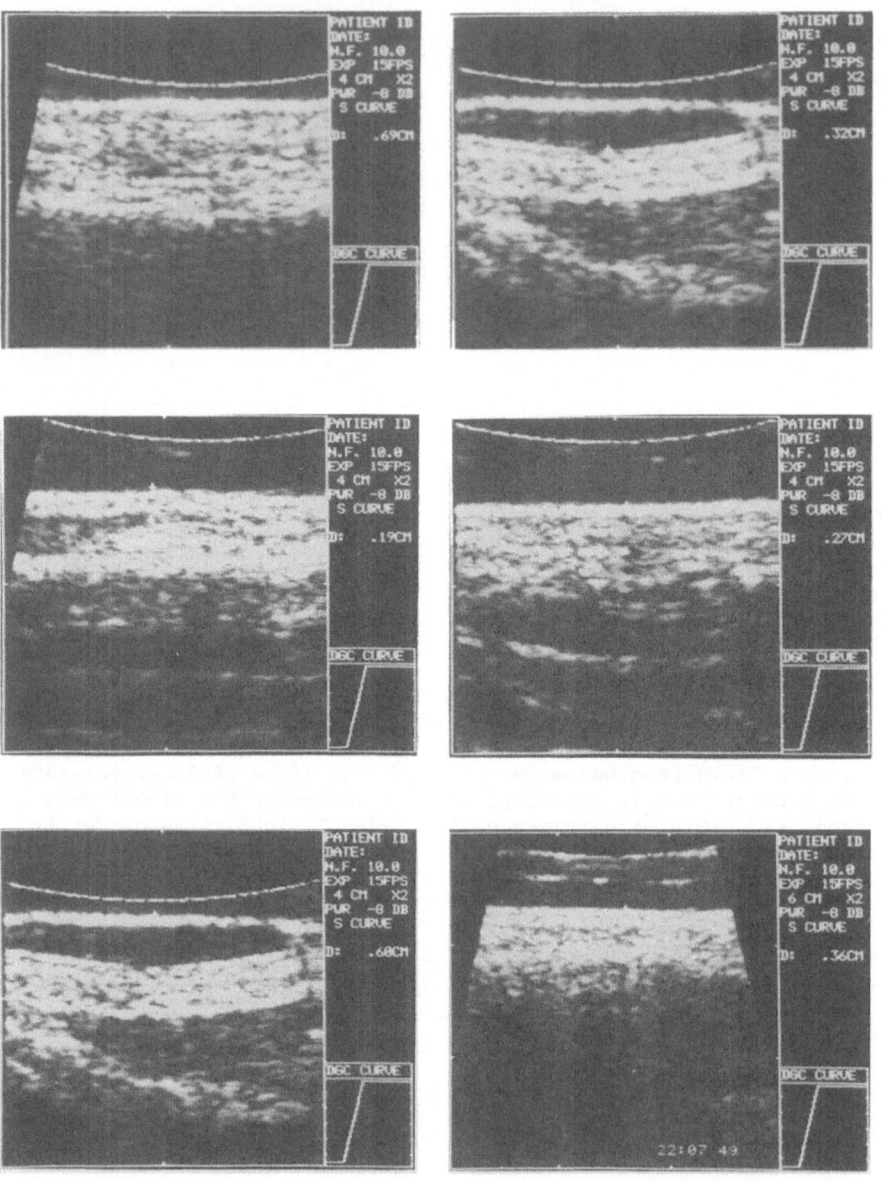

Fig. 1.   Sonograms of blister; the first sonogram shows a blister in
statu nascendi (1/2 h post burning)

RESULTS

Cysteinyl leukotriens were not measurable. Very high levels of LTB$_4$ were found in the 8 patients before CWT. 7 of these patients surely had no CWT immediately after buring, one patient applied a probably insufficient CWT at home. In the 10 patients becoming CWT before removal blister fluid low LTB$_4$ levels wer found : CWT caused a decrease of LTB$_4$. The "late comers" become no CWT, and low levels were measured too: This group becomes probably a physiological decrease of LTB$_4$ content. Our findings cannot be explained by a local leukotriene induction regulated by glutathion release from the liver (1), because only cysteinyl LT synthesis is connected with glutathion release.

INITIALS	BIRTHDATE	LOCALISATION OF BURNS	TIMEINTERVAL IN MIN	LTB 4 (PG/ML)
GROUP 1: EARLY COMERS, FLUID REMOVAL BEFORE CWT				
1) T. CH. ♀	17.12.19	BOTH HANDS AND FOREARMS	100	953
2) E.V. ♀	10.12.10	RIGHT HAND AND ARM	45	359
3) L.E. ♂	08.04.38	FACE, HANDS AND THORAX	10	398
4) H.G. ♂	07.08.52	LEFT LEG AND ABDOMEN	60	481
5) SCH.G. ♂	01.11.63	NECK AND BACK	35	1189
6) R.E. ♀	28.11.28	RIGHT HAND AN FOREARM	40	666
7) R.E. ♀	24.07.06	LEFT ARM AND SHOULDER	65	670
8) R.L. ♂	24.05.59	BREAST	40	888
GROUP 2: EARLY COMERS, FLUID REMOVAL AFTER CWT				
1) D.A. ♀	01.07.19	FACE AND HANDS	20	97
2) K.J. ♂	14.09.45	HANDS AND LEFT SHOULDER	15	292
3) K.D. ♂	09.02.59	RIGHT FEET AND LEG	50	240
4) B.R. ♂	16.05.61	LEFT ARM, LEFT LEG	30	221
5) D.J. ♂	10.02.43	LEFT LEG AND HIP	35	156
6) W.CHR. ♀	26.12.59	BOTH THIGHS	60	232
7) C.L. ♀	19.09.48	RIGHT LEG	30	378
8) H.W. ♂	20.08.65	LEFT HAND AND FOREARM	85	350
9) K.E. ♀	17.02.38	LEFT HAND AND LEFT LEG	70	283
10) J.D. ♀	01.02.45	LEFT BREAST, RIGHT HAND AND RIGHT ARM	40	345

Fig. 2. List of our patients with personal data, time interval, and LTB4 level in pg/ml.   (continued)

INITIALS	BIRTHDATE	LOCALISATION OF BURNS	TIMEINTERVAL IN MIN	LTB 4 (PG/ML)
GROUP 3: LATE COMERS, NO CWT				
1) L.E. ♀	16.07.23	BOTH HANDS AND FOREARMS	365	208
2) W.U. ♀	23.09.59	RIGHT LEG	155	214
3) B.R. ♂	16.03.49	BOTH LEGS	240	180
4) A.TH.♂	08.02.65	BACK AND SEAT	195	234
5) S.E. ♀	14.10.09	RIGHT HAND	120	318
6) SCH.J. ♀	03.05.02	RIGHT HAND AND FOREARM	1020	194
7) W.W. ♂	17.07.51	RIGHT HAND AND FOREARM	510	198
8) B.Z. ♀	16.05.56	LEFT HAND AND FOREARM	570	210
9) W.E. ♀	22.08.44	NECK AND SHOULDERS	915	247
10) G.W. ♀	07.05.38	RIGHT ARM	ONE DAY	296
11) F.G. ♀	02.12.63	ABDOMEN, VULVA AND RIGHT HAND	55	253
12) P.H. ♂	29.04.49	LEFT LEG	460	319
13) P.E. ♀	13.11.30	BOTH LEGS	60	253

Fig. 2 (Continued)

CONCLUSION

LTB_4 appearance in human burn blisters is time dependent. The interval is very short. Our findings correspond to the time dependent appearance (after 30 min.) of chemotactic activity in serum (2). Any therapeutic approach to the burn disease must be considered in further drug research also.

REFERENCES

1. Sies, H., and Cadenas, E.
   Oxidative stress: damage to intact cells and organs.
   Phil Trans. R. Soc. (London) B 311:617 (19-5).
2. Till, G.O., Beauchamp, Ch., Menapace, D.., Tourtellotte, W., Kundel, R., Johnson, K.J., and Ward, P.A.
   Oxygen radical dependent lung damage following thermal injury of rat skin.
   J. Traums, 23:269 (1983).

PLATELET-ACTIVATING FACTOR : AN INFLAMMATORY MEDIATOR INVOLVED IN THE PATHOPHYSIOLOGY OF SHOCK

M. Sanchez Crespo, S. Fernandez-Gallardo and E. Cano

Servicio de Nefrologia, Fundacion Jimenez Diaz

Avenida Reyes Catolicos 2, 28040-Madrid, Spain

INTRODUCTION

The shock state is an important clinical condition characterized by the occurrence of profound hemodynamic disturbances and the subsequent development of functional failure of many organs. The pathophysiology of shock is very complex, and many important aspects remain to be elucidated. Hemodynamic changes such as reduction of arterial pressure and cardiac output, a decrease in the count of platelets and leukocytes, an elevation of pulmonary arterial pressure and the stimulation of blood coagulation, kinin and complement system have been repeatedly observed in the process of shock evolution. At the present time, many chemical mediators and activation systems have been implicated in the pathogenesis of these disturbances, and this has surely opened the door to a vista of new therapeutic trends for this severe condition. Histamine, kinin, serotonin and endorphins have been found to be released during shock evolution, and most recently metabolites of arachidonic acid have been recognized to play a central role in the pathophysiology of shock in view of their potent actions on the cardiovascular system.

Platelet-activating factor (PAF, PAF-acter, AGEPC) was initially described by Benveniste, Henson and Cochrane[1] in their 1972 report as a substance released from IgE-stimulated basophils that could trigger the secretory response of rabbit platelets. However, the role of this mediator in pathology grew preatly after the elucidation of its chemical structure as 1-O-alkyl-2acetyl-sn-glycero-3-phosphocholine[2,3,4]. In addition to its well known action on platelets, PAF-acether has been found to be a potent stimulator of polymorphonuclear leukocytes[5,6,7] and also causes diverse etfects such as systemic hypotension,[4,8,9] extravasation of protein-rich plasma,[8,10] pulmonary hypertension,[11] bronchoconstriction[12] and glycogenolysis. This wide spectrum of actions for a unique molecule, in combination with its ability to promote the activation of mamany different types of cells and activation systems have suggested that PAF-acether has an important function in the "entangled web" of humoral and cellular effectors of inflammation and in vascular homeostasis. In this review we analyse the available data in order to ascertain whether PAF-acether could be recognized as a shock factor in both humans and experimental animals.

---

* On sabbatical leave, Departamento de Farmacognosia, Facultad de Farmacia, Universidad de Santiago de Compostela, Spain.

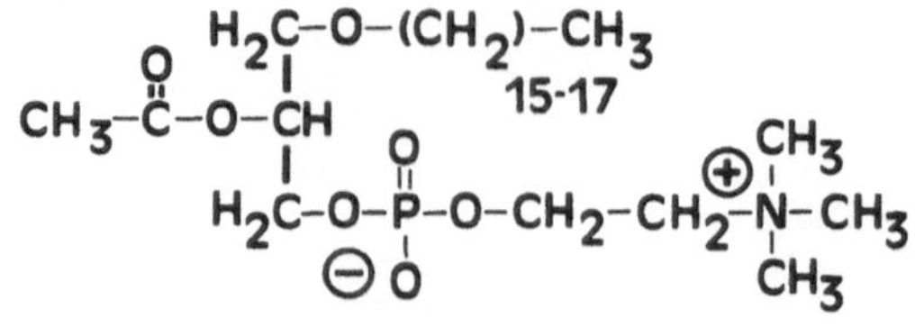

$$CH_3-\overset{\overset{\text{O}}{\|}}{C}-O-\underset{\underset{\overset{\text{O}}{\|}}{\overset{|}{C}H_2-O-\overset{\overset{\text{O}}{\ominus}}{\underset{}{P}}-O-CH_2-CH_2-\overset{\oplus}{\underset{\underset{CH_3}{|}}{\overset{\overset{CH_3}{|}}{N}}}-CH_3}}{\overset{|}{C}H}\overset{\overset{H_2C-O-(CH_2)-CH_3}{\quad 15\text{-}17}}{}$$

Fig. 1. Chemical structure of platelet-
        activating factor: 1-O-hexadecyl/
        octadecyl-2-acetyl-sn-glycero-3-
        phosphocholine.

EFFECT OF THE INFUSION OF PLATELET ACTIVATING FACTOR IN EXPERIMENTAL ANIMALS

The first criterium to be fulfilled by a substance for it to be considered as a possible mediator of a physiological event, must be the ability of this substance to reproduce that physiological event. This has been fulfilled with regard to PAF-acether and shock state by a number of different laboratories. Initial studies with the synthetic molecule were carried out in the rabbit by McManus et al.[14] These authors showed that the compound was extremely potent and caused the death of the animals at very low doses. In these rabbits the clinical picture mimicked systemic anaphylaxis and it was considered that the activation of rabbit platelets by PAF-acether played a central role in this process, since the mortality rate was significantly reduced by platelet depletion.[15]

Initial experiments from our group[8] were carried out in normal Sprague-Dawley rats using a racemic standard of paf-acether synthetized by Prof. Godfroid[16] (University of Paris). This seems important, because the available standard at that time did not only contained the active enantiomer, and therefore, it possessed a lower specific activity than those employed at the present time. Under the above mentioned conditions, the infusion of PAF-acether at doses as low as 0.05 ug induced a significant reduction in blood pressure. This effect was instantaneous and disappeared in a few minutes. With doses higher than 0.5 ug, there was no significant differences in the reduction of mean arterial pressure in mm Hg, but the effect was long-lasting. The effect of PAF-acether on peripheral vascular resistance and cardiac output was explored by using plastic microspheres labeled with ^{57}Co and ^{113}Sn. In these experiments, PAF-acether was found to cause a profound reduction of peripheral vascular resistance "pari passu" with a fall of mean arterial pressure. Cardiac output showed similar values before and after PAF-acether infusion. The ability of PAF-acether to induce the extravasation of protein-rich plasma was studied by

Table 1. Biological Actions of Paf-acether

Activation of platelets
Activation of polymorphonuclear leukocytes
Systemic hypotension
Extravasation of protein-rich plasma
Pulmonary hypertension
Bronchoconstriction
Glycogenolysis
Cardiodepression and arrythmias
Glomerular filtration rate reduction
Mesangial cell contraction
Gastric ulcerogenesis
Release of tissue-type plasminogen activator

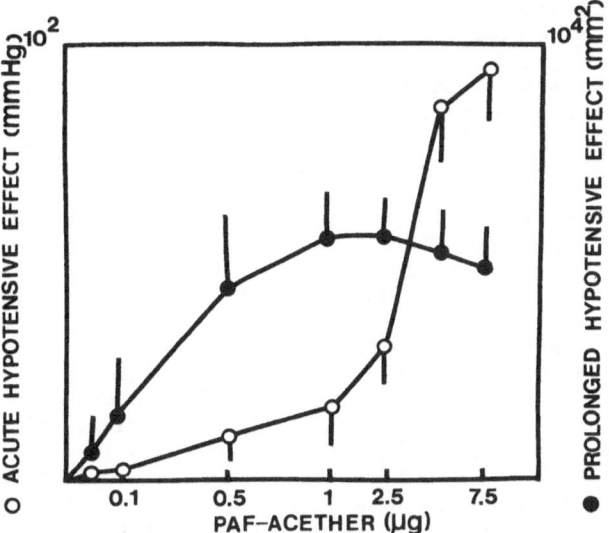

Fig. 2. Effect of the infusion of PAF-acether
on blood pressure. The acute hypotensive effect of PAF-acether represents
the fall in mm Hg of the mean arterial
pressure. The prolonged hypotensive effect represents the area of the right
triangle under the mean arterial pressure tracing. From Sanchez Crespo et
al.[8], with permission.

measuring the changes of the hematocrit value and the clearance from the
circulation of ^{125}I-HSA. The intravenous infusion of amounts of PAF-acether
higher than 1 ug, induced a marked increase of the vascular permeability,
as judged from a reduction of up to 70 % of the starting plasma level of
^{125}I-HSA and an increase of the hematocrit value of up to 20 %. This extravasation was dose-dependent and paralleled hematocrit changes. A good correlation was found between the prolonged hypotensive effect of PAF-acether
(measured by the area of the right triangles under the mean arterial pressure tracing) and the variation of ^{125}I-HSA plasma levels. The effect of
PAF-acether on platelets and leukocytes was studied by performing peripheral blood cell counts. In no case a significant reduction of the count was
observed, and this suggested to us that both platelets and neutrophils are
not important targets of PAF-acether actions in this particular model. These initial findings have been confirmed and extended by other authors in
many other systems. Bessin et al.[9] have shown that thromboxane A_2 is generated in dogs following the infusion of PAF-acether, and the same authors
have demonstrated the occurrence of pulmonary hypertension, impairment of
myocardial metabolism and severe metabolic acidosis under these experimental conditions. Halonen et al.[17] have also showed that the hemodynamic actions of PAF-acether are platelet-independent even in those animal species
whose platelets are fully responsive to paf-acether. The cardiac actions
of PAF-acether consist of early alterations of the electrocardiogram, characterized by variations in heart rate and rhythm associated with a consistent depression of the ST segment, a sign of subendocardial ischemia, and
a negative inotropic effect which is observed in the isolated guinea-pig
heart with very low concentrations of paf-acether[18].

Studies on isolated lungs have provided controversial data with regard to the mechanisms involved in the action of PAF-acether. On the one hand, Hamasaki el al.[19] have observed an elevation in the airway and vascular pressures in guinea pig lungs perfused with platelet-free Krebs-Ringer solution. These effects were associated with stimulated synthesis of thromboxane A_2, but the mechanism of their production was not determined. On the other hand, Heffner et al.[20] have found that the potential of PAF-acether to produce pulmonary hypertension and edema in isolated rabbit lungs was dependent on the ability of rabbit platelets to generate thromboxane A_2. To make this question even more complex, Vargaftig and coworkers[12] have provided evidence suggesting that PAF-acether possibly releases from platelets a bronchoconstrictor component, distinguishable from thromboxane A_2 that can be depleted by reserpine administration to the animals.

The effect of paf-acether on kidney circulation has been studied by several groups of authors[21,22,23,24]. Most results agree as to the vasoconstrictor action of this compound, which results in dose-dependent reductions in renal blood flow, glomerular filtration rate, urine volume and urinary sodium excretion. Once more, the possible involvement of arachidonate metabolites in the mediation or in the modulation of the action of PAF-acether has been suggested and most results agree as to the appearance of these changes in the absence of both systemic hypotension and significant platelet-count reduction. In addition, PAF-acether has been found to be an active agonist at the glomerular level, since it can stimulate the contraction of mesangial cells influencing their function[25] and increasing the permeability of the glomerular capillary wall[26].

A recent report by Rosam et al.[27] has shown potent ulcerogenic actions of PAF-acether on the gastric mucosa. These actions do not seem to be attributable solely to the hypotensive effect of PAF-acether and, apparently, are not mediated via effects on platelets or cycloxygenase products. This finding is a new data linking PAF-acether and the pathogenesis of shock, since there is a close association between septic shock and acute gastric damage.

THE PLACE OF PAF-ACETHER IN THE "ENTANGLED WEB" OF CELLULAR AND HUMORAL EFFECTORS OF SHOCK

The study of PAF-acether in the last years has shown that this mediator displays almost universal agonistic properties. This is probably due to the ability of PAF-acether to stimulate calcium entry and the degradation of inositol phospholipids[28]. This wide spectrum of pharmacological actions makes it likely the involvement of PAF-acether in a clinical condition in which these pathophysiological changes are observed. The shock state is probably the only clinical condition in which all the pathophysiological changes induced by PAF-acether infusion can be simultaneously observed.

Another important aspect is the extremely elevated capacity of PAF acether to promote the generation and release of secondary mediators. The relationship between PAF-acether and eicosanoids has been observed in platelets[29], neutrophils[30] and endothelial cells[31]. In all these different types of cells, PAF-acether has been found to initiate the generation of arachidonate metabolites and this has suggested that all the pharmacological actions of PAF-acether are the consequence of the generation of these secondary mediators. Recent studies with pharmacological antagonists of eicosanoids and inhibitors of the enzymes involved in their generation have provided controversial results, but the present opinion is that the generation of these secondary mediators is an additional property of PAF-acether and not its unique way of inducing biological responses. As an example of this assertion, it can be mentioned that inhibition of platelet-cyclooxygenase by aspirin only modifies the response to low concentrations of PAF-acether, whereas the response to high concentrations is unaffected[32]. This probably suggests that the generation of thromboxane A_2 by PAF-acether stimulated platelets could potentiate the response to suboptimum concentrations of the agonist, but it is not required when the agonist is fully acting.

Doebber et al.[33] have recently published that PAF-acether provokes the intravascular secretion of lysosomal hydrolases, and Emeis and Kluft[34] have reported that PAF-acether stimulates fibrinolysis in rats by promoting the generation of tissue-type plasminogen activator from endothelial cells. Hartung and Hadding[35] have recently published that PAF-acether promotes the synthesis of complement factors by macrophages "in vitro". The ability of PAF-acether to initiate the production of oxygen toxic metabolites by neutrophils and other cell types[36] has also been reported.

The question of whether PAF-acether is an important factor in the recruitment of all these mediators in shock could be revolved only by the use of specific antagonists of the PAF-acether receptor.

GENERATION OF PAF-ACETHER DURING THE SHOCK STATE

The next hypothesis to be tested was to ascertain whether PAF-acether could be generated when experimental animals are challenged with some of the agonists which can initiate the shock state. The first description of the presence of PAF-acether in blood from rabbits with systemic anaphylaxis was published by Pinckard et al.[37] in 1980. These authors developed a procedure which allowed the inactivation of serum acetylhydrolase and permitted the isolation and assay of PAF-acether.

Soluble aggregates of immunoglobulin G can initiate a number of responses in experimental animals which mimic the pathophysiology of the shock state, and they are a potent stimulus for the generation of PAF-acether "in vitro"[38,39,40]. This suggested to us that PAF-acether could be a likely endogenous mediator of the physiological responses initiated by soluble aggregates of immunoglobulin G[41]. A set of experiments was carried out in normal mice in which the variations of the intravascular volume were measured by using ^{51}Cr-labeled homologous red cells. Simultaneously, the liver and the spleen of these animals were removed and placed in Potter-glass homogenizers containing methanol and kept on ice. These organs were mechanically disrupted and the methanolic extract centrifuged at 1000 x g. Phase formation was achieved by adding chloroform and water, and the chloroform phase collected, evaporated to dryness and treated by thin layer chromatography on silica gel plates. The lipid fraction migrating as PAF-acether was eluted and tested on (^3H) serotonin-labeled rabbit platelets. Under these conditions, we were able to show that a lipid fraction analogous to PAF-acether could be obtained from the liver and the spleen of the animals which had received an intravenous challenge with soluble aggregates of immunoglobulin G. Furthermore, the apperance of this substance preceded the occurence of blood volume depletion due to extravasation of protein-rich plasma. Previous treatment of the animals with quinacrine or depletion of mononuclear phagocytes by total irradiation induced both the disappearance of the extravasation induced by the aggregates and the abrogation/diminution of the PAF-acether generated by the spleen and the liver. Buxton et al.[12] have used a model of "in situ" perfused rat liver challenged with soluble aggregates of immunoglobulin G. Under these conditions, a compound analogous to PAF-acether could be found in the effluent, and, interestingly, this compound was able to initiate glycogenolysis in the hepatic cell.

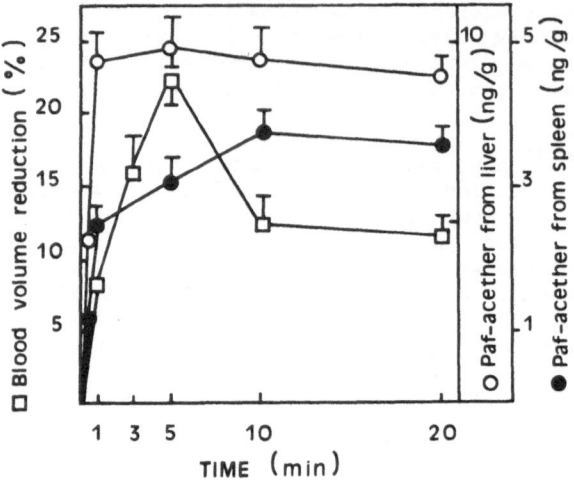

Fig. 3. Generation of PAF-acether from the mononuclear phagocytic system in mice
challenged with soluble aggregates of
IgG. Liver and spleen were removed after intravenous infusion of 2mg of aggregates. Another group of animals was
processed in parallel to study blood
volume reduction. From Iñarrea et al.,
ref. 41. With permission.

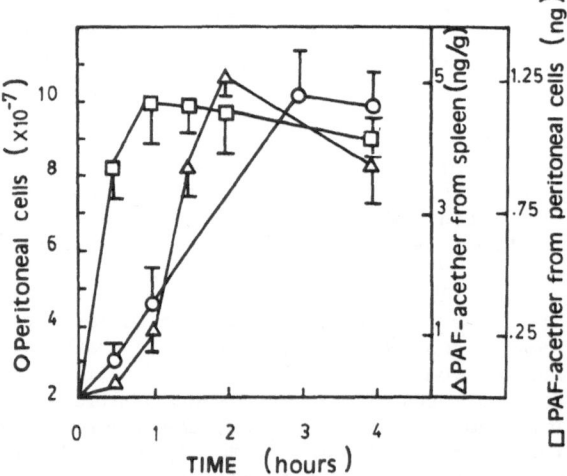

Fig. 4. Time course of PAF-acether formation
and peritoneal infiltration of cells.
Rats were injected with $2x10^8$ CFU of
E. coli and at the times indicated
peritoneal cells were counted and
paf-acether measured in the peritoneal exudate and in the spleen. From
Iñarrea et al.[42], with permission.

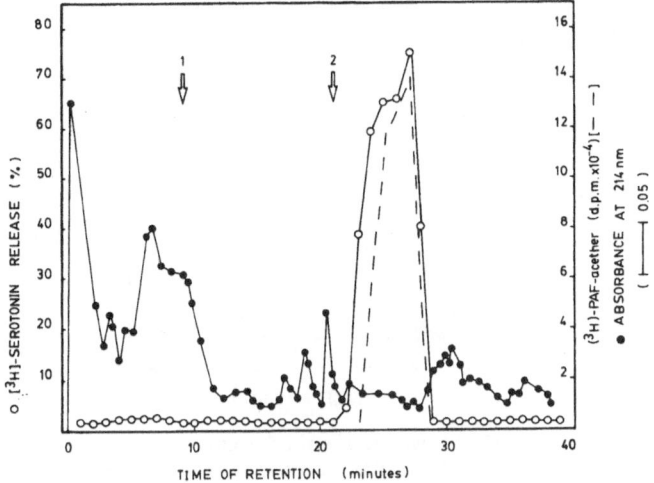

Fig. 5. HPLC separation of fractions of lipid ex-
        tracts from peritoneal cells which migra-
        ted in thin layer chromatography as paf-
        acether. UV absorbance at 214 nm was mo-
        nitored and the ability to release (^{3}H)
        serotonin from rabbit platelets assayed
        in each fraction. Numbered arrows indi-
        cate the position of the standards emplo-
        yed: 1, phosphatidylethanolamine; 2, phos-
        phatidylcholine. The dotted line repre-
        sents the chromatographic behaviour of
        a (^{3}H) PAF-acether standard. From Iñarrea
        et al.[42], with permission.

Gram-negative sepsis is a major cause of severe circulatory shock, and
this condition may be mimicked in laboratory animals by the infusion of either living bacteria or bacterial endotoxin. Gram-negative sepsis was induced in rats by intraperitoneal injection of E. coli[42]. Under these circumstances septicemia occurred and mortality reached a 50 % rate when rats
were injected with 2 x 10^8 CFU (colony forming units). The animals inocu-
lated with the amounts of bacteria which induced mortality showed a time-
and dose-dependent increase of vascular permeability, as judged from the
presence of abundant peritoneal exudate and by the depletion of the circulating volume. At the same time, PAF-acether was found in the peritoneal
exudate and in the spleen of these animals. The generation of PAF-acether
at both levels preceded the appearance of blood volume depletion, and, apparently, the generation and release of PAF-acether seemed to be the trigger for the development of extravasation and vascular volume depletion.
Interestingly, in a recent report by Doebber et al.[43], PAF-acether was isolated from the blood of experimental rats which developed systemic hypotension in response to the intravenous injection of bacterial endotoxin.

ANTAGONISM OF PAF-ACETHER AND THE SHOCK STATE

The recent development of specific antagonists of the PAF-acether receptor has provided the opportunity to fulfil a primary criterium as to

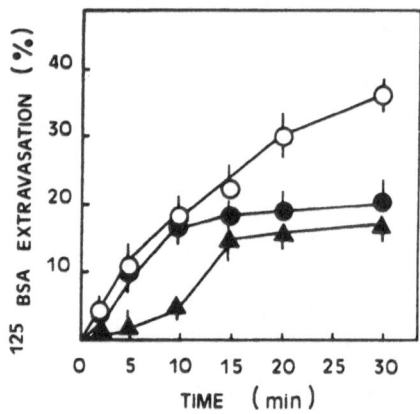

Fig. 6. The time course of extravasation
of ^{125}I-BSA in response to soluble aggregates of IgG and its
inhibition by BN 52021. Rats were pretreated with vehicle (○) or
5 mg/kg (▲) or 1 mg/kg (●) of BN
52021. At time 0 the challenge
was performed with 40 mg/kg of
aggregated IgG. From Sanchez Crespo et al.[46], with permission.

the involvement of this potent mediator in the pathophysiology of the shock
state. These studies have been carried out in different shock types and
also with different compounds. Since PAF-acether has been initially associated with anaphylaxis, some experiments have been performed in sensitized animals challenged with the specific antigen[44,45]. Under these conditions, two chemically unrelated PAF-acether antagonists, BN 52021 and CV
3988, have been found to specifically antagonize both systemic hypotension[45] and bronchoconstriction in response to the antigen[44].

Since the possible role of PAF-acether in the pathophysiology of the
circulatory shock has been enlarged to include other etiological factors,
the effect of a previous treatment with PAF-acether antagonists has been
tested in normal rats challenged with an intravenous bolus of soluble aggregates of immunoglobulin G[46]. In these experiments, the animals received an intravenous infusion of 1-5 mg/kg of the specific PAF-acether antagonist BN 52021 (IHB-IPSEN, Le Plessis Robinson, France), a compound
extracted from the Ginkgo biloba tree, and 10 min later an intravenous
challenge with soluble aggregates of IgG. Under these conditions, a significant reduction of the extravasation and of the hypotensive response
to the challenge was observed. Some experiments were also performed injec-
ting the aggregates first, and the PAF-acether antagonist some minutes after the completion of the hypotensive response. In this case, the infusion
of the drug reversed the hypotension.

In another study by Terashita and coworkers[47], the specific antagonist of the PAF-acether receptor CV 3988 was found to inhibit the hypotensive response to bacterial endotoxin. Again, when CV 3988 was injected 7-10

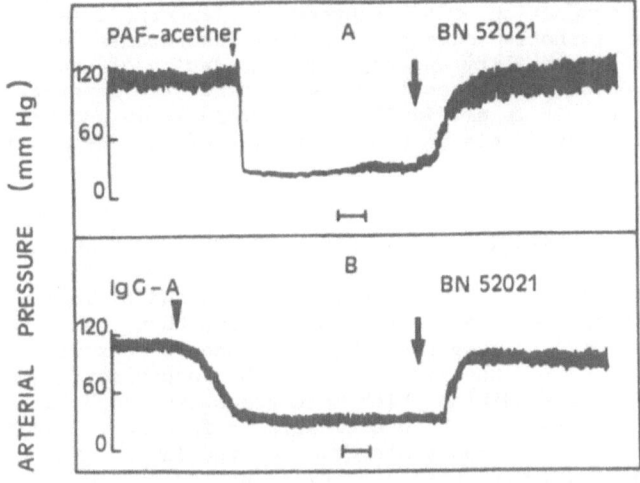

Fig. 7. Reversal of the hypotension induced with
paf-acether and IgG aggregates by intravenous injection of BN 52021. (A) A normal rat received a bolus dose of 2.5 ug
per kg of paf-acether, and at the time
indicated 5 mg per kg of BN 52021. (B)
Another rat received a bolus dose of
40 mg per kg of aggregated IgG and at
the time indicated 5 mg per kg of BN
52021. The bar indicates 1 min. From
Sanchez Crespo et al[46], with permission.

minutes after endotoxin, the hypotension was rapidly reversed. In a
recent report, Doebber et al.[43] have been able to inhibit the hypotensive
response to bacterial endotoxin with kadsurenone, another specific PAF-acether receptor antagonist. In this study the reversal of the hypotensive
response was also observed when kadsurenone infusion followed the intravenous administration of endotoxin. Similar findings have been reported
by Adnot et al.[48] using endotoxin from S. tiphymurium and BN 52021. Etienne et al.[49] have also shown that BN 52021 diminishes the mortality rate
in response to the same endotoxin. In this study BN 52021 was also found
to reduce the elevation of rectal temperature. Preliminary data from Doebber et al. have shown that L 652,731, a compound with PAF-acether antagonistic properties, blunts the responses to intravenously infused immuno complexes. Feuerstein et al.[50] have reported that thyrotropin-releasing hormone (TRH), a tripeptide with potent antihypotensive activity in experimental shock, is a potent PAF-acether antagonist in the unanesthetized guinea
pig.

In a recent study from this laboratory[51], the presence of high levels
of PAF-acether has been shown in samples of blood from cirrhotic rats.
These animals were made cirrhotic by the combined administration of oral
phenobarbital and inhaled carbon tetrachloride and they showed a hyperdynamic state with enhanced cardiac output, decreased mean arterial pressure
and reduced peripheral vascular resistance. When these animals were intravenously treated with BN 52021, a decrease in cardiac output with increase
in peripheral vascular resistance was also observed. In contrast, mean arterial pressure increased slightly, but not significantly. The reason for

an increased production of PAF-acether in cirrhotic animals is not yet clear. Since cirrhotic animals often have endotoxemia because of alterations of gut permeability and hepatic degradation of endotoxin, and since bacterial endotoxin has been found to initiate the generation of PAF-acether in rats, it is difficult to ascertain if in addition to endotoxin other stimuli play a role in the generation of PAF-acether in cirrhotic animals.

THE ROLE OF PAF-ACETHER IN HUMAN DISEASES RELATED TO SHOCK

The measurement of PAF-acether in fluids and biological samples from patients with the diverse clinical conditions that could lead to the shock state has not been an easy task. This had been succesfully attempted by Pinckard et al.[38] for the first time in the anaphylactic shock of the rabbit. The clinical condition which most resembles anaphylaxis in which PAF-acether has been measured , is urticaria "a frigore". Grandel et al.[52] have been able to show a rapid elevation of the levels of PAF-acether in blood from patients with this condition, preceding the appearance of edema after immersion of the hand in a mixture of ice and water. The method employed by these authors was similar to that described in this laboratory[53] to extract and assay PAF-acether in human blood, the onli significant difference being a rapid centrifugation step to separate plasma from blood cells. Grandel and coworkers[54] have also been able to measure high amounts of PAF-acether in sputum from patients suffering from clinical conditions related to asthma. In another study from this laboratory[55], we found high levels of PAF-acether in blood and ascitic fluid from patients with cirrhosis of the liver. The highest levels were found in decompensated patients. Compensated cirrhotics showed lower blood values, but higher than controls. PAF-acether levels in ascitic fluid were similar to those of blood. Acetylhydrolase activity, the main catabolic enzyme for PAF-acether showed similar values in both patients and controls which probably indicates that a reduced catabolism is not the reason why PAF-acether levels are increased in cirrhotic patients. These findings can be relevant to shock pathophysiology, since pathophysiological disturbances similar to those existing in shock are usually observed in cirrhotic patients and this analogy is even more marked in the patients with the hepato-renal syndrome. In fact, cirrhotic patients show peripheral vasodilatation and hypotension, renal perfusion presents considerable lability and changes in vascular permeability seem to be present in cirrhosis.

Table 2. Effect of BN 52021 on the hemodynamics of cirrhotic rats

	CONTROL		CIRRHOTIC	
	BASAL	BN 52021	BASAL	BN 52021
CO ml/min.100g	41.1+8.8	47.9+12.4	53.4+5.4*	35.9+3.6**
PVR mmHg.min.100g/ml	1.38+0.26	1.22+0.18	0.86+0.11*	1.34+0.17**
MAP	128+7.9	132+7.2	97.5+5.6*	100+5.4*

*Indicates a p<0.05 as compared to control rats.**Indicates p<0.05 as compared to the basal period. Figures of CO and PVR have been corrected for the weight of the animals and are expressed as the values corresponding to 100 g body weight. CO indicates cardiac output, PVR indicates peripheral vascular resistance, MAP indicates mean arterial pressure.From Villamediana et al.[51], with permission.

In a preliminary report of a patient with a clinical condition known as Atkinson syndrome[56], and characterized by the sudden occurrence of hypotension, edema due to the extravasation of protein-rich plasma, monoclonal gammapathy and hypocomplementemia, it has been observed that the use of a Ginkgo biloba extract enriched in BN 52021 has significant beneficial effect on all the clinical manifestations. This has suggested the involvement of PAF-acether in the pathophysiology of this condition, but no other data apart from the effect of PAF-acether antagonist substantiate the involvemen t of this mediator in this interesting clinical situation.

CONCLUDING REMARKS

PAF-acether, a recently described inflammatory mediator seems to be a shock factor in experimental animals. Its antagonization in experimental animals has shown promising reults in acute models. With regard to more chronic models and human diseases, more studies are needed.

ACKNOWLEDGEMENTS

S. Fernandez-Gallardo is a fellow from the "Conchita Rabago Foundation". This study has been supported by grants from "CAICYT", "Fondo de Investigaciones Sanitarias" and "Fundacion Alvarez de Toledo".

REFERENCES

1. J. Benveniste, P. M. Henson and C. G. Cochrane, Leukocyte-dependent histamine release from rabbit platelets. The role of IgE, basophils and a platelet-activating factor, J. Exp. Med. 136:1356 (1972).
2. C. A. Demopoulos, R. N. Pinckard and D. J. Hanahan, Platelet-activating factor. Evidence for 1-alkyl-2-acetyl-sn-glyceryl-3-phosphorylcholine as the active component (a new class of lipid chemical mediators), J. Biol. Chem. 254:9355 (1979).
3. J. Benveniste, M. Tence, J. P. LeCouedic, P. Varenne, C. Bidault and J. Polonski, Semisynthese et structure proposee du facteur activant les plaquettes (PAF): PAF-acether un alkyl ether analogue de la lysophosphatidylcholine, C.R. Acad. (Paris) 289:1037 (1979).
4. M. L. Blank, F. Snyder, L. W. Byers, B. Brooks and E. E. Muirhead, Antihypertensive activity of an alkyl ether analog of phosphatidylcholine, Biochem. Biophys. Res. Commun. 90:1194 (1979).
5. E. J. Goetzl, C. K. Derian, A. I. Tauber and F. H. Valone, Novel effects of 1-O-hexadecyl-2-acetyl-sn-glycero-3-phosphorylcholine mediators in human leukocyte function: delineation of the specific actions of the acyl substituents, Biochem. Biophys. Res. Commun. 94:881 (1980).
6. J. O. Shaw, R. N. Pinckard, K. S. Ferrigni, L. M. McManus and D. J. Hanahan, Activation of human neutrophils with 1-O-hexadecyl/octadecyl-2-acetyl-sn-glyceryl-3-phosphorylcholine (platelet-acetivating factor), J. Immunol. 124:1482 (1981).
7. J. T. O'Flaherty, R. L. Wykle, C. H. Miller, J. C. Lewis, M. Waite, D. A. Bass, C. E. McCall and L. R. DeChatelet, 1-O-alkyl-2-acety-sn-glyceryl-3-phosphorylcholine: A novel class of neutrophil stimulants, Am. J. Pathol. 103:70 (1981).
8. M. Sanchez Crespo, F. Alonso, P. Iñarrea, V. Alvarez and J. Egido, Vascular actions of synthetic paf-acether (A synthetic platelet-activating factor) in the rat: Evidence for a platelet-independent mechanism, Immunopharmacology 4:173 (1982).

9. P. Bessin, J. Bonnet, D. Apfell, C. Soulard, L. Desgroux, J. Pelas and J. Benveniste, Acute circulatory collapse caused by platelet-activating factor (paf-acether) in dogs, Eur. J. Pharmacol. 86:406 (1983).
10. D. M. Humphrey, D. J. Hanahan and R. N. Pinckard, Induction of leukocytic infiltrates in rabbit skin by acetyl glyceryl ether phosphorylcholine, Lab. Invest. 47:227 (1982).
11. N. F. Voelkel, S. Worthen, J. T. Reeves, P. M. Henson and R. C. Murphy, Nonimmunological production of leukotrienes induced by platelet-activating factor, Science. 218:286 (1982).
12. B. B. Vargaftig, J. Lefort, M. Chignard and J. Benveniste, PLatelet-activating factor induces a platelet-dependent bronchoconstriction unrelated to the formation of prostaglandins derivatives, Eur. J. Pharmacol. 65:185 (1980).
13. D. B. Buxton, D. J. Hanahan and M. L. Olson, Stimulation of glycogenolysis and platelet-activating factor production by heat-aggregated immunoglobulin G in the perfused rat liver, J. Biol. Chem. 259:13758 (1984).
14. L. M. McManus, D. J. Hanahan, C. A. Demopoulos and R. N. Pinckard, Pathobiology of the intravenous infusion of acetyl glyceryl ether phosphorylcholine (AGEPC) a synthetic platelet-activating factor (PAF), in the rabbit, J. Immunol. 124:2919 (1980).
15. R. N. Pinckard, M. Halonen, D. J. Palmer, C. Butler, J. O. Shaw and P. M. Henson, Intravasacular aggregation and pulmonary sequestration of platelets during IgE-induced anaphylaxis in the rabbit. Abrogation of lethal anaphylactic shock by platelet depletion, J. Immunol. 119:2185 (1977).
16. J. J. Godfroid, F. Heymans, E. Michel, C. Redeuilh, E. Steiner and J. Benveniste, Platelet-activating factor (PAF-acether): total synthesis of 1-O-octadecyl-2-O-acetyl-sn-glycero-3-phosphorylcholine, FEBS Lett. 116:161 (1980).
17. M. Halonen, J. D. Palmer, I. C. Lohman, L. M. McManus and R. N. Pinckard, Respiratory and circulatory alterations induced by acetyl glyceryl ether phosphorylcholine (AGEPC), a mediator of IgE anaphylaxis in the rabbit, Am. Rev. Respir. Dis. 122:915 (1980).
18. R. Levi, J. A. Burke, Z. Guo, Y. Hattori, C. M. Hoppens, L. M. McManus, D.J. Hanahan and R. N. Pinckard, Acetyl glyceryl ether phosphorylcholine (AGEPC): a putative mediator of cardiac anaphylaxis in the guinea-pig, Cir. Res. 54:117 (1984).
19. Y. Hamasaki, M. Mojarad, T. Saga, H-H. Tai and S. I. Said, PLatelet-activating factor raises airway and vascular pressures and induces edema in lungs perfused with platelet-free solution, Am. Rev. Respir. Dis. 129:742 (1984).
20. J. E. Heffner, S. A. Shoemaker, E. M. Canham, M. Patel, I. F. McMurtry, H. G. Morris and J. E. Repine, Acetyl glyceryl ether phosphorylcholine-stimulated human platelets cause pulmonary hypertension and edema in rabbit isolated rabbit lungs, J. Clin. Invest. 71:351 (1983).
21. G. Plante, R. L. Hebert, P. Braquet and P. Sirois, Effect of platelet-activating factor on renal hemodynamics and sodium excretion,Prostaglandins. 30:708 (1985).
22. S. Vemulapalli, P. J. S. Chiu and A. Barnett, cardiovascular and renal actions of platelet-activating factor in anesthetized dogs, Hypertension. 6:489 (1984).
23. H. Scherf, A. S. Nies, U. Schwertschlag, M. Hughes and J. G. Gerber, Hemodynamic effects of platelet activating factor in dog kidney "in vivo", 6th International Conference on Prostaglandins and Related Compounds, Florence 1986, Abstract.
24. P. G. Baer and L. M. Cagen, Alprazolam inhibits the vasoconstrictor action of platelet-activating factor in dog kidney, Southwestern Symposium on Pulmonary and Cardiovascular Effects of Eicosanoids, Gainesville, Florida 1986, Abstract.
25. D. Schlondorff, J. A. Satrians, J. Hagege, J. Perez and L. Baud, Effect of platelet-activating factor and serum treated zymosan on prostaglandin E_2 synthesis, arachidonic acid release and contraction of cultured mesan-

gial cells, J. Clin. Invest. 74:1227 (1984).

26. G. Camussi, C. Tetta, R. Coda, G. Segoloni and A. Vercellone, Platelet activating factor-induced loss of glomerular anionic charges, Kidney Int. 25:73 (1984).

27. A. C. Rosam, J. L. Wallace and B. J. R. Whittle, Potent ulcerogenic actions of platelet-activating factor on the stomach, Nature. 319:54 (1986).

28. E. G. Lapetina, Platelet-activating factor stimulates the phosphatidylinositol cycle, J. Biol. Chem. 257:7314 (1982)

29. J. O. Shaw, S. J. Klusick and D. J. Hanahan, Activation of rabbit platelet phospholipase and thromboxane synthesis by 1-O-hexadecyl/octadecyl-2-acety-phosphorylcholine (platelet-activating factor), Biochim. Biophys. Acta. 663:222 (1981).

30. F. H. Chilton, J. T. O'Flaherty, C. E. Walsh, M. J. Thomas, R. L. Wykle, L. R. DeChatelet and B. M. Waite, Platelet-activating factor. Stimulation of the lipoxygenase pathway in polymorphonuclear leukocytes, J. Biol. Chem. 257:5402 (1982).

31. F. Bussolino, M. Aglietta, F. Sanavio, A. Stachini, D. Lauri and G. Camussi, Alkyl-ether phosphoglycerides influence calcium fluxes into human endothelial cells, J. Immunol. 135:2748 (1985).

32. B. B. Vargaftig, M. Chignard, J. Benveniste, J. Lefort and F. Wahl, Background and present status of research on platelet-activating factor (PAF-acether), Ann. N. Y. Acad. Sci. USA. 370:119 (1981).

33. T. W. Doebber, M. S. Wu and T. Y. Shen, Platelet-activating factor intravenous infusion in rats stimulates vascular lysosomal hydrolase secretion independent of blood neutrophils, Biochem.Biophys. Res. Commun. 125:980 (1984).

34. J. J. Emeis and J. Kluft, Paf-acether-induced release of tissue type plasminogen activator from vessel walls, Blood. 66:86 (1985).

35. H. P. Hartung and U. Hadding, Phorbol myristate acetate and platelet-activating factor promote synthesis of complement by macrophages in vitro, in: "Inflammatory Mediators", G. A. Higgs and T. J. Williams, ed., IUPHAR 9th International Congress of Pharmacology, VCH Publishers, Deerfield Beach, Florida (1984).

36. B. Poitevin, R. Rubin and J. Benveniste, Paf-acether generates chemiluminiscence in human neutrophils in the absence of cytochalasin B, Immunopharmacology. 7:135 (1984).

37. R. N. Pinckard, R. S. Farr and D. J. Hanahan, Physiochemical and functional identity of rabbit platelet-activating factor (PAF) released in vivo during IgE anaphylaxis with PAF released in vitro from IgE sensitized basophils, J. Immunol. 123:84 (1979).

38. J. M. Mencia Huerta and J. Benveniste, Platelet-activating factor and macrophages. II. Phagocytosis associated release of paf-acether from rat and mouse macrophages and not from mastocytes, Cell. Immunol. 57:281 (1981).

39. M. Sanchez Crespo, F. Alonso and J. Egido, Platelet-activating factor in anaphylaxis and phagocytosis. Release from human peripheral polymorphonuclears and monocytes during the stimulation by ionophore A23187 and phagocytosis but not from degranulating basophils, Immunology. 40:645 (1980).

40. G. Virella, M. F. L. Lopes-Virella, C. Shuler, T. Sherwood, T. Spinosa, P. Winocour and J. A. Cowell, Release of PAF by human polymorphonuclear leukocytes stimulated by immunocomplexes bound to Sepharose particles and human erythrocytes, Immunology. 50:43 (1983).

41. P. Iñarrea, F. Alonso and M. Sanchez Crespo, Platelet-activating factor and effector substance of the vasopermeability changes induced by the infusion of immune aggregates in the mouse, Immunopharmacology. 6:7 (1983).

42. P. Iñarrea, J. Gomez-Cambronero, J. Pascual, M. C. Ponte, L. Hernando and M. Sanchez Crespo, Synthesis of paf-acether and blood volume changes in Gram-negative sepsis, Immunopharmacology.9:45 (1985).

43. T. W. Doebber, M.S. Wu, J. C. Robbins, B. Ma Choy, M. N. Chang and T. W. Shen, Platelet-activating factor involvement in endotoxin-induced hypotension in rats: Studies with PAF-receptor antagonist kadsurenone, Biochem.

Biophys. Res. Commun. 127:799 (1985).

44. P. Braquet, A. Etienne, C. Touvay, R. H. Bourgain, J. Lefort and B. B. Vargaftig, Involvement of platelet-activating factor in respiratory anaphylaxis, demonstrated by paf-acether inhibitor BN 52021, Lancet. I:1501 (1981).

45. Z. Terashita, Y. Imura, K. Nisnikawa and S. Sumida, Beneficial effect of (RS)-2-methoxy-3-(octadecylcarbamoyloxy)propyl-2-(3-thiazolis)ethyl phosphate (CV 3988), a paf antagonist, in endotoxin and anaphylactic shock. Kioto Conference in Prostaglandins. Kioto 1984, Abstract.

46. M. Sanchez Crespo, S. Fernandez-Gallardo, M. L. Nieto, J. Baranes and P. Braquet, Inhibition of the vascular actions of immunoglobulin G aggregates by BN 52021, a highly specific antagonist of paf-acether, Immunopharmacology. 10:69 (1985).

47. Z. I. Terashita, Y. Imura, K. Nishikawa and S. Sumida, Is platelet activating factor (PAF) a mediator of endotoxin shock?, Eur. J. Pharmacol. 109: 173 (1985).

48. S. Adnot, J. Lefort, V. Lagente, P. Braquet and B. B. Vargaftig, Interference of BN 52021, a PAF-acether antagonist, with endotoxin induced hypotension in the guinea-pig. Ninth Annual Conference on Shock, Arizona 1986, Abstract.

49. A. Etienne, F. Hecquet, C. Soulard, B. Spinnewyn, F. Clostre and P. Braquet, "In vivo" inhibition of plasma protein leakage and salmonella enteriditis-induced mortality in the rat by a specific PAF-acether antagonist: BN 52021, Agents Actions 17: In press (1986).

50. G. Feuerstein, W. E. Lux, D. Ezra, E. C. Hayes, F. Snyder and A.I. Faden, Thyrotropin releasing hormone blocks the hypotensive effects of platelet activating factor in the unanesthetized guinea-pig, J. Cardiovas. Pharmacol. 7:335 (1985).

51. L. M. Villamediana, E. Sanz, S. Fernandez-Gallardo, C. Caramelo, M.Sanchez Crespo, P. Braquet and J. M. Lopez Novoa, Effects of the platelet-activating factor antagonist BN 52021 on the hemodynamics of rats with experimental cirrhosis of the liver, Life Sci. 39:201 (1986)

52. K. E. Grandel, R. S. Farr, A. A. Wanderer, T. C. Eisenstadt and S. I. Wasserman, Association of platelet-activating factor with primary acquired cold urticaria, New Engl. J. Med. 313:405 (1985).

53. C. Caramelo, S. Fernandez Gallardo, D. Marin Cao, P. Iñarrea, J. C. Santos, J. M. Lopez Novoa and M. Sanchez Crespo, Presence of platelet-activating factor in blood from humans and experimental animals. Its absence in anephric individuals, Biochem. Biophys.Res. Commun. 120:789 (1984).

54. K. E. Grandel, R. S. Farr and M. L. Wardlow, Platelet-activating factor in sputum of patients with asthma and COPD, J. Allergy Clin Immunol. 75:184 (1985) Abstract.

55. C. Caramelo, S. Fernandez-Gallardo, J. C. Santos, P. Iñarrea, M. Sanchez Crespo and J. M. Lopez Novoa,Increased levels of platelet-activating factor in blood from patients with cirrhosis of the liver, Eur. J. Clin. Invest. In press

56. G. Lagrue, K. rahbar, A. Sobel, P. Braquet, P. F. Michel and A. C. Yeung-Laiwah, Cyclic shock with monoclonal gammapathy: treatment in acute and chronic phase by IPS 200, a standarized Ginkgo biloba extract, 2nd World Conference on Inflammation, Monte-Carlo 1986, Abstract.

MECHANISMS OF CIRCULATORY COLLAPSE INDUCED BY PAF-ACETHER

R. E. Goldstein, F. R. M. Laurindo, D. Ezra and G. Z. Feuerstein
Divisions of Cardiology and Clinical Pharmacology and
Neurobiology Research Unit, Departments of Medicine,
Pharmacology, and Neurology Uniformed Services University of
Health Sciences Bethesda, Maryland 20814

PAF-acether (1-0-alkyl-2-acetyl-sn-glyceryl-3-phosphorylcholine) is an inflammatory mediator with potent circulatory actions (1). Release of this agent by activated leukocytes, macrophages, platelets, and thrombin-stimulated endothelial cells is likely to occur during Ig-E related anaphylaxis or sepsis. Adverse effects of PAF-acether may contribute significantly to circulatory deterioration and collapse that can accompany these conditions.

The cardiovascular actions of PAF-acether have been examined in several different experimental contexts. When studied in vitro, PAF-acether can alter vascular smooth muscle tone in addition to its potent capacity to promote platelet aggregation (2,3). When administered to isolated, perfused guinea pig hearts, PAF-acether reduces coronary blood flow and diminishes contractile performance of the left ventricular myocardium (4,5). When given to intact animals, PAF-acether can induce profound systemic arterial hypotension and even sudden death (6-9). The origins of this abrupt circulatory collapse are unclear. During more prolonged PAF-induced hypotension, cardiac output and left ventricular contractility are decreased and both systemic and pulmonary vascular resistances are increased (7). In addition, effective arterial blood volume falls and hematocrit rises, suggesting extravascular plasma sequestration (6). Previous data do not clarify which of these multiple alterations is causal and which a secondary consequence of poor cardiovascular function. Less attention has been focused on early hemodynamic changes during onset of PAF-induced circulatory collapse. In this phase, PAF-acether causes severe rises in pulmonary vascular resistance accompanied by elevated plasma levels of thromboxane A_2 (TXA_2) (7-9). In both domestic pigs and rabbits, PAF-initiated systemic arterial hypotension parallels TXA_2 release and prior cyclooxygenase blockade largely prevents the hypotensive response (8,10,11).

To clarify the complex circulatory actions of PAF-acether we administered this substance to in situ, blood-perfused hearts of domestic pigs. Intracoronary administration was utilized to evaluate direct influence on coronary perfusion. Intravenous administration was employed to assess PAF-acether actions on integrated circulatory function.

Methods   Domestic pigs (9–12 weeks old) weighing 25–35 kg were sedated
with ketamine hydrochloride (20 mg/kg per hr i.v.) and anesthetized with
pentobarbital sodium (2–4 mg/kg per hr i.v.). The pigs were ventilated with
a Harvard respirator modified to deliver 3 cm H_2O end-expiratory pressure.
Ventilatory rate and inspired oxygen concentration were adjusted to maintain
arterial blood gas concentrations within physiologic limits. Hematocrit was
25–32% (normal for pigs of this age). Rectal temperature was maintained at
37.5–38.0°C by an external heating pad. Catheters filled with heparinized
saline were placed in the jugular vein and internal mammary artery. The
left ventricular cavity was catheterized via a carotid artery. Mean
systemic arterial pressure, left ventricular pressure, and the surface
electrocardiogram were monitored continuously.

A left lateral thoracotomy was performed and the heart suspended in a
pericardial cradle. A circumferential electromagnetic flow probe was placed
around a proximal portion of the left anterior descending (LAD) coronary
artery and attached to a Model 501D Carolina square-wave flowmeter (Carolina
Medical Electronics, King, NC). A fine Tygon catheter used for administration of PAF-acether was introduced into the LAD several millimeters distal
to the flow probe. Pure synthetic PAF-acether (1-0-hexadecyl-2-acetyl-sn-
glyceryl-3-phosphorylcholine) was kindly provided by Dr. F. Snyder, Oak
Ridge Associated Universities (Oak Ridge, TN). Each dose was dissolved in
0.1 ml sterile, pyrogen-free 0.9% NaCl (vehicle), loaded into the LAD
catheter, and followed by a 0.5 ml vehicle flush (catheter volume 0.2 ml).
The pH of all injection solutions was 7.15–7.20. After baseline recordings
of mean LAD coronary blood flow (CBF) and other measurements, a 0.5 ml bolus
of vehicle was injected into the LAD. This uniformly failed to affect base-
lines. Individual bolus doses of PAF-acether were then injected over 20 sec
in an ascending order: 0.03, 0.1, 0.3, 1, 3, and 10 nmol. After each
injection, all parameters were continuously monitored for 10–15 min or until
return to baseline.

In further experiments to study effects of continuous infusion, PAF-
acether was administered at 1–6 nmol/min for 5–8 min. Regional contractility was assessed in these studies by a pair of 2 mm piezoelectric crystals
inserted into the midwall of the left ventricle in the distribution of the
LAD, oriented parallel to the diagonal branches, and attached to a sonomicrometer (Triton Technology, Inc., San Diego, CA). This allowed repeated
assessment of myocardial fiber length in systole and diastole as well as
myocardial shortening fraction [(diastolic length – systolic length)/dia-
stolic length].

To evaluate the biochemical consequences of PAF-acether, blood was
withdrawn rapidly from a catheter in the left ventricle (arterial sample)
and from the coronary vein draining the territory of the distal LAD. Plasma
was separated by rapid centrifugation (Beckman Microfuge B) and immediately
frozen on dry ice. Thromboxane B_2 (TXB_2) was determined by radioimmunoassay. Plasma lactate was measured by routine enzymatic methods.

Data are reported as mean values ± standard error.

Results   Intracoronary bolus administration of PAF-acether (7 pigs)
resulted in a consistent biphasic change in CBF (Fig. 1). An initial, brief
rise was followed immediately by a somewhat more long-lasting decline. The
initial rise was not accompanied by systemic effects, implying a rapidly
transient coronary vasodilator action of PAF-acether. This dose-dependent
effect increased CBF up to 50% (10 nmol dose) and was not influenced by

prior cyclooxygenase blockade with indomethacin, 6 mg/kg i.v. The later fall in CBF was also dose-dependent, lowering CBF by as much as 90% of pre-treatment baseline value (10 nmol dose). Larger doses of PAF-acether (1-10 nmol) produced hypotension and electrocardiographic evidence of myocardial ischemia along with steep declines in CBF. Calculated coronary vascular resistance rose in this phase from a baseline of 2.6±0.3 mm Hg/ml per min at baseline to 18.4±0.7 after 10 nmol of PAF-acether. Unlike the initial CBF rise, the subsequent PAF-induced decline in CBF and its concomitants were markedly diminished by pretreatment with indomethacin.

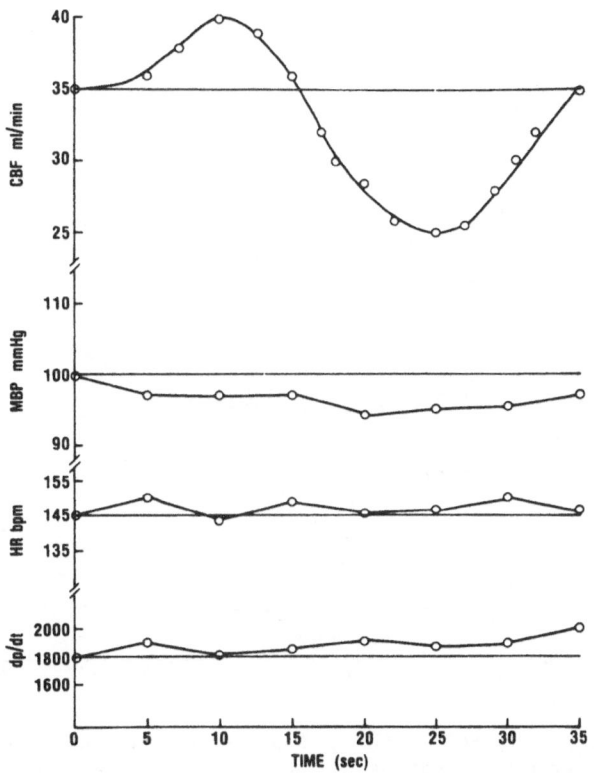

Fig. 1. Effect of intracoronary PAF-acether, 0.3 nmol, on coronary blood flow (CBF) and systemic hemodynamic variables in a domestic pig. MBP, mean blood pressure; HR, heart rate; dp/dt, peak rate of rise of left ventricular pressure (in mmHg/sec); abcissa, time after PAF-acether administration into the left anterior descending coronary artery. Reproduced, with permission, from Prostaglandins, Leukotrienes, and Lipoxins (JM Bailey, editor). Plenum (NY) p 304 (1985).

During PAF-induced periods of ischemic electrocardiographic abnormality in 4 pigs, coronary venous lactate rose (p<0.05) from 5.0±0.3 mg% at baseline to 16±6 after PAF-acether with the development of net myocardial release of lactate. Simultaneously, TXB_2 levels rose about equally in arterial and coronary venous samples: arterial TXB_2 increased from 0.7±0.4 ng/ml to 2.3±0.8 (p<0.05) while venous TXB_2 rose from 0.7±0.3 to 2.9±0.5 (p<0.05).

In contrast to results after bolus administration, steady intracoronary infusion of PAF-acether at 1 nmol/min produced no sustained change in CBF, coronary vascular resistance, or mean arterial pressure. Heart rate rose slightly. However, end-diastolic myocardial segment length and systolic shortening in the LAD territory and left ventricular end-diastolic pressure decreased throughout the period of infusion. Similar but more marked changes in left ventricular performance were seen with infusion of 3 nmol/min (Fig. 2). Although CBF did not change, systemic hypotension did occur at this higher infusion rate. All hemodynamic changes reversed rapidly at the end of infusion.

<u>Discussion</u>. Bolus intracoronary administration of PAF-acether revealed the capability of this inflammatory mediator to alter coronary vascular

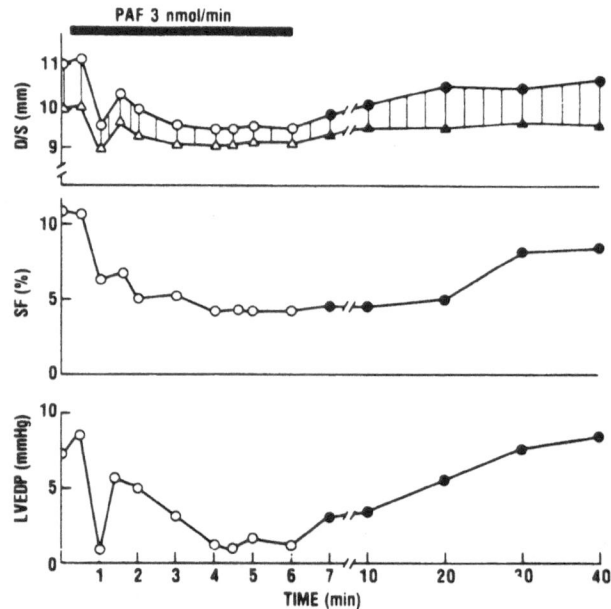

Fig. 2.   Effect of PAF-acether, 3 nmol/min, infused into the left anterior descending coronary artery of a domestic pig. D/S, distance between crystals in end-diastole (circles)/end-systole (triangles); verticals connecting circles and triangles (D-S), extent of systolic shortening (mm) in region between crystals; SF, shortening fraction (D-S)/D, shown as a percentage. Black horizontal bar represents the duration of the infusion. Closed symbols are data obtained after infusion was halted. Reproduced, with permission, from Prostaglandins, Leukotrienes, and Lipoxins (JM Bailey, editor). Plenum (NY) p 307 (1985).

resistance, first in a downward and then in an upward direction. The latter phase can result in marked coronary underperfusion with electrocardiographic and metabolic evidence of transient myocardial ischemia. Although the exact mechanism of this action of PAF-acether is not known, our data suggest that cyclooxygenase products, possibly TXA_2, mediate this latter phase. The earlier coronary vasodilation appears independent of cyclooxygenase metabolites and may reflect the effects of a specific platelet polypeptide released by PAF-acether (12).

The effects of continuous intracoronary infusion of PAF-acether are quite distinct from those seen after bolus administration. CBF changes are minimal, possibly due to buffering of coronary constrictor action by the vasodilator polypeptide. However, left ventricular changes are seen that are not characteristic of either myocardial ischemia or a primary myocardial depressant action of PAF-acether. Rather, the declines in end-diastolic pressure and myocardial segment length are most suggestive of underfilling of the left ventricle. These phenomena can also account for decreased systolic shortening via the Frank-Starling mechanism. PAF-induced left ventricular underfilling might be explained by peripheral venous pooling or extravascular sequestration of plasma volume (6). However, administration of even large volumes of intravenous fluids failed to reverse these changes. The unexplained nature of these findings led to further study of the influence of PAF-acether on pulmonary vessels and right ventricular function.

It should be noted that the coronary and myocardial effects of PAF-acether, even during infusions, are transient. Thus, if PAF-induced platelet or leukocyte aggregation figure in these phenomena, the impact of such aggregation on the heart appears fully reversible.

PULMONARY VASCULAR AND VENTRICULAR FUNCTION STUDIES

Methods - The initial phases and general care of this preparation were the same as those described in preceding paragraphs. After thoracotomy 9 domestic pigs were instrumented with a circumferential electromagnetic flowmeter cuff around the ascending aorta. A catheter-tip manometer (Millar Instruments, Houston, TX) was placed in the left ventricle via the left carotid artery. Fluid-filled catheters were positioned for pressure measurement in main pulmonary artery, right atrium, and a systemic artery (the left internal mammary). An infusion catheter was placed in the LAD of 6 pigs; in the remaining 3, drugs were given intravenously. A pair of piezoelectric crystals was inserted in the left ventricular myocardium, as described above. A second pair was inserted into the free wall of the right ventricle at minimum depth needed to bury each 2 mm crystal. The right ventricular intercrystal axis was positioned parallel to the LAD; its midpoint was located approximately 2 cm lateral to the midpoint of the LAD and 3 cm caudal to the pulmonic valve ring. Both crystal pairs were attached to the multichannel sonomicrometer mentioned previously.

Once baseline values were obtained, 7 pigs were infused with PAF-acether, 40-280 pmol/kg/min. Doses were begun at the low end of this range and increased every 2 min until the appearance of systemic hypotension (mean arterial pressure decrease in excess of 50%), at which point the infusion was halted. Observations were continued until baseline values returned or, in 2 cases, until death. Six animals received infusions of the stable thromboxane A_2 analog, U-46619 (300-3000 pmol/kg/min), for comparison with the actions of PAF-acether. A suitable interval (45-60 min) was interposed between successive infusions to allow for return and stabilization of hemodynamic parameters.

Data are reported as means ± standard error and analyzed by Student's t test (two-tailed) and ANOVA.

Results. Hemodynamic findings were the same in subgroups receiving intracoronary and intravenous PAF-acether except for the transient changes with intracoronary infusion described previously. Findings from each of these subsets are pooled in subsequent analyses.

Infusion of PAF-acether produced a progressive rise in pulmonary vascular resistance (PVR), eventually reaching 5-120 (mean 44) x baseline values (Fig. 3). This increase was initially associated with a rise in mean pulmonary artery pressure from baseline value of 19±1 mmHg to a peak of 45±2. However, pulmonary artery pressure actually fell to 25±4 mmHg with further progression of hemodynamic abnormality. Cardiac output and stroke

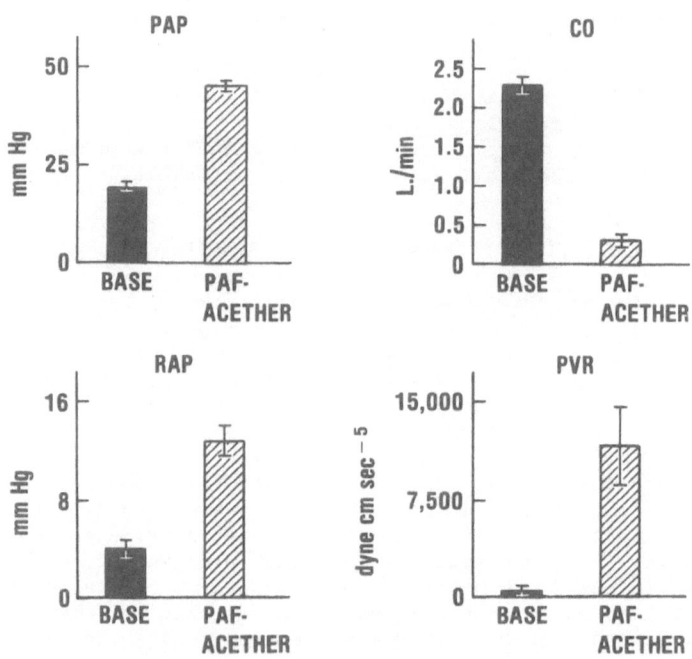

Fig. 3.   Effects of PAF-acether, 40-280 pmol/kg/min i.v., in 7 domestic pigs are shown by comparison of pretreatment baseline mean value (left bar) and mean value at peak effect (right bar). Verticals denote standard errors. PAF-acether caused a progressive rise in pulmonary vascular resistance (PVR), initially accompanied by increased pulmonary artery pressure (PAP). Right ventricular dysfunction, developing with marked rises in PVR, featured elevation in mean right atrial pressure (RAP) and severe decline in cardiac output (CO). With peak effects on CO, systemic arterial pressures also fell to shock levels. Reproduced, with permission, from Pharmacol Res Commun, in press (1986).

182

volume declined steadily during PAF-acether infusion to values 75 - 98%
below baseline. Systemic arterial pressure also fell from a baseline value
of 94±3 mmHg to an endpoint value of 34±5. However, this pressure decrement
only reached statistical significance at extremes of cardiac output reduction. Systemic vascular resistance was little changed at first but later
rose by 144% with falling systemic arterial pressure.

These PAF-induced alterations were accompanied by signs of right
ventricular overload and failure: mean right atrial pressure rose from 3±1
mmHg to 13±1 with evidence of tricuspid regurgitation on the right atrial
pressure tracings; right ventricular end-diastolic fiber length increased
from 9.7±1.0 mm to 11.0±1.0 mm (p<0.01); and right ventricular shortening
fraction fell from 10±1% to 5±1%. Left ventricular shortening fraction was
also severely reduced from 13±1% to 1±2%. However, left ventricular indices
showed significant underloading: left ventricular end-diastolic pressure
fell from 12±1 mmHg to 2±1 mmHg and left ventricular end-diastolic fiber
length fell from 11±1 mm to 8.0±0.4. Each of these effects of PAF-acether
was reproduced in 5 pigs by gradually progressive, mechanically-induced
occlusion of the main pulmonary artery.

Confirmatory studies were performed in 12 closed-chest chloralose-anes-
thetized domestic pigs. Intravenous bolus administration of either 3 or
10 nmol PAF-acether produced the same effects on cardiac output and pressures in cardiac chambers and great vessels. On average, PAF-acether,
10 nmol, raised PVR 24-fold, reduced systemic arterial pressure from
110±3 mmHg to 45±5, augmented systemic vascular resistance by 170%,
increased right atrial pressure from 4±1 mmHg to 17±2 with signs of tricuspid regurgitation, and lowered left ventricular end-diastolic pressure from
10±2 mmHg to 2±1. Radioimmunoassay performed on simultaneous mixed venous
plasma samples showed that PAF-acether raised TXB_2 levels from 0.7±0.1 ng/ml
to 2.5±0.7 (3 nmol) or 3.0±0.3 (10 nmol). Pretreatment with indomethacin,
6 mg/kg i.v., eliminated this rise. Indomethacin pretreatment also
eliminated all hemodynamic changes produced by PAF-acether, 3 nmol, and
significantly attenuated changes caused by 10 nmol. Nevertheless, 10 nmol
PAF-acether did raise PVR 10-fold, lower systemic arterial pressure from
113±6 mmHg to 79±11, increase systemic vascular resistance by 58% and raise
right atrial pressure from 4±1 mmHg to 7±1 in the absence of any change in
TXB_2 levels (0.74±0.18 ng/ml before and 0.76±0.06 after PAF-acether).

Intravenous or intracoronary infusion of the stable thromboxane A_2
analog, U46619, in 6 open-chest pigs produced effects that closely resembled
those of PAF-acether infusion. A progressive, dose-related ris in PVR and
fall in cardiac output culminated in severe systemic arterial hypotension
and signs of right ventricular overload and failure. Reduction in left
ventricular shortening fraction also occurred with infusion of U-46619
accompanied by declining left ventricular end-diastolic pressure and fiber
length. As in the case of PAF-acether, circulatory effects waned and
usually reversed completely in the 30 min after discontinuation of infusion.

Discussion. Our results confirm the capacity of PAF-acether to
produce profound shock, manifested as severe systemic arterial hypotension,
markedly reduced cardiac output, and increased systemic vascular resistance.
These changes were accompanied and even anticipated by progressive and very
severe rises in PVR. Both right and left ventricles showed signs of reduced
contractile function (diminished shortening fraction). The right ventricle
exhibited signs of acute myocardial failure: increased filling pressure and
end-diastolic fiber length along with decreased cardiac output, stroke
volume, and systolic shortening. The severe and abrupt rise in afterload
imposed by PAF-induced elevation of PVR was very likely a major factor in
precipitation of right ventricular failure. Reduced coronary perfusion,
particularly during systemic arterial hypotension and right-sided diastolic

pressure rise, and primary PAF-induced depression of inotropic state may have interfered with the capacity of the right ventricle to cope with enormously increased afterload. Our data suggest that the diminished pumping action of the right ventricle was further compromised by tricuspid regurgitation, probably due to right ventricular dilation. Our results exclude venous pooling or plasma sequestration as a significant factor contributing to right ventricular dysfunction occurring soon after administration of PAF-acether.

In contrast to the situation in the distended right ventricle, reduced contractile function in the left ventricle was accompanied by evidence of underfilling—decreased left ventricular end-diastolic pressure and fiber length. Although diminution in the inotropic state of the myocardium might contribute, reduced left ventricular contractile performance during PAF-acether infusion is adequately explained by insufficient preload (stretching of the myocardium just prior to systole). This insufficiency in left ventricular preload probably reflects right ventricular incapacity to sustain adequate blood flow through pulmonary vessels severely obstructed by PAF-acether. Plausibility of this analysis is demonstrated by the capacity of mechanical constriction of the main pulmonary artery to reproduce all early hemodynamic consequences of PAF-acether infusion. If it were possible to eliminate PAF-induced pulmonary vascular obstruction, the normally-filled left ventricle might still be dysfunctional due to an unmasking of PAF-induced contractile depression. This theoretical possibility, however, is likely to be of little consequence. Measures that alleviate the influence of PAF-acether on the circulation would probably act in the heart as well as the lungs.

Results in open-chest and closed-chest animals show similar circulatory effects of PAF-acether. PAF-induced changes observed in pulmonary vascular function were not dependent upon concomitant thoracotomy, abnormalities in blood gas concentrations, or a particular type of anesthesia, nor were they exclusively related to either bolus administration or steady infusion. The similarity of our data in domestic pigs and results of Kenzora et al. in intact dogs (7) suggest that PAF-acether may have an obstructive influence on in situ pulmonary vessels of many species.

Indomethacin blocked much of the pulmonary vascular response and other consequences of PAF-acether. This finding plus the similar circulatory actions of PAF-acether and the TXA_2 analog, U-46619, suggest that PAF-induced release of TXA_2 may be a major mechanism by which PAF-acether causes early circulatory collapse. This concept is favored by the substantial TXA_2 release associated with PAF-acether administration. However, this does not appear to be the sole mechanism of PAF-acether action in our model. Even when PAF-induced TXA_2 release was fully blocked, PAF-acether was able to cause substantial rises in PVR and pulmonary artery pressure.

CONCLUSIONS

Release of PAF-acether by activated leukocytes or platelets can exert profound deleterious effects on the circulation. Bolus administration of PAF-acether directly into the coronary circulation causes a dose-related, biphasic flow response—a brief fall in coronary vascular resistance followed by a more prolonged rise. When severe, the latter phase is associated with systemic hypotension and evidence of myocardial ischemia. All aspects of this latter phase are substantially diminished by cyclooxygenase blockade.

More steady and prolonged administration of PAF-acether, either by intracoronary or intravenous route, discloses a different pattern of circulatory response. High doses rapidly cause profound systemic hypotension, circulatory collapse, and death. Examination of a broader dosage range discloses a progressive rise in PVR with increasing exposure to PAF-acether. Initially, this is manifested chiefly by a rise in pulmonary artery pressure. Greater rise in PVR initiates signs of acute right ventricular failure: increase in right ventricular filling pressures and dimensions and decrease in systolic shortening, stroke volume and cardiac output. By contrast, the left ventricle shows signs of underfilling, most likely the effect of reduced flow delivery from the pulmonary circulation. In this phase of steady PAF-acether exposure, coronary flow changes are modest and, in any event, cannot fully explain hemodynamic changes. When progressively increasing PAF-acether administration results in systemic hypotension, the right ventricle simultaneously exhibits signs of far-advanced failure, including tricuspid regurgitation. Our data suggest that the rise in PVR and resultant acute right ventricular failure are predominant factors producing circulatory collapse associated with PAF-acether administration to domestic pigs.

ACKNOWLEDGEMENT

Research was supported by the Uniformed Services University of the Health Sciences, Protocols R08346 and G19214. Dr. Ezra was an International Fellow in Clinical Pharmacology of the Merck Foundation, and Dr. Laurindo is an International Fellow of the Fogarty Foundation, National Institutes of Health. The authors are grateful to Mr. John Czaja for technical assistance and to Mrs. Joan McMillen for help in manuscript preparation.

REFERENCES

1. Benveniste J and Chignard M. Circulation 72:713 (1985)
2. Kamitani T, Katamoto M, Tasumi M, Katsuta K, Ono T, Kikuchi H, and Kumada S. Eur J Pharmacol 98:357 (1984)
3. Cervoni P, Herzlinger HE, Lai FM, and Tanikella TE. Br J Pharmacol 79:667 (1983)
4. Benveniste J, Boullet C, Brink C, and Labat C. Br J Pharmacol 80:81 (1983)
5. Levi R, Burke JA, Guo LG, Hattori Y, Hoppens CM, McManus LM, Hanahan DJ, and Pinckard PN. Circ Res 54:117 (1984)
6. Bessin P, Bonnet J, Apffel D, Soulard C, Desgroux L, Pelas I, and Benveniste J. Eur J Pharmacol 86:403 (1983)
7. Kenzora JL, Perez JE, Bergmann SR, and Lange LG. J Clin Invest 74:1193 (1984)
8. Lefer AM, Mueller HF, and Smith JB. Br J Pharmacol 83:125 (1984)
9. Sybertz EJ, Watkins RW, Baum T, Pula K, and Rivelli M. J Pharmacol Exp Ther 232:156 (1985)
10. Feuerstein GZ, Boyd LM, Ezra D, and Goldstein RE. Am J Physiol 246:H466 (1984)
11. Feuerstein GZ, Ezra D, Ramwell PW, Letts G, and Goldstein RE in: Prostaglandins, Leukotrienes, and Lipoxins (editor: JM Bailey). Plenum (NY) p 301 (1985)
12. Jackson CV, Schumacher WA, Kunkel SL, Driscoll EM, and Lucchesi BR. Circ Res 58:218 (1986)

LIPID MEDIATORS IN LUNG ANAPHYLAXIS : KINETICS OF THEIR RELEASE AND MODULATION BY SELECTED DRUGS

P. Sirois[1], M. Harczy[1], J. Maclouf[2], P. Pradelles[3], P. Braquet[4], P. Borgeat[5]

[1] Depts. Pediatrics and Pharmacology, Faculty of Medicine, University of Sherbrooke, Sherbrooke, P.Q. Canada
[2] Hôpital Lariboisière, Paris, France. [3] CEN-Saclay, France
[4] IHB, 17 avenue Descartes, 92350 Le Plessis Robinson, France
[5] CHUL, Québec, Canada

ABSTRACT

Various arachidonic acid metabolites are released from the lungs during hypersensitivity reactions such as anaphylaxis or asthma. Thromboxane A_2 and leukotrienes were shown to play a key role in the bronchoconstriction associated with these conditions. In our experiments, Reverse Phase High Performance Liquid Chromatography (RP-HPLC) and Enzyme Immunoassay (EIA) techniques were use (a) to study the profile of cyclooxygenase and lipoxygenase products released during guinea pig anaphylaxis, (b) to characterize their time-course of release, and (c) to investigate the complex interactions regulating their synthesis. The lungs of guinea pigs sensitized to ovalbumin (100 mg intraperitonealy and 100 mg subcutaneously) were perfused with Krebs solution. Anaphylaxis was induced by specific challenge (ovalbumin, 100 ug/ml) and the effluent was collected at 1 min intervals for the measure of prostaglandin E_2 (PGE$_2$), thromboxane B_2 (TxB$_2$), leukotrienes B_4 (LTB$_4$) and D_4 (LTD$_4$). In a set of experiments, the 12-hydroxyheptadecatrienoic acid (HHT) and 12-keto-heptadecatrienoic acid (12-keto-HT) were analysed. Our results showed that the time-course of release of all the arachidonic acid metabolites were maximal approximately 5-6 min following the onset of the challenge. Leukotriene D_4 and to a higher extent LTB$_4$ were the major lipoxygenase products detected during anaphylaxis. Perfusion of the lungs with aspirin and indomethacin decreased the formation of PGE$_2$ and TxB$_2$ but did not modify the release of leukotrienes. On the contrary, BW755C and eicosatetraynoic acid (ETYA) reduced the release of all icosanoids from the lungs. FPL-55712, a selective leukotriene antagonist, significantly reduced the release of PGE$_2$, TxB$_2$, LTB$_4$ and LTD$_4$ at the high concentration (20 uM). In summary, this study made use of novel techniques to quantify arachidonic acid metabolites in the effluent of perfused lungs. New metabolites were described. Our study also stresses the possible significance of LTB$_4$ as a mediator of anaphylaxis and asthma.

INTRODUCTION

The release of "Slow Reacting Substance" ("SRS" – a mixture of leukotriene C_4, D_4 and E_4)[1-3] by perfused guinea pig lung following specific antigen challenge was first reported by Kellaway and Trethewie[4]. Brocklehurst[5] used the guinea pig ileum, a pharmacological preparation which is exquisitely sensitive to SRS, to study the time-course of release of SRS-A (Slow Reacting Substance of Anaphylaxis) in the same experimental model. Piper and Vane[6] described the release of "additional factors" namely RCS (Rabbit aorta contracting substance – now known to be a mixture of thromboxane A_2 and prostaglandin endoperoxides) and other prostaglandin-like substances in addition to the previously identified mediators SRS-A and histamine from guinea pig lung during anaphylaxis. These mediators account for the symptoms of immediate hypersensitivity reactions. Pharmacological evidence points to a role for leukotrienes in the bronchoconstriction associated with anaphylaxis and asthma. For instance, the administration of chemically pure leukotrienes was shown to mimick the symptoms of asthma crisis[7,8]. However measurements of the various arachidonic acid metabolites and especially lipoxygenase products in biological fluids are relatively difficult and remain a handicap for assessing the full significance of these compounds in health and disease.

The aims of the following study were:
(a) to use novel Reverse Phase High Performance Liquid Chromatography (RP-HPLC)[9] and Enzyme Immunoassay (EIA)[10] techniques to study the profile cyclooxygenase and lipoxygenase products during guinea pig anaphylaxis,
(b) to characterize the kinetics of their release following antigen challenge, and
(c) to investigate the interactions between the various mediators by modulating their release by selected drugs.

MATERIAL AND METHODS

Guinea pig sensitization and lung perfusion

Guinea pig of either sex were sensitized with two initial doses of ovalbumin (Sigma, Grade II, 100 mg, intraperitonealy and 100 mg subcutaneously) and a subsequent recall (10 mg intraperitoneally) one week later[11,12]. Three weeks after the initial injections, the animals were sacrificed and the heart and lungs were excised and perfused via the pulmonary artery with oxygenated Krebs solution (15 ml/min) for 2 min. The flow rate was subsequently decreased to 2 ml/min.

Sample collection and icosanoid measurements

Aliquots of effluents (0.5 ml) were collected before and 0, 2, 4, 6, 8, 10, 15, 20, 25 and 30 min following antigen challenge (ovalbumin, Sigma, Grade IV, 100 ug/ml for 30 min) for the determination of prostaglandin E_2 (PGE_2) and thromboxane B_2 (TxB_2). Aliquots corresponding to one minute of perfusion (2 ml) were collected for 25 min after the antigen challenge for the study of the time-course of release of leukotrienes. However for the experiments using various drugs, the collection of the complete 30 min effluent was done for the measurement of leukotrienes. All samples were collected in tubes containing equal volumes of cold methanol.

RP-HPLC analyses were performed with a C_{18} Radial-Pak cartridge

(100 x 8 mm; 10 um particle size) as described before[12,13]. PGB$_2$ (200 ng) was added as internal standard and each sample were centrifuged (3000 g; 30 min). Supernatants were acidified and injected (4 ml) without further treatment or in the cases where the volumes were 60 ml, the samples were initially extracted with methanol before RP-HPLC analysis.

Elution of the arachidonic acid metabolites was monitored by ultra-violet spectrophotometry at 229 nm (5-HETE; 5-hydroxyeicosatetraenoic acid and HHT; 12-hydroxyheptadecatrienoic acid) and at 280 nm (Leuko-trienes, diHETE, PGB$_2$ and 12-keto-HT;12-keto-heptadecatrienoic acid). Quantification of these metabolites was done by comparison of the peak areas with the internal standard. Corrections for differences in molar extinction coefficients and recorder settings were performed. Recoveries of the arachidonic acid metabolites in our experimental conditions were estimated to be over 90%.

Prostaglandin E$_2$ (PGE$_2$) and thromboxane B$_2$ were measured with a novel EIA technique as described before[10,14].

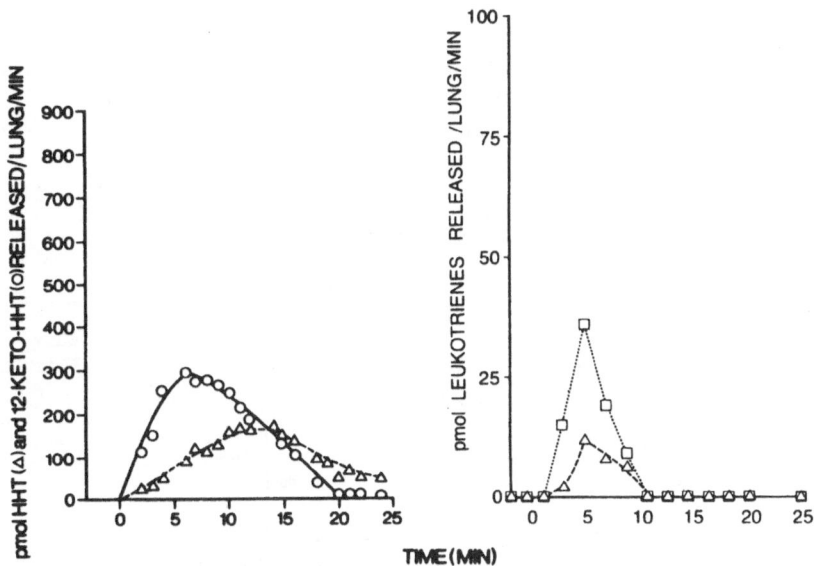

Figure 1. (Right Panel) Time-course of LTB$_4$ (squares) and LTD$_4$ (triangles) release from perfused guinea pig lungs during anaphylaxis. Figure 1. (Left Panel) Time-course of HHT (triangles and 12-keto-HT (circles) release from perfused guinea pig lungs during anaphylaxis. Aliquots corresponding to 1 min of perfusion (2 ml) were analysed by RP-HPLC. Experiments were repeated at least 6 times and the figure is a representative one.

Drugs used

The following drugs were used: aspirin (Sigma Chem., St-Louis, U.S.A.); indomethacin (Merck Frosst, Montreal, Canada); BW755C (Welcome Labs., Beckenham, U.K.); eicosatetraynoic acid (ETYA; Hoffman Laroche, Nutley, U.S.A.). FPL-55712 and the leukotrienes have been kindly provided by Dr P. Sheard of Fisons Ltd (U.K.) and Dr J. Rokach of Merck Frosst (Canada) respectively.

Time-course of the release of arachidonic acid metabolites

When perfused guinea pig lungs were challenged with ovalbumin, LTB_4 and LTD_4 were the only two 5-lipoxygenase products which could be detected by RP-HPLC techniques in the two ml effluents. Maximal concentrations of LTB_4 (40 pmole/2 ml) and LTD_4 (12 pmole/2 ml) were

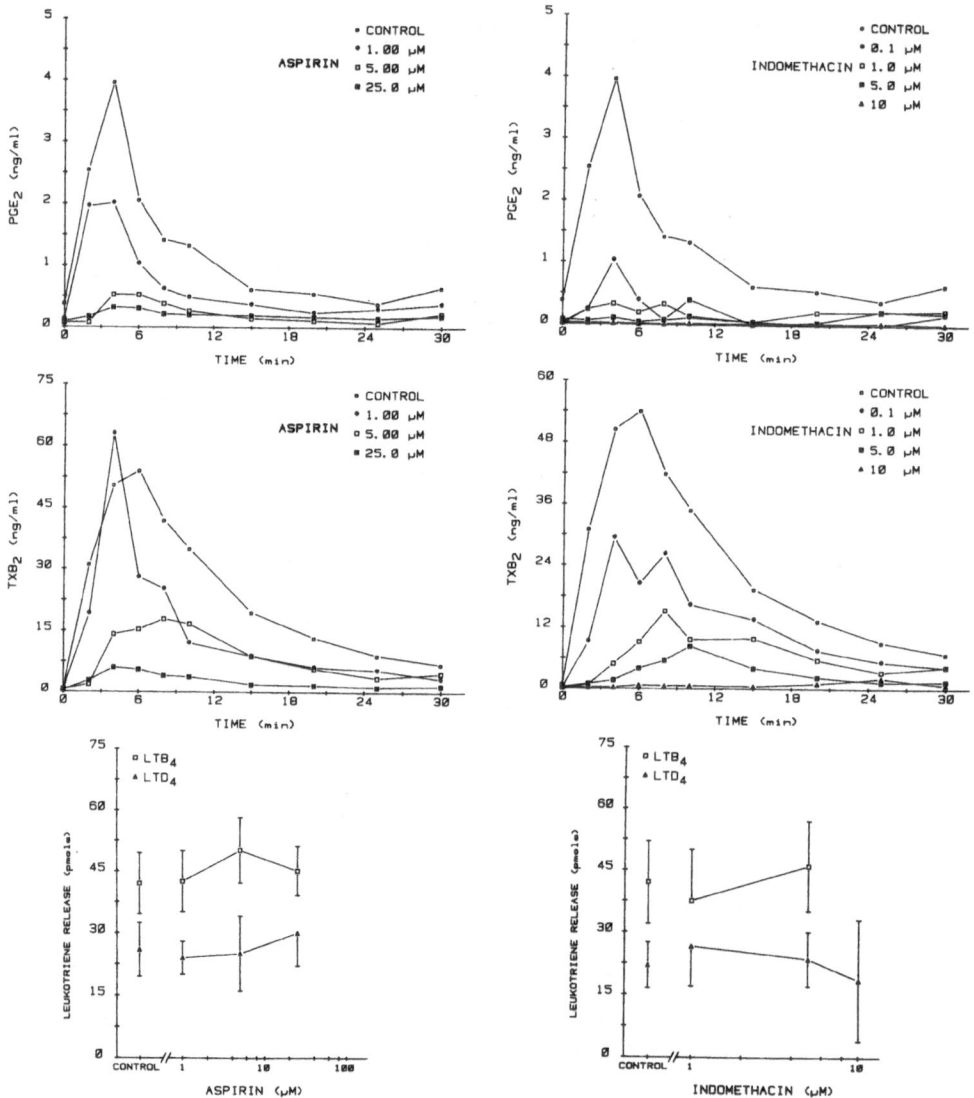

Figure 2. Effects of aspirin on the release of PGE_2 (upper panel), TxB_2 (center panel), LTB_4 and LTD_4 (lower panel) from perfused guinea pig lungs during anaphylaxis. Values are means $\pm$ SEM of 14, 5, 5, 7 (control, 1.0, 5.0, 25.0 uM) observations for PGE_2; 16, 5, 5, 7 observations for TxB_2 and the leukotrienes.

Figure 3. Effects of indomethacin on the release of PGE_2 (upper panel), TxB_2 (center panel), LTB_4 and LTD_4 (lower panel) from perfused guinea pig lungs during anaphylaxis. Values are means $\pm$ SEM of 14, 2, 3, 3, 3 (control, 0.1, 1.0, 5.0, 10 uM) observations for PGE_2; 16, 2, 3, 3, 3 for TxB_2 and the leukotrienes.

reached approximately 5 min after the beginning of the challenge. The amounts of leukotrienes rapidly decreased afterwards and were no longer detectable after 10 min of perfusion. As shown clearly in Fig 1 (right panel), the amount of LTB_4 released was approximately three times the amount of LTD_4. However the presence of 5-HETE, LTB_4, LTC_4, LTE_4 and the ω-oxidation products of LTB_4 were not detectable in most experiments.

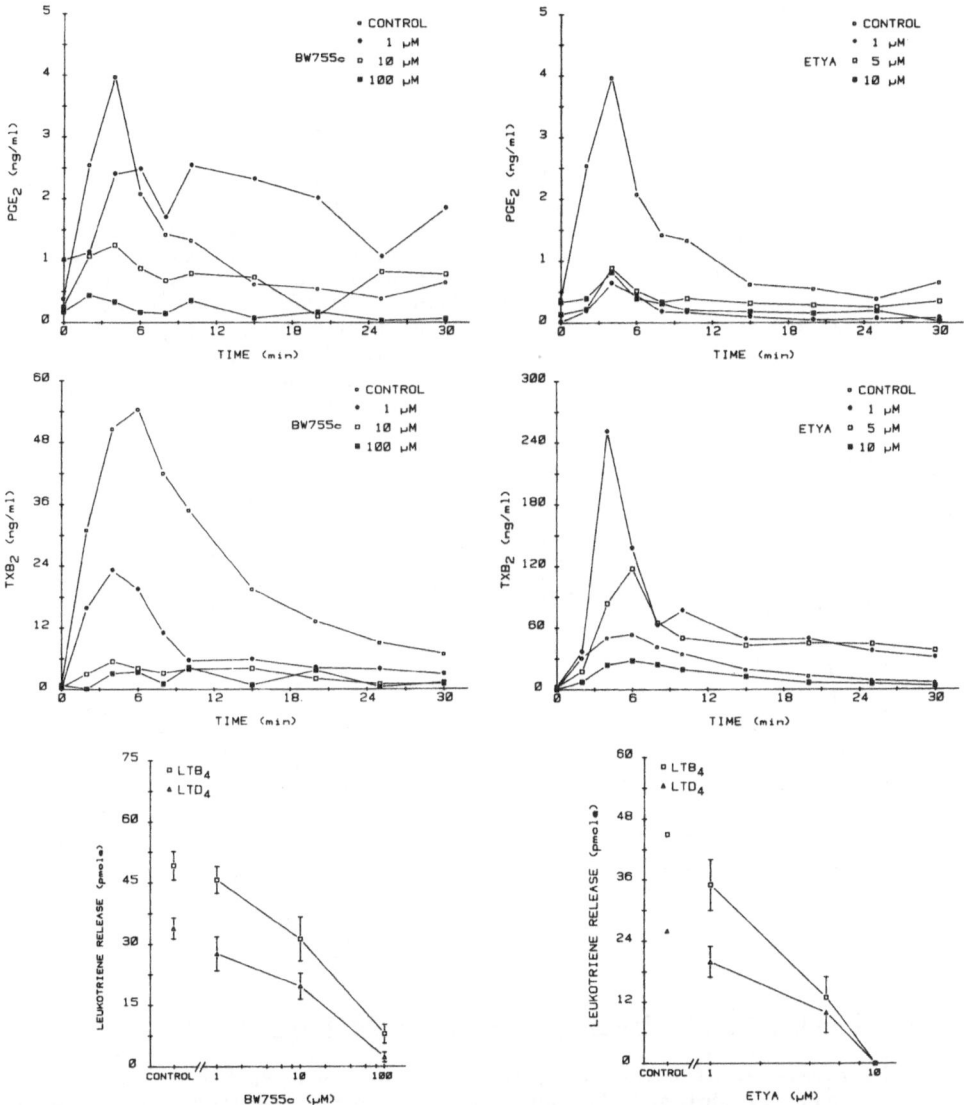

Figure 4. Effects of BW755C on the release of PGE_2 (upper panel), TxB_2 (center panel), LTB_4 and LTD_4 (lower panel) from perfused guinea pig lungs during anaphylaxis. Values are means $\pm$ SEM of 14, 3, 2, 3 (control, 1, 10, 100 uM) for PGE_2; 16, 3, 3, 3 observations for TxB_2 and the leukotrienes.

Figure 5. Effects of eicosatetraynoic acid (ETYA) on the release of PGE_2 (upper panel), TxB_2 (center panel), LTB_4 and LTD_4 (lower panel) from perfused guinea pig lungs during anaphylaxis. Values are means $\pm$ SEM of 14, 3, 3, 2 (control, 1, 3, 2 uM) observations for PGE_2; 16, 3, 3, 2 observations for TxB_2 and leukotrienes.

The left panel of Fig 1 shows the time-course of release of two cyclooxygenase products, HHT and 12-keto-HT. The presence of these compounds in the lung perfusates was confirmed by Gas ChromatographicMass spectrometric analyses. The amount of the two products detected in each of the sample analysed was similar and the maximal level (approximately 250 pmole/2 ml) was reached slightly after the peak level of leukotrienes (respectively 5-6 min for 12-keto-HT and 12 min for HHT).

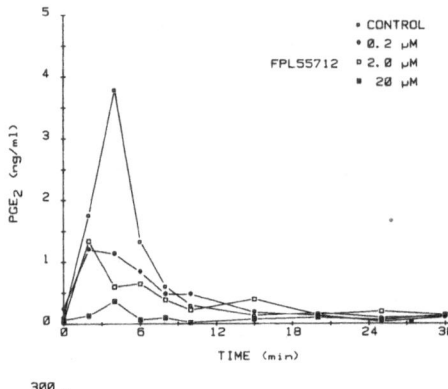

Figure 6. Effects of FPL-55712 on the release of PGE_2 (upper panel), TxB_2 (center panel), LTB_4 and LTD_4 (lower panel) from perfused guinea pig lungs during anaphylaxis. Values are means $\pm$ SEM of 20, 2, 2, 6 observations (respectively for controls, 0.2, 2 and 20 uM) for PGE_2, TxB_2 and the leukotrienes.

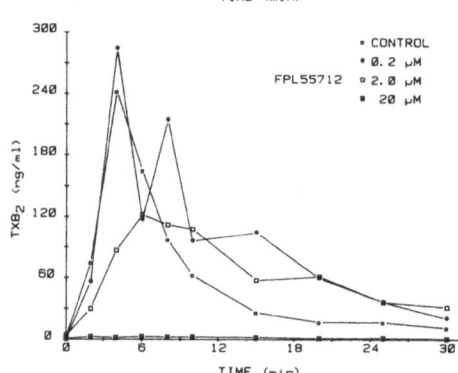

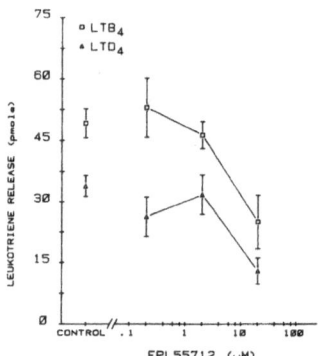

Effect of selected inhibitors of arachidonic acid metabolism

The treatment of perfused guinea pig lung with aspirin 15 min before and continuously during the antigen challenge resulted in a major dose-dependent decrease of both PGE_2 and TxB_2. As shown in Fig 2 (upper and center panels), the lowest concentration of aspirin used (1 uM) had little effect on the release of mediators. However increasing the concentration to 5 uM strongly reduced the release of cyclooxygenase products, and the concentration of 25 uM almost abolished it. The peaks of release of the mediators were always reached 5 min after the beginning of the challenge.

The same concentrations of aspirin had no significant effects on LTB$_4$ and LTD$_4$ released from the anaphylactic lungs. As shown in the lower panel of Fig 2, control anaphylactic lungs released approximately 25 pmole of LTD$_4$ and 40 pmole of LTB$_4$.

Indomethacin (0.1, 1.0, 5.0 and 10 uM) had effects very similar to those of aspirin on the release of mediators of anaphylaxis. At the lowest concentration used (0.1 uM), it decreased both PGE$_2$ and TxB$_2$ release by more than 50%. As shown in Fig 3 (upper and center panels), the inhibition of PGE$_2$ appeared more marked than that of TxB$_2$. Increasing the concentrations to 1.0 and 5.0 uM completely abolished the appearance of PGE$_2$ in lung effluent and diminished by approximately 80% the release of TxB$_2$. At the concentration of 10 uM, both mediators were almost undetectable in the effluent.

The effect of indomethacin on the C-5 lipoxygenase was also similar to that of aspirin. In brief, the concentration of indomethacin used (0.1 - 10 uM) did not affect either positively or negatively the release of LTB$_4$ or LTD$_4$ (Fig 3, lower panel). It is important to note that the effect of indomethacin (10 uM) on the release of LTB$_4$ could not be evaluated because of the retention time of indomethacin which is similar to that of LTB$_4$ in our RP-HPLC system; at this high concentration, the large peak of indomethacin completely covered the peak of LTB$_4$.

Fig 4 shows the effects of BW755C, an inhibitor of cyclooxygenase and lipoxygenases, on the time-course of release of PGE$_2$ (upper panel), TxB$_2$ (center panel) and leukotrienes B$_4$ and D$_4$ (lower panel) by the anaphylactic lungs. This compound produced a concentration-dependent inhibition of the release of all arachidonic acid metabolites. At the opposite of indomethacin, BW755C appeared to be more potent against TxB$_2$ than against PGE$_2$. As shown on the upper and center panels of Fig 4, the lowest concentration of BW755C (1 uM) did not change significantly the release of PGE$_2$ whereas, the release of TxB$_2$ was reduced by more than 50%. At the concentration of 100 uM, BW755C almost abolished the release of both cyclooxygenase products.

The release of leukotrienes B$_4$ and D$_4$ which averaged about 50 and 33 pmole in control lungs was not affected significantly at the concentration of BW755C of 1 uM but was decreased slightly at the concentration of 10 uM and abolished at the concentration of 100 uM (Fig 4, lower panel).

The effects of eicosatetraynoic acid (ETYA) (1-10 uM) on the release of anaphylactic mediators by the perfused lungs is shown in Fig 5. This compound strongly decreased the release of PGE$_2$ at each concentration used. No significant difference were seen between the lowest and the highest concentration of the drug (Fig 5, upper panel). The effect on the thromboxane release was quite different. ETYA potentiated the release of TxB$_2$ from the lungs at concentrations of 1 and 5 uM. The highest concentration (10 uM) diminished by approximately 50% the release of TxB$_2$ (Fig 5, center panel).

ETYA was a potent inhibitor of leukotriene release (Fig 5, lower panel). At the concentation of 1 uM, ETYA had no significant effect on the content of leukotrienes in the effluent but at the concentration of 5 uM, the levels were decreased by two thirds and leukotrienes were undetectable in the effluent of lungs treated with 10 uM of the drug.

Effect of FPL-55712, a LTD$_4$ antagonist

In this set of experiments, we investigated the effects of FPL-55712, a peptido-leukotriene antagonist, on the release of anaphylactic mediators. As shown in Fig 6, this compounds produced a clear inhibition of lung mediator release. The release of PGE$_2$ (Fig 6, upper panel) was decreased by more than 50% at the concentrations of 0.2 and 2.0 uM and was very low at the concentration of 20 uM. This high concentration of the drug also completely abolished the release of thromboxane B$_2$. However the concentration of 0.2 and 2 uM did not affect significantly the release of TxB$_2$.

The release of LTB$_4$ and LTD$_4$ (Fig 6, lower panel) was not affected by the concentrations of FPL-55712 of 0.2 and 2.0 uM. However, both mediators were inhibited by around 50% when the lungs have been treated with FPL-55712 (20 uM).

DISCUSSION

This study shows the time-course of release of arachidonic acid metabolites by the perfused lungs of sensitized guinea pigs. Maximal amounts of these metabolites were detected in aliquots of the effluent approximately 5 min after the beginning of the antigen challenge. Significant amounts were release up to 30 min after the onset of anaphylaxis although for some of these substances, namely the leukotrienes, the sensitivity of the RP-HPLC technique do not allow to measure significant amounts in the effluent 10 min after the onset of the reaction. Other C-5 lipoxygenase products were only observed occasionally. It is interesting to note that LTB$_4$ was release in larger quantities than LTD$_4$ from the anaphylactic lungs. This evidence together with previous observations which showed that (a) the lungs bear specific receptors for LTB$_4$, (b) LTB$_4$ is nearly as potent a bronchoconstrictor as LTD$_4$ in the lung[16], (c) the bronchoconstriction induced by Platelet Activating Factor (PAF) appears to be mediated by LTB$_4$[17], and (d) LTB$_4$ may be involved in the cell accumulation and the increased vascular permeability in the lung during immediate hypersensitivity reactions[18], strongly support a major role for LTB$_4$ in anaphylaxis and asthma.

Our results also showed that HHT and 12-keto-HT were release in signification amounts from the anaphylactic lung. These two compounds which have a ultraviolet chromophores, were detected respectively at 229 and 280 nm. They constitute good markers of the cyclooxygenase, the products of which, namely the prostaglandins, cannot be detected by ultraviolet detectors. HHT is released together with TxB$_2$ during lung anaphylaxis. Since the lungs contain a 15-OH-prostaglandin dehydrogenase, our results suggest that 15-keto-HT is the metabolite resulting from the activity of this enzyme on HHT. This product was recently reported to be formed by the action of purified kidney 15-hydroxy-prostaglandin dehydrogenase on HHT[19].

Our results also extended previous studies of the effects of aspirin and indomethacin on the release of arachidonic acid metabolites from anaphylactic lungs. Using novel sensitive EIA techniques which allowed the measure of PGE$_2$ and TxB$_2$ in tiny amounts of lung effluents, it has been possible to carefully analyse the time-course of release of a few mediators from the same aliquots. The time course of release of cyclooxygenase products as measured with the EIA techniques was the same as that measured with RP-HPLC.

The two cyclooxygenase inhibitors produced concentration-dependent inhibition of the release of PGE_2 and TxB_2 but did not modify the release of LTB_4 and LTD_4. The results on PGE_2 and TxB_2 are in agreement with previous studies. However inhibition of prostaglandin synthesis in anaphylaxis was shown by various groups to potentiate leukotriene release[11]. Our present results do not support this finding. Further studies will clarify this question.

Our results also showed that BW755C and ETYA, two inhibitors of arachidonic acid metabolism, inhibited the release of PGE_2, TxB_2, LTB_4 and LTD_4 from the perfused anaphylactic lungs. These findings confirm previous studies using isolated cell preparations[20]. The inhibition produced by ETYA was different from that of BW755C which was concentration-dependent. Low concentrations of ETYA appeared to potentiate TxB_2 release whereas it was inhibitory on PGE_2 and on leukotrienes.

The inhibition of the release of icosanoids from the anaphylactic lungs by FPL-55712 is interesting and raises a number of questions. FPL-55712 was described in 1973 as a selective SRS-A antagonist[21] and various groups have shown that FPL-55712 was an inhibitor of lung cyclooxygenase[22]. More recently this antagonist was also shown to inhibit the formation of leukotrienes. Our data confirm that FPL-55712 is a relatively potent inhibitor of both pathways of metabolism of arachidonic acid and may suggest that there are structural similarities between leukotriene receptors and the binding and/or catalytic sites of arachidonic acid on both the C-5 lipoxygenase and the cyclooxygenase.

In summary, this study analyses the time-course of release of PGE_2, TxB_2, LTB_4 and LTD_4 from the anaphylactic guinea pig lungs using novel specific and sensitive methods. The effects of selected inhibitors are also studied. The results presented confirm the major role of leukotrienes in lung anaphylaxis. The possible significance of the high amounts of LTB_4 detected in lung effluent is also emphasized.

REFERENCES

1.- Hammarström S, Murphy RC, Samuelsson B, Clark DA, Mioskowski C, Corey EJ. Biochem. Biophys. Res. Commun, 91: 1266 (1979).

2.- Murphy RC, Hammarström S, Samuelsson B. Proc. Natl. Acad. Sci., U.S.A., 76: 4279 (1979).

3.- Morris HR, Taylor GW, Piper PJ, Samhoun MN, Tippins JR. Prostaglandins, 19: 185 (1980).

4.- Kellaway CH, Trethewie ER. Q. J. Exp. Physiol., 30: 185 (1980).

5.- Brocklehurst WE. In: P Kallos, BH Waksman. Progress in Allergy, Basel, Karger, p: 539 (1962).

6.- Piper PJ, Vane JR. Nature, Lond. 223: 29 (1969).

7.- Drazen JM, Austen KF, Lewis RA, Clark DA, Goto G, Marfat A, Corey EJ. Proc. Natl. Acad. Sci. U.S.A., 77: 4354 (1980).

8.- Woodrow Weiss J, Drazen JM, Coles N, McFadden RR, Weller PF, Corey EJ, Lewis RA, Austen KF. Science, 216: 196 (1982).

9.- Borgeat P, Fruteau de Laclos B, Rabinovitch H, Picard S, Braquet P, Hébert J, Laviolette M. J. All. Clin. Immunol., 74: 310 (1984).

10.- Pradelles P, Grassi J, Maclouf, J. Anal. Chem. 57: 1170 (1985).

11.- Engineer DM, Niederhauser U, Piper PJ, Sirois P. Br. J. Pharmacol., 62: 61 (1978).

12.- Salari H, Borgeat P, Braquet P and Sirois P. J. All. Clin. Immunol., 77: 720 (1986).

13.- Sirois P, Brousseau Y, Chagnon M, Gentile J, Gladu M, Salari H and Borgeat P. Exp. Lung Res. 9: 17 (1985).

14.- Harczy M, Maclouf P, Prádelles P, Braquet P, Borgeat P and Sirois P. Pharmacol. Res. Commun. (in press).

15.- Sirois P, Roy S, Borgeat P, Picard S, Corey EJ. Biochem. Biophys. Res. Commun., 99: 385 (1981).

16.- Sirois P, Roy S, Tétrault JP, Borgeat P, Picard S, Corey EJ. Prostaglandins and Med., 7: 327 (1981).

17.- Sirois P, Harczy M, Braquet P, Borgeat P, Maclouf J and Pradelles P. Advances in Prostaglandin, Thromboxane and Leukotriene Research (submitted)

18.- Malik AB, Noonan TC, Selig WM and Garcia JGN. IN: Leukotrienes in Cardiovascular and Pulmonary function. AM Lefer and MH Gee. Alan R Liss, Inc., New York, p: 221 (1985).

19.- Liu Y, Yoden K, Shen RF, Tai HH. Biochem. Biophys. Res. Commun., 129: 268 (1985).

20.- Salari H, Braquet P and Borgeat P. Prostagl. Leukotr. and Med., 13: 53 (1984).

21.- Krell RD, Osborn R, Falcone K, Vickery L. Prostaglandins, 22: 423 (1981).

A ROLE FOR PLATELET ACTIVATING FACTOR IN SHOCK AND ACUTE LUNG INJURY

C.P. Page and D.N. Robertson
Dept. of Pharmacology, King's College, London University
Chelsea Campus, Mansera Road, London SW3

INTRODUCTION

Acute lung injury is associated with a number of clinical conditions including endotoxic and septic shock, acute respiratory distress syndrome (ARDS) and pulmonary embolism. Despite improved therapy with steroids, aprotinin and dobutamine, mortality figures still remain high, being between 50 and 75% in patients with ARDS (1). The pathophysiology of shock and acute lung injury is very complex and still remains to be elucidated. The clinical symptoms characterising shock and acute lung injury are profound haemodynamic disturbances, in particular systemic hypotension associated with a fall in peripheral vascular resistance, pulmonary hypertension, bronchoconstriction and pulmonary oedema. Additionally, there are major alterations in circulating blood elements such as thrombocytopenia, leukopenia and systemic vascular permeability. Several animal models have been utilised in attempts to further understand the pathogenesis of lung injury. These models include exposure of animals to thrombin or fibrin derived peptides (2), hypoxia (3), and more commonly, endotoxin shock (4). In particular, infusion of Escherichia Coli into unanaesthetised sheep is a widely used model of endotoxic shock which reproduces several features of the human clinical syndrome. There is an early transient phase of pulmonary hypertension and increased resistence to air flow, followed by a prolonged phase with high flow of protein-rich lung lymph indicating increased permeability and pulmonary oedema (5). Following infusion of endotoxin there is an increased reactivity of the airways to inhaled bronchoconstrictor stimuli (6), and increased airway reactivity has also been reported in patients with ARDS (7). Accompanying these changes is accumulation and disruption of lymphocytes and granulocytes in the pulmonary microcirculation, followed by migration into the interstitium (8). A massive influx of neutrophils into the lungs has also been found in patients with ARDS (9), which is thought to be due to the presence of extravascular chemotactic factors (10). Leukocytes have therefore been implicated in the pathogenesis of acute lung injury. This is further supported by the observation that neutrophil depletion in sheep significantly reduces the increase in pulmonary vascular permeability seen after endotoxin infusion (11). The exact role of these inflammatory cells in the sequence of lung injury is unclear, but the release of their toxic contents undoubtedly is involved. Platelet activation and subsequent thrombocytopenia is also a feature of shock and acute lung injury clinically. In some animal species, such as dogs and monkeys, infusion of endotoxin results in thrombocytopenia (12,13), but in the sheep, platelets do not seem to participate in the response to endotoxin, as thrombocytopenic sheep respond identically to sheep with normal circulating platelet counts (14).

Many chemical mediators are released during shock including prostanoids (15), complement-derived peptides (4), histamine, 5HT (16), kinins and endorphins (17,18). There has been a considerable amount of interest in the possible contribution of arachidonic acid metabolites to acute lung injury and shock. During the early phase of pulmonary hypertension, increased concentrations of the metabolites of both thromboxane and prostacyclin (TXB_2 and $6ketoPGF_{1\alpha}$) have been observed in both the lung lymph and plasma (19,20). Cyclooxygenase inhibitors inhibit both the pulmonary hypertension and the increase in thromboxane B_2 (21), and both thromboxane synthetase inhibitors (22), and thromboxane receptor antagonists (23), have been shown to afford protection against the acute pulmonary vascular response to endotoxin. Kubo & Kobashi (24), however, have shown that the late phase of increased vascular permeability is unaffected by inhibition of thromboxane synthesis. These results suggest that thromboxane plays a crucial role in the early transient pulmonary hypertension induced by endotoxin in sheep, but does not mediate the late permeability changes. The source of thromboxane synthesis during endotoxemia is unknown. Platelets are generally thought to be the major source, but this seems unlikely in the sheep in light of the inability of thrombocytopenia affording protection against endotoxin-induced lung injury. The role of thromboxane however remains controversial as infusion of thromboxane mimetics does not increase lung vascular permeability (25), and inhibition of thromboxane has no effect upon the sustained pulmonary hypertension that accompanies the increased vascular permeability. This has led to the investigation of the involvement of other vasoactive agents.

Several products of the lipoxygenase pathway of arachidonic acid metabolism, namely 5-HETE and 12-HETE, have been found to be released later in the endotoxin response during the period of increased vascular permeability (26,27). The peptidoleukotrienes (LTC_4, D_4, E_4) which are products of 5-lipoxygenase metabolism, exert a variety of biological actions that could contribute to shock, including bronchoconstriction and vasoconstriction. The precise role of lipoxygenase products of arachidonic acid metabolism awaits the availability of good lipoxygenase inhibitors and antagonists of leukotrienes.

5HT is another material that is released from platelets which can produce bronchoconstriction and vasoconstriction and which may play a role in a number of forms of acute respiratory failure (28,29). The $5HT_2$ antagonist, ketanserin, resulted in a decrease in the degree of pulmonary hypertension in patients with sepsis-induced ARDS (28). Demling et al. (16) found increased 5HT levels in lung lymph but not plasma in the late phase of endotoxin-induced lung injury, and ketanserin attenuated both the pulmonary hypertension and hypoxia, but did not influence the lung vascular permeability. The increased vascular permeability is thought to be largely due to endothelial cell damage from activated neutrophils. These cells release toxic oxygen radicals which have also been implicated in endotoxin-induced lung injury (4). N-acetylcysteine, a free radical scavenger, reduces both the early and late phase changes induced by endotoxin, when given in large doses to sheep.

The many studies of endotoxin-induced shock in animals has resulted in major advances in the understanding of mechanisms of lung injury, but has not led to novel therapeutic advances or an explanation of existing treatments for these clinical conditions. The inability of classical mediators to fully account for the pathology of shock and acute lung injury has prompted investigation into other mediators.

Platelet activating factor (PAF) is an ether linked phospholipid having the structure 1-0-alkyl-2-acetyl-sn-glyceryl-3-phosphorylcholine (30,31,32), and has a number of synonyms: Paf-acether, AGEPC and APRL. PAF was first identified as a product of IgE sensitised rabbit basophils that was able to elicit platelet activation (33). It is now recognised as being a product of a variety of cell types including platelets (34), neutrophils (35), macrophages (36), eosinophils (37) and endothelial cells (38).

In 1983 Myers et al.,(39) reported that i.v. administration of PAF to mice induced a lethal shock-like response which could be blocked by glucocorticoids but not by cyclooxygenase inhibitors, thus having a pharmacological sensitivity reminiscent of the clinical condition. This

Table 1. Evidence for PAF as a mediator of shock and acute lung injury

---

1) PAF can mimic the major features of shock and lung injury

> systemic hypotension
> pulmonary hypertension
> thrombocytopenia
> neutropenia
> bronchoconstriction and increased bronchial reactivity
> pulmonary oedema

2) PAF is released in models of shock and acute lung injury

3) PAF antagonists inhibit and in some cases reverse shock and acute lung injury in animal models.

---

has led to considerable interest in PAF as a possible mediator of shock and lung injury. This review examines the evidence supporting a role for PAF as a mediator of shock and lung injury adopted from the criteria for identification of a mediator proposed by Dale in 1933 (40). Table 1 summarises this evidence.

I.v. administration of PAF to experimental animals produces a range of biological effects that are similar to that observed following administration of endotoxin or gram negative bacteria and accompanying ARDS or trauma.

Haemodynamic disturbances

PAF will induce a pronounced and often fatal systemic hypotension in experimental animals (31,41,42) which has also been observed in human subjects with brain stem death (43). In experimental animals, hypotension can be induced following either i.v. or oral administration indicating that PAF generated extravascularly is able to enter the circulation and exert its hypotensive effects (44). This seems to occur in man as well, since a recent report has indicated that inhalation of small concentrations of PAF, by normal healthy volunteers, can result in facial flushing and a fall in systemic blood pressure indicative of PAF exerting an action on the cardiovascular system (45). The mechanism of PAF-induced hypotension seems likely to be due to a direct action on vascular smooth muscle since it cannot be abrogated by prior depletion of circulating blood elements such as the platelet (46,47). Futhermore, PAF is able to induce systemic hypotension in the rat, an animal species whose platelets are totally insensitive to PAF (48,49). In the dog, systemic hypotension induced by PAF cannot be significantly affected by theophylline, indomethacin, BW755c or following sympathetic denervation, suggesting that this particular cardiovascular parameter is not secondary to adenosine acting via purogenic (Pl) receptors, cyclooxygenase or lipoxygenase products of arachidonic acid metabolism or any involvement of the sympathetic nervous system (50). Direct administration of PAF into the lateral ventricles of the CNS is without effect on systemic hypotension indicating that this phenomenon does not involve central nervous pathways (51). Systemic hypotension is associated with a fall in blood flow to a number of vital organs including the brain, the liver and the kidney, and this may well contribute to the fall in total peripheral resistance following PAF administration (52).

In a variety of experimental animals, PAF induces an increase in pulmonary vascular resistance associated with an increase in pulmonary wedge pressure. Such changes are concomitant with the fall in systemic blood pressure.

Changes in blood elements

Another striking feature associated with systemic administration of PAF is the marked change in blood and plasma volume. There is a dose-related decrease in both plasma and blood volume associated with an increase in red cell volume and haematocrit (53,49). Plasma loss can be detected throughout the vasculature reflecting the capacity of PAF to increase vascular permeability (54). A variety of plasma proteins can be shown to leave the blood following systemic administration of PAF including albumin, low density lipoproteins and very low density lipoproteins (55). Utilising blue dye as a marker for plasma proteins, increased vascular permeability can be visualised throughout the systemic and pulmonary vasculature following PAF administration, a phenomenon also observed following systemic administation of endotoxin (56).

The systemic administration of endotoxin or PAF to experimental animals is associated with a variety of changes in other circulating blood elements. There is both thrombocytopenia and leukopenia with subsequent accumulation of microemboli of activated blood elements in a variety of organs. In particular the pulmonary, renal and hepatic circulations have evidence of microembolisation.

Gamma imaging has allowed detection of platelet accumulation within the pulmonary vasculature following systemic administration of PAF (57) or endotoxin (58) in animals receiving autologous 111-indium labelled platelets. Histopathologically, lung tissue removed from PAF treated animals reveals aggregates of platelets in close association with polymorphonuclear leukocytes at all levels of the pulmonary vasculature (59,60).

Pulmonary oedema

One of the major clinical features of acute lung injury complicting sepsis or shock is pulmonary oedema. Lungs taken from PAF-treated animals reveal oedematous changes both by an increase in the wet:dry ratio and by evidence of accumulation of plasma markers such as Evans blue dye or radiolabelled plasma proteins (61,62). In isolated perfused lungs, oedema formation can be induced by PAF in the absence of circulating blood elements (22), although at least in the rabbit, the addition of platelets to the perfusate increases both the pulmonary pressure and the extent of oedemaformation (63). In other tissues, including human skin, PAF has been shown to be one of the most potent mediators eliciting increased vascular permeability with associated oedema formation which seems to be in a large part due to a direct action of PAF on vascular endothelium (64). Histopathologically, lungs taken from PAF treated animals also reveal frank endothelial cell damage in addition to evidence of vascular permeability (66).

Bronchospasm and increased bronchial hyperreactivity

Bronchoconstriction is often a feature of shock syndromes, particularly anaphylaxis and acute respiratory distress syndrome. PAF is now recognised as the most potent endogenous spasmogenic agent yet identified in both experimental animals and man (67,45). As little as 10 ng/kg i.v. will induce changes in airways resistance and dynamic compliance indicative of changes in both large and peripheral airways. PAF is interesting in that it does not cause the contraction of airway smooth muscle directly, but only in the presence of platelets suggesting the release of secondary mediators from activated platelets. In vivo in rabbits and guinea pigs, selective platelet depletion will inhibit PAF induced bronchoconstriction supporting a platelet dependent process (47,67). Furthermore, a number of anti-platelet agents have been shown to reduce PAF-induced bronchoconstriction in vivo, secondary to an inhibitory effect on the platelet release reaction (46). PAF-induced bronchoconstriction is associated with an accumulation of platelet aggregates in the pulmonary vasculature but kinetic studies monitoring the accumulation of radiolabelled platelets in the pulmonary vasculature reveal that the platelet embolisation is maximal after the peak bronchoconstriction, again suggestive of some other aspect of platelet activation than aggregation contributing to the airways obstruction (57). As yet the spasmogen(s) derived from activated platelets resposible for PAF-induced bronchospasm remain to be identified. Although some investigators have suggested that arachidonic acid metabolites of both cyclooxygenase (65) and lipoxygenase metabolism (68) may play a role, this view is by no means supported by all investigators (46).

Altered bronchial reactivity to other spasmogens (both endogenous and exogenous) can be a clinical feature accompanying acute lung injury and shock. It is therefore of considerable interest that PAF is one of the only chemically defined endogenous substances capable of eliciting a non-specific increase in bronchial reactivity in both experimental animals and man (69,70,45).

EVIDENCE FOR PAF RELEASE IN SHOCK

The important question as to whether PAF is released in any given clinical or experimental situation has been hampered by the lack of a reliable specific radioimmunoassay. It had been demonstrated by a number of groups that exposure of platelets to PAF render them specifically desensitised to futher stimulation by PAF (71,72,73). This desensitisation bioassay has been utilised to detect the release of PAF as an alternative to radioimmunoassay in a variety of circumstances. Henson &Pinckard (74) first utilised this technique to demonstrate the release of PAF during IgE-induced anaphylaxis in sensitised rabbits and recently PAF release has been detected in this way in human allergic asthma (75). Very recently PAF has been demonstrated to be released into the blood of animals undergoing endotoxic shock (76), and in the lavage fluid obtained from experimental animals following acute lung injury induced by hypoxia (77,78). In the latter circumstance, PAF release has been shown to be demonstrated by chromatographic analysis of samples.

The PAF receptor has now been isolated and this has been utilised successfully as the basis of a radioreceptor assay for detecting PAF in biological fluids (79,80,81,82).

EFFECT OF SELECTIVE PAF ANTAGONISTS IN MODELS OF ACUTE LUNG INJURY AND SHOCK

In the last few years a number of selective PAF receptor antagonists have been described derived from both natural and synthetic origins (83,84,85). All of these PAF antagonists have been demonstrated to inhibit the wide range of biological activities induced by PAF in vitro and in vivo. It is now recognised that there may be at least 2 distinct PAF receptor sub-types and that the existing PAF antagonists may have preferential activity on one particular receptor sub-type. For example, the PAF antagonist kadsurenone has been shown to inhibit the PAF receptor 1 found on platelets and neutrophils preferentially to the PAF receptor 2 found on macrophages (86). This classification of the PAF receptor into distinct populations will hopefully lead to more selective compounds and allow a better understanding of the role of PAF in pathophysiological mechanisms.

A number of PAF antagonists have recently been observed to inhibit endotoxic shock in experimental animals, both the cardiovascular complications and the acute lung injury (87,88,89,90). Of particular note is the ability of PAF antagonists to reverse the cardiovascular changes induced by endotoxin once they have become established offering the possibility of this class of drugs being both preventive and curative therapy for shock related disorders (89). PAF antagonists have been successfully utilised to prevent ischaemic bowel necrosis secondary to an endotoxin induced hypotension in the mesenteric circulation and to prevent the respiratory changes associated with anaphylactic shock (91,92,93). It remains to be seen whether such PAF antagonists will be useful in the treatment of acute lung injury and cardiovascular complications of shock clinically, but the preliminary animal experiments look encouraging.

202

CONCLUSION

Dale (40) originally proposed a number of criteria for establishing the validity of a substance as a mediator of nervous transmission and the same criteria have been utilised to assess mediators of other processes. The cells and organs implicated in endotoxic shock have been shown to release PAF and release of this phospholipid has now been demonstrated in vivo in a model of endotoxin-induced lung injury. PAF is able to mimic all the features of shock and lung injury following systemic administration to experimental animals making it a plausible candidate as a mediator for these pathological conditions. Furthermore, the ability of a number of PAF antagonists to prevent and reverse certain aspects of endotoxin induced pathology in experimental animals suggests that PAF satisfies the criteria established by Dale (40) for serving the role of a mediator. It is hoped that the availability of PAF antagonists will allow a better understanding and in time provide a novel therapeutic approach to the treatment of this important clinical problem.

REFERENCES

1. Bell, R.C., Coalson, J.J., Smith, J.D., Johanson, W.G. Ann.Intern.Med 99 : 293 (1983).

2. Gerdin. B., Hogstorp, H., Lindquist, O., Saldeen, T., Svensjo, E. and Wallin, R. Forens Sci 10 : 156 (1977).

3. Voelkel, N. Am. Rev. Resp. Dis 133 : 1186 (1986).

4. Brigham, K. and Meyrick, B. Am. Rev. Resp. Dis 133 : 913 (1986).

5. Brigham, K., Bowers, R. and Haynes, J. Circ. Res 45 : 292 (1979).

6. Hutchinson, A., Hinson, J., Brigham, K. and Snapper, J. J.Appl.Physiol 54 : 1463 (1983).

7. Newman, J. Clin. Chest Med 6 : 371 (1985).

8. Meyrick, B. and Brigham, K. Lab. Invest 48 : 458 (1983).

9. Fowler, A., Walchak, S., Giclas, P., Henson, P. and Hyers, T. Chest 81 (suppl) : 50 (1982)

10. Parsons, P., Fowler, A., Hyers, T. and Henson, P. Am.Rev.Resp.Dis 132 : 490 (1985).

11. Heflin, A. and Brigham, K. J.Clin.Invest 68 : 1253 (1981).

12. Bredenberg, C., Taylor, G. and Webb, W. Surgery 87 : 59 (1980).

13. McKay, D., Margarelten, W. and Csarossy, I. Surg.Gynecol.Obstet 125 : 825 (1967).

14. Snapper, J., Hinson, J., Hutchinson, A., Lefferts, P., Ogletree, M. and Brigham, K. J.Clin.Invest 74: 1782 (1984).

15. Lefer, A. Fed.Proc 44 : 275 (1985).

16. Demling, R., Smith, M., Guenther, R., Flynn, J. and Gee, M. Am.J.Physiol 240 : H348 (1981).

17. Holaday, J. and Faden, A. _Nature_ 275 : 450 (1978).

18. Carr, D., Bergland, R., Hamilton, A., Blume, H., Kasting, N., Arnold, M., Martin, J. and Rosenblatt, M. _Science_ 217: 845 (1982).

19. Frolich, J., Ogletree, M. and Brigham, K. _Advances in Prostaglandin and Thromboxane Research_ (editors: Samuelson, B., Ramwell, P. and Paoletti, R.). Raven Press (New York) 7: 745 (1980).

20. Demling, R., Wong, C., Fox, R., Hechtman, H. and Huval, W. _Am.Rev.Resp.Dis_ 132 : 1257 (1985).

21. Ogletree, M. and Brigham, K. _Prostaglandins, Leukotrienes and Medicine_ 8 : 489 (1982).

22. Halushka, P., Cook, J. and Wise, W. _Br.J.Clin.Pharmac_ 15 : 1335 (1983).

23. Armstrong, R., Jones, R., & Wilson, N. _Prostaglandins_ 29 : 703 (1985).

24. Kubo, K., and Kobayashi, T. _Am.Rev.Resp.Dis_ 132 : 494 (1985).

25. Bowers, R., Ellis, E., Brigham, K. and Oates, J. _J.Clin.Invest_ 63 : 131 (1979).

26. Ogletree, M., Oates, J., Brigham, K. and Hubbard, W. _Prostaglandins_ 23 : 459 (1982).

27. Ogletree, M., Begley, C., King, G. and Brigham, K. _Am.Rev.Resp.Dis_ 133 : 55 (1986).

28. Huval, W., Lelaik, S., Shepro, D. and Hechtman, H. _Ann.Surg_ 200 : 166 (1984).

29. Sibbald, W., Peters, S. and Lindsay, R. _Crit.Care Med_ 8 : 490 (1980).

30. Benveniste, J., Tence, M., Varenne, P., Bidault, J., Boullet, C. and Polonski, J. _C.R.Acad.Sci_ (Paris) 289 : 1037(1979).

31. Blank, M., Snyder, F., Byers, W., Brooks, B. and Muirhead, E. _Biochem.Biophys.Res.Comm_ 90 : 523 (1979).

32. Demopolous, C., Pinckard, R. and Hanahan, D. _J.Biol.Chem_ 254: 9355 (1979).

33. Benveniste, J., Henson, P. and Cochrane, C. _J.Exp.Med_ 136 : 1356 (1972).

34. Benveniste, J., Jouvin, E., Pirotzky, E., Arnoux, B., Mencia-Huerta, J., Roubin, R. and Vargaftig, B.B. _Int.Arch.All.Appl.Immunol_ (suppl.1) 121 : (1981).

35. Lotner, G., Lynch, J., Betz, S. and Henson, P. _J.Immunol_ 124 : 676 (1980).

36. Arnoux, D., Duval, D. and Benveniste, J. _Eur.J.Clin.Invest_ 10 : 437 (1980).

37. Lee, T.C., Malone, B., Wasserman, S., Fitzgerald, V. and Snyder, F. Biochem.Biophys.Res.Comm. 105 : 1303 (1982).

38. Camussi, G., Agletta, M., Malauasi, F., Bussolino, F., Piacibello, W. Sanavio, F. and Tetta, C. In: Platelet Activating Factor Inserm Symposium No.23 (Eds: J.Benveiste and B. Arnoux) Elsevier Science Publishers p.83 (1983).

39. Myers, A., Ramsey, E. and Ramwell, P. Br.J.Pharmac 79 : 595 (1983).

40. Dale, H.H., J. Hopkins. Med.J 53 : 207 (1933).

41. McManus, L., Hanahan, D., Demopolous, C. and Pinckard, R. J.Immunol 124 : 2919 (1980).

42. Bessin, P., Bonnet, J., Apffel, D., Soulard, C., Desgroux, L., Pelas, J. and Benveniste, J. Eur. J. Pharmacol 86 : 406 (1983).

43. Gateau, O., Arnoux, B., Deriaz, H., Viars, P. and Benveniste, J. Am.Rev.Resp.Dis 129 : A3 (1984).

44. Muirhead, E., Byers, L., Desiderio, D., Smith, K., Prewitt, R. and Brooks, B. Hypertension 3 : I-107 (1981).

45. Cuss, F., Dixon, C. and Barnes, P. Am.Rev.Resp.Dis A212 (1986).

46. Vargaftig, B., Lefort, J., chignard, M. and Benveniste, J. Eur.J.Pharmacol 65 : 185 (1980).

47. Halonen, M., Palmer, J., Lohman, I., McManus, L. and Pinckard, R. Am.Rev.Resp.Dis 124 : 416 (1981).

48. Namm, D., Tadepalli, A. and High, J. Thromb.Res 25 : 341 (1982).

49. Sanchez-Crespo, M., Alonso, F., Inarrea, P., Alvarez, V. and Egido, J. Immunopharmacol 4 : 173 (1982).

50. Sybertz, E., Watkins, W., Baum, T., Pula, K. and Rivels, M. J.Pharm.Exp.Ther 232 : 156 (1985).

51. Saeki, S., Masugi, F., Ogihara, T., Otsuka, A., Koyama, Y., Nagano, M., Kumagai, A. and Kumahara, Y. Med.J.Osaka Univ 34 : 57 (1984).

52. Goldstein, B., Gabel, R., Huggins, F., Cervoni, P. and Crandall, D. Life Sciences 53 : 1513 (1984).

53. McManus, L., Pinckard, R., Fitzpatrick, F., O'Rourke, R., Crawford, M. and Hanahan, D. Lab.Invest 45 : 303 (1981).

54. Handley, D., Van Valen, R., Melden, M. and Saunders, R. Thromb.Haemostas 52 : 34 (1982).

55. Handley, D., Arbeeny, C., Lee, M., Van Valen, R. and Saunders, R. Immunopharmcol 8 : 137 (1984).

56. Etienne, A.m Hecquet, F., Soulard, C., Spinnewyn, B., Clostre, F. and Braquet, P. Agents and Actions (in press).

57. Page, C., Paul, W. and Morley, J. Int.Archs.Allergy Appl.Immunol 74 : 347 (1984).

58. Poskitt, K., Irwin, J., Mathews, J., Lane, I., Fox, B. and McCollum, C. J.Med.Micro 18 : X11 (1984).

59. Dewar, A., Archer, C.B., Paul, W., Page, C.P., MacDonald,D.M. and Morley, J. J.Pathol 144 : 25 (1984).

60. Lellouch-Tubiana, A., Lefort, J., Pirotsky, E., Vargaftig, B. and Pfister, A. Br.J.Exp.Path 66 : 345 (1985).

61. Mojarad, M., Hamasaki, Y. and Said, S. Bull.Eur.Physiopath.Resp 19 : 253 (1983).

62. Page, C., Paul, W., Dewar, A., Wood, L., Basran, G.S. and Morley, J. Agents and Actions 13 : 177 (1983).

63. Heffner, J., Shoemaker, S., Carham, E., Patel, M., McMurphy, I., Morris, H. and Repine, J. J.Clin.Invest 71 : 351 (1983).

64. Archer, C., MacDonald, D., Morley, J., Page, C., Paul, W. and Sanjar, S. Br.J.Pharmac 85 : 109 (1985).

65. Lewis, A., Derviis, A. and Chang, J. Agents and Actions 15 : 636 (1984).

66. Lewis, J.C. O'Flaherty, J.T., McCall, C.E., Wykle, R.L. and Bond, M.G. Exp.Mol.Pathol 38 : 100 (1983).

67. Vargaftig, B., Lefort, J., Wal, F., Chignard, M., and Medeiros, M. Eur.J.Pharmacol 82 : 121 (1982).

68. Bonnet, J., Thibaudeau, D. and Bessin, P. Prostaglandins 26 : 457 (1983).

69. Mazzoni, L., Morley, J., Page, C. and Sanjar, S. J.Physiol 369 : 107P (1985).

70. Chung, K., Aizawa, H., Leikauf, G., Ueki, I., Evans, T. and Nadel, J. J.Pharmac.Exp.Ther 236 : 580 (1986).

71. Henson, P. J.Exp.Med 143 : 937 (1976).

72. Lalau Keraly, C. and Benveniste, J. Br.J.Haematol 51 : 313 (1982).

73. Cargill, D., Cohen, D., Van Valen, R., Klimck, J. and Levine, R. Thromb.Haemostas 49 : 204 (1983).

74. Henson, P. and Pinckard, R. J.Immunol 119 : 2179 (1977).

75. Thompson, P., Hanson, J., Bilani, H., Turner-Warwick, M. and Morley, J. Am.Rev.Resp.Dis 129 : A3 (1984).

76. Chang, S., Henson, P. and Voelkel, N. Am.Rev.Resp.Dis 133 : A279 (1986).

77. Fasules, J., Stenmark, K., Henson, P., Voelkel, N., Tucker, A. and Reeves, J. Am.Rev.Resp.Dis 133 : A227 (1986).

78. Prevost, M., Carivers, C., Simon, M., Chap, H. and Douste-Blazy, L. Biochem.Biophys Res.Comm 119 : 58 (1984).

79. Valone, F., Coles, E., Reinhold, V. and Goetzl, E. J.Immunol 129 : 1637 (1982).

80. Hwang, S., Lee, C.S., Cheah, M. and Shen, T.Y. Biochem 22 : 4756 (1983).

81. Inarrea, P., Gomez-Cambronero, J., Nieto, M. and Sanchez-Crespo, M. Eur.J.Pharmacol 105 : 309 (1984).

82. Tuffin, D., Davey, P., Dyer, R., Lunt, D. and Wade, P. In: Mechanisms of stimulus-response coupling in platelets (Eds: J. Westwick, M. Scully, D. MacIntyre and V. Kakker) Plenum Publishing Corporation p. 83, (1985).

83. Terashita Z., Tsushima. S., Yoshioka, Y., Nomura, H., Inada, Y. and Nishikawa, K. Life Sci 32 : 1975 (1983).

84. Hwang. S., Lam, M-H., Biftu, T., Beattie, T. and Shen, T.Y. Biol.Chem 260 : 15639 (1985).

85. Braquet, P., G.B Patent 8, 418, 424, (1984) ; US Patent (1984).

86. Lambrecht, G. and Parnham, M. Br.J.Pharmac 87 : 287 (1986).

87. Doebber, T., W, M., Robbins, J., Choy, B., Chang, M. and Shen, T. Biochem.Biophys.Res.Comm 127 : 799 (1985).

88. Terashita, Z., Imura, Y., Nishikawa,K. and Sumida, S. Eur.J.Pharmacol 109 : 257 (1985).

89. Adnot, S., Lefort, J., Lagente, V., Braquet, P. and Vargaftig B.B. Lancet (in press).

90. Toyofuku, T., Kubo, K., Kobayashi, T. and Kusama, S. Prostaglandins 31 : 271 (1986).

91. Hsueh, W., Gonzlaez-Crussi, F., Arroyave, J., Anderson, R., Lee, M. and Houlihan, W. Eur.J.Pharmacol 1230 : 79 (1986).

92. Darius, H., Lefer, D.J., Smith, J.B. and Lefer, A.M. Science 232 : 58 (1986)

93. Braquet. P., Etienne, A., Touvay, C., Bourgain, R., Lefort, J. and Vargaftig, B.B. Lancet i : 1501 (1985).

MORPHOLOGICAL AND PHARMACOLOGICAL EVIDENCES FOR THE PARTICIPATION OF PAF-ACETHER IN ANAPHYLACTIC SHOCK IN THE GUINEA-PIG

A. Lellouch-Tubiana*, J. Lefort, V. Lagente, M. Cirino,
M. and B.B. Vargaftig

Institut Pasteur, Unité associée Institut Pasteur / INSERM 285
* Faculté de Médecine Necker-Enfants Malades Paris, 25 rue du
Dr. Roux, 75015 Paris, France

INTRODUCTION

Anaphylaxis in the guinea-pig is widely used as a model for human asthma. Nevertheless, no ultrastructural morphological studies of the anaphylactic shock in this animal were reported until now. Beside histamine, alternative mediators accounting for asthma have been incriminated, the peptido-leukotrienes first, and, more recently, PAF-acether was added to the list (Braquet et al, 1985; Page and Morley, 1986). This ether-linked phospholipid is known to induce various biological actions. It is also released during allergic reactions (Benveniste et al., 1972; Pinckard et al., 1979). PAF-acether administrated i.v. to the guinea-pig induces bronchoconstriction, which is histamine and cyclooxygenase-independent and platelet-dependent (Vargaftig et al., 1980). This is consistent with the fact that non-steroïdal anti-inflammatory drugs are ineffective in clinical asthma.

I.v. administration of PAF-acether induces effects similar to those of anaphylaxis, including bronchoconstriction, systemic hypotension, thrombocytopenia, neutropenia and vasopermeation.

We previously described the ultrastructural effects induced by i.v. PAF-acether in the guinea-pig lungs (Lellouch-Tubiana et al, 1985). The analogy between PAF-acether and anaphylactic shock effects led us now to compare the histological alterations induced by both anaphylactic shock and PAF-acether in the guinea-pig lungs.

MATERIALS AND METHODS

Drugs employed

Pentobarbitone (Nembutal, Lathévet, France); ovalbumin (antigen) (Worthington, New Jersey); Al $(OH)_3$; pancuronium (Pavulon, Organon, France); mepyramine maleate (Rhône-Poulenc, France); FPL 55712 (a gift from Dr. M. Sheard, from Fisons, U.K.); aspirin (lysine acetylsalicylate; Egic Laboratories, France); PAF-acether (Platelet-Activating-Factor, PAF), a gift from Prof.J.J. Godfroid (Université de Paris VII).

Sensitization procedure

30 Hartley guinea-pigs of both sexes (300-500 g) were injected s.c. with 0.5 ml of 0.9 % NaCl (saline) containing 10 μg of ovalbumin and 1 mg of Al (OH)$_3$ (Andersson and Bergstrand, 1981). This injection was repeated 14 days later and 7 days after the 2nd injection, the animals were bled to provide serum for passive sensitization. 1 ml of a pooled serum was injected i.p. to naïve animals, which were used 10-14 days later.

In vivo studies

Passively sensitized guinea-pigs were anaesthesized with 30 mg/kg i.p. of pentobarbitone, treated with the neuro-muscular blocking agent pancuronium (2mg/kg i.v.) and prepared for the recording of bronchial resistance to inflation (Lefort and Vargaftig, 1978). After controlling bronchial reactivity with serotonin (5HT, 1-3 g/kg i.v.), shock was triggered with a 1 minute i.v. infusion of 1 mg/kg of ovalbumin. For comparison purpose 66 ng/kg of PAF-acether was injected i.v. to separate animals. The potential antagonists were given i.v. 5 minutes before antigen. In separate experiments, ovalbumine or PAF-acether were administered by aerosol as described (Cirino et al., in press; Lefort et al., 1984).

Histological techniques

Lungs were removed 1 minute and 1 hour after the end of ovalbumin or PAF-acether administration. For light microscopy, they were fixed by 10% formalin, embedded in parafin and stained with hematoxylin-eosin-safranin. For electron microscopic study, 1 mm^3 fragments of lungs were fixed with 2.5% glutaraldehyde in phosphate buffer, at 4°C during 24 hours. The lung samples were then rinsed in the same buffer, post-fixed in 1% osmic acid, dehydrated and embedded in Epon. Semi-thin sections of 1 μm thickness stained with toluidine blue were examined by light microscopy. Ultra-thin sections were then prepared with a LKB ultramicrotome, stained with uranyl acetate and lead citrate, and examined with a Philips EM 300 electron microscope.

RESULTS

Systemic passive anaphylaxis (ovalbumin i.v.)

1 minute after the end of the intravenous infusion of ovalbumin, lung parenchyma showed: congested alveolar capillaries, constricted bronchi, bronchioles and arterioles. Massive platelet aggregates obstructed the lumen of small pulmonary arteries, peri-bronchial veins and veinules and occasionally alveolar capillaries (Figure 1a). These aggregates, which contained neutrophils, were formed by platelets partially or totally de-granulated (Figure 2). Unlike what was observed after i.v. PAF-acether, platelet aggregates were seldom found in the capillaries and were less disseminated during shock. (Figure 1b). Marked lesions of the vascular walls were seen in the aggregated vessels: endothelial disruption and dissociation of the basement membrane. Neutrophils remained attached to adjacent areas of damaged endothelium (Figure 3). A peri-vascular oedema dissociated the collagen fibers in the bronchial submucosa.

Images of platelet diapedesis were also observed. At the vicinity of the bronchial smooth muscle, degranulated platelets were noted containing tubules and glycogen granules (Figure 4). Neutrophils accumulated in the vessels and migrated in broncho-vascular spaces and in the septa interstitium.

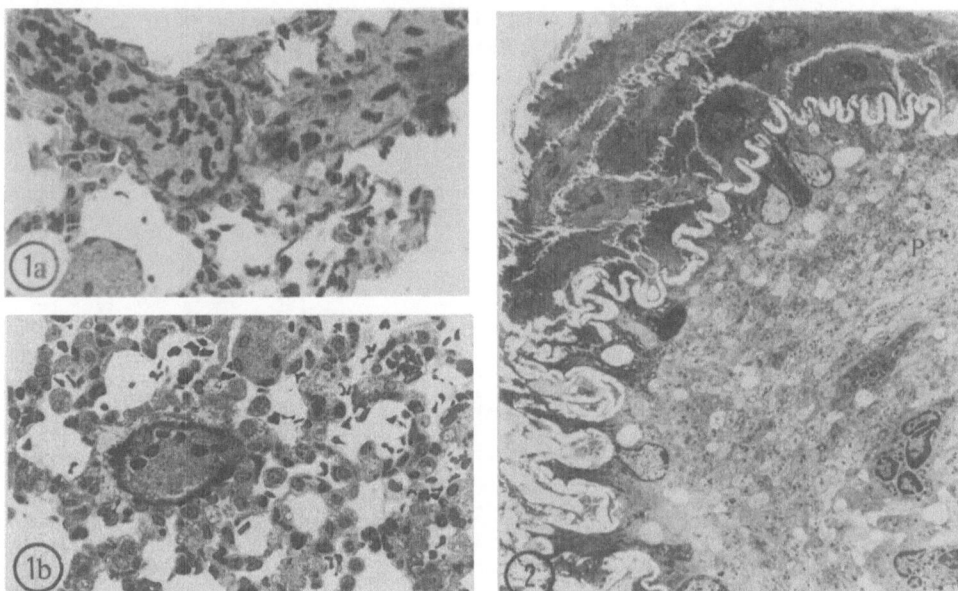

Fig. 1a. Light micrograph of the lung parenchyma 1 min after the i.v. injection of ovalbumin showing large platelet thrombi in pulmonary veinules. X200

Fig. 1b. 1 min after i.v. PAF-acether, similar platelets thrombi obstruct alveolar capillaries lumens and some veinules. X200

Fig. 2. Electron micrograph (EM) of an arterial lumen 1 min after ovalbumin i.v. The endothelium (En) is ruptured at several points. Neutrophils (Pmn) are attached to adjacent areas of damage endothelium. X17 000.

Eosinophils were present in the bronchial walls, in the septa and in the respiratory alveoli. Isolated eosinophilic granules were found in the bronchial submucosa, at the vicinity of the smooth muscle. The same infiltration of eosinophils was observed after injection of PAF-acether.

One hour after the end of the i.v. perfusion of ovalbumin, bronchoconstriction was over and platelet aggregates had disappeared. The respiratory epithelium was denuded and necrosed cells accumulated in the bronchial lumen. Numerous inflammatory cells (mastocytes, plasmocytes, eosinophils and macrophages) were found in the bronchial mucosa and sub-mucosa.

Antigen and PAF-acether aerosolization

Two minutes after the end of ovalbumin aerosolization bronchoconstriction and arterial vasoconstriction were observed.

Sequestration of neutrophils occurred in the alveolar capillaries and free neutrophils granules were noted in the capillaries lumen (Figure 5 and 6). Another important finding was the presence of activated macrophages, showing numerous pseudopods in the process of phagocytosis of surfactant in the alveolar lumen, associated with abundant oodematous alveolitis (figure 6). Septal and peribronchial mast cells were also seen, sometimes degranulated. No platelet aggregates were present.

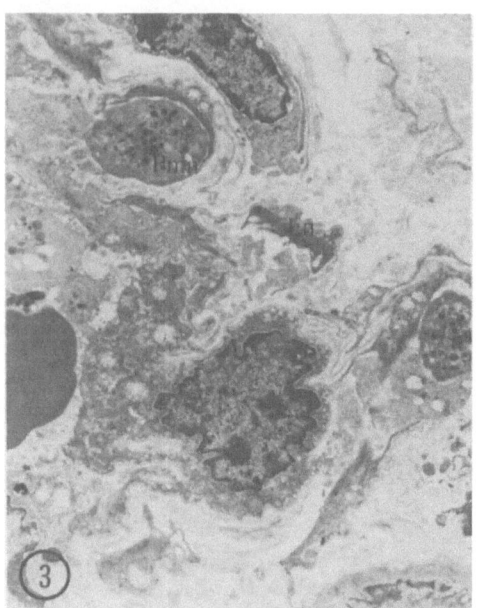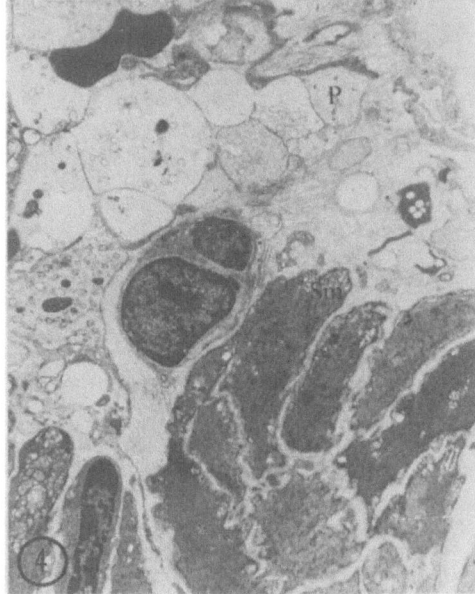

Fig. 3. EM of a submucosal bronchial veinule 1 min after ovalbumin i.v. The endothelium (En) is ruptured at several points. Neutrophils (Pmn) are attached to adjacent areas of damage endothelium. X17 000.

Fig. 4. EM of a bronchiolar submucosa showing degranulated platelets containing glycogen granules (P) in contact with the bronchial smooth muscle (Sm). X5600.

One hour after the ovalbumin aerosolization, the bronchial wall was infiltrated by numerous inflammatory cells, including plasma cells, eosinophils, mast cells and macrophages. Neutrophils remained accumulated in vascular lumens. An important finding was the presence of eosinophils in the different layers of the bronchial wall; in the epithelium, in the lamina propria and in the submucosa (Figure 7). PAF-acether inhalation induced similar effects: neutrophils accumulated in and beyond the vessels whereas no platelet aggregates were observed. One hour after PAF-acether aerosolization, neutrophils aggregates persisted. In analogy with shock, eosinophils and macrophages were present in the bronchial submucosa.

Pharmacological modulation of shock

Pre-treatment of the animals by various antagonists (the anti-histamine mepyramine, the selective leukotriene antagonist FPL 55712, the cyclooxygenase inhibitor aspirin) failed to prevent the appearance of the described histological alterations.

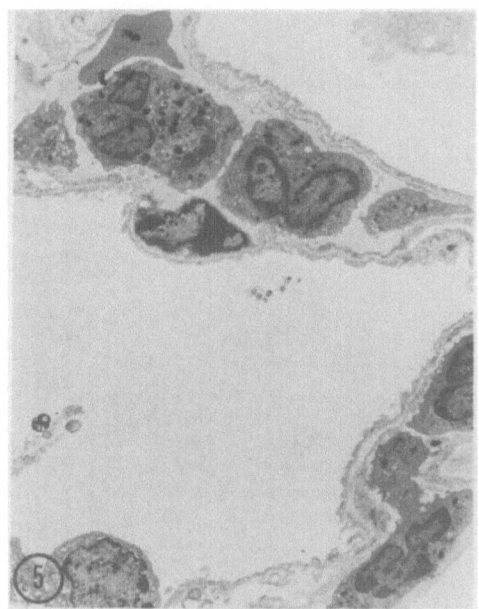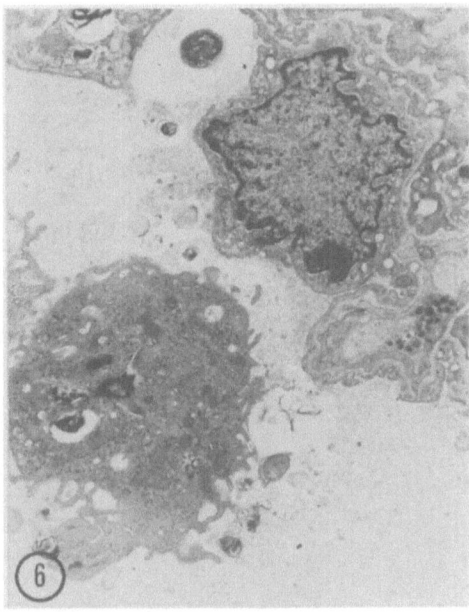

Fig. 5. EM of an alveolar wall illustrating the accumulation of neutrophils in the capillary lumen 1 min after the challenge with ovalbumin by aerosol. X3920

Fig. 6. EM of lung parenchyma 1 min after ovalbumin aerosolization. The alveolar lumen is oedematous and contains an activated macrophage with numerous pseudopods. Note the presence in the alveolar capillaries lumen of free neutrophil granules. X27900

DISCUSSION

In this study, we found out that the pulmonary histological alterations induced by passive anaphylactic shock in the guinea-pig bear a striking resemblance to those observed after PAF-acether administration. Indeed, in both cases, were observed BC, platelet aggregates and neutrophis accumulation, eosinophils infiltration, inflammatory cells diapedesis and vascular lesions.

Arterial vasoconstriction following ovalbumin i.v. was also present after PAF-acether (Lellouch-Tubiana et al., 1985), confirming the findings of Basran et al. (11). Bronchoconstriction, which is a major feature of shock and i.v. PAF-acether (Vargaftig et al., 1980) was clearly seen in our experiments 1 minute after ovalbumin injection or aerosolization, and resolved one hour later. Bronchial epithelium damage was also found after i.v. ovalbumin and PAF-acether, and the presence of free eosinophilic granules containing major basic protein in the bronchial wall suggest that they account for these lesions. Infiltration of eosinophils in the brochial wall was also noted after antigen and PAF-acether inhalation, which highlights the role of these cells in passive anaphylatic shock. Here again, the analogy with the effects of PAF-acether is striking. Indeed, infiltration of eosinophils is a well known feature of asthma (Kallos and Kallos, 1984; Frigas and Gleich, 1986; Page and Morley, 1986). The presence of large intravascular platelet aggregates in systemic passive anaphylaxis contrasts with the absence of platelet activation after antigen aerosolization. The intravascular distribution of platelet aggregates differs in passive systemic shock and after PAF-acether i.v.

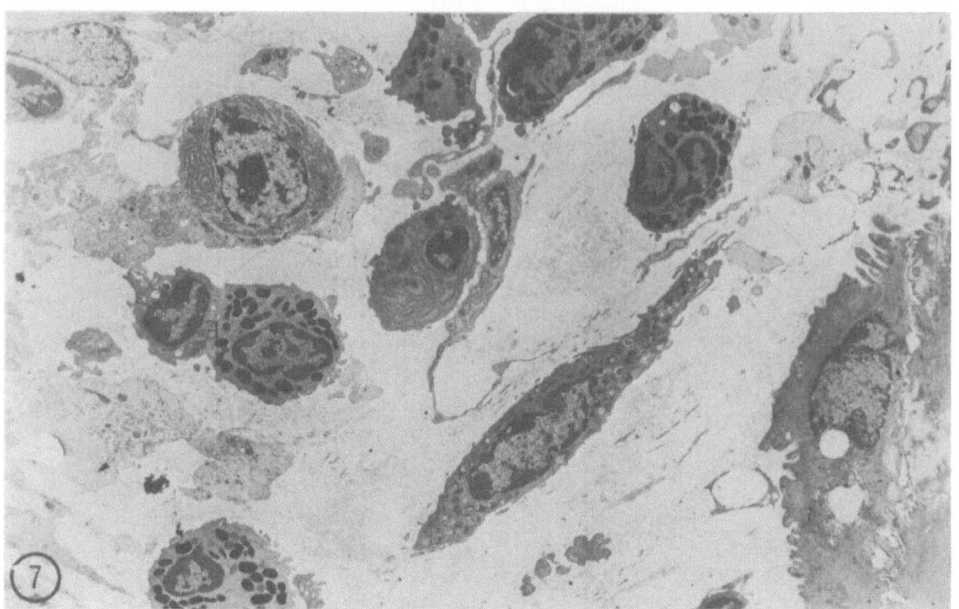

Fig. 7. EM of a bronchial wall 1 hour after the challenge with ovalbumin by aerosol showing the presence of eosinophils (E) in the submucosa. X4800

administration since in the former case, the aggregates were mostly located in small arteries and arterioles, as well as in bronchial veinules. In contrast, when PAF-acether was used, the aggregates were mostly disseminated in alveolar capillaries. One possible explanation accounting for these differences is that, with PAF-acether, platelets are a very early target and aggregates are formed in the pulmonary microvasculature as soon as PAF-acether is injected. This is clearly demonstrated by thrombocytopenia which follows in 10 seconds this injection (Vargaftig et al., 1980). On the other hand, in systemic passive anaphylactic shock, thrombocytopenia is less pronounced and develops in a few minutes. Thus, the hypothesis of a primary platelet target is unlikely. In this case, platelet aggregation may result from a secondary effect, initiated by the damaged pulmonary endothelium and/or leukocytes.

Endothelial disruption in the pulmonary veinules were observed after i.v. ovalbumin, but not after aerosol. Leukocytes were attached to adjacent areas of damaged endothelium, suggesting their involvement with those lesions, specially since free neutrophils granules were present in vascular lumens. Endothelial lesions were also reported after superfusion of PAF-acether onto the guinea-pig exposed mesentery (Bourgain et al., 1985).

Activation and clumping of alveolar macrophages was found after antigen aerosolization. Here again, the analogy with PAF-acether is striking (Maridonneau-Parini et al., 1985).

CONCLUSION

The analogies observed between the bronchopulmonary lesions induced by passive anaphylactic shock in the guinea-pig and those caused by PAF-acether are in favour of the suggestion that PAF-acether is involved in anaphylaxis and specially in asthma. The role of eosinophils is highlighted in our study, since they were found at all levels of the bronchial wall.

Eosinophils present IgE receptors and may possibly act as effector cells in immediate hypersensitivity reactions (Capron et al., 1985). Furthermore, they are very sensitive to the chemotactic effect of PAF-acether in vitro (Wardlaw and Kay, 1986) and in vivo (Henocq and Vargaftig, 1986). Activated eosinophils release different substances, including the Major Basic Protein, leukotrienes and PAF-acether (Spry, 1985). The former is found in asthma and may be toxic for the bronchial epithelium (Kay et al. in press). The lesions observed by us agree with this hypothesis.

ACKNOWLEDGEMENTS

Miss Marie-Thérèse SIMON is gratefully acknowledged for competent technical assistance.

REFERENCES

1. Braquet, B., Etienne, A., Touvay, C., Bourgain, R.H., Lefort, J. and Vargaftig, B.B. Lancet 1(8444):1501 (1985).
2. Page, C.P. and Morley, Pharmacol. Res. Com. 18, Suppl: 217 (1986).
3. Benveniste, J., Henson, P.M. and Cochrane, C.G. J. Exp. Med. 136: 1356 (1972).
4. Pinkard, R.N., Farr, R.S. and Hanahan, D.J. J. Immunol. 123: 1847 (1979).
5. Vargaftig, B.B., Lefort, J., Chignard, M. and Benveniste, J. Eur. J. Pharmacol. 65: 185 (1980).
6. Lellouch-Tubiana, A., Lefort, J., Pirotzky, E., Vargaftig, B.B. and Pfister, A. Brit. J. exp. Path. 66: 345 (1985).
7. Andersson, P. and Bergstrand, H. Br. J. Pharmacol. 74: 601 (1981).
8. Lefort, J. and Vargaftig, B.B. Br. J. Pharmacol. 63: 35 (1978).
9. Cirino, M., Lagente, V., Lefort, J. and Vargaftig, B.B. Prostaglandins, in press.
10. Lefort, J., Rotilio, D. and Vargaftig, B.B., Br. J. Pharmacol. 82:565 (1984).
11. Basran, G.S., Dewar, A., Morley,J., Page, C.P., Paul, W., Wood, L.E. Br. J. Pharmacol. 77: 437P (1982).
12. Kallos, P. and Kallos, L. Int. Archs. Allergy & appl. Immunol. 73:77 (1984).
13. Frigas, E. and Gleich, G.J. J. Allergy Clin. Immunol. 77: 527 (1986)
14. Bourgain, R.H., Andries, R., Braquet, P. and Deby, C. Prostaglandins 30: 915 (1985).
15. Maridonneau-Parini, I., Lagente, V., Lefort, J., Randon, J., Russo-Marie, F. and Vargaftig, B.B. Biochem. Biophys. Res. Commun. 131: 42 (1985).
16. Capron, M., Kusnierz, J.P., Prin, L., Spegelley, H.L., Klahife, J., Tonnel, A.B. and Capron, A. Int. Archs. Allergy appli. Immunol. 77: 246 (1985).
17. Wardlaw, A. J. and Kay, A.B. J. Allergy clin. Immunol. 77: 236 (1986).
18. Henocq, E. and Vargaftig, B.B. Lancet 11: 1378 (1986).
19. Spry, C.J.F. Immun. Today 6: 332 (1985).
20. Kay,A.B., Wardlaw, A.F., Mogbel, R., Buchanan, D.R. and Cromwell, O. in Proceedings of 1986 Symposium on "Allergy and Inflammation" (Academic Press).

BIOGENIC AMINES : MICROCIRCULATORY ASPECTS IN SHOCK, SEPSIS AND TRAUMA

D. H. Lewis
Clinical Research Center, University Hospital
S-581 85 Linköping, Sweden

INTRODUCTION

The purpose of this communication is to give a concise over-view of the effect of shock, sepsis and trauma on the handling of biogenic amines in the body and also how these substances affect the organism exposed to these pathophysiologies. I will deal primarily with some of the microcirculatory phenomena and their mechanisms, and will not deal with the use of biogenic amines or their blockers in the therapy of these disorders. Further, I will consider only the monoamines, i.e. catecholamines (including epi-nephrine, norepinephrine and dopamine), serotonin (i.e.5-hydroxy-tryptamine) and histamine. To my knowledge, the effect of shock, sepsis and trauma on polyamines has not been studied. These substances are involved in cell differentiation and growth, and it is conceivable that disorders here could have long-term effects. For a review of these substances see Heby (1986). Since much more is known about the reactions of catecholamines than about serotonin and histamine, most emphasis will be placed on catecholamines.

NORMAL FUNCTION

Catecholamines serve as nerve transmitters and hormones, both local and general. They have important normal functions, relating to regulation of microvascular tone, blood flow distribution, distribution of fluid volume amongst the various bodily compart-ments and regulation of myocardial activity. Their actions depend on a great many factors, including, the balance between a_1 and a_2 receptor activity(Alabaster and Davey, 1984) and the receptor dens-ity of the series-coupled microvessels (Burnstock et al, 1984).For serotonin and histamine physiological functions have been more difficult to establish. In the lung, serotonin increases vascular pressures, probably more post-capillary than precapillary, without appearing to alter vascular permeability (Brigham and Owen, 1975a), while histamine has a major effect on vascular permeability with less of an effect on vascular pressures (Brigham and Owen, 1975b).

Improvements in measuring techniques have meant a great deal for research in this area. At present high-performance liquid chromatography with electro-chemical detection appears to be the method of choice for catecholamine determinations (Allenmark, 1986; Oka et al, 1984). For all biogenic amines, the availability of specific blockers has also led to rapid advancement in our understanding of the actions in both animals and man.

An additional complication is the probable interplay between the actions of the various biogenic amines as well as interactions with other mediators. Thus, serotonin and histamine may well be involved in feed-back regulation of, among other things, sympathetic nerve transmission and histamine is involved in catecholamine release from the adrenal medulla (Fahmy et al, 1983). For a review of the actions of serotonin on blood vessels see Vanhoutte et al, (1984). Adenosine 5'-triphosphate (ATP) is a co-transmitter with norepinephrine (Burnstock and Kennedy, 1986) and prostaglandins of the E series inhibit norepinephrine release from nerve terminals. "It is reasonable to asume that NE release from the adrenergic terminals is controlled by a local PGE-mediated feedback mechanism, which operates through restriction of availability of calcium for the NE release process, and which seems to be particularly effic- ient within the 'physiological' frequency range of nerve impulses." (Hedqvist, 1977). Kopin (1985) has recently reviewed catecholamine metabolism and the importance of aging on adrenergic mechanisms was the subject of a symposium led by Roberts and Steinberg (1986).

EFFECTS OF ANESTHESIA

This important subject, especially for the experimentalist, has been reviewed in detail by Longnecker and Harris (1980).Studies both *in vivo* and *in vitro* indicate that anesthetics in general cause a vasodilatation. Altura et al, (1980) suggest that this is due to inhibition of normal vasomotion of vascular smooth muscle as well as a depression of the contractile responses to those endogen- ous neurohumoral substances that play a role in maintaining vasc- ular tone. This may be due to interference with calcium ion move- ment. "This may help to explain why some of the commonly used anesthetics, in particular, compromise the control of the cardio- vascular system, produce hypotension and venous pooling, and exacerbate the effects of circulatory shock." (Altura et al, 1980).

A number of studies have investigated in detail the effect of pentobarbital on the circulation both with and without shock. "In the absence of anesthesia, hemorrhage activates mainly the cardiac sympathetics (and possibly also the sympathetics innervating the capacitance vessels) with little sympathetic modifications of the overall vascular resistance. Under pentobarbital, however, hemorrhage cannot further increase the discharge of the cardiac sympathetics and the major sympathetic response is to elicit constriction of the resistance vessels." (Chien, 1971).Pentobarb- ital depresses norepinephrine output from sympathetic nerves (Göthert and Rieckesmann, 1978). It depresses the catecholamine response to hemorrhage (Farnebo et al, 1979; Adamicza et al, 1985) and has a central depressant action (Carrier and Holland, 1966).

EFFECT OF SHOCK AND TRAUMA ON BIOGENIC AMINES

Chien and Simchon (1983) have reviewed the effects on the

sympathetic and central nervous systems and Parratt (1983) has reviewed the release of neurohumoral agents.The normal action of the sympatho-adrenal system is essential for survival in shock and trauma, at least in the early stage, but may well contribute to demise of the organism in the late phase. Thus, we showed both with reduced skeletal muscle blood flow (Lewis and Mellander, 1962) and hemorrhagic shock (Mellander and Lewis, 1963) that the early action of the sympathetics was to transfer fluid from the extravascular space to the intravascular space, but that this was just the opposite later on, due to the more rapid fall off of pre-capillary responsiveness than post-capillary responsiveness. In trauma there was an immediate, marked vasodilatation (Sandegård, 1974) as well as a decreased responsiveness to sympatho-adrenal stimulation (Lewis and Lim, 1970; Lewis and Kerstein, 1970).

Vital microscopic studies have shown that the hallmark of irreversibility is the hyposensitivity of the arterioles (Zhao et al, 1985). One important factor here could be the membrane depolarizing action of catecholamines, as shown for epinephrine (Clemens et al, 1985). Even before irreversibility is reached, there is a difference in the sensitivity of arterioles depending on their size. Thus, in hemorrhagic shock Flint et al, (1984) have shown that larger arterioles become less sensitive to topically applied norepinephrine, while smaller arterioles do not. Here again, interrelationships with other mediators make the picture very complex. PGE_2 stimulates release of catecholamines (Feuerstein et al, 1981) and steroids potentiate them (Hellman, 1980). Tyrosine, a catecholamine precursor, can stimulate catecholamine synthesis (Conlay et al, 1985), thus possibly acting in a compensatory manner, but the synthetic pathway can lead instead to tyramine and may then contribute to sympathetic nerve system exhaustion (Hamburger and Henry, 1986). In the brain, both catecholamines and serotonin show an early decrease followed by a subsequent partial recovery (Laborit et al, 1984a), and the situation can be improved with a combination of tyrosine, glucose, insulin and vitamin C (Laborit et al, 1984b).

In Table 1, are summarized some of the important microcirculatory aspects of the early, compensatory actions of the sympatho-adrenal system to hemorrhage and shock. Note that all of these actions are compensatory in nature and are truly essential for a proper response to the injurious stimulus. It can be suggested that one important function of vasoconstriction is to keep the microcirculatory hematocrit at its normal low level. Insulin resistance, which develops in this state, has often been looked upon as an inappropriate response. Ware (1982) has, however, pointed out that this allows the body to use the sympathoadrenally-induced hyperglycemia to pull water out of the cell, thus aiding in the refilling of both the extra- and intra-vascular spaces, at the expense of the much larger intracellular space.

One of the major differences between shock and trauma is that the traumatized tissue has a depressed reactivity to catecholamines, which means that the involved tissue cannot take part in the compensatory reactions. Catecholamines are also released into the injured tissues, and at least for the nervous system have been shown to participate in the injury, possibly by the production of oxygen free radicals (Kurihara, 1985). The possible importance of this for the patient has been emphasized by Young and Becker(1982): "The fact remains that the spinal cord is often not transected by trauma and that a delayed detrimental pathophysiological event

Table 1. Acute Response of Microcirculation to Hemorrhage & Shock

Vascular Response	Physiological Effect	Significance for Homeostasis
generalized vaso-constriction, both pre- and post-capillary	increased peripheral vascular resistance	maintained arterial blood pressure in the face of low cardiac output
	decreased (or maintained) microcirculatory hematocrit	decreased (or maintained) microcirculatory blood viscosity
increased tone of capacitance vessels	venoconstriction with decrease of blood in peripheral veins	adjusted size of vascular bed to reduced blood volume; movement of blood volume centrally
greater increase in precapillary resistance than in post-capillary resistance	decrease in capillary hydrostatic pressure with net inward movement of extravascular fluid	restoration of circulating blood volume
opening up of "pre-capillary sphincters"	increase in size of capillary bed available for fluid exchange	restoration of circulating blood volume more even distribution of available capillary blood flow
metabolic response sympatho-adrenal effect on liver, pancreas and adrenal medulla	release of glucose from liver--> hyper-glycemia--> increased osmolality of plasma and extravascular space	restoration of circulating blood volume and extra-vascular volume

occurs that can be reversed. It is possible that a secondary injury of the spinal cord may be prevented by proper immobilization, manipulation of blood pressure and *alteration of biogenic amines or other metabolites at the trauma site*." (Italics are mine).

Serotonin appears to be involved in the hypoxia-induced pulmonary vasoconstriction (Lauweryns and Cokelaere, 1973), possibly mediated by the serotonin found in the pulmonary neuroepithelial bodies (Lauweryns et al, 1986). This response is important in preventing blood from traversing unoxygenated alveoli. In the peripheral vessels, anoxia potentiates the constricting action of serotonin (Shepherd and Vanhoutte, 1985). The puzzling array of both constrictor and dilator actions of serotonin can perhaps be explained in part by the possibility that it modifies adrenergic transmission (Vanhoutte et al, 1984). Finally, of interest clinically is the observation that serotonin constricts collateral blood vessels (Verheyen et al, 1984).

Histamine, coming mainly from mast cells, has a major action on vascular permeability. The original proposal by Majno et al, (1967), that this is accomplished by endothelial cell contraction with opening up of the interendothelial junction is still the subject of debate. Histamine also plays a role in the hyperglycemia seen with shock. It stimulates release of catecholamines from the adrenal medulla by a H_1-receptor mechanism (Dimlich, 1985).

EFFECT OF ENDOTOXIN AND SEPSIS ON BIOGENIC AMINES

Nagler (1980) has reviewed the effects of endotoxin on the circulation and microcirculation and Schrauwen's doctoral thesis (1986) explores in detail effects on the mesenteric circulation of the pig. The effect of endotoxin on the microcirculation has been studied by vital microscopy (Urbaschek and Urbaschek, 1975; Baker and Wilmoth, 1984). Table 2 taken from the data of Baker and Wilmoth (1984) shows that in rats given endotoxin intravenously, there is a marked depression in the ability of both arterioles and venules of all sizes of the cremaster muscle to respond to norepinephrine applied topically. Similarly, McKenna et al (1985) have observed that sepsis decreases vascular contractility to norepinephrine. The explanation for these observations is not clear, but there are changes in receptor density (Shepherd et al, 1986) and in receptor sub-types (Winbery et al, 1986) on exposure of tissue to endotoxin. In addition, septic plasma affects the ability of rat myocytes *in vitro* to produce cAMP upon stimulation by epinephrine (Carmona et al, 1985).

Endotoxin causes marked elevations of plasma catecholamines in conscious rats. "It would appear that the elevation in plasma catecholamines may be a direct indication of the severity of endo-

Table 2. Videomicroscopy of rat cremaster muscle. Endotoxin: 6 mg/kg iv during a 1 hr period. Norepinephrine applied topically to muscle. Threshold dose (-log molar conc) necessary to produce vasoconstriction. Data from Baker and Wilmoth (1984)

Type of microvessel	Approx. diam in μm	Control	Post-endotoxin max. change	at time,min
large order arteriole	140	8.5	4.1	120
first order branch arteriole	80	8.4	4.1	120
second order branch arteriole	55	8.8	5.2	120
large order venule	170	8.2	4.0	120
first order branch venule	80	7.9	4.5	90
second order branch venule	60	8.1	4.7	90

toxicosis and under certain conditions have potential as a predictor of ultimate shock and death." (Jones and Romano, 1984).
The observations of Benedict and Grahame-Smith (1978) indicate that
the same is true for humans. They found higher and longer lasting
levels of catecholamines in non-survivors than in survivors.

The studies of McKechnie et al, (1985) confirmed the data of
Jones and Romano (1984) and explored in more detail the role of the
sympatho-adrenal system in this response. In addition to studying
the effect of endotoxin on intact rats, they noted the effects of
"sympathectomy" with guanethidine and/or adrenal demedullation. It
was shown that "sympathectomy" had some effect on the levels of
plasma catecholamines and on mortality, while adrenal demedullation
blocked both the catecholamine and hyperglycemic responses and had
a devastating effect on mortality, as shown in Table 3.

Studies in sepsis and septic models have revealed myocardial
dysfunction (Spitzer et al, 1986) as well as dysfunction of the
peripheral vasculature in septic patients. Thus, Siegel et al
(1967) found a decreased peripheral vascular resistance independent
of the flow level in patients in septic shock as compared to those
suffering from shock of other etiologies. This picture resembles
the peripheral vascular status seen in hepatic failure, suggesting
the possible production of false transmitters (Siegel, 1986), as
suggested by Fischer and Baldessarini (1971). They reasoned that
amines absorbed from the gut, normally metabolized in the liver,
would not be in the presence of hepatic failure. These amines
would then overflow into the circulation and accumulate in nerves.
The same could well be true in shock, in view of the decreased
barrier function of the gut in this situation. This would then be
a slight modification of the original hypothesis put forward by
Fine and colleagues as to the mechanism of irreversibility in shock
(see Fine, 1970).

Table 3. Effect of "Sympathectomy" (SY) and/or Adrenal Demedullation (AD) on plasma catecholamines and glucose, and on mortality
in rats. Data taken from McKechnie et al, 1985.

Endotoxin-induced plasma increases	SY	AD	SY+AD
Epinephrine	early-no effect late-reduced	blocked	blocked
Norepinephrine	early-no effect late-reduced	reduced	markedly reduced
Glucose	no effect	blocked	blocked

Mortality	Endotoxin alone	SY	AD	SY+AD
1 hr	0/10	0/7	0/8	0/8
4 hr	0/10	1/7	6/8	7/8
8 hr	3/10	3/7	8/8	8/8

As Thijs et al, (1984) have pointed out, the hemodynamic
problems in patients with septic shock involve 4 main factors.
These include peripheral pooling, peripheral vascular failure,
increased permeability and myocardial failure. It is not difficult
to see that most of these items can be related to alterations in
the reactivity to catecholamines and the excess presence of the
other biogenic amines, as well as other mediators.

Serotonin and histamine are undoubtedly involved in the pathophysiological changes seen in endotoxemia and sepsis, but their
exact roles have been difficult to establish. Since there are great
species variations in these substances, care must be taken in
interpreting experimental data. In sheep, Emau et al, (1984) found
elevated serotonin levels with endotoxin, and no change in
histamine levels. In rats, Schauer (1975) found elevated histamine
levels upon exposure to endotoxin, higher in non-survivors than in
survivors. Serotonin, coming to a large extent from platelets and
from the gut may (Stein and Thomas, 1967) or may not (Allison et
al, 1981) play a role in the pulmonary changes seen with endotoxin.
In dogs, Murphy et al, (1981) concluded that endotoxin did not
cause immediate, marked platelet aggregation and that serotonin
from these cells did not seem to play a major role in the immediate
pulmonary response. Using the H1-blocking agent, diphenhydramine,
Brigham et al, (1980) concluded that endogenous histamine was
responsible only in part for the increase in pulmonary vascular
permeability seen with endotoxin and was not involved in the pulmonary hypertension.

ROLE OF ENDOTHELIUM

The metabolism of biogenic amines by the endothelium has
recently been reviewed by Shepro and Dunham (1986). It is becoming
increasingly clear that the endothelial cells play a major role in
many important metabolic processes, not the least involving biogenic amines. Of great importance for the control of the microcirculation is the realization of the role of the endothelium in
mediating vasodilator responses (Furchgott et al, 1984). For a long
list of vasoactive substances, including for example, serotonin,
the ability of the substance to produce vasodilatation depends on a
functioning endothelium, which via an endothelial-derived relaxant
factor (or factors) causes the smooth muscle to relax. In the
absence of a properly functioning endothelium, or if the substance
reaches the smooth muscle directly from the outside of the vessel,
there will then be vasoconstriction. Shepherd and Vanhoutte (1985)
point out how this may well be one of the explanations for coronary
artery vasospasm. It is not necessarily the only mechanism for the
production of vasospasm, however, as Svendgaard et al, (1985) have
shown that cerebral vasospasm after subarachnoid hemorrhage in the
rat is mediated via intracerebral catecholamine pathways.

Since much of the endothelium is found in the lung, which
receives all of the venous drainage before blood is delivered to
the systemic circulation, the function of the pulmonary endothelium
is of paramount importance. It can be considered to protect the
body from an overflow of local hormones, by removing them from the
circulating blood. Thus, epinephrine, a general hormone, is not
metabolized by pulmonary endothelium, whereas norepinephrine, a
local hormone, is removed. We have suggested (Post and Lewis,
1979)that failure of this function could result in toxic actions by
the body's own mediators. Thus, cardiac arrhythmias could be due

in part to failure of the pulmonary endothelium to remove norepinephrine from the venous return. The role of endothelium in the metabolism of serotonin and the problems that can arise when this function fails has been described in detail by Vanhoutte et al, (1984).

FINAL COMMENTS

Catecholamines, functioning as nerve transmitters and both local and general hormones, play a central role in the effects of shock, sepsis and trauma on the microcirculation. They are vital for survival in the early state, but may contribute to death in the later stages. When and how this about-face occurs is not clear. It may well be that many of the other mediators also important in these pathophysiological states have their effect by modulating catecholamine effects. The reverse is also a very good possibility.

Since the normal functions of serotonin and histamine are not really understood, it is unclear how these are altered in shock, sepsis and trauma. What is clear is that they are produced in excess and appear to have, in general, detrimental effects, since blockade of their effects does not worsen the organism's status, as is the case for catecholamines. In some cases there is an improvement.

What we know about the biogenic amines in these states is much less than what we do not know. Much work remains to be done.

ACKNOWLEDGEMENTS

The original results reported in this communication were supported in part by grants-in-aid from the Swedish Medical Research Council (Project Nr. 02042), the Swedish National Defence Research Institute (Project Nr. D 60), the County Council of Östergötland, and Linköping University.

REFERENCES

1. Adamicza A., Tarnoky K., Nagy A. and Nagy S. Acta Physiol Hung 65: 239 (1985).
2. Alabaster V. and Davey M. J Cardiovasc Pharmacol 6: S365 (1984).
3. Allenmark S. in: HPLC and Its Application in Endocrinology (Editors: H.L.J. Makin and R. Newton), Springer-Verlag (Wien, New York) (In Press)
4. Allison R.C., Murphy T.L., Weisman I.M., McCaffree D.R. and Gray B.A. J Appl Physiol 50: 185 (1981).
5. Altura B.M., Altura B.T., Carella A., Turlapaty P.D.M.V. and Weinberg J. Fed Proc 39: 1584 (1980).
6. Baker C.H. and Wilmoth F.R. Circ Shock 12: 165 (1984).
7. Benedict C.R. and Grahame-Smith D.G. Quart J Med 47: 1 (1978).
8. Brigham K.L. and Owen P.J. Circ Res 36: 761 (1975a).
9. Brigham K.L. and Owen P.J. Circ Res 37: 647 (1975b).
10. Brigham K.L., Padove S.J., Bryant, D., McKeen, C.R., Bowers, R.E. J Appl Physiol 49: 516 (1980).
11. Burnstock G., Griffith S.G. and Sneddon, P. J Cardiovasc Pharmacol 6: S344 (1984).

12. Burnstock G. and Kennedy C. Circ Res 58: 319 (1986).
13. Carmona R.H., Tsao T., Dae M. and Trunkey D.D. Arch Surg 120: 30 (1985).
14. Carrier O. Jr. and Holland W.C. Can J Physiol Pharmacol 44: 176 (1966).
15. Chien S. Proc Soc Exp Biol Med 136: 271 (1971).
16. Chien S. and Simchon S. in: Handbook of Shock and Trauma. Vol.I: Basic Science. (Editors: B.M. Altura, A.M. Lefer and W. Schumer) Raven Press (New York) p. 149 (1983).
17. Clemens M.G., Chaudry I.H. and Baue A.E. Circ Shock 16: 55 (1985) (abstract).
18. Conlay L.A., Maher T.J. and Wurtman, R.J. Brain Res 333: 81 (1985)
19. Dimlich R.V.W. Ann Emerg Med 14: 91 (1985).
20. Emau P., Giri S.N. and Bruss M.L. Circ Shock 12: 47 (1984).
21. Fahmy N.R., Sunder N. and Soter N.A. Clin Pharmacol Ther 33: 615 (1983).
22. Farnebo L.-O., Hallman H., Hamberger B. and Jonsson, G. Circ Shock 6: 109 (1979)
23. Feuerstein G, Jimerson D.C. and Kopin, I.J. Amer J Physiol 240: R166 (1981)
24. Fine J. Gastroenterology 59: 301 (1970).
25. Fischer J.E. and Baldessarini R.J. Lancet 2: 75 (1971)
26. Flint L.M., Cryer H.M., Simpson C.J. and Harris, P.D. Surgery 96: 240 (1984)
27. Furchgott R.F., Cherry P.D., Zawadzki J.V. and Jothianandan D. J Cardiovasc Pharmacol 6: S336 (1984).
28. Göthert M. and Rieckesmann J.-M. Experientia 34: 382 (1978).
29. Hamburger S.A. and Henry D.P. Circ Shock 18: 344 (1986) (abstract)
30. Heby O. News in Physiol Sci (In Press)
31. Hedqvist P. Annu Rev Pharmacol Toxicol 17: 259 (1977).
32. Hellman A. Thesis, University of Göteborg, Sweden, (1980) (In Swedish)
33. Jones S.B. and Romano F.D. Circ Shock 14: 189 (1984).
34. Kopin I.J. Pharmacol Rev 37: 333 (1985)
35. Kurihara M. J Neurosurg 62: 743 (1985)
36. Laborit H., Baron C., Ferran C. and Henriet I. Res Commun Chem Pathol Pharmacol 43: 425 (1984a).
37. Laborit H., Baron C., Ferran C. and Henriet I. Res Commun Chem Pathol Pharmacol 43: 435 (1984b).
38. Lauweryns J.M. and Cokelaere M. Z Zellforsch 145: 521 (1973)
39. Lauweryns J.M., van Ranst L. and Verhofstad A.A.J. Cell Tissue Res 243: 455 (1986).
40. Lewis D.H. And Kerstein M.D. Eur surg Res 2: 12 (1970).
41. Lewis D.H. and Lim R.C.Jr. Acta orthop scand 41: 17 (1970).
42. Lewis D.H. and Mellander S. Acta physiol scand 56: 162 (1962).
43. Longnecker D.E. and Harris P.D. in : Microcirculation. Volume III.(Editors: G. Kaley and B.M. Altura) University Park Press (Baltimore) p. 333 (1980).
44. Majno G., Gilmore V. and Leventhal M. Circ Res 21: 833 (1967).
45. McKechnie K., Dean H.G., Furman B.L. and Parratt J.R. Circ Shock 17: 85 (1985).
46. McKenna T.M., Briglia F.A., Chernow B. and Roth B.L. Circ Shock 16: 81 (1985) (abstract).
47. Mellander S. and Lewis D.H. Circ Res 13: 105 (1963).
48. Murphy T.L., Allison R.C., Weisman I.M. McCaffree, D.R. and Gray, B.A. J Appl Physiol 50: 178 (1981).
49. Nagler A.L. in: Microcirculation. Volume III.(Editors: G. Kaley and B.M. Altura) University Park Press (Baltimore) p. 107 (1980).

50. Oka K., Kojima K., Togari A. and Nagatsu T. _J Chromatogr_ 308: 43 (1984).
51. Parratt J.R. in: Handbook of Shock and Trauma. Vol. I: Basic Science.(Editors: B.M. Altura, A.M. Lefer, and W. Schumer) Raven Press (New York) p. 311 (1983).
52. Post C. and Lewis D.H. _Acta pharmacol toxicol_ 45: 218 (1979).
53. Roberts J. and Steinberg G.M. _Fed Proc_ 45: 40 (1986)
54. Sandegård J. _Acta chir scand_, Suppl 447 (1974).
55. Schauer A. in: Gram-Negative Bacterial Infections and Mode of Endotoxin Action.(Editors: B.Urbaschek, R.Urbaschek and E. Neter). Springer-Verlag (Wien, New York) p.315 (1975).
56. Schrauwe, E.M.J. Thesis, University of Utrecht, The Netherlands, 1986.
57. Shepherd J.T. and Vanhoutte P.M. _Mayo Clin Proc_ 60:33 (1985).
58. Shepherd R.E., Lang C.H. and McDonough K.H. _Circ Shock_ 18: 368 (1986) (abstract).
59. Shepro D. and Dunham B. _Annu Rev Physiol_ 48: 335 (1986).
60. Siegel J.H. First Vienna Shock Forum (In Press)
61. Siegel J.H., Greenspan M. and Del Guercio L.R.M. _Ann Surg_ 165: 504 (1967).
62. Spitzer J.J., Smith L.W. and McDonough K. First Vienna Shock Forum (In Press).
63. Stein M. and Thomas D.P. _J Appl Physiol_ 23: 47 (1967).
64. Svendgaard N.A., Brismar J., Delgado T.J. and Rosengren E. _Stroke_ 16: 602 (1985).
65. Thijs L.G., Teule G.J.J. and Bronsveld W. _Resuscitation_ 11: 147 (1984).
66. Urbaschek B. and Urbaschek R. in: Gram-Negative Bacterial Infections and Mode of Endotoxin Action.(Editors: B.Urbaschek, R. Urbaschek and E. Neter). Springer-Verlag (Wien, New York) p.323 (1975).
67. Vanhoutte P.M., Cohen R.A. and van Nueten J.M. _J Cardiovasc Pharmacol_ 6: S421 (1984)
68. Verheyen A., Vlaminckx E., Lauwers F., van den Broeck C. and Wouters, L. _Arch int Pharmacodyn_ 270: 280 (1984).
69. Ware J. _Acta chir scand_, Suppl 511 (1982).
70. Winbery S.L., Smith L.W., McDonough K.H. and Barker L.A. _Circ Shock_ 18: 343 (1986) (abstract)
71. Young H.F. and Becker D.P. in: Pathophysiology of Shock, Anoxia, and Ischemia. (Editors: R. A. Cowley and B. F. Trump) Williams & Wilkins (Baltimore/London). p. 613 (1982)
72. Zhao K.-S., Junker D., Delano F.A., and Zweifach B.W. _Microvascular Res_ 30: 143 (1985).

TRYPSIN-LIKE ACTIVITY IN BURN AND SEPSIS

G. Deby-Dupont[1], M.E. Faymonville[2], J. Pincemail[3], A. Adam[4],
R. Goutier[3], M. Lamy[2], P. Franchimont[1]

[1] Laboratory of Radioimmunology, CHU, Sart-Tilman, Liège, Belgium
[2] Dept. of Anesthesiology, University of Liège, Bavière Hospital,
66 Bd de la Constitution, B-4020 Liège, Belgium. [3] Laboratory of
Applied Biochemistry, University of Liège, Sart-Tilman, 4000
Liège 1, Belgium. [4] Laboratory of Clinical Biology,
Centre Hospitalier de Saint-Ode, B-6970, Baconfov, Belgium

INTRODUCTION

A variety of insults (such as trauma, burns, acute pancreatitis and
so on...) are accompanied by shock and often lead to the development of
Adult Respiratory Distress Syndrome (ARDS) which defines a clinical and
physiological sydrome characterized by an increase in pulmonary permeability with interstitial edema, subsequent hypoxemia and severe respiratory
failure and by a high rate of mortality (50 to 70 % of patients) (1,2).

Sepsis is largely involved in either the etiology or the complication
of shock and ARDS. It appears more and more that "respiratory failure is
not the major reason for death" (3) and that "most of the deaths occuring
more than 72 hours after the events or illness that resulted in ARDS are
related to sepsis syndrome" (4).

Shock and ARDS also result in multi-organ system failure. Like the
kidney, the pancreas is highly vulnerable during shock, particularly when
pancreatic ischemia is possible due to hypoperfusion. This could be a
critical factor leading to pancreatitis with proteinases release (5).

These enzymes are able to activate complement leading to polymorphonuclear cells activation with production of activated oxygen species. They can
destroy proteins and membranes and activate many other zymogens such as
prekallikrein, prothrombin, plasminogene, and so on... (6,7). One of these
proteinases is known as particularly dangerous : it is trypsin. In acute
pancreatitis patients, shock and adult respiratory distress syndrome are
frequent (8). But, until now, the presence of this enzyme in blood has
never been described except for acute pancreatitis.

We have previously reported the results of trypsin measurements in
ARDS patients : a significant correlation appeared between ARDS and
abnormal immunoreactive trypsin (IRT) and more particularly between sepsis
and IRT (9,10). These results were confirmed by Nicod et al. who found a

19-fold increase in IRT in ARDS patients after a mean of six days evolution (11).

Further studies were pursued in multiple injured patients, in patients who underwent major surgical procedure and in severely burned patients, at the admission to the Intensive Care Unit (ICU) and during several days of hospitalisation. We studied the evolution of trypsin release in blood in parallel with amylase and lipase and tried to correlate our observations with the development of ARDS and sepsis.

MATERIEL AND METHODS

Patients

123 patients admitted to the Intensive Care Unit (ICU) of the University Hospital of Liege have been studied (Table I). They were divided in three groups : the first group consisted of 89 patients : 65 multiple injured patients with at least three major injuries (head, chest, pelvis or limbs) leading to severe shock and 24 patients who underwent major abdominal surgical procedures on digestive tract or abdominal aorta and needed long duration intensive care post-operatively. 53 (59,5%) of these patients presented septic phenomena. 48 (54%) patients of this group developed ARDS.

The second group included 13 multiple injured patients who were studied immediately after their arrival in the emergency unit. Diagnosis of shock was established in 11 out of these patients. Eight of them developed ARDS and 7 presented septic phenomena.

The third group of patients consisted of 21 burned patients classified in heavy and benign burned patients according to the measurements of the Burn Skin Area (BSA) and the Units of Burned Skin (UBS) as proposed by Sachs and Watson (12). The 13 heavy burned patients had more than 30% BSA and more than 40% UBS. Cause of burn was essentially flam (16 patients). Out of these 21 patients, 10 developed ARDS generally with sepsis.

Table I : Classification of the 123 studied patients (pts).

Group 1 n=89	65 multiple injured patients with sepsis : 35 pts	24 abdominal surgery patients with sepsis : 18 pts
Group 2 n=13	13 multiple injured patients	
	with shock : 11 pts	with shock and ARDS } 8 pts
Group 3 n=21	21 burned patients heavy burns : 8 pts	benign burns : 13 pts

Sepsis and ARDS criteria

Sepsis was considered as present when there were fever, white blood cells count greater than 12.000 cells/mm^3 and local signs of infection.

The criteria for ARDS were a non cardiogenic edema without an associated increase in capillary wedge pressure and severe arterial hypoxemia requiring mechanical ventilation as previously described (9,10). The severity of this syndrome was estimated by using the respiratory failure score established by Morel et al. (13). In the ARDS patients, this respiratory failure score remained above 2 during four to five consecutive days.

Sampling procedure

Blood sampling started within 24 hours of admission to the ICU for all the patients of the first group. Arterial or venous samples were drawn twice daily, until the patient died or left the ICU.

In group 2, the first blood sample was taken immediately after arrival in the ICU. The following samples were taken every 6 hours during the first 30 hours after injury, twice a day the second and third days and once a day during 10 days.

In group 3, the first blood sample was taken within the first six hours after thermal injury and the second between the 6th and 16th hours. After this period, blood samples were taken twice a day, in the morning and in the afternoon. Some patients were studied over a twelve days period.

Blood was drawn form short catheters in polystyrene tubes containing heparin for trypsin determination. For amylase, lipase, α_1 proteinase inhibitor (α_1PI) and α_2 macroglobulin (α_2M) determinations, serum was used. After centrifugation (10 min.,1500xg), plasma or serum samples were stored at -30° C until assay.

Plasma or serum assay

Immunoreactive trypsin (free trypsin, trypsinogen and the complex trypsin - α_1-proteinase inhibitor) was measured by radio-immunoassay technique using a rabbit antiserum to human cathodic trypsin, non labelled human cathodic trypsin as standards and ^{125}I-labelled human cathodic trypsin as tracer.

The lower limit of sensitivity is 5 µg/L and the mean normal value 33.0 ± 12.3 µg/L. The value of 70 µg/L (mean value plus three standard deviations) was taken as the upper limit of normal. Enzymatic methods were used for lipase and amylase measurements and laser nephelometry for α_1 PI and α_2M. Upper normal values for amylase and lipase were respectively 140 and 290 U/L. Mean normal values was 2.6 g/L for α_1PI and 2.4 g/L for α_2M.

RESULTS

1. Correlation between immunoreactive trypsin and sepsis in multiple injured and abdominal surgery patients

Fig. 1 presents the mean of highest IRT values measured in the 89 patients of group 1. The mean values are high in septic as well as in non septic patients when compared to the normal mean value of 33 µg/L. In non septic patients, the mean of highest IRT values was 96 µg/L of plasma. Eleven patients presented value above 70 µg/L (upper normal value), but IRT values higher than 500 µg/L were never measured. In 25 patients, IRT always remained within the normal range.

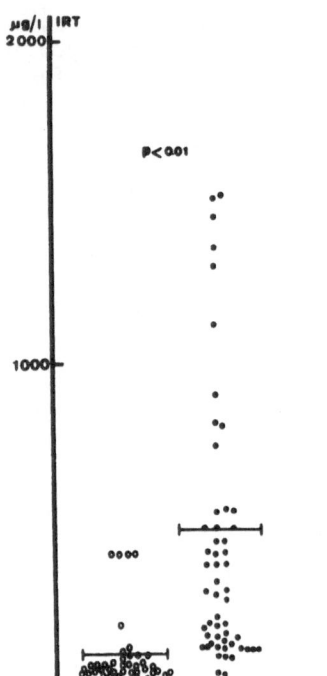

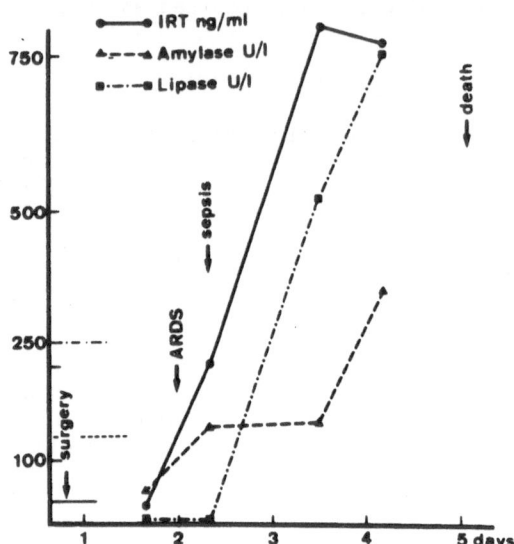

Fig. 1. Highest IRT values measured in septic (•) and non septic (o) patients of group 1. (⊢——⊣) indicates the mean value in µg/L plasma

Fig. 2. Evidence of pancreatic failure in an abdominal surgery male patient. (——) : mean normal value for IRT (33ng/ml); (----) : upper normal value for amylase (140 U/L); (-.-.-) : upper normal value for lipase (290 U/L).

High IRT values were correlated with the presence of septic phenomena (p < 0.01). In septic patients, the mean IRT value reached 476 µg/L and only two patients always remained below the upper limit of normal. Values higher than 500 µg/L were found in 13 patients. These high IRT values were measured during several consecutive days and were thus not a transitory phenomenon. Plasmatic IRT started rising one or two days before the clinical signs of sepsis became evident and reached its maximal value when sepsis was well established. Then it slowly declined with improvement of patient or remained high until death.

ARDS occurred on days 1 to 2 in multiple injured patients and in abdominal vascular surgery patients. It occurred on days 2 to 5 for the digestive tract operation patients. In accord with previous studies, mortality was higher in patients who developed ARDS (2,4). IRT rises were concomitant or occurred one or two days after the onset of ARDS.

Fig. 2 illustrates these observations in a male patient who underwent a surgical procedure on digestive tract. On day 2, ARDS was present. Septic phenomena were evident in the afternoon of the same day. At this moment, IRT was already high (200 µg/L). It continued rising the following day and remained high until death. Amylase and lipase presented a similar but delayed evolution.

2. Simultaneous release of IRT, amylase and lipase.

In 13 multiple injured patients with major injuries and severe shock (group 2) IRT, amylase and lipase were measured on blood samples taken at well defined times after injury.

Fig. 3 shows the mean values obtained for IRT, amylase and lipase at each sampling time. Until day 5th, these mean values were obtained on the 13 patients; from day 6th to day 13th, the mean values were established on 7 remaining patients with sepsis and ARDS (one patient died and the five others had left the ICU, after improvement). The mean IRT value was abnormal on day 4th and always remained high in the following days, reaching a maximal mean value on day 11th. Amylase and lipase presented a similar evolution. The drop of mean values observed on day 13th accompanied an improvement of patients.

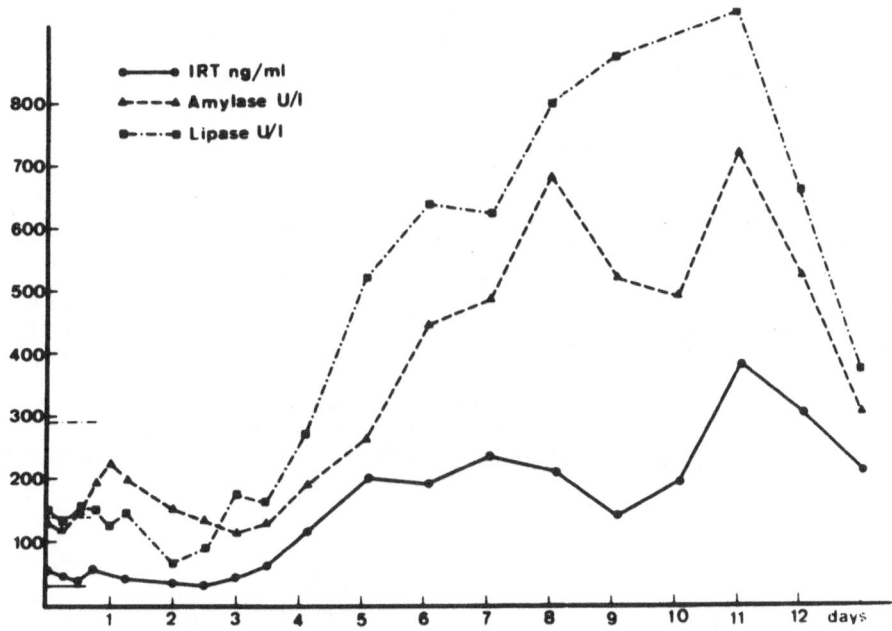

Fig. 3. Simultaneous release of IRT, amylase and lipase in the 13 multiple injured patients of group 2.

——— : mean normal value for IRT

---- : upper normal value for amylase

-.-. : upper normal value for lipase

3. Pancreatic failure in burned patients

The mean value for IRT, amylase and lipase were calculated for each sampling time. As shown on fig. 4, all along the study, the means of the three measured substances were found normal. However, mean IRT and mean amylase raised slowly in the last days particularly when compared to the values obtained on day one. On day 7th, the mean amylase value reached the upper limit of normal.

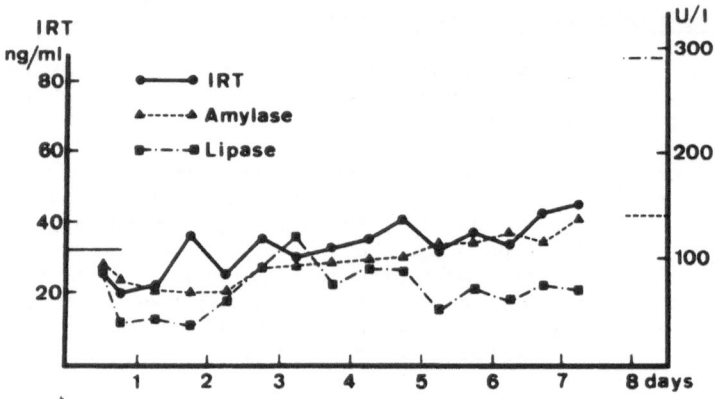

Fig. 4 Evolution of pancreatic markers in
        burned patients (group 3). Left ordi-
        nate : immunoreactive trypsin in µg/L.
        ———— : mean value for IRT
        right ordinate : amylase and lipase in
        U/L.
        ---- : upper limit of normal value for
              amylase
        -.-. : upper limit of normal value for
              lipase

It seemed thus that burned patients, unlike multiple injured and
abdominal surgery patients, did not developed a pancreatic failure.
However, surprisingly, in four heavy cases with ARDS and sepsis, the
markers of a pancreatic suffering were present : IRT, amylase and lipase
became abnormal on day 3th, and remained high for several days. Two of
these heavy burned patients died on day 7th with IRT values above 500 µg/L.

Fig. 5 illustrates one of these particular cases. On fig. 5A, are
presented the evolution of IRT, amylase and lipase and on fig. 5B, the
evolution of the two anti-proteinases (α_1PI and α_2M) in the same patient.

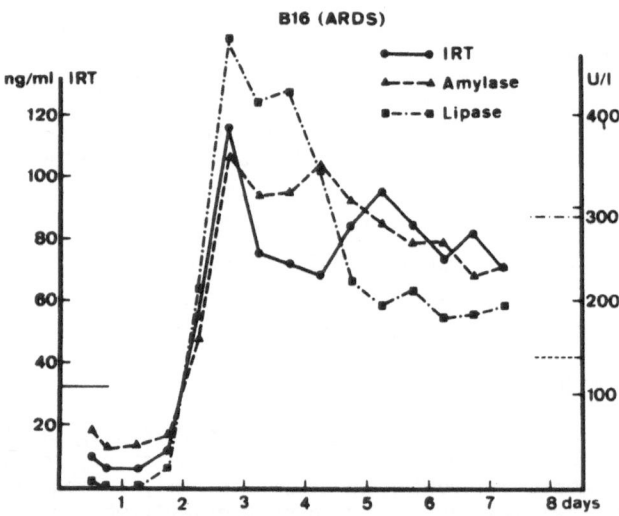

Fig. 5A. Pancreatic failure in a 24 years
         male patient burned at 42% UBS.
         Legend : see fig. 4.

Trypsin, amylase and lipase were abnormal from day 3th; the curves drawn for these three substances presented a parallel evolution; however, lipase quickly returned to normal value (day 4th) while trypsin and amylase remained either abnormal or at limit of the upper normal value. Curiously, we observed abnormally low values for the three measured substances on the first day, 6 to 24 hours after thermal injury.

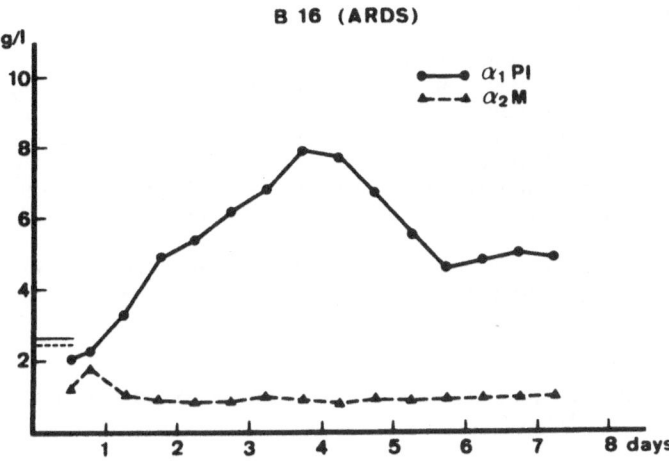

Fig. 5B Evolution of the two main plasmatic
proteinase-inhibitors in the same patient
as fig. 5A.
α_1PI : α_1 proteinase inhibitor
———— : mean normal value for α_1PI
α_2M : α_2 macroglobulin
———— : mean normal value for α_2M

As concerned α_1PI and α_2M, their evolution was similar to what had been previously observed in patients at risk of developing ARDS (14). α_2M dropped dramatically after 24 hours and remained very low (below 1 g/L) all along the study; α_1PI behaved as an acute phase protein, reaching aroung 8 g/L on day 4th remaining above 4 g/L until the end of blood sampling.

DISCUSSION

1. Pancreatic failure in critically ill patients

We observed very important releases of immunoreactive trypsin (reaching 10 to 50 - fold the mean normal value of 33 µg/L) in multiple injured and abdominal surgery patients in correlation with sepsis. In burned patients, we observed moderate releases of the same enzyme. At this point, it is important to note that our technique measured trypsinogen, potential free trypsin and trypsin bound to α_1PI, but not the trypsin bound to α_2M. In complex with this large molecule, trypsin remains enzymatically active and thus very dangerous. So, by our radioimmunological technique, we do not obtain any information about the presence of trypsin- α_2 macroglobulin complexes which represent the most important part of trypsin complexes in plasma (15,16). When immunoreactive trypsin was measured in patients plasma, it appeared late after injury (4 to 5 days) but persisted for several days. Its appearance was generally concomitant or a little delayed to the onset of ARDS : trypsin can thus not be considered as a marker of ARDS useful for an early diagnosis of this syndrome. On the contrary, the release of trypsin in plasma always preceded (by at least 24 hours) the clinical signs of sepsis; so, this enzyme appears as an early marker of sepsis syndrome which is a frequent cause of death in ARDS and burned patients (4).

We also observed a release of amylase and lipase in plasma of our patients. In most cases, this release accompanied the release of trypsin; this simultaneous increase of IRT, lipase and amylase is classically associated with the diagnosis of pancreatic disorder. It seems thus that pancreatic failure is frequent after severe injury and shock despite the fact that a diagnosis of pancreatitis was never established in our patients. In a previous paper, we already mentioned the simultaneous release of trypsin and amylase in ARDS and septic patients (9); however, the origin of amylase may be extra-pancreatic and Weaver et al. have underlined the presence of non pancreatic amylase in critically ill patients plasma (17).

With our new observations about the simultaneous release of the three substances (trypsin, amylase and lipase), we are now able to confirm that pancreatic failure occurs during shock.

2. What are the ways of penetration of trypsin in blood and how could it accumulate in plasma ?

During shock, there is a circulatory stagnation : trypsin can be absorbed from gastrointestinal tract. During shock, a pancreatic ischemia is possible due to hypoperfusion (5) : this is a critical factor leading to pancreatitis as we have observed in our patients. During this pancreatic failure, trypsin and trypsinogen escaping from pancreas accumulate in abdominal cavity. From there, these pro-enzyme and enzyme can flows out in blood via the lymph of thoracic duct.

In plasma, it seems that trypsin does not remain free : it is quickly bound by the two main antiproteinases, α_1PI and α_2M. However, in our patients, an unbalance between proteinases and antiproteinases cannot be excluded : α_1PI levels were elevated, but it appears more and more that this antiproteinase is rapidly inactivated (at level of a methionine residue) by toxic forms of oxygen and the chlorinated derivatives which are produced by activated leucocytes during ARDS and shock (18,19). On the contrary, plasma levels of α_2M were low (below 50% of normal value) in our patients. A similar fall in α_2M plasmatic levels was also observed in septic shock patients by Witte et al. (20). α_2M is normally present in high concentration in plasma and is able to entrap nearly all known proteinases, particularly serine proteinases, facilitating their removal via reticulo endothelial system.

Our observations in severely ill patients suggest that there are a consumption and clearance of this inhibitor indicating a saturation by proteinases released not only by leucocytes but also by pancreas.

3. What are the consequences of this pancreatic failure in severely ill patients ?

The main consequence of this pancreatic failure will be a release of numerous enzymes such as phospholipases, elastase, lipase ... (22). Among all these enzymes, trypsin appears as the most dangerous.

Trypsinogen, released from pancreas and appearing in blood is inactive.

However, it will be rapidly activated to trypsin by autoactivation or by action of other proteinases released, for example, by activated polymorphonuclear leucocytes (21). This active trypsin is able to activate all the other zymogens of pancreatic origin such as prophospholipases, prokallikrein. It can also activate plasmatic prothrombin and plasminogen leading to coagulation, fibrinolysis disorders and bradykinin release (7,10). It activates complement with a degradation of complement fraction C_3 (6). This complement activation will entertain the granulocytes activation in patients. Trypsin also destroys proteins and tissues, and when bound to α_2M, remains a potent endopeptidase. In this way, pancreas failure may contribute to the occurence and to the degree of severity of shock and acute respiratory failure. The released enzymes will be secondary mediators enhancing tissues and membranes destruction and increasing the endothelial permeability responsible for lung dysfunction.

CONCLUSIONS

Trypsin-like activity was detected in critically ill patients, particularly in correlation with sepsis. Trypsin release in plasma was accompanied by abnormal levels of amylase and lipase, suggesting a pancreatic failure in these patients. This pancreatic failure may contribute to the degree of severity of shock and acute respiratory failure.

REFERENCES

1. Blaisdell FW, and Lewis FR. in : Major problems in clinical surgery, vol. 21. (ed : P.A. Ebert) Saunders (Philadelphia) p 85 (1977).
2. Zapol W, Snider MT, Hill JD et al. JAMA 242 : 2193 (1979).
3. Petty TL. Am Rev Respir Dis, 132 : 471 (1985).
4. Montgomery B, Stager MA, Carrico CJ, and Hudson LD. Am Rev Respir Dis, 132 : 485 (1985).
5. Warshaw AL, and O'Hara PJ. Ann Surg, 188 : 197 (1978).
6. Lasson A, and Ohlsson K. Biochim Biophys Acta, 709 : 227 (1982).
7. Walsh KA, and Wilcox PE. Methods Enzymol, 19 : 31 (1970).
8. Boumghar M, and Cavin R. Schweiz Rundschau Med, 38 : 1394 (1978).
9. Deby-Dupont G, Haas M, Pincemail J, Braun M, Lamy M, Deby C, and Franchimont P. Intensive Care Med, 10 : 7 (1984).
10. Lamy M, Deby-Dupont G, Pincemail J, Braun M, Duchateau J, Deby C, Van Erck J, Bodson L, Damas P, and Franchimont P. Bull Eur Physiopathol Respir, 21 : 221 (1985).
11. Nicod L, Leuenberger C, Seydoux C, Rey F, Van Melle G, and Perret Cl. Am Rev Respir Dis, 131 : 696 (1985).
12. Sachs A, and Watson T. in : The immune consequences of thermal injury (ed : J. Ninnemann). Williams & Wilkins (Baltimore) p 1 (1981).
13. Morel DR, Dargent F, Bachmann M, Suter PM, and Junod A. Am Rev Respir Dis, 132 : 479 (1985).
14. Lamy M, Faymonville ME, Adam A, et al. (paper submitted for publication).
15. Barrett AJ, and Starkey PM. Biochem J, 133 : 709 (1973).
16. Travis J, and Salvesen GS. Annu Rev Biochem, 52 : 655 (1983).
17. Weaver DW, Busuito MJ, Bouwman DL, Wilson RF. Crit Care Med 13 : 532 (1985).
18. Carp H, and Janoff A. J Clin Invest, 63 : 793 (1979).
19. Cohen AB. Am Rev Respir Dis, 119 : 953 (1979).
20. Witte J, Jochum M, Scherer R, Schramm W, Hochstrasser K, and Fritz H. Intensive Care Med 8 : 215 (1982).
21. Janoff A, White R, Carp H, Harel S, Dearing R, and Lee D. Am J Pathol, 97 : 111 (1979).
22. Nevalainen TJ. Scan J Gastroenterol 15 : 641 (1980).

SIGNIFICANCE OF THE ACTIVATION OF THE KININ SYSTEM FOLLOWING THERMAL INJURY

J. Damas[1], A, Adam[2], J. Marichi[3] and J. Lecomte[1]

[1] Institut Léon Fredericq, Physiologie Humaine,
   Université de Liège, Liège, 4020 Belgium
[2] Centre Hospitalier de Sainte-Ode, Baconfoy-Tenneville, Belgium
[3] Centre des Brûlés, Hôpital Herriot, Lyon, France

INTRODUCTION

Bradykinin and related kinins have been shown to produce all four of the cardinal signs of the inflammatory response, vasodilatation, increased vascular permeability, pain and, in higher doses, accumulation of leucocytes (1). Some observations have described the appearance of kinins in inflammatory exudates induced by thermal injury. Roche e Silva and Antonio (2) observed that immersing rat paw in hot water induced the appearance of a kinin-like activity in the perfusate of the subcutaneous spaces. This observation was confirmed several times by the same group (3, 4, 5). Edery and Lewis (6) detected a kinin-forming activity in the lymph draining the burned limb of the dog. Kinins could thus participate in the inflammatory response to burn.

Kinins are released from plasma proteins, the kininogens by several specific (plasma or tissue kallikreins) or nonspecific (e.g. trypsin) enzymes. Human plasma contains two kininogens, low molecular weight (LMW) and high molecular weight (HMW) kininogens. Both kininogens are formed by two chains surrounding bradykinin. The heavy chain is similar for both kininogens, while the light chain is characteristic for each kininogen. The light chain of HMW-kininogen is a cofactor for the initiation of the contact system of blood coagulation, and exists in complexes with plasma prekallikrein and coagulation factor XI. HMW-kininogen is the preferential kinin-forming substrate for plasma kallikrein (7, 8). Recently, the heavy chain of both kininogens was shown to contain two potential reactive site sequences, Gln-Val-Val-Ala-Gly, that inhibit thiol-proteinases (9, 10). Kininogens thus apparently have two functions: they release pro-inflammatory kinins and are anti-inflammatory α2-cysteine proteinase inhibitors.

The Plasma Kinin System in Burned Patients

We have developed specific radioimmunoassays for the heavy chain of kininogens and for the light chain of HMW-kininogen (11).These assays can be used to estimate the plasma content in total kininogens and in HMW-kininogen directly. The difference between total kininogen and HMW-kininogen gives the level of LMW-kininogen (11).Whe showed that "in vitro",a release of kinins accompanies a decrease of the level of kininogens estimated by radioimmunoassays (12). Moreover, we developed an automated method for the measurement of plasma prekallikrein enzyme activity (13). For both kinds of assays, we established reference values in a large population of normal subjects (11, 13). We showed that the plasma level in kininogens and prekallikrein is similar whatever the sex or the age of the subject (11, 13).

Very few results about the involvement of kinins in man after thermal injury have been reported (14). Thus in the first place, we quantified the plasma levels of prekallikrein and kininogens in 36 severely burned patients admitted to the "Centre des Brûlés", of Herriot Hospital in Lyon. Upon admission to the intensive care unit, the plasma level of proteins in the patients was reduced by 13,5 %, that of HMW-kininogen was diminished in the same range (- 14 %), while the plasma levels of prekallikrein (- 26 %) and of total kininogens (-26 %) were more drasticaly reduced. The plasma content in kininogens and in prekallikrein stayed at these low values during the first three days following thermal injury (15).

We have shown (15) that reduction in total kininogens is directly proportional to the extent of the burn injury estimated by the Burn Skin Area (BSA) or by the Burn Unit Skin (BSU) of Sachs and Watson (16). We have also observed, as others (17), a reduction in complement plasma level. In our patients, there was a positive correlation between the plasma level in total kininogens on the one hand and on the other hand the plasma level in C_{3c} and C_4 (15).

A decrease of prekallikrein activity in plasma suggests that this enzyme is activated by the injured surface and bradykinin released from HMW-kininogen. However HMW-kininogen was slightly reduced in our patients. This apparent discrepancy could be explained by the protection of the immunogenicity of the light chain of this kininogen in spite of the release of kinins as we have shown "in vitro" (12).

The Kinin System in Blister Fluids

The decrease of total kininogens can be explained, at least, in part by an extravasation through the burned skin. Indeed, in six other patients, we were able to obtained blister fluids (18). These fluids contained $23 \pm 4.1 \ g.L^{-1}$ of proteins, or 30 % plasma level. They also contained prekallikrein and kininogens. Prekallikrein was present in the same proportion as plasma proteins while total kininogen level was relatively 2.5 times higher than plasma proteins (18). Jacobsen and Waaler (19) previously observed large increase of kininogens in lymph after thermal injury. Like ourselves, they showed that scalding limbs of the dog or of the rabbit did not induce the appearance of active kallikrein in lymph (19). Similarly, Armstrong et al. (20) found no active kallikrein but prekallikrein in blister fluid induced by other means. Thus, on the surface of the skin, there is no activation of kinins but a loss of prekallikrein and mainly of kininogens. This loss would explain part of the decrease of the plasma level.

The Inflammatory Response to Burn in Kininogen-Deficient Animals

To estimate the importance of the involvement of kinins in the inflammatory response to burn, we compared the edemas induced by heating (30 sec; 60°C) the paw of normal Wistar rats and of kininogen-deficient Brown Norway rats. The plasma of our Brown Norway rats from the strain BN/May Pfd f contains only T-kininogen but not the two usual kininogens and does not form kinins by the usual ways (21, 22). T-kininogen is not a kinin-forming substrate for kallikreins and releases kinins with very large concentrations of trypsin (23). It is an acute phase plasma protein and is identical with α_1-cysteine proteinase inhibitor (10). We observed that this inflammatory response induced by heating the paw has similar pattern and extent in both strains of rats (24). Other factors would thus be responsible for the inflammatory response. Indeed, Green (25) has described some differences between the exudates induced by bradykinin and by thermal injury. In the experiments of Rocha e Silva and his collaborators (2, 4, 5), the thermal insult was milder than in our experiments.

CONCLUSIONS

Following thermal injury, there is an activation of the kinin system as demonstrated by the decrease of plasma prekallikrein level. The role of kinins in the inflammatory response is apparently minor as the edema induced by heating is not modified by the lack of these polypeptides, as shown in a specific strain of rats. There is also an accumulation of kininogens on the skin and in blister fluids. In blister fluids, kininogens could inhibit cysteine proteinases.

REFERENCES

1. Lewis, G. In : Handbook of experimental Pharmacology (editor: E. Erdos). Springer Verlag (Berlin, Heidelberg, New York) 25 : p 516 (1970).
2. Rocha e Silva, M. and Antonio, A. Med. Exp. 7 : 371 (1960).
3. Rocha e Silva, M. and Rosenthal, S. J. Pharmacol. exp. Ther. 132 : 110 (1961).
4. Garcia Lemme, J., Hamamura, L. and Rocha e Silva, M. Brit. J. Pharmacol. 40 : 294 (1970).
5. Limaos, F., Borges, D., Souza-Pinto, J., Gordon, A. and Prado, J. Br. J. exp. Pathol. 62 : 591 (1981).
6. Edery, M. and Lewis, P. J. Physiol. (London) 169 : 568 (1963).
7. Movat, H. In : Handbook of experimental Pharmacology (editor: E. Erdos). Springer Verlag (Berlin, Heidelberg, New York) 25 suppl. p. 1(1979).
8. Colman, R. J. Clin. Invest. 73 : 1249 (1984).

9. Ohkubo, I., Kurachi, K., Takasawa, T., Shikawa, H. and Sasaki, M. Biochemistry 23 : 5691 (1984).
10. Muller-Esterl, W., Fritz, H., Kellermann, J., Lottspeich, E., Machleidt, M. and Turk, V. FEBS Letters 191 : 221 (1986).
11. Adam, A., Albert, A., Calay, G., Closset, J., Damas, J. and Franchimont, P. Clin. Chem. 31 : 423 (1985).
12. Adam, A., Damas, J., Ers, P., Albert, A., Stas, J. and Lecomte, J. Pathol. Biol. 34 : 19 (1986).
13. Adam, A., Azzouzi, M., Boulanger, J., Ers, P., Albert, A., Damas, J. and Faymonville, M. Clin. Chem. Clin. Biochem. 23 : 203 (1985).
14. Colman, R. and Wong, P. In : Handbook of experimental Pharmacology (editor: E/ Erdos). Springer Verlag (Berlin, Heidelberg, New York) 25 suppl. p. 569 (1979).
15. Adam, A., Damas, J., Albert, A., Ers, P., Marichi, J., Calay, G. and Laurent, P. Thrombosis Res. 41 : 537 (1986).
16. Sachs, A. and Watson, T. In : The immune consequences of thermal injury (editor: J. Ninnemann). Williams and Wilkins (New York) p. 1 (1981).
17. Moore, F., Davis, C., Rodrick, M., Mannick, J. and Fearon, D. New Engl. J. Med. 314 : 948 (1986).
18. Marichi, J., Adam, A., Damas, J. and Lecomte, J. C.R. Soc. Biol. 179 : 748 (1985).
19. Jacobsen, S. and Waaler, B. Brit. J. Pharmacol. 27 : 222 (1966).
20. Armstrong, D., Jepson, J., Keele, C. and Stewart, J. J. Physiol. (London) 135 : 350 (1957).
21. Damas, J. and Adam, A. Experientia 36 : 586 (1980).
22. Damas, J. and Adam, A. Mol. Physiol. 8 : 307 (1985).
23. Greenbaum, L. Biochem. Pharmacol. 33 : 2943 (1984).
24. Damas, J., Remacle-Volon, G. and Adam, A. Int. J. Tissue Reactions 6 : 391 (1984).
25. Green, K. Br. J. exp. Pathol. 59 : 38 (1978).

THE ROLE OF LIPID MEDIATORS IN BLOOD FIBRINOLYSIS

J.J. Emeis
Gaubius Institute for Cardiovascular Research Herenstraat 5d
2313 AD Leiden, The Netherlands

INTRODUCTION : THE FIBRINOLYTIC SYSTEM

The fibrinolytic system is the physiological counterpart of the coagulation system. Formed from fibrinogen by thrombin after activation of the coagulation cascade, fibrin is not meant to be a permanent structure in the body, but a temporary one bound to be removed by proteolytic enzymes. The most important of these enzymes is plasmin that, by limited proteolytic cleavage, degrades the insoluble polymer fibrin into soluble fragments, the fibrin degradation products. As plasmin is a broad spectrum protease and can degrade many plasma proteins, it will not normally be present in the blood in its active form, but as an inactive pro-enzyme (plasminogen). Plasminogen is activated into plasmin by other specific proteases : the plasminogen activators. Of the three types of plasminogen activators present in the blood, tissue-type plasminogen activator (t-PA) is considered to be the most important for intravascular fibrinolysis. The role of the other two plasma plasminogen activators in intravascular fibrinolysis (plasma pro-urokinase, and the Factor XII-dependent plasminogen pro-activator, both circulating in blood as pro-enzymes) is still obscure. Not only is fibrin the substrate for plasmin, but it also dramatically accelerates plasminogen activation by t-PA, thus virtually limiting plasmin formation to the surface of its substrate fibrin. Moreover, fibrin protects plasmin from inactivation by protease inhibitor α_2-antiplasmin, which rapidly inactivates plasmin in the fluid phase. Plasminogen activator activity is regulated by other specific protease inhibitors, the plasminogen activator inhibitors (PA inhibitors). The fibrinolytic system is schematically depicted in Fig. 1. For more detailed descriptions of the fibrinolytic system, the reader is referred to recent reviews (1-5).

In the present paper the -relatively few - data regarding the role of lipid mediators (especially the eicosanoids and platelet-activating factor) in fibrinolysis will be reviewed. Also, a brief survey will be given of the changes in the fibrinolytic system in trauma and sepsis.

LIPID MEDIATORS AND CELLULAR PLASMINOGEN ACTIVATOR SYNTHESIS

The cellular sources of the plasma plasminogen activators (PA) are only partly defined. Tissue-type PA, the major plasma PA involved in

OUTLINE OF THE FIBRINOLYTIC SYSTEM

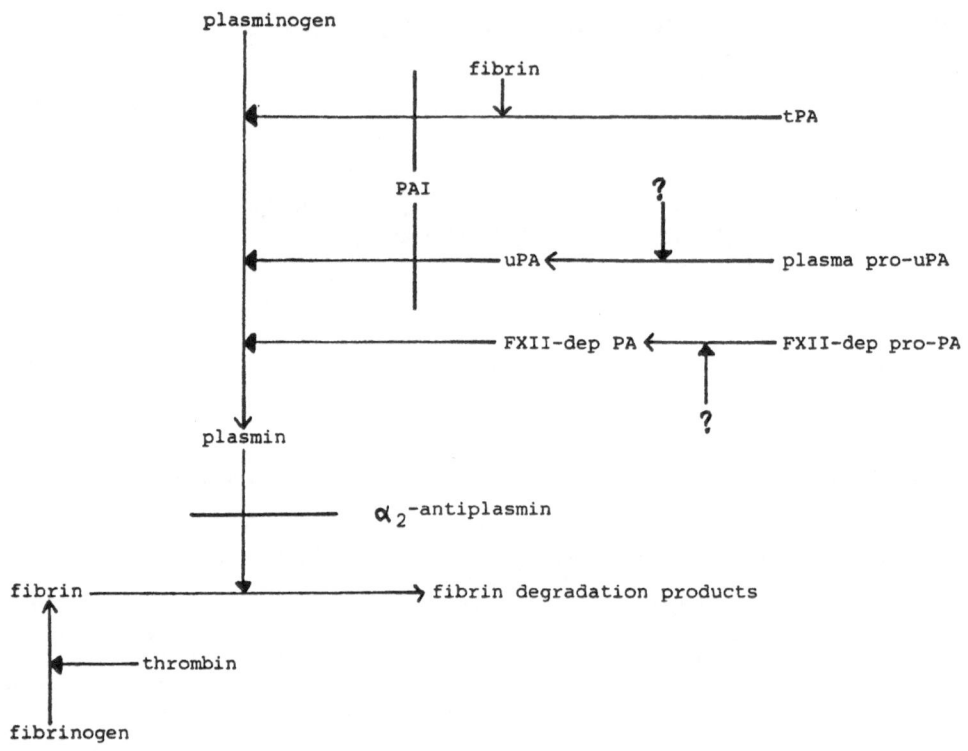

Abbreviations used:

PA   :plasminogen activator

tPA   :tissue-type plasminogen activator

u-PA   :urokinase-type plasminogen activator

pro-uPA   :pro-urokinase (single-chain urokinase)

FXII-dep   :coagulation factor XII-dependent

PAI   :plasminogen activator inhibitor(s)

Fig. 1

fibrinolysis, is synthesized by vascular endothelial cells _in vitro_ (6,7) and is present in endothelial cells _in vivo_, as determined by histochemical (8,9) and immunohistochemical (10) techniques. From said cells it is released into the blood (11). At present, no data are available on the possible effects of lipid mediators on endothelial cell t-PA synthesis, either _in vivo_ or _in vitro_.

Urokinase-type PA immunoreactivity is widely distributed in many cell types throughout the body (12) ; which of these contribute to the level of circulating plasma u-PA is unknown.

Many cell types synthesize and secrete PA's _in vitro_ (for recent reviews, see 13,14) ; depending on the type of cell, the activator synthesized can be either u-PA (generally as its inactive precursor pro-uPA), t-PA, or both activators. Since methods to differentiate between these two types of activators have only recently become available, the interpretation of previously reported data on the effects of lipid mediators on PA synthesis is in many cases hampered by incomplete knowledge of the type of activator studied. Moreover, in may studies PA _activity_ in cell extracts or conditioned media was assayed, and as many cells are now known to produce PA inhibitors, activity data should be interpreted with caution since changes in PA inhibitor synthesis will also lead to changes in the PA activity measured.

In HeLa cells (which produce t-PA, ref. 15,16) Crutchley and co-workers (17,18) showed that phorbol myristate acetate (PMA) induced increased synthesis of both E- and F-type prostaglandins, and of t-PA (17). As increased prostaglandin synthesis preceded increased t-PA synthesis by some hours, these authors hypothesized that prostaglandins might mediate the increase in t-PA synthesis. The cyclo-oxygenase inhibitor indomethacin abolished prostaglandin production in these cells, but had no effect on the induction of increased t-PA synthesis (17). However, two inhibitors of arachidonate metabolism via the lipoxygenase pathway, eicosatetraynoic acid (ETYA) and nor-dihydroguaiaretic acid (NDGA), dose-dependently inhibited t-PA induction by PMA (18). Also, 5-hydroxyeicosatetraynoic acid directly stimulated t-PA synthesis in HeLa cells (18). Similarly, in rat RBL-1 basophilic leukemia cells, calcium ionophore-induced increase in PA-synthesis proved insensitive to indomethacin, but was abolished by ETYA and by NDGA (18). The inhibitors had little effect on basal rates of PA synthesis in these cells. As PG's E_1, E_2 and I_2 induce increased PA activity (19), prostaglandins, in human skin fibroblasts, are involved in the modulation of PA synthesis (likely t-PA). The mechanism of this induction presumably involves activation of adenyl cyclase, as the effect of PGE_1 was potentiated by phosphodiesterase inhibitors. Nevertheless, the PMA-induced increase in PA activity in these fibroblasts was not influenced by indomethacin (19).

Increased PA activity was also induced by PGE_1 and PGE_2 in mouse C1300 neuroblastoma cells (20) - the activator in this cell line being predominantly t-PA (21) -, by PGE_1 in human embryonic long cells (22), and by PGE_2 in rat osteoblasts (23). In cultured rat granulosa cells, which produce t-PA as the sole activator (24), again PGE_1 and PGE_2 induced increased PA activity whereas $PGF_{1\alpha}$ and $PGE_{2\alpha}$ had no effect (25). In cultured macrophages (which produce pro-uPA, ref. 26) PGE_1 and PGE_2 reduced PA activity ; $PGF_{1\alpha}$ and $PGF_{2\alpha}$ proved inactive, as well as indomethacin and aspirin (27,28). In both granulosa cells, human lung cells and macrophages cAMP had similar effects as PGE_1 and E_2.

In human synovial fibroblasts, cyclo-endogenous oxygenase products could enhance PA activity in cells treated with supernatants of activated mononuclear cells (possibly interleukin-1), but did not enhance PA-activity in retinoic acid-treated cells (28). Prostanoids by themselves were poor activators of basal synovial cell PA activity (29, see also 30).

The above data strongly suggest that eicosanoids are involved in the regulation of the fibrinolytic activity of cells, especially after PA synthesis stimulation. The mechanisms involved are still poorly understood, but might be related to changes in cAMP metabolism (19,20,22,25,27,29). It is also clear that the effects of prostanoids on cellular fibrinolytic activity vary with the cell type, and possibly with the activator type, studied. As already mentioned above, possible concomitant changes in PA inhibitor synthesis have to be taken into account before changes in fibrinolytic (plasminogen activator) activity can be interpreted as changes in PA synthesis. To date, no studies on the effects of eicosanoids on PA inhibitor synthesis are available, except for one study in macrophages (31).

RELEASE OF TISSUE-TYPE PLASMINOGEN ACTIVATOR FROM ENDOTHELIAL CELLS

The fibrinolytic activity of blood is to a large extent determined by the plasma level of t-PA. This t-PA is synthesized and stored by vascular endothelial cells, and can be acutely released from these cells into the circulation. As the half-life of t-PA in the circulation is very short (5-7 min in man, ref. 32), the plasma level of t-PA is highly variable, and can be subject to rapid, short-term changes (see refs. 11,33,34 for recent reviews). Lipid mediators are involved in the t-PA release reaction, both as inducers of the release reaction, and as mediators of said release. Experimental studies demonstrated that platelet-activating factor (PAF-acether) was a very potent inducer of t-PA release in the rat in vivo (35). Studies in isolated perfused vascular beds in rats (35) and pigs (36) showed that PAF-acether induced the release of t-PA by a direct effect on vascular endothelial cells. Compared to other compounds that induce the release of t-PA, such as acetylcholine, histamine, calcium ionophore A-23187, bradykinin, eledoisin and thrombin, PAF-acether proved equipotent to thrombin, and more potent than other inducing agents (11,35,37). (A recent report suggests that PAF-acether may be less potent in mice, ref. 38). Studies on the mechanism of PAF-acether-induced t-PA release in perfused vascular beds then demonstrated that the release reaction was calcium-dependent, and could be inhibited by the phospholipase inhibitors mepacrine and p-bromophenacylbromide, by the lipoxygenase inhibitors NDGA, AA-861 and ETYA, and by the leukotriene synthesis inhibitor diethylcarbamazine (35,36,39). In the rat system, the cyclo-oxygenase inhibitors acetylsalicylic acid and indomethacin were unable to inhibit PAF-acether induced t-PA release (35), though in the pig system acetylsalicylic acid was inhibitory (36). A similar inhibitory profile was found in the rat perfusion system when A-23187, bradykinin, carbachol or thrombin was used to induce the release of t-PA, and when mepacrine, NDGA or calcium-free solution was used to inhibit (39). The data suggest that t-PA release involves the (calcium-dependent) activation of phospholipase(s), resulting in increased availability of arachidonic acid for further metabolization to eicosanoids. The metabolite involved might be a lipoxygenase product, although a cytochrome P_{450}-dependent reaction cannot be excluded (11,40). Still in the perfused rat vascular system, leukotriene D4 was able to directly induce the release of t-PA (39).

Whether this same reaction sequence is also involved in t-PA release in vivo is still under investigation.

Although, as mentioned, PAF-acether is a potent inducer of the release reaction in vivo, it is not indispensible : in rats, t-PA release induced by bradykinin or by venous occlusion was not inhibited by the PAF-antagonist BN 52021, which would inhibit the t-PA release, in vivo as well as in perfused vascular beds, induced by PAF-acether (39).

Lipid mediators therefore in two different ways may be involved in the acute release of t-PA from endothelial cells into the circulation : as inducer of the release reaction (PAF-acether and leukotriene D4), and as intracellular mediators (a still non identified arachidonate metabolite).

ASPIRIN, PROSTACYCLIN AND FIBRINOLYSIS

The widespread use of aspirin and related cyclo-oxygenase inhibitors in the prevention and treatment of thromboembolic disease has led to a number of studies on the effects of aspirin on blood fibrinolytic activity. High-dose aspirin not only inhibits thromboxane synthesis by platelets, but also prostacyclin synthesis by endothelial cells (41), and prostacyclin has been reported to increase fibrinolytic activity in patients with ischemic arterial disease (42-45), and to increase t-PA synthesis in fibroblasts (see above). High-dose aspirin, by decreasing fibrinolysis, might influence the balance between coagulation and fibrinolysis unfavourably. The first studies, however, reported a favourable effect of aspirin on humoral (46) and blood leukocyte (47) fibrinolytic activity. Subsequently, Levin et al. (48) reported that aspirin (650 mg/day x 2) reduced the increase in t-PA activity in plasma obtained after venous occlusion, even though it did not change plasma t-PA activity under resting conditions. (Venous occlusion for 10-20 min, at a pressure halfway between systolic and diastolic pressure, results in the release of t-PA from the endothelium into the blood vessels of the occluded limb, and is widely used in fibrinolysis studies to assess the fibrinolytic capacity of patients). The data of Levin et al. therefore suggested that high-dose aspirin reduced the t-PA release capacity of endothelial cells. These observations were not confirmed by Bounameaux et al. (49), who found no effect on plasma t-PA activity, t-PA antigen or PA inhibitor levels of aspirin, indomethacin or dazoxiben (a thromboxane synthase inhitor), either before or after venous occlusion. Similarly, Keber and Keber (50) found no effects of chronic aspirin treatment on blood fibrinolytic activity, either before or after venous occlusion. Indeed, Hammouda and Moroz (51) and Korninger et al. (52) noted increased t-PA release after venous occlusion in aspirin-treated volunteers.

Mussoni et al. (53,54) found no effect of low-dose aspirin (20 mg/day x 7), but confirmed Levin's observation that high-dose aspirin and indobufen (200 mg/day x 2) significantly reduced the t-PA activity increase after venous occlusion, without however influencing the t-PA antigen increase. Similar data, using similar aspirin dosages, were reported by Keber et al. for venous occlusion-induced changes, but not for exercise-induced increases in t-PA antigen or activity (55). Besides, Brommer et al. (56) reported that aspirin did not affect the release of t-PA induced by Desmopressin (a vasopressin derivative) infusion.

In combination with the observations that prostacyclin infusion results in a (transient) increase in blood fibrinolytic activity (42-45), these studies suggest that in some situations (e.g. venous occlusion) cyclo-oxygenase products may favourably affect plasma fibrinolytic activity by a mechanism, as yet unexplained, that is not operative during exercise - or Desmopressin - induced t-PA release, and presumably does not involve t-PA synthesis rate or the endothelial release mechanism.

SEPSIS, TRAUMA AND FIBRINOLYSIS

Trauma and sepsis are associated with profound changes in the fibrinolytic system (for reviews, see 57-60). A well-known example of such a change is the post-operative fibrinolytic shut-down, i.e. decreased plasma fibrinolytic activity during the immediate post-operative period. Before the discovery of plasma PA inhibitors, it was generally thought that decreased plasma fibrinolytic activity was due to decreased plasma levels of t-PA. However, it soon became clear that fibrinolytic activity was the result of both t-PA and PA inhibitor (PAI) activities, and that decreased plasma fibrinolytic activity could be present in the face of increased t-PA antigen levels, provided the PAI level was even more increased.

Increased levels of PAI are found after, and even already during, surgery (61-64). Despite the concomitant increase in t-PA antigen in the same period, the plasma fibrinolytic activity drops, resulting in the post-operative (or even per-operative) fibrinolytic shut-down. A similar increase in PAI is seen in traumatized patients (63). In all these cases the increased PAI levels rapidly revert to normal (even faster than the acute-phase protein C-reactive protein, ref. 65) suggesting that PAI behaves like a (very) acute phase-type reacting protein. The mechanisms leading to increased plasma levels of PAI are still unknown. Recently, we could demonstrate that the infusion of interleukin-1 into rats resulted in increased plasma levels of PAI (65). Trauma-induced increases in plasma interleukin-1 may thus be one of the causes leading to increased PAI levels, probably by an effect of interleukin-1 on vascular endothelial cells, since these cells, in vitro, respond to interleukin-1 by increased synthesis of PAI (65-67). Moreover, the PAI increasing after interleukin-1 infusion is of the same type as the PAI produced by endothelial cells (65).

A second component that induces increased levels of (endothelial cell-type) PAI is endotoxin, as first reported by Colucci et al. (68). Even minute amounts of endotoxin gave rise to large increases in plasma PAI levels within a few hours in experimental animals (65,68) and endotoxin induced increased PAI synthesis in cultured endothelial cells (65,68-70). Very high levels of PAI were found in septic patients (68) and in patients with meningococcal disease (71). In the latter group of patients, PAI levels normalized parallel with plasma endotoxin levels during penicillin treatment (71).

Whether lipid mediators are in any way involved in the regulation of plasma PAI levels is unknown ; in rats, in any case, the increase in PAI level after endotoxin injection was not influenced by pre-treating the animals with aspirin, indomethacin, or the PAF-antagonist BN 52021 (39).

CONCLUSION

Lipid mediators are in various ways involved in fibrinolysis. Eicosanoids can, in vitro, influence the synthesis of plasminogen activators (and possibly plasminogen activator inhibitors) by cultured cells. The release of t-PA from endothelial cells into the blood stream is induced both by PAF-acether and leukotriene D_4, which activate an endothelial release mechanism presumably involving a lipoxygenase metabolite. The role of lipid mediators (if any) in the induction of plasminogen activator inhibitor synthesis induced by interleukin-1 or endotoxin, remains to be defined.

REFERENCES

1. Aoki N. Sem. Thromb. Haemostas. 10 : 1 (1984).
2. Collen D., Lijnen H.R. and Verstraete M. Thrombolysis, Churchill Livingstone, Edinburgh (1985).
3. Emeis J.J., Brommer E.J.P., Kluft C. and Brakman P. In : Recent Advances in Blood Coagulation (Poller, L. ed), Churchill Livingstone, Edinburgh. 4 : 11 (1985).
4. Kluft C. Tissue-type Plasminogen Activator (t-PA), Physiological and Clinical Aspects, CRC Press, Boca Raton (in press).
5. Verstraete M. and Collen D. Blood 67 : 1529 (1986).
6. Erickson L.A., Schleef R.R., Ny T. and Loskutoff D.J. Clin. Haematol. 14 : 513 (1985).
7. Van Hinsbergh V.W.M. In : Tissue-type Plasminogen Activator (t-PA) Physiological and Clinical Aspects (Kluft C., ed) CRC Press, Boca Raton (in press).
8. Todd A.S. Bibl. Anat. 12 : 18 (1973).
9. Rijken D.C., Wijngaards G. and Welbergen J. Thromb. Res. 18 : 815 (1980).
10. Kristensen P., Larsson L.-I., Nielsen L.S., Grondahl-Hansen J., Andreasen P.A. and Dano K. FEBS Lett. 168 : 33 (1984).
11. Emeis J.J. In : Tissue-type Plasminogen Activator (t-PA), Physiological and Clinical Aspects (Kluft C., ed) CRC Press, Boca Raton (in press).
12. Larsson L.-I., Skriver L., Nielsen L.S., Grondahl-Hansen J., Kristensen P. and Dano K. J. Cell Biol. 98 : 894 (1984).
13. Dano K., Andreasen P.A., Grondahl-Hansen J., Kristensen P., Nielsen L.S. and Skriver L. Adv. Cancer Res. 44 : 139 (1985).
14. Kadouri A. and Bohak Z. Adv. Biotechn. Proc. 5 : 275 (1985).
15. Waller E.K. and Schleuning W.-D. Biol. Chem. 260 : 6354 (1985).
16. Crutchley D.J. and Smariga P.E. Biochim. Biophys. Acta 886 : 26 (1986).
17. Crutchley D.J., Conanan L.B. and Maynard J.R. Cancer Res. 40 : 849 (1980).
18. Crutchley D.J. and Maynard J.R. Biochim. Biophys Acta. 762 : 76 (1983).
19. Crutchley D.J., Conanan L.B. and Maynard J.R. J. Pharmacol. Exp. Ther. 222 : 544 (1982).
20. Laug W.E., Jones P.A., Nye C.A. and Benedict W.F. Biochem. Biophys. Res. Commun. 68 : 114 (1976).
21. Soreq H., Miskin R., Zutra A. and Littauer U.Z. Dev. Brain Res. 7 : 257 (1983).
22. Rifkin D.B. J. Cell. Physiol. 97 : 421 (1978).
23. Hamilton J.A., Lingelbach S.R., Partridge N.C. and Martin T.J. Biochem. Biophys. Res. Commun. 122 : 230 (1984).
24. Canipari R. and Strickland S. J. Biol. Chem. 260 : 5121 (1985).
25. Strickland S. and Beers W.H. J. Biol. Chem. 251 : 5694 (1976).
26. Vassalli J.-D., Dayer J.-M., Wohlwend A. and Belin D. J. Exp. Med. 159 : 1653 (1984).
27. Vassalli J.-D., Hamilton J.A. and Reich E. Cell 8 : 271 (1976).
28. Hamilton J.A. J. Reticuloend. Soc. 30 : 115 (1981).
29. Hamilton J.A., Leizer T. and Lingelbach S.R. Biochim. Biophys. Acta 886 : 195 (1986).
30. Hamilton J.A. J. Rheumatol. 10 : 872 (1983).
31. Drapier J.-C. and Petit J.-F. Int. J. Immunopharmacol. 6 : 345 (1984).
32. Bounameaux H., Verstraete M. and Collen D. In : Thrombolysis (Collen D., Lijnen H.R. and Verstraete M. eds), Churchill Livingstone, Edinburgh, p. 85 (1985).
33. Prowse C.V. and Cash J.D. Sem. Thromb. Haemostas. 10 : 51 (1984).

34. Prowse C.V. and MacGregor I.R. In : Tissue-type Plasminogen Activator (t-PA) Physiological and Clinical Aspects (Kluft C., ed) CRC Press, Boca Raton (in press).

35. Emeis J.J. and Kluft C. Blood 66 : 86 (1985).

36. Klöcking H.-P., Markwardt F. and Hoffmann A. Thromb. Res. 38 : 413 (1985).

37. Emeis J.J. Thromb. Res. 30 : 195 (1983).

38. Jansen J.W.C.M. and Olieberg H.H. Fibrinolysis 1, suppl. 1 : 72 (1986).

39. Emeis J.J. Unpublished observations (1986).

40. Singer H.A., Saye J.A. and Peach M.J. Blood Vessels 21 : 223 (1984).

41. Bertele V. and Salzman E.W. Arteriosclerosis 5 : 119 (1985).

42. Dembinska-Kiec A., Kostka-Trabka E., Zmuda A. et al. Pharmacol. Res. Commun. 14 : 485 (1982).

43. Korbut R., Byrska-Danek A. and Gryglewski R.J. Thromb. Haemostas. 50 : 893 (1983).

44. Musial J., Wilczynska M., Sladek K. et al. Prostaglandins 31 : 61 (1986).

45. Szczeklik A., Kopec M., Sladek K. et al. Thromb. Res. 29 : 655 (1983).

46. Menon I.S. Lancet i : 364 (1970).

47. Moroz L.A. New Engl. J. Med. 296 : 525 (1977).

48. Levin R.I., Harpel P.C., Weil D., Chang T.-S. and Rifkin D.B. J. Clin. Invest. 74 : 571 (1984).

49. Bounameaux H., Gresele P., Hanss M., De Cock F., Vermylen J. and Collen D. Thromb. Res. 40 : 161 (1985).

50. Keber I. and Keber D. Thromb. Res. 39 : 761 (1985).

51. Hammouda M.W. and Moroz L.A. Thromb. Res. 42 : 73 (1986).

52. Korninger C., Kircheimer J., Christ G., Schwaiger N. and Binder B.R. Thromb. Haemostas. 54 : 175 (1985).

53. Mussoni L., Carriero M.R., Cerletti C. and De Gaetano G. Blood 66 : 352a (1985).

54. Mussoni L., Carriero M.R., Cerletti C. and De Gaetano G. Fibrinolysis 1, Suppl. 1 : 171 (1986).

55. Keber I., Jereb M. and Keber D. Fibrinolysis 1, Suppl. 1 : 175 (1986).

56. Brommer E.J.P., Derkx, F.H.M., Barrett-Bergshoeff M.M. and Schalekamp M.A.D.H. Thromb. Haemostas. 51 : 42 (1984).

57. Innes D. and Sevitt S. J. Clin. Pathol. 17 : 1 (1964).

58. Risberg B. Eur. Surg. Res. 10 : 373 (1978).

59. Risberg B. J. Surg. Res. 26 : 698 (1979).

60. Saldeen T. The microembolism syndrome, Almquist and Wiksell, Stockholm (1979).

61. Aillaud M.F., Juhan-Vague I., Alessi M.C. et al. Thromb. Haemostas. 54 : 466 (1985).

62. D'Angelo A., Kluft C., Verheijen J.H. et al. Eur. J. Clin. Invest. 15 : 308 (1985).

63. Kluft C., Verheijen J.H., Jie A.F.H. et al. Scand. J. Clin. Lab. Invest. 45 : 605 (1985).

64. Mellbring G., Dahlgren S. and Wiman B. Acta Chir. Scand. 151 : 109 (1985).

65. Emeis J.J. and Kooistra T. J. Exp. Med. 163 : 1260 (1986).

66. Bevilacqua M.P., Schleef R.R., Gimbrone M.A. and Loskutoff D.J. J. Clin. Invest. 78 : 587 (1986).

67. Nachman R.L., Hajjar K.A., Silverstein R.L. and Dinarello C.A. J. Exp. Med. 163 : 1595 (1986).

68. Colucci M., Paramo J.A. and Collen D. J. Clin. Invest. 75 : 818 (1985).

69. Dubor F., Dosne A.M. and Chedid L.A. Inf. Imm. 52 : 725 (1986).

70. Crutchley D.J. and Conanan L.B. J. Biol. Chem. 261 : 154 (1986).

71. Engebretsen L.F., Kierulf P. and Brandtzaeg P. Thromb. Res. 42 : 713 (1986).

FIBRINOLYSIS IN ENDOTOXEMIA

A-M. Dosne, F. Dubor and L. Chedid
UA-579 Immunothérapie Expérimentale, Institut Pasteur
Paris, France

INTRODUCTION

The fibrinolytic process includes plasminogen cleavage by different types of plasminogen activators leading to plasmin formation. Plasmin degrades fibrin but also cartilage proteoglycan[1] and activates latent collagenase[2]. Besides its role in hemostasis, fibrinolysis may play a role in tissue remodeling and metastasis of tumor cells[3,4]. The fibrinolytic process is controlled by inhibitors acting either at the plasminogen activation step or by antiplasmins. An alteration in fibrinolysis was first observed following the injection of certain vaccines[5]. These changes were reproducible with purified lipopolysaccharides (LPS) from the walls of gram-negative bacteria. After experimental endotoxemia there is first an acceleration of fibrinolysis followed by a late depression. Some of the mechanisms underlying these sequential modifications have been recently clarified by the identification of mediators induced by endotoxin and the ability to study the responses of isolated cells.

ACTIVATION OF FIBRINOLYSIS FOLLOWING ENDOTOXIN INJECTION

Thirty years ago, it was observed that blood clots dissolved spontaneously in 60 min when blood was collected 100 min after injection of small doses of LPS from Salmonella abortus equi in humans[6]. Fibrinolytic activity was more easily detected in the plasma euglobulin fraction, which lysed in about 6 min[7]. The acceleration of fibrinolysis declined rapidly and was no longer observed 4 hr after endotoxin injection. A parallel between activation of fibrinolysis and fever rise was reported but a causal relationship was excluded since pretreatment with antipyretics suppressed fever but not the fibrinolytic reaction[7]. At that time, these results led to the proposal that purified LPS could be used as thrombolytic therapy.

Several mechanisms could produce increase in blood fibrinolytic activity. Activation of Hageman factor was reported, as reflected by decrease in prekallicrein[8,9] and generation of bradykinin[10] during endotoxemia in man. In vitro studies provided evidence for a direct interaction between the Lipid A moiety of the endotoxin and Hageman factor leading to its activation[11]. A Hageman factor-dependent plasminogen activator has been demonstrated in blood but its function is not well defined[12]. Activation of the Hageman factor-kinin system during endotoxemia has been mainly studied with respect to coagulation activation and hypotension and its role in fibrinolysis activation was not established. The release of plasminogen activator from vascular endothelium has received much attention since endothelial cells contain and produce tissue plasminogen activator[13,14]. As LPS does not directly induce plasminogen activator release from cultured endothelial cells, mediators are likely to be involved. Adrenalin release subsequent to endotoxin injection could play a role since catecholamines induce in vivo a certain but weak fibrinolytic reaction[15]. The fact that adrenalectomy suppresses the fibrinolytic response to LPS in the rat supports the role of adrenals[16]. Release of vasopressin from pituitary gland might be an important factor since this hormone is a powerful inducer of plasminogen activator release in vivo[17] and in perfused vessels[18]. A peak in vasopressin activity occurs in blood 1-2 hr after endotoxin injection. This appears to be independent of known physiological stimuli such as hypotension and hypovolaemia[19]. As the blood brain barrier does not allow the penetration of endotoxin, further investigations are needed to understand the mechanism of this vasopressin release. In addition to these factors, platelet aggregating factor could be also involved. Some data suggest that this lipid mediator plays a role in endotoxemia[20] and this factor is endowed with a potent tissue plasminogen activator releasing activity[21].

Besides endothelial cell production of tissue-type plasminogen activator, in vitro findings suggest that lymphokines produced by T cells exposed to LPS, increase the expression in macrophages of a urokinase-type plasminogen activator which is partly membrane-bound[22,23,24]. The importance of this phenomenon during endotoxemia is presently unknown.

The biological importance of this transient hyperfibrinolysis has been generally evaluated in relation with the activation of coagulation and the occurrence of intravascular coagulation after endotoxin injection. Administration of fibrinolytic inhibitors favorizes fibrin deposition in the kidneys[25,26] and treatment with the plasminogen activator streptokinase prevents the lesions of the generalized Shwartzman reactions[27]. These data would suggest that the hyperfibrinolytic response to endotoxin has a protective effect in the pathogenesis of Shwartzman reaction.

Table 1. Decreased Fibrinolytic Activity in Blood Collected 4 hr after
Endotoxin Injection in the Rat

Endotoxin (μg/kg)	0	.03	.3	3	30
Euglobulin clot lysis time (min)	52	137	193	231	> 240
Plasma inhibitor anti-urokinase (U/ml)	1.2	4.7	12	21	41

Wistar rats (5-10 per group) received i.v. injection of LPS from S.enteritidis and were bled 4 hr later. Plasma euglobulin were precipitated at pH 5.9 with 0.01 % acetic acid. Plasma inhibitor was estimated by adding 0.15-10 U of urokinase to 100 μl of plasma followed by euglobulin precipitation. Residual lytic activity of urokinase was determined on fibrin plate rich in plasminogen. One unit of inhibitor was defined as the amount neutralizing one unit of urokinase.

It has been observed in humans that hyperfibrinolytic state is followed by a decreased blood fibrinolytic activity which could last up to 24 hr[7]. In rats which receive 30 ng to 30 µg endotoxin per kg we have noticed a prolongation of the euglobulin clot lysis time which is depedent upon doses of endotoxin (Table 1).

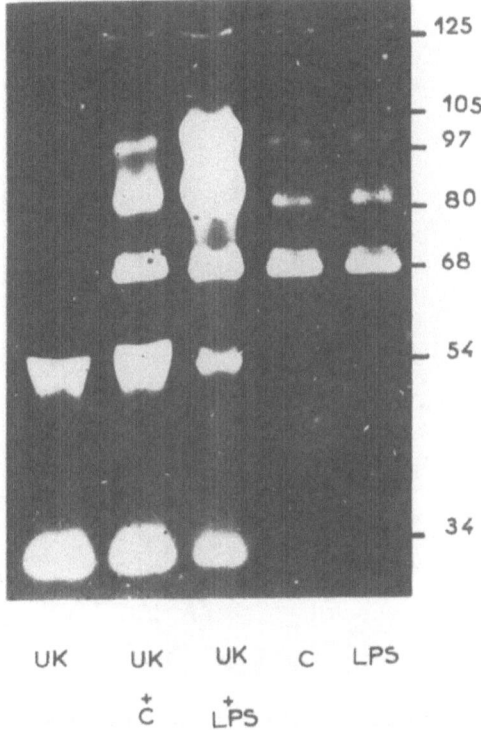

Fig. 1. Fibrinoenzymatographic pattern of rat plasma euglobulin 4 hr after injection of LPS (30 µg/kg). Euglobulins (15 µl) from control (C) or LPS-treated rats were submitted to SDS-PAGE in the presence or absence of 90 mU of urokinase (UK). Acrylamide gel was applied on a fibrin agarose gel rich in plasminogen. Lysis areas (clear zones) correspond to the different enzymatic activities whose molecular weight were calculated by reference to calibrated proteins. Without UK supplementation enzymatic profiles of control or post-LPS samples were not very different. When UK (34 and 54 kd) was added to control sample, these lytic bands were not reduced but a 80 kd lytic area increased. When UK was added to post-LPS sample, activities corresponding to 34 and 54 kd were decreased and a major lytic area extended from 80 to 105 kd. This shift in the molecular weight of UK indicates the presence of PAI which binds to UK and forms higher molecular weight complexes.

Recent findings have shown that a plasminogen activator inhibitor (PAI) increases in plasma with a peak between 3 and 6 hr after endotoxin injection[28,29]. When plasma from endotoxin-treated rabbits is passed on a gel filtration column the relative molecular weight of tissue plasminogen activator-related antigen is 100 kd as opposed to 66 kd when normal rabbit plasma is used[28]. This study concluded that PAI binds to plasminogen activator yielding high molecular complexes devoid of enzymatic activity.

Such complexes could be also detected when plasma supplemented with plasminogen activator was submitted to polyacrylamide gel electrophoresis in the presence of sodium dodecylsulfate (SDS-PAGE) since this detergent renders the complexes fibrinolytically active[30]. Utilizing this method, we have found that the molecular weight of urokinase added to post-endotoxin rat plasma is augmented about 50 kd. This increase in relative molecular weight is an estimation of the molecular weight of PAI (Fig. 1).

Several mechanisms are involved in PAI generation induced by bacterial lipopolysaccharides. In vitro experiments have demonstrated that cultured endothelial cells produce PAI[14,31,32] and that endotoxin stimulates this synthesis[28,33,34]. Figure 2 shows that supernatant from human endothelial cells exposed to LPS neutralizes urokinase and forms higher molecular weight complexes.

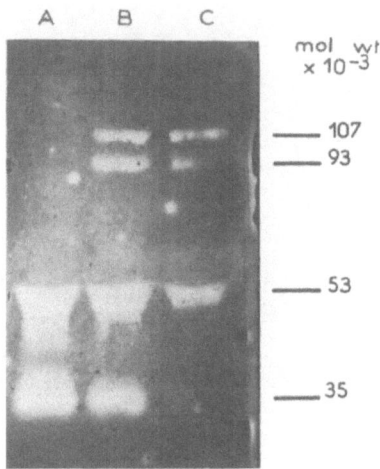

Fig. 2. Fibrinoenzymatographic analysis of urokinase neutralization by endothelial cell supernatants. Urokinase (25 µl of 6 U/ml) was mixed with 25 µl of unconditioned medium (lane A) or supernatants from control endothelial cells (lane B) or from cells incubated with LPS (1 µg/ml) for 24 hr (lane C). SDS-PAGE and development of fibrinolytic activities were performed. Control supernatant decreased the 35 and 53 kd lytic bands of urokinase and led to the formation of 93 and 107 kd complexes. Supernatant from LPS-treated cells suppressed the 35 kd lytic band and further reduced the 53 kd band.

Pulmonary macrophages also generate PAI in response to endotoxin[35,36]. These reactions could be amplified in vivo by interleukin-1 which is a potent inflammatory and immunological mediator released by macrophages[37] and also a powerful PAI inducer in endothelial cells and in vivo[29,38]. Although PAI secreted by different cells shows a molecular weight in the range of 50 kd, this inhbitory activity might be supported by various proteins whose specificity towards different types of plasminogen activators has to be established. Such information would help to clarify the biological role of PAI(s). The inactivation of tissue plasminogen activator which was demonstrated in vivo[28] might contribute to the development of intravascular coagulation which is frequently observed in septicemic patients. On the other hand, PAI has been reported to inhibit Hageman factor[39]. The consequence of the neutralization of urokinase-type plasminogen activators expressed by inflammatory macrophages[40],[41] or tumoral cells[42] is still speculative.

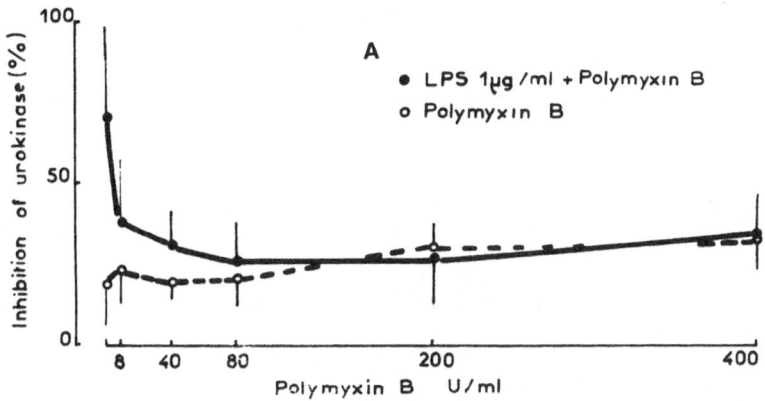

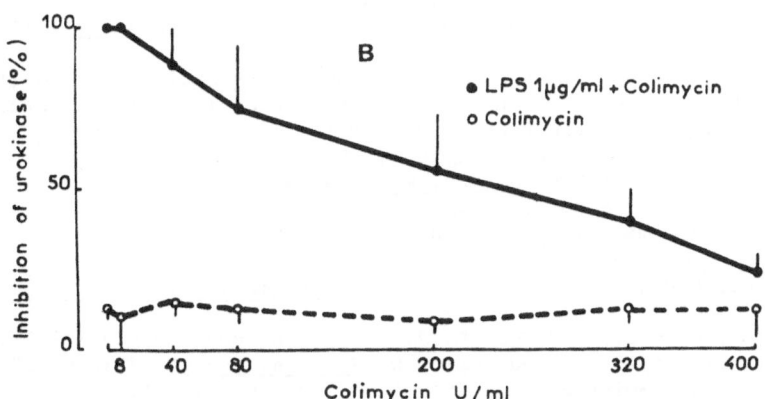

Fig. 3. Prevention by polymyxin B and colimycin of the LPS-induced inhibition of urokinase in endothelial cell cultures.

Some attempts at preventing generation of LPS-induced PAI have been done with the help of polymyxin B since this antibiotic binds to the Lipid A moiety[43] and blocks several LPS reactions[44],[45],[46]. In cultured endothelial cells, polymyxin B and colimycin suppress PAI generation induced by LPS (Fig. 3). When assessed in vivo in the rat, we have found these compounds to be effective only after low doses of LPS.

In conclusion, alteration in blood fibrinolysis induced by endotoxemia consists of transient increase in plasminogen activator(s) followed by generation of plasminogen activator inhibitor(s). Some of these reactions are due to direct effect of endotoxin on endothelial cells and macrophages and others appear to be regulated by endotoxin-induced mediators. These alterations are probably relevant to the risk of intravascular coagulation.

ACKNOWLEDGEMENTS

We wish to thank Dr. H.S. Warren for helpful advice during the preparation of the manuscript and C. de Champs for skillful secretarial assistance.

REFERENCES

1. Mochan E., and Keler T. Biochim. Biophys. Acta 800:312 (1984).
2. Paranjpe M., Engle L., Young N., and Liotta L. Life Sci. 26:1223 (1980).
3. Ng R., Kellen J., and Wong A. Invasion Metastasis 3:243 (1983).
4. Carlssen S., Ramshaw I., and Warrington R. Cancer Res. 44:3012 (1984).
5. Meneghuini P. Acta Haemat. 19:65 (1958).
6. Eichenberger E. Acta Neuroveg. 11:201 (1955).
7. Von Kaulla K. Circulation 17:187 (1958).
8. Robinson J., Klodnycky M., Loeb H., Racic M., and Gunnar R. Am. J. Med. 59:61 (1975).
9. Aasen A., Smith-Ericksen N., and Amundsen E. Arch. Surg. 118:343 (1983).
10. Kimball H., Melmon K., and Wolff S. Proc. Soc. Exp. Biol. Med. 139:1078 (1972).
11. Morrison D., and Cochrane C. J. Exp. Med. 140:247 (1974).
12. Kluft C., Trumpi-Kalshoven M., Jie A., and Veldhuyzen-Stolk E. Thromb. Haemost. 41:756 (1979).
13. Kristensen P., Larsson L., Nielsen L., Grondahl-Hansen J., Andreasen P., and Dano K. Febs 168:33 (1984).
14. Philips M., Juul A., and Thorsen S. Biochim. Biophys. Acta 802:99 (1984).
15. Dosne A.M. Path. Biol. 23(suppl.):63 (1975).
16. Fracasso J., and Rotschild A. Throm. Haemost. 50:557 (1983).
17. Gader A., DaCosta J., and Cash J. Lancet II:1417 (1973).
18. Jablonovski C., Klöcking H., and Markwardt F. Folia Haematol. 111:567 (1984).
19. Kasting N., Mazurek M., and Martin J. Am. J. Physiol. 248:E420 (1985).
20. Braquet P. in : Prostaglandins and Leukotrienes in Health Disorders (editor: U. Zor). Raven Press (New York), (1986), in press.
21. Emeis J., and Kluft C. Blood 66:86 (1985).
22. Vassali J., and Reich E. J. Exp. Med. 145:429 (1977).
23. Klimetzek V., and Sorg C. Eur. J. Immunol. 7:185 (1977).
24. Tiku M., and Tomasi T. Transplantation 40:293 (1985).
25. Lee L. J. Exp. Med. 115:1065 (1962).
26. Beller F., Mitchell P., and Gorstein G. Thrombos. Diathes. Haemorrh. 17:427 (1966).
27. Condie R., Hong C., and Good R. J. Lab. Clin. Med. 50:803 (1957).
28. Colucci M., Paramo J., and Collen D. J. Clin. Invest. 75:818 (1985).
29. Emeis J., and Kooistra J. J. Exp. Med. 163:1260 (1986).
30. Granelli-Piperno A., and Reich A. J. Exp. Med. 148:223 (1978).
31. Loskutoff D., and Edgington T. Proc. Natl. Acad. Sci. USA 74:3903 (1977).
32. Dosne A., Dupuy E., Bodevin E. Thromb. Res. 12:377 (1978).
33. Dubor F., Dosne A., and Chedid L. Infect. Immun. 52:725 (1986).
34. Crutchley D., and Conanan L. J. Biol. Chem. 261:154 (1986).
35. Chapman H., and Stone O. Biochem. J. 230:109 (1985).
36. Chapman H., and Stone O. Am. Rev. Resp. Dis. 132:569 (1985).
37. Dinarello C. Rev. Infect. Dis. 6:51 (1984).
38. Nachman R., Hajjar K., Silverstein R., and Dinarello C. J. Exp. Med. 163:1595 (1986).
39. Hedner U., and Martinsson G. Throm. Res. 12:1015 (1978).
40. Lemaire G., Drapier J., and Petit J. Biochim. Biophys. Acta 755:332 (1983).
41. Chapman H., Stone O., and Vavrin Z. J. Clin. Invest. 73:806 (1984).
42. Saksela O., Vaheri A., Schleuning W., Mignatti P., and Barlati S. Int. J. Cancer 33:609 (1984).
43. Morrison D., and Jacob D. Immunochemistry 13:813 (1976).
44. Rifkind D., and Hills R. J. Immunol. 99:564 (1967).
45. Corrigan J., and Bell B. J. Lab. Clin. Med. 77:802 (1971).
46. Van Miert A., and Van Duin C. Arzneim-Forsch./Drug Res. 28:2246 (1978).

Chapter 3

The immune response
in
critically ill patients

ORIGINS OF IMMUNOLOGICAL IMPAIRMENTS IN BURNS

M. Braquet

Unité de Recherche Clinique, Centre de Traitement des Brûlés
Hôpital d'Instruction des Armées Percy, F-92141 Clamart, France

One of the best examples of the immunodepressed host is afforded by burn patients since the natural protection barriers are broken and the host defenses are no longer operative. In this respect a frequent complication of serious burn injury is sepsis that leads to death in a large percentage of patients surviving the shock phase of such injury (1). In the 1940s, Gram-positive cocci were the main source of the infections but with the discovery of antimicrobial therapy longer the case. Infections are now caused mainly by gram-negative bacilli such as pseudomonas (2) and methicillin-resistant Staphylococcus aureus, Aspergillus Mucor, Herpes virus and Candida (3-5). Pneumonia is also extremely frequent in patients with inhalation injury or shock lung (6). Burn patients may furthermore present suppurative chondritis and thrombophlebitis, pyelonephritis and endocarditis (2,7).

The total body surface area affected (TBSA) and depth of burn are the factors influencing the extent of abnormality in immunologic functions. Prolonged skin allograft survival may be noted after burn (8). The immunological impairments in thermal injury and the suspected mechanisms of immunosuppression are reviewed in this paper.

IMMUNOLOGIC ALTERATIONS IN BURN PATIENTS

Delayed hypersensitivity skin tests point to the occurrence of anergy in burn patients. There is a high survival rate in patients displaying a consistent positive reaction or in those who regain reactivity in the second phase. On the contrary, there are high mortality rates in individuals with continuously negative skin reactions (9). This fall in the skin immune response is often linked with the activities of the T cell system which are reduced to a greater extent than those of the B cell system (10-13). Table 1 presents a summary of the principal T cell impairments.

Burn patients often develop T cell lymphopenia as evidenced by the early fall in OKT3+ cells. This disorder becomes clearly apparent at 48 hours after thermal injury. OKT3+ cell number then slowly falls until D17 when it attains the minimal value. A slow rise towards normal values then occurs. The alteration of helper T cells (OKT4+) is the main cause of this biphasic pattern (10). On the contrary, the number of suppressor cells (OKT8+) remains appreciably the same. The OKT4/OKT8 ratio is thus decreased and reaches a minimal value between D10 and D15 (10). After D20 after severe burn two types of response have been observed as a general rule : (a) the OKT4/OKT8 ratio stayed low in patients who were critically ill and died ; and (b) the OKT4/OKT8 ratio rose slowly in patients who recovered (10,12). Antonacci et al. (14) explains the fall in OKT4/OKT8+

ratio by a "sequestration" phenomenon, perhaps triggered by corticosteroids, in which OKT4 cells in particular would be sequestered to lymphoid compartments other than the peripheral blood.

Table 1. Abnormalities of T-cell functions after severe thermal injury

- OKT3 + cell number decreased
- OKT4 + cell number decreased
- OKT8 + cell slightly or not impaired
- Decreased OKT4 +/OKT8 + ratio
- Reduced in vitro cytotoxicity
- Decreased lymphoproliferative responses in vitro to nonspecific mitogens, soluble antigens, and allogenic cells
- Cutaneous anergy displayed by negative delayed hypersensitivity skin reactivity
- Prolonged survival of skin allograft

Although the possibility exists that impaired OKT4/OKT8 ratio may denote a relative predominance of suppressor cells, when mixed lymphocyte responses are analyzed it appears that the fundamental alteration of significance is deficient T cell "help" characterized by a fall in the number of IL2-producing cells and/or a concomitant dysfunction of these cells (13).

T cells are also affected by qualitative impairments in addition to the above-mentioned quantitative alterations. Their cytotoxicity is lowered in burn patients, possibily deriving from lower release of lymphokines (15). In addition, T cell proliferative responses to nonspecific mitogens, specific antigens, and histoincompatible cells (mixed lymphocyte reactions) are lowered after burn injury (16-20). The T cell impairments are summarized in Table 1.

As for the B cell system, no alteration was noted in the total number of circulating cells (12). There was a transient fall in immunoglobulin concentration early after thermal injury with the lowest point being situated around D5. Generally, immunoglobulin G(IgG) levels are weakened the most, with little early change in immunoglobulin M (IgM) levels (21-23). Burn blister fluid contains all the serum immunoglobulins, one cause of the fall in immunoglobulin levels possibly being the extravasation of previously intravascular IgG into tissue through the increasingly permeable capillary endothelium. Some works also report a higher tissue levels of IgG in deeper layers of burned skin (24).

In general the lower serum antibody concentrations and the incidence of infection are not correlated (25).

Burn injury significantly affects natural killer (NK) cells too. These cells trigger a cell-mediated cytotoxic reaction, are IgG-Fc receptor positive and have been reported to be a key mechanism in host anti-viral defense. After burn injury NK-dependent cytotoxicity is lowered (26,27). In addition, NK activity in burn patients was not increased by either interferon, (IFN)-γ or α (27). This lack of effect may add to burn patients' susceptibilities to viral infections, especially cytomegalovirus and herpes simplex.

Apart from alterations in T cells and NK activity, a set of disorders act upon the nonspecific cellular immunity in patients with severe thermal injury, leading to abnormalities of phagocytic cells (macrophages and neutrophils), as shown in Table 2. Burn patients thus present highly altered monocyte and neutrophil functions.

Neutrophils display a lower microbicidal activity (28-30), impaired chemotaxis (31-34), a fall in lysosomal enzyme content (34-35), oxygen consumption (36), nitroblue tetrazolium reduction (37), chemiluminescence (38) and glucose oxidation (39), and impaired NADH-NADPH oxidase activity (40). Neutrophils may be degranulated and present pseudopodes. Lower inflammatory responsiveness has been reported, as measured by skin windows (41). Thermal injury also alters A 23187-stimulated arachidonic acid (AA) metabolism, a biphasic sequential release of the different metabolites (5-HETE, LTB$_4$, 20-OH LTB$_4$) being recorded (see accompanying paper) (10). The capacity for AA metabolite production is highly altered after an early but transient rise, the fall being associated with the clinical outcome of the patients. After the third week, two patterns were generally noted : (a) there was an extremely low calcimycine-stimulated AA metabolism in patients who died ; (b) on the contrary, the stimulated release rose progressively back to normal in patients who recovered. These functional alterations are correlated with lower superoxide production (10), and a higher susceptibility to severe infection by various microorganisms. Pulmonary alveolar phagocytes (42) and the reticuloendothelial system (RES) have also been reported to present similar alterations.

Table 2. Impairments of phagocytic cells after severe thermal injury

- Diminished chemotactic responsiveness
- Diminished superoxide and hydrogen peroxide release
- Diminished arachidonate metabolite production
- Impairment of bacterial activity
- Granulocytopenia (early phase)
- Partial degranulation
- PMN leukocyte aggregation
- Decreased serum opsonic activity

Phagocytic depression of neutrophils from burn patients is usually not seen until more than 5 days following injury.

A variety of cells are comprised in the RES including noncirculating phagocytic cells located close to the vascular system throughout the body. The RES takes part in the action of host-defense phagocyting bacteria, cellular debris... A considerable depression of RES has been noted after thermal injury, apparently in relation with a deficit of fibronectin (43). This glycoprotein is able to bind to a variety of materials leading to their phagocytosis mainly by Kupffer cells in the liver.

Abnormal activation of the complement system is another potential dysfunction of the host defense. Lowbury and Ricketts (44) were the first to report a defect in the alternative complement pathway in 1957 (44). Many researchers have since reported serious alterations of complement functions and levels after burn injury. A pronounced fall in hemolytic serum complement activity (CH50) has been observed, mainly in septic burn patients with burn injury over more than 30 % of the TBSA (45-48). Falls in CH50 units were accompanied by falls in C3 activity (45) and in concentrations of C3 and C4 (49). Sepsis or a potential risk of sepsis appear to be reflected by very low levels of hemolytic complement activity. Complement titers remain decreased until death in fatally burn injured patients (50). CH 50 titers and protein concentrations of individual complement components (C4, C2, C3) are found to be normal or raised in minor thermal injuries (45,47,51). To date the only complement components that have been decreased, whatever the size and severity of the thermal injury, are C1 esterase activity and C1 γ protein (46,51). At present researchers have only a poor understanding of the cause or consequences of depressed C1 activity and the mechanism of hemolytic complement activation in thermal injury. Sera from burn injured patients has consistently presented depressed complement activity of the alternate activation pathway, as assessed by the C3 conversion assays (52,53). The depression was seen in both bacteremic and nonseptic burn patients and correlated with TBSA. Some evidence was presented by Bjornson et al. (46,52) suggesting that sera from burn injured patients contain an unknown inhibitor of C3 conversion. Holder and Ablin (54) have also reported inhibitory serum activity of the alternative complement pathway ; on the basis of their findings, they suggest that an inhibitor of the C3 proactivator may have been produced as a result of burn injury. While classic pathway activity is normal or supranormal, the alternative pathway depletion continues for at least a week (55). Complement activation results in the formation of various components (C5a, C3a, C4a) that promote various pathologic events (e.g. chemotaxis, release of lysosomal enzymes, superoxide production, T/B cell interactions, histamine release, immune adherence).

MECHANISMS OF IMMUNOSUPPRESSION IN BURN PATIENTS

The exact mechanisms underlying anergy in the burn patient are still not yet assessed. The first possible hypothesis may involve an increase in suppressive cells. An enhancement in the relative suppressor activity of T cells has been proposed (10). In addition, the involvement of inhibitory macrophages has been advanced (56,57) the suppressor activity resulting in this case form exaggerated production of prostaglandins (58). On the other hand, it has been proposed that the major defect in cellular immunity is the failure of lymphocytes from burn patients to produce adequate quantities of IL 2 in response to standard stimuli (59). Failure of IL 2 production is correlated early after injury with a marked reduction in the percentage of circulating T lymphocytes of the helper/inducer phenotype. Late after injury, diminished IL 2 production persists, despite recovery of the helper T cell subset (59).

Detection of immunoregulatory factors has been increasingly reported in serum of patients with thermal or traumatic injuries, as detailed in Table 3 ; the second hypothesis takes account of this fact. The nature of these suppressive factors is not yet elucidated, although various hypotheses have been advanced (Table 3). These factors affect both the specific and nonspecific host defense mechanisms. Thus, soluble plasma factors from burn patients have been reported to inhibit chemotaxis of normal, polymorphonuclear (PMN) leukocytes (61,62). This inhibitor is heat-stable and nondialyzable (61). The binding of abnormal proteins in the leukocyte membrane has been evidenced by the appearance of antileukocytic autoantibodies (63). The incubation of normal PMN with plasma from recently burned patients suppresses or at least depresses A 23187-stimulated AA metabolite synthesis and phorbol myristate acetate-activated superoxide production (10). This phenomenon was observed with plasma at D2 and D3 and lasts for about 20 days, although it varies among patients. The suppressive activity disappears after the plasma is heated, and the use of FPLC showed that the molecular weight of this factor was lower than 50,000 daltons (D) (10).

Similar findings were obtained by Baxter's group (44), which isolated two circulating protein moieties exhibiting superoxide inhibitory activity (mol wt 26,000 D and 52,000 D). Inhibition produced by the large mol wt protein can be blocked by anti-C3D antibodies, while the lesser inhibition exhibited by low mol wt protein cannot (64).

Bjornson et al. (65) demonstrated an inhibitory effect of sera from septic burn patients on the phagocytosis of Staphylococcus aureus and Escherichia coli by normal neutrophils. Inhibition of phagocytosis by sera from burn patients was neither due to cytotoxic effects on neutrophils nor

Table 3. Possible plasmatic factors related to immunodepression in thermal injury

Nature	Mol/wt (D)	Suppressive Effect	Reference
Acute phase proteins	—	Decreased PMN arachidonate and superoxide release	27, 11
Protein(s)	<50,000	Decreases PMN arachidonate metabolism	11
Proteins	26,000 52,000	Decreases PMN superoxide poduction	64
Nondialyzable	>12,000	Decreases PMN phagocytosis	65
Fluid phase C3b	—	Decreases PMN phagocytosis and killing	66
Nondialyzable, heat stable	—	Decreases chemotactics	61
SAP*	3,654	Suppresses PHA response	60, 68
Nondialyzable	5,000	Suppresses IL2 production	59

The suppressive factors are generally not present in the normal serum.

*Suppressor active factor comprising peptide, carbohydrate, and fatty acid components

to defective opsonization but was caused by a direct and irreversible interaction of the sera with the neutrophils. Further studies by the same authors revealed the presence in sera from bacteremic burn patients of an inhibitor of phagocytosis that was nondialyzable, suggesting a substance with a mol wt of more than 12,000 D.

Ogle et al. (66) demonstrated that fluid phase C3b, a degradation product of C3 from normal human serum, inhibits phagocytosis and killing of E coli by neutrophils. At a concentration that is similar to the C3 concentration in normal human serum, over 99 % inhibition of neutrophil function could be observed. Although the exact mechanism of the inhibition of neutrophil function is not yet known, fluid phase C3b may play an important role in thermal injury, which is accompanied by intravascular complement activation. The correlation of the appearance of the phagocytosis inhibitor after D5 following thermal injury with the appearance in the circulation of C5a can be seen as an indicator of complement activation. Whether fluid phase C3b also has an inhibitory effect on neutrophil locomotion deserves further investigation.

The suppressive factors also impair lymphocyte function. The first demonstration was provided by Ninnemann et al. (60) and Ninnemann and Stein (67) who showed that burn patient serum suppresses the PHA response of normal human lymphocytes. The suppressive factors are not present in the normal serum ; are heat stable ; pH stable ; unaffected by trypsin, proteinase K, DNase or RNase ; and are related to the presence of endotoxin and cutaneous burn toxin (68,69). After a complex procedure including Amicon ultrafiltration, ion-exchange chromatography, and Sephadex C-25 gel filtration, a suppressor active factor (SAP) was isolated (68,69). The mol wt of SAP was estimated to be 3,654 D and its isolectric point between 3.2 and 3.6. Characterization of SAP revealed a complex structure comprised of (a) a peptide component rich in glycine, serine, and alanine ; (b) a carbohydrate component containing sialic acid; and (c) a fatty acid component tentatively identified as prostaglandin E. In addition to the immunosuppressive activity on lymphocytes, SAP also causes inhibition of PMN chemotaxis and hemolytic activity.

Circulating factor(s) suppressing IL 2 production and IL 2 activity are found in the serum of burn patients and may play a role in the protracted impairment of IL 2 production by lymphocytes from these individuals. A major portion of this serum suppressive activity appears once again to reside in a low mol wt (5000 D) polypeptide fraction (59).

Interestingly, the presence of plasmatic suppressive factors has also been reported in other pathologic states, including cancer and acquired immunodeficiency syndrome (AIDS). In cancer, macrophage cytotoxicity is inhibited by plasma (70), and the nature of these factors has been reported to be immune complex (71) or low mol wt factors (1,400 D) that inhibit proliferation of mitogen-stimulated T cells as a function of the concentration of cAMP (72). Concerning AIDS, plasmapheresis and selective immunoadsorption have been proposed as a treatment, suggesting the involvement of plasma suppressive activity (73).

CONCLUSION

The main risk of death in burn patients is still sepsis deriving from anergy. Appearance in the plasma of potent suppressive factors may account for the immune depression. Some therapeutic modalities may therefore be used in combating these negative effects : (a) plasma exchange that provides a significant beneficial effect in restoring lymphocyte function in burn patients (74) and at the same time improves allograft survival times (75) ; (b) drugs inhibiting or at least controlling the early massive inflammatory phase, such as leukotrienes or platelet-activating factor (PAF) antagonists. Products of lipoxygenase pathways and PAF (see accompanying papers) are indeed important modulators of lymphocyte functions (76). Above all LTB_4 activates cytotoxic effector cells of NK type by through increased lytic efficiency. LTB_4 also stimulates IL 1 production by large granular lymphocyte and suppressor cells that may need the participation of monocytes and cyclooxygenase products. LTB_4 may furthermore induce the appearance of T8+ cells from T8-depleted lymphocyte cultures. Since PAF produces LTB_4 in various cell models it could have similar effects as LTB_4. Any drug that could control and decrease the first phase may thus reduce the relative intense negative feed back and, as a consequence, anergy. This assumption is at present being investigated in our laboratory.

REFERENCES

1. Pruitt B.A. and Moylan J.A. Adv. Surg. 6 : 237 (1972).
2. Curreri P.W. In Simmons R.L., Howard R.J. (eds) : Infections in Surgery. New York, Appleton-Century-Crofts, p. 1125 (1982).
3. Bruck H.M., Hash G., Foley F.D. et al. Arch. Surg. 102 : 476 (1971).
4. Bruck H.M., Nash G., Stein J.M. et al. Surg. 176 : 108 (1972).
5. Salisbury R.E., Silverstein P., Goodwin M.N. Plast. Reconstr. Surg 54: 654 (1974).
6. Pruitt B.A. Jr, DiVincenti F.C., Mason A.D. Jr et al. J. Trauma 10 : 519 (1970).
7. Stein J.M., Pruitt B.A. Jr. N. Engl. J. Med. 282 : 1452 (1970).
8. Ninnemann J.L., Fisher J.C., Frank H.A. Transplantation 25 : 69 (1978).
9. Meakins J.L., Christou N.V., Shizgal H.M., et al. Ann. Surg. 190 : 286 (1979).
10. Braquet, M., Lavaud, P., Dormont, D., et al. Prostaglandins 29 : 747 (1985).
11. Braquet, M., Dormand D., Garay R., et al. Adv. Prostaglandin Thromboxane and Leukotriene Res. 15 : 677 (1985).
12. Antonacci A.C., Reaves L.E., Calvano S.E., et al. Surg. Gynecol Obstet 159 : 1 (1984) (suppl. 1).
13. Braquet M., Dormont D., Ducousso R., et al. Agents Actions 17 : 385 (1986) (Suppl. 3-4).
14. Antonacci A.C., Calvano S.E., Organ B.C., et al. Lipid Mediators in Immunology of Burn and Sepsis. Nato Adv. Res. Workshop, Helsingor, Denmark, July 20-25. New York, Plenum (in press) (1986).
15. Markely K, Smallman E.T. Proc. Soc. Exp. Biol. Med. 160 : 468 (1979).

16. Warden G.D. and Ninnemann J.L. In Ninnemann J.L. (ed.) : The Immune Consequences of Thermal Injury. Baltimore, Williams & Wilkins p. 1. (1981).
17. Wood G.W., Volence F.J., Manni M.M. et al. Clin. Exp. Immunol. 31 : 291 (1978).
18. Munster A.M., Winchurch R.A., Birmingham W.J. et al. Ann. Surg. 192 : 772 (1980).
19. Baker C.C., Miller C.L., Trunkey D.D. J. Trauma 19 : 641 (1979).
20. Miller C.L. and Baker C.C. J. Clin. Invest. 63 : 202 (1979).
21. Arturson G., Hogman C.F., Johansson S.G.O., et al. Lancet 1 : 546 (1969).
22. Munster A.M., Hoagland H.C., Pruitt B.A. Jr. Ann. Surg. 172 : 965 (1970).
23. Daniels J.C., Larson D.L., Abtson S. et al. Trauma 14 : 137 (1974).
24. Daniels J.C., Fukushima M., Larson D.L. et al. J. Trauma 11 : 699 (1971).
25. Alexander J.W., Moncrief J.A. Arch. Surg. 93 : 75 (1966).
26. Antonacci A.C., Gupta S., Good A.R. et al. Curr. Surg. Jan-Feb 24 : 20 (1983).
27. Stein M.D., Gamble D.N., Klimpel K.D. et al. Cell Immunol. 86 : 551 (1984).
28. Alexander J.W., Dionigi R., Meakins J.L. Ann. Surg. 173 : 206 (1971).
29. Grogan J.B. and Miller R.C. Surg. Gynecol. Obstet. 137 : 784 (1973).
30. Grogan J.B. J. Trauma 16 : 734 (1976).
31. Warden G.D., Mason A.D. Jr., Pruitt B.A. Jr. J. Clin. Invest. 54 : 1001 (1974).
32. Grogan, J.B. J. Trauma 16 : 985 (1976).
33. Fikrig, S.M., Karl, S.C. and Suntharalingam, K. Ann. Surg. 186 : 746 (1977).
34. Davis J.M., Dineen P. and Galin, J.I. J. Immunol. 124 : 1467 (1980).
35. Alexander J.W. Arch. Surg. 95 : 482 (1967).
36. Heck E.L., Browne L., Curreri P.W. et al. J. Trauma 15 : 486 (1975).
37. Curreri P.W., Heck E.L., Browne L. et al. Surgery 74 : 6 (1973).
38. Howes R.M., Allen R.C., Su C.T. et al. Surg. Forum 27 : 558 (1976).
39. Canocico P.G., McManus A.T.N. and Powanda M.C. In Dingle J.T., Jacques P.J., Shaw I.H. (eds) : Lysosomes in Applied Biology and Therapeutics, Amsterdam, North Holland 6 : 287 (1979).
40. Heck E.L., Edgar M.A., Masters B.S. et al. J. Trauma 19 : 49 (1979).
41. McCabe W.P., Rebuck, J.W., Kelly A.P. Jr. et al. Arch. Surg. 106 : 155 (1973).
42. Loose L.D., Turinsky, J. Infect. Immun. 26 : 157 (1979).
43. Saba T.M. and Jaffe E. Am. J. Med. 68 : 577 (1980).
44. Lowbury E.J.L. and Ricketts C.R. J. Hyg. 55 : 266 (1957).
45. Bjornson A.B. and Alexander J.A. J. Lab. Clin. Med. 83 : 372 (1974).
46. Bjornson A.B., Altemeier W.A. and Bjornson H.S. Ann. Surg. 191 : 323 (1980).
47. Dhennin C., Pinon G., Greco J.M. J. Trauma 18 : 129 (1978).
48. Gelfand J.A., Donelan M. and Burke J.F. Ann. Surg. 198 : 58 (1983).
49. Zuckerman L., Caprini J.A., Lipp V. et al. J. Trauma 18 : 432 (1978).
50. Fjellstrom K.E., Arturson G. Acta Pathol. Microbiol. Immunol. Scand. 59 : 257 (1963).
51. Erbs F., Muller F.E., Opferkuch W. Fortschr. Med. 98 : 397 (1980).
52. Bjornson A.B., Altemeier W.A., Bjornson H.S. J. Trauma 16 : 905 (1976).
53. Bjornson A.B., Altemeier W.A. and Bjornson H.S. Ann. Surg. 186 : 88 (1977).
54. Holder I.A. and Ablin, R.J. In Ninnemann J.L. (ed) : The Immune Consequences of Thermal Injury. Baltimore Williams & Wilkins, p. 134 (1980).
55. Gelfand J.A., Donelan M. and Burke J.F. Am. Surg. 198 : 58 (1983)

(suppl 1).

56. Munster A.M. Lancet 1 : 1329 (1976).
57. Baker C.C., Miller C.L., Trunkey D.D. et al. J. Surg. Res. 26 : 478 (1979).
58. Ninnemann J.L., Stockland A.E. and Condie J.T. J. Clin. Immunol. 3 : 142 (1983).
59. Mannick J.A. In Immunology of Burn and Sepsis. Nato Adv. Res. Workshop, Helsingor, Denmark, July 20-25. New York, Plenum (in press) (1986).
60. Ninnemann J.L., Condie T., Davis S.E. et al. J. Trauma 22 : 837 (1982).
61. Altman L.C., Furukawa C.T. and Klebanoff S.J. Clin. Res. 25 : 117A (1977).
62. Warden G.D., Masson A.D. and Pruitt B.A. Ann. Surg. 181 : 363 (1975).
63. Wolfe J.H., Saporoschetz A.E., Youns A.D. et al. Ann. Surg. 193 : 513 (1981) (suppl. 4).
64. Baxter C.R. In Immunology of Burn and Sepsis. Nato Adv. Res. Workshop, Helsingor, Denmark, July 20-25. New York, Plenum (in press) (1986).
65. Bjornson A.B., Bjornson, H.S. and Altemeier W.A. Ann. Surg. 194 : 568 (1981).
66. Ogle J.O., Ogle C.K. and Alexander J.W. Infect. Immun. 40 : 967 (1983).
67. Ninnemann J.L. and Stein, M.D. J. Trauma 20 : 959 (1980) (suppl. 11).
68. Ninnemann J.L. and Ozkan A.N. J. Trauma 25 : 113 (1985) (suppl. 2).
69. Ozkan A.N. and Ninnemann J.L. J. Clin. Immunol. 5 : 172 (1985) (suppl. 3).
70. Cameron D.J. and Collawn S.S. Int. J. Immunopharmacol. 5 : 55 (1983) (suppl. 1).
71. Nagel G.A., Beyer J.H., Schuff-Werner P., et al. Cancer Treat. Rev. 11 : 139 (1984) (suppl. A).
72. Farmer J.L. and Prager, M.D. Cancer Immunol. Immunother 18 : 101 (1984).
73. Kiprov D.D., Lippert R., Sandstrom E. et al. J. Clin. Apher. 2 : 427 (1985).
74. Warden G.D., Ninnemann, J., Stratta R.J. et al. Surgery 96 : 321 (1984) (suppl. 2).
75. Shelby J. and Krob M.J. J. Trauma 26 : 54 (1986) (suppl. 1).
76. Parnham M.J. and Engleberger, W. In Morley J., Bray M.A. (eds). Handbook of Exp. Pharmacol. The Pharmacology of Lymphocytes (in press).

IMPORTANCE OF IMMUNE FUNCTION IN TRAUMA FOR SURVIVAL

J. L. Meakins
Professor of Surgery and Microbiology Mc Gill University
Montreal, Quebec, Canada

INTRODUCTION

Trauma may manifest itself in a number of different
ways, including thermal injury, blunt injury, and the injury
associated with elective surgery. Each of these forms of
trauma has been demonstrated to produce important changes in
the immune response which are clearly secondary and which in
the absence of compounding variables such as the complications
of infection, shock, malnutrition etc. have a recovery period
which is a function of the magnitude of the injury and the age
of the patient. This chapter will review data on the immune
system following trauma and integrate these acquired immune
defects with the increased likelihood of developing infection
as well as the influence of compounding variables on the
evolution of these defects.

THE EFFECT OF SURGERY

Surgery is a controlled form of trauma and as such
stratification of its magnitude permits an evaluation of its
influence upon the immune response. While there have been
many studies done in this area, the first comprehensive
approach was that of Slade et al.[1] who found that a standard
donor nephrectomy, lasting approximately three hours and
twenty minutes, led to depression of the delayed
hypersensitivity skin response in the immediate post-operative
period, which could last up to two weeks. In vitro tests of
T and B lymphocyte function were also depressed, following
induction of anesthesia and stayed depressed up to five days.
Correlation of different parameters was difficult to confirm
but there was overall decreased immune function following the
donor nephrectomy. McLouglin et al. demonstrated an excellent
correlation between the appearance of immuno-suppressive
activity in the serum to PHA and skin test anergy in patients
following major vascular surgical procedures specifically
those on the aorta.[2] It is of interest that those patients
who demonstrated the altered skin test responses and the
presence of the immuno-suppressive serum were that group of

patients who developed an increased incidence of infectious complications following surgery.

Our initial studies of the influence of surgery failed to demonstrate any influence of surgery upon skin testing[3] but the timing was incorrectly selected and subsequent studies[4] have demonstrated clearly that skin testing, neutrophil chemotaxis and neutrophil adherence may be altered by any surgical procedure. Furthermore, the magnitude of the operation was an important contributing factor to the degree of abnormality of the immune parameter measured. Most functions we studied had returned to normal by seven days.

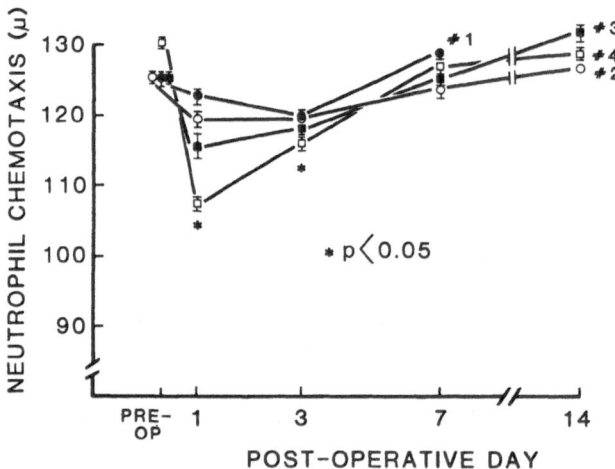

Fig.1. The effect of surgery on neutrophil chemotaxis.

Figure I demonstrates the influence of surgery upon neutrophil chemotaxis and in addition reinforces the idea that the magnitude of the procedure influences the degree of abnormality of neutrophil chemotaxis.

As part of an attempt to correct these abnormalities following surgery, a previously demonstrated protein-sparing nutritional regimen ,[5] was instituted on the hypothesis that the neuroendocrine response which produced the changes in body composition would be that same response which influenced the immune system; a study of 85 patients randomly allocated to an isocaloric protein-sparing regimen versus a regular 10% glucose solution, demonstrated there were no differences in either outcome or benefit to the immune system. In both groups of patients the cell-mediated and neutrophil functions measured were equivalently decreased. Serum albumin was, however, higher in those patients receiving the protein-sparing regimen. There is a study confirming this result from

O'Mahony and his collegues who looked at PHA responsiveness and a descriptive review of T-lymphocytes and their subsets.[6] They demonstrated no influence upon the immune changes following surgery, of a similar protein-sparing regimen. It therefore seems clear that the alterations produced by pure injury following surgery are not mediated by any of the factors related to nutritional or body composition measures may be mediated by some factors unrelated to the neuroendocrine response.

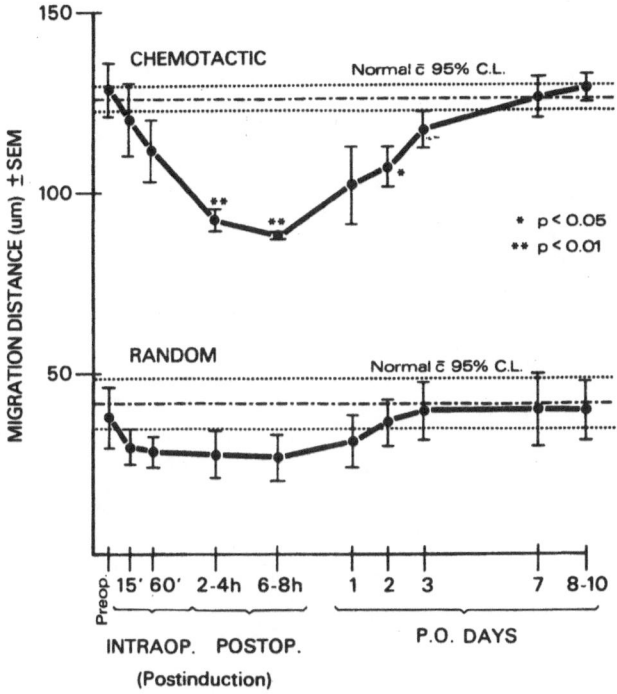

Fig.2.

This is demonstrated in figure 2 where the influence of donor nephrectomy starts to produce abnormal chemotaxis extremely early in the procedure. The complexity of these issues is demonstrated by Mochizuki et al. who ablated the neuroendocrine response in burned guinea pigs with enteral but not parenteral feeding.[7]

THE INFLUENCE OF TRAUMA

burn injury has been demonstrated to affect all aspects of the immune response.[8,9] These abnormalities occur shortly after burn injury and are initially caused by the trauma but in this setting compounding factors such as complications, the state of the patient's health, patient age, infection and other uncertain factors contribute to the persistance of the acquired immunodefiency. The importance of humoral mediators is perhaps best seen in Warden's approach to the use of plasma pheresis in the infected and failing thermally injured patient.[10]

Blunt injury has important influences upon the immune response. Our data [11,12] demonstrate that skin test abnormalities occur promptly after the injury and appear to be a function of the age of the patient and the magnitude of the injury.

In addition to the abnormalities of skin tests there are clear changes in neutrophil chemotaxis and neutrophil adherence.

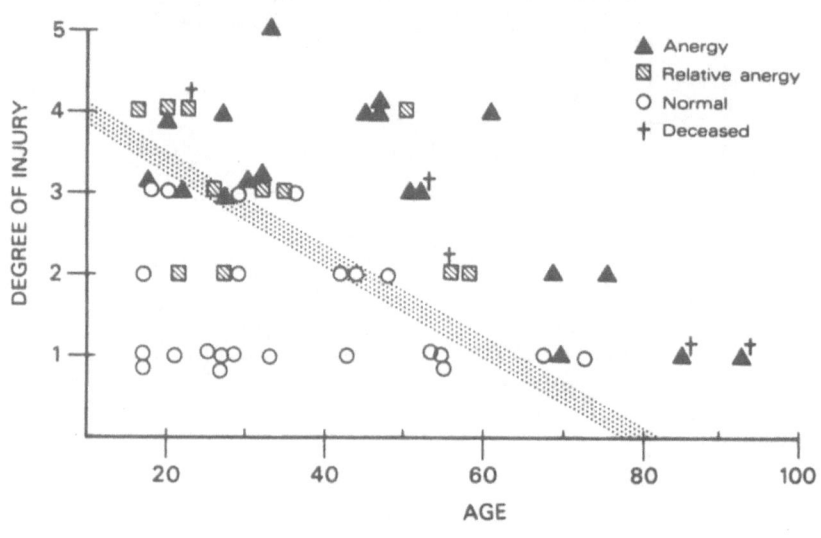

Fig. 3.

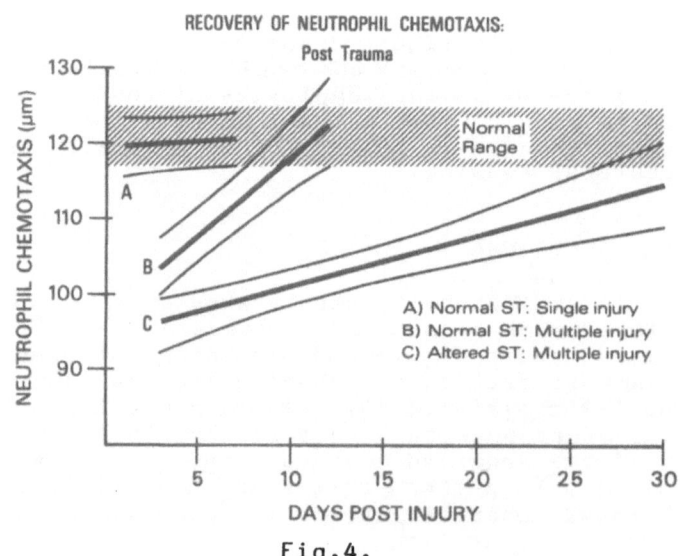

Fig.4.

Figure 4 demonstrates the influence of trauma on three subgroups of injured patients. Those patients whose neutrophil
chemotaxis remains normal throughout their course in hospital
have a minor injury of no greater magnitude than a fracture.
Those patients who have more than one system injured have an
initial decrease in their neutrophil chemotaxis but their
injury, not being sufficient to alter skin test reactivity, is
associated with a rapid return to normal of neutrophil
chemotaxis. The third group are those patients who had
multiple injuries sufficient to alter skin test reactivity and
in this population at least four weeks are required before
skin test returns to normal. Neutrophil adherence is
similarly abnormal and there is a close correlation between
the presence of abnormal chemotaxis and abnormal adherence [12].

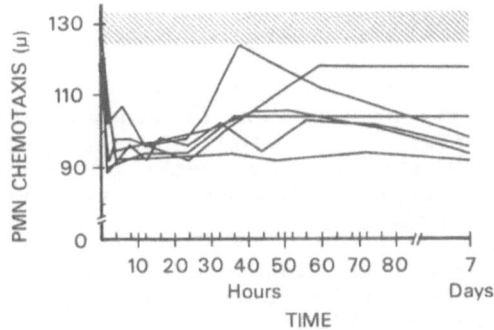

Fig.5. Neutrophil chemotaxis studied upon arrival in emergency
       following trauma.

The patients who became anergic following injury had a higher
incidence of infection; however some did not have any
infectious complication. Determination of time at risk, that
is the period of time during which patients were immunocompromised demonstrated a substantially higher rate of
infection and septicemia, which increased with duration of
risk. The altered immune responses were always associated
with increased mortality from all causes but particularly
those of an infectious nature.

    The timing of these abnormalities is almost instantly
following the injury.

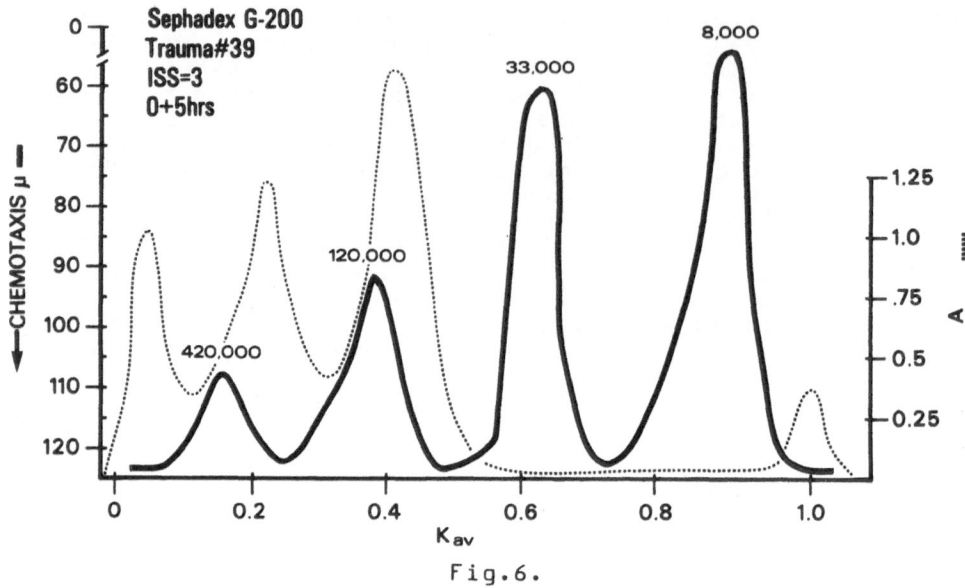

Fig.6.

Patients studied two hours after trauma can be seen to have altered PMN chemotaxis with serum inhibitors of cell migration. Figure 6 showing four peaks of serum inhibitors of neutrophil chemotaxis is from a patient who was studied five hours after their injury.

These data suggest the instantaneous production of mediators which can influence the immune response. The four peaks of chemotactic inhibitors seen in figure 6 are remarkably similar to those abnormalities found by Ninnemann in his burn population.[14] Because of technical difficulties with the chemotaxis assay we have been unable to further fractionate or identify these peaks of inhibitory activity.

Christou's data in trauma patients suggest strongly that the immune response and nutritional factors and body composition are separate in the trauma patient. Immune measures (skin test reactivity and chemotaxis) have returned to normal in patients whose anthropometric studies (tricep's skin-fold thickness, mid-arm circumference and serum albumin) are increasingly abnormal.[12]

COMPOUNDING FACTORS

The normal course of the changes following trauma or surgery is one of recovery, and is a function of the magnitude of the trauma. However, when compounding factors such as complications of the injury itself, the development of infection, or other factors intervene in the recovery following trauma, the abnormal host defense mechanisms persist as a result of the compounding variable, at the level to which they had been depressed by the initial injury. It goes without saying that this is an extremely complex and difficult area to unravel particularly in a patient population where clinical investigation is controlled more by the evolution of the patient than by other factors. It is only in animal models that a truly controlled situation can be created and permits the possibility of dissecting out the various factors that are involved in the development of these abnormalities.

The development of intraabdominal sepsis following either surgery or basic intraabdominal disease is very much associated with abnormalities of host defense. It is also apparent that the drainage of the last abscess, that is control of the intraperitoneal infection surgically, does lead to improvement in host defenses. Our own studies in drainage of abscesses point this out[15]. Solomkin et al. in studies of neutrophil chemiluminescence and chemotaxis has demonstrated exactly the same thing, in particular, that it is the drainage of the last abscess that is of most importance.[16] His neutrophil chemotaxis and chemiluminescence can be seen to improve temporarily following drainage but never completely to normal until infection is controlled. Heideman[17] looking at complement levels (C-3, C-4, and C-5) has shown that resection of necrotic tissue, drainage of abscesses, control of trauma are all clearly associated with the return to normal of circulating complement factors. The failure to correct these basic disease processes is similarily associated with the persistence of their abnormality. In the same vein Richards et al.[18] have demonstrated with regard to fibronectin that those patients whose intraabdominal abscess is drained without accompanying multiple organ failure have a routine return to normal of fibronectin levels (along the same course that would be expected following recovery following intraabdominal surgery). Those patients with a more complex course, that is multiple organ failure, do not have the same return to normal; their fibronectin levels remain remarkably abnormal.

Chrohn's disease and its complications have been associated with abnormalities of the immune response[19]. In an attempt to determine whether this was an intrinsic defect associated with Crohn's disease itself and therefore present in all patients, or a function of the patient's illness, we utilized the Crohn's disease activity index (CDAI) and correlated   with a variety of immune parameters.[20] It is apparent that while not all of these measures go perfectly with the Crohn's disease activity index that the symptomatic patients with a high CDAI had abnormalities of their skin test reactivity, neutrophil chemotaxis and the ability to deliver neutrophils to an inflammatory focus.

| | ST | CTX(u)* | CDAI* | PMN Delivery (x10^6) | |
				6H*	24H*
Normal HD	All N	126.6±2.3	63.7±23.1	11.4±2.0	144.4±23.1
	2 N				
Abnormal HD	7RA	103.7±3.2	341.2±33.0	0.5±0.2	25.2±12.5

The correlation of both the CDAI and neutrophil chemotaxis and cell delivery into skin windows was statistically significant. When in the same patients their CDAI returned to lower levels and their disease was in a state of remission all of their immune measures were normal. The patients still had Crohn's disease but it was quiescent suggesting strongly that it is

271

the state of the patient's illness in this setting that is
creating the abnormalities of the immune response rather than
the specific disease itself.

The implications are such that one can imagine a
traumatized patient having the same abnormalities with the
complications of his trauma once the influence of the injury
itself had passed. As an extension of these ideas Tchervenkov
et al.[21] working with rats has demonstrated that there is a
marked decrease in skin test reactivity following burn injury
which starts to return to normal on day 5 following the burn.
These burns universally became infected on about the 12th
post-burn day at which time the skin tests again deteriorated
markedly. This second drop in skin test reactivity was a
result of the infection of the burn wound as the animals had
already demonstrated that they were partially recovered from
the burn injury itself. The response to KLH remained abnormal
for the remainder of the experiment, Fig.7. The complexity of
cause and effect is apparent and its resolution will only be
managed in burn models such as this.

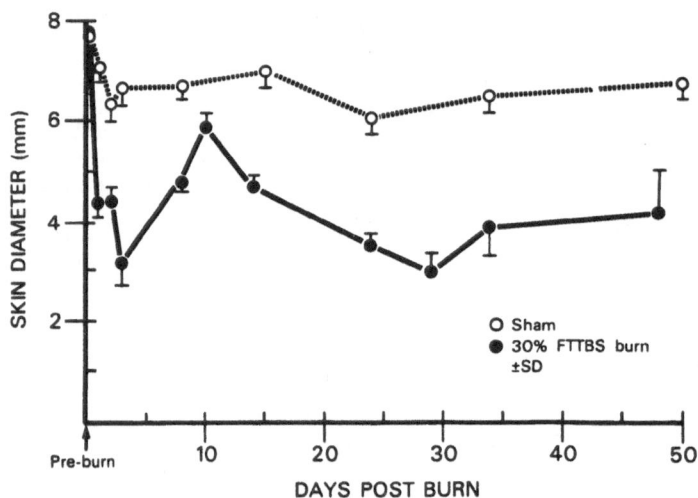

Fig.7. DTH response following a 30% burn in the rat.

Utilizing the Staphylococcus abscess model of Miles, Miles
and Burke[22], the DTH response to KLH and size of the
staphylococcal abscess were correlated. Throughout the
evolution of the burned rat model described above the skin
test reaction predicted the size of the abscess: the smaller
the DTH response, the larger the abscess (fig.8) and vice-
versa.

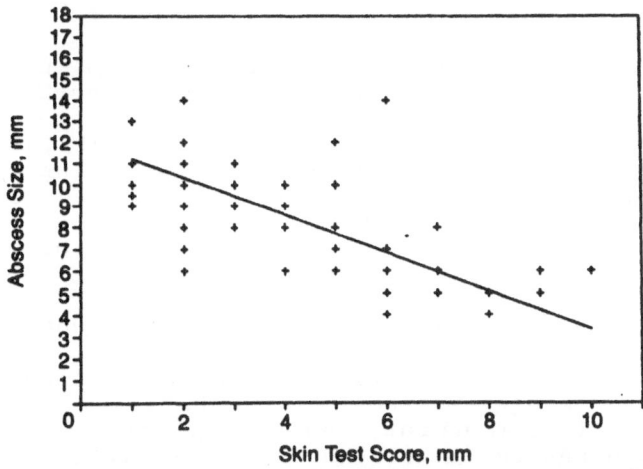

—Linear regression correlation curve showing negative correlation between skin test score (delayed-type hypersensitivity response) and abscess size (bacterial lesions). Abscess size = [12.07 − (0.87 × skin test score)]; $r = .73$ ($P < .001$). Smaller delayed-type hypersensitivity response corresponded to larger bacterial lesion.

Fig.8. Correlation of abscess size with DTH response.

The relationship between a cell mediated immune response and a bacterial infection is striking and suggests that compartmentalization of the immune response and control of infection may not be as clear as previously thought. McPhee et al.[23] have shown that the product of a mixed lymphocyte culture reaction is effective in markedly reducing mortality in intraabdominal infection in rats.

SUMMARY

All forms of injury and perhaps even stress lead to changes in the immune system. It is likely that age has an influence on the degree of these abnormalities. These changes almost certainly have important influences on outcome. In the absence of compounding factors the natural history of these changes is recovery which is related to the initial injury and the changes in physiology.

Development of any complications will lead to prolongation of this state of acquired immuno-deficiency. Compounding factors such as infection, malnutrition or organ failure create their own problems and become so intertwined with the effects of the initial injury that they are inseparable in their effects upon the immune system and consequent susceptibility to infections.

REFERENCES

1. M.S. Slade, R.D. Simmons, E. Yunis, and L.J. Greenberg, Immunodepression after major surgery in normal patients, Surgery 78:363-372 (1975).

2. G.A. McLoughlin, A.V. Wu, and I. Saporoschetz,et al., Correlation between anergy and a circulating immunosuppressive factor following major surgical trauma, Ann. Surg. 190:297-304 (1979).

3. J.B. Pietsch, J.L. Meakins, D. Gotto, and L.D. MacLean, Delayed hypersensitivity responses: The effect of surgery, J. Surg. Res. 22:228-230 (1977).

4. N.V. Christou, R. Superina, M. Broadhead, and J.L. Meakins, Post operative depression of host resistance; determinants and effects of peripheral protein sparing therapy, Surgery 92:786-792 (1982).

5. H.M. Shizgal, C.A. Milne, and A.H. Spanier, The effect of nitrogen-sparing intravenously administered fluids on postoperative body composition, Surgery 85:496-503 (1979).

6. J.B. O'Mahony, A.J. McIrvine, S.B. Palder, et al., The effect of short term postoperative intravenous feeding upon cell-mediated immunity and serum suppressive activity in well nourished patients, Surg. Gyn. Obs. 159:27-32 (1984).

7. H. Mochizuki, O. Trocki, L. Don, et al., Mechanism of prevention of post burn hypermetabolism and catabolism by early enteral feeding, Ann. Surg. 200:297-310 (1984).

8. J.W. Alexander, C.K. Ogle, J.D. Stinnett, et al, A comparison of immunologic profiles and their influence on bacteremia in surgical patients with a high risk of infection, Surgery 86:343 (1979).

9. J.W. Alexander, and J.L. Meakins, A physiologic basis for opportunistic infection, Ann. Surg. 176:273-387 (1972).

10. G.L. Warden, The use of plasma exchange in the management of thermally injured patients, chapter 14, in: "Traumatic injury: infection and other immunologic sequelae," J.L. Neinnemann, ed., University Park Press, Baltimore (1983).

11. J.L. Meakins, A.P.H. McLean, R. Kelly, et al, Delayed hypersensitivity response and neutrophil, J. Trauma 18:240-247 (1978).

12. N.V. Christou, A.P.H. McLean, and J.L. Meakins, Host defence in blunt trauma: Interrelationships of kinetics of anergy and depressed neutrophil function, nutritional status and sepsis, J. Trauma 20:833-841 (1980).

13. N.V. Christou, and J.L. Meakins, Partial analysis and purification of the polymorphonuclear neutrophil chemotactic inhibitors in serum from anergic patients, Arch. Surg. 118:156-160 (1983).

14. J.L. Ninnemann, Immune depression in burn and trauma patients: the role of circulating suppressors, chapter 3, in: "Traumatic injury: infection and other immunologic sequelae", J.L. Ninnemann, ed., University Park Press, Baltimore (1983).

15. J.L. Meakins, N.V. Christou, H.M. Shizgal and L.D. MacLean, Therapeutic approaches to anergy in surgical patients: Surgery and Levamisole, Ann. Surg. 190:285-296 (1979).

16. J.S. Solomkin, L.A. Coha, J.K. Brodt, et al., Neutrophil dysfunction in Sepsis: III Degranulation as a mechanism for non specific deactivation. <u>J. Surg. Res</u>. 36:407-412 (1984).
17. M. Heideman, Bacteremia and multiple organ failure: The role of injured tissue and abscess, chapter 8, <u>in:</u> Traumatic injury: infection and other immunologic sequelae", J.L. Ninnemann, ed., University Park Press, Baltimore (1983).
18. W.O. Richards, W. Scovill, B. Shin, W. Reed, Oprsonic fibro-nectin deficiency in patients with intraabdominal infection. <u>Surgery</u> 94:210-7 (1983).
19. A.W. Segal, and G. Loewi, Neutrophil dysfunction in Crohn's disease, <u>Lancet</u> ii:219-221 (1976).
20. R.A. Superina, N.V. Christou, M. Broadhead, et al, Relationship between disease activity index and host defense in patients with inflammatory bowel disease, Ann. Royal Coll. Phys. Surg. Canada 14 (abst) (1981).
21. J.I. Tchervenkov, E. Diano, J.L. Meakins, and N.V. Christou, The delayed type hypersensitivity (DTH) skin test score (STS) accurately measures susceptibility to bacterial sepsis, <u>Arch. Surg</u>. (1986).
22. A.A. Miles, E.M. Miles and J. Burke, The value and duration of defense reactions of the skin to the primary lodgement of bacteria, <u>Br. J. Exp. Path.</u>, 24:95-107 (1957).
23. M. McPhee, I. Zakaluzny, J. Marshall, et al., Mixed lymphocyte culture (MLC) supernatants provide effective immunotherapy for accute peritonitis in immunosuppressed rats, <u>Surgical Forum</u> (in press) (1986).

ANERGY IN CRITICALLY ILL PATIENTS

A. Engquist and S. H. Johansen
Department of ICU and Anesthesia
Bispebjerg and Herlev Hospitals Copenhagen, Denmark

INTRODUCTION

The failure of delayed hypersensitivity skin responses
to recall antigens or anergy has been widely used in order to
document a possible correlation to the incidence of postoperative
sepsis and mortality. Patients with anergy developed significantly
more infectious complications postoperatively than the others[1,2].
Additionally, the degree of surgical trauma determined the extent
of the postoperative decrease of skin test responses. In contrast,
Bancewicz et al.[3] found that delayed hypersensitivity skin
testing was not of general value as a predictive test in surgical
patients, since their data demonstrated an increasing incidence
of anergy with age. Furthermore, abnormal skin tests followed
clinical detection of sepsis, and therefore did not contribute
to management.

Factors implicated as causes of anergy are sepsis, shock,
major trauma, malnutrition and high age. All these factors are
commonly found in patients treated in the intensive care unit
and these patients carry a high mortality.

The purpose of the present study was to evaluate the
usefulnes of intradermal tests of immunity for determining
survival in critically ill medical or surgical patients
admitted to an intensive care department[4].

MATERIAL AND METHODS

Out of a total of 952 patients admitted to the intensive
care unit (ICU) at Herlev Hospital during a 3 year period,
290 consecutive patients were tested by intradermal injection
of 4 antigens: candida, mumps, tuberculin and streptokinase –
streptodornase (Varidase(R)) and a control solution containing
solvents as described by Meakins et al[5].

The skin testing was performed within the first 24 h
after admission to the ICU as described previously[4]. The
induration was measured 48 h after injection. A positive

reaction was defined as a diameter of at least 5 mm. Two or more positive reactions represent normal immunity. One positive reaction indicates relative anergy. If no positive reactions are present, anergy exists. Sequential testing was performed at weekly intervals.

The 290 patients were representative concerning distribution of age, sex, diagnoses and mortality rates.

The definition of sepsis was a quickly rising core temperature and shivering, the presence of a well-known focus or a suspicion hereof with or without a positive blood culture. In no case was the treatment altered as a consequence of the skin test response, since the results were excluded from the patient's chart and, therefore, unknown to the clinicians.

RESULTS

Table 1 shows the distribution of the main diagnoses in the total group of 290 patients, consisting of 165 surgical and 125 medical patients, respectively.

Table 1. Main diagnoses

Diagnoses	No.	%
Respiratory failure	161	55.5
Cardiovascular failure	42	14.5
Gastrointestinal disease	21	7.2
Renal failure	11	3.8
Endocrine disease	8	2.8
Other[a]	47	16.2

[a]Poisoning, sepsis of unknown origin, fluid-electrolyte disorders.

Twenty-seven of the patients died within 24 h after skin testing or were transferred to other hospitals. The results from these patients are excluded which leaves 263 patients. In Table 2, the distribution of the 263 patients and their responses to skin testing are shown including mortality rates.

Of the 263 patients, 55.1% were anergic, 23.6% were relative anergic, and 21.2% showed normal responses. The difference in skin test responses between surgical and medical patients was insignificant. The mortality rate was 45.5% in all anergic patients, being significantly higher in anergic surgical patients (50.0%) as compared to anergic

Table 2. Skin test responses and mortality

Response			Death	
	No.	%	No.	%
Surgical patients n = 148				
Anergic	82	55.4	41[a]	50.0
Relative anergic	35	23.6	9	25.7
Normal	31	20.9	7	22.5
Medical patients n = 115				
Anergic	63	54.8	25[a]	39.7
Relative anergic	27	23.5	6	22.2
Normal	25	21.7	3	12.0
Total n = 263				
Anergic	145	55.1	66[a]	45.5
Relative anergic	62	23.6	15	24.2
Normal	56	21.2	11	17.9

[a] Mortality rate was significantly higher in anergic patients as compared to relative anergic and normal patients, respectively (p < 0.001).

medical patients (39.7%)(p < 0.001). In relative anergic patients the total mortality rate was 24.2%. In patients showing normal skin test responses the total mortality rate was 17.9% (surgical patients 22.5%; medical patients 12.0%). The total mortality rate in anergic patients was significantly higher (p < 0.001) than the mortality rates in relative anergic and normal patients. The difference between the latter two groups of patients differed insignificantly.

Table 3. Sepsis rates and mortality in septic patients

Response	Sepsis		Death	
	No.	%	No.	%
Surgical patients n = 148				
Anergic	24	29.3	15	62.5
Relative anergic	11	31.4	3	27.3
Normal	3	9.7	2	66.6
Medical patients n = 115				
Anergic	16	25.4	7	43.8
Relative anergic	3	11.1	2	66.6
Normal	3	12.0	0	0.0
Total n = 263				
Anergic	40[b]	27.6	22	55.0[a]
Relative anergic	14[b]	22.6	5	35.7
Normal	6	10.7	2	33.3

[a]
Mortality rate was significantly higher in anergic patients as compared to relative anergic and normal patients, respectively (p < 0.001).

[b]
Sepsis rate was significantly higher in anergic and relative anergic patients (p < 0.001).

The total group of anergic and relative anergic patients showed sepsis rates of 27.6% and 22.6%, respectively, which were significantly higher than the sepsis rate in normal patients (10.7%) (p < 0.001). The sepsis rates between surgical and medical patients differed insignificantly.

In the total group of 40 anergic septic patients the mortality rate was 55.0% as compared to 35.7% in relative anergic and 33.3% in normal patients, respectively, being significantly higher (p < 0.001). The total group of anergic patients thus showed significantly higher mortality and sepsis rates. Additionally, the mortality rate in anergic septic patients was significantly higher as compared to mortality in septic patients with relative anergy or normal skin test responses.

Table 4. Sequential skin testing

Response	Unaltered		Improved		Worse	
	No.	Mortality%	No.	Mortality%	No.	Mortality%
Anergic	19	89.5[a]	20	25.0	–	–
Relative anergic	5	60.0	3	33.3	9	55.5
Normal	6	33.3	–	–	0	0

a
Mortality rate is significantly higher in patients who remain anergic as compared to anergic patients with improved responses (p < 0.001).

Table 4 shows the results from sequential skin testing in 62 patients. 19 patients remained anergic and 17 of these died. In contrast, 20 patients showed improved skin test responses. Of these patients only 6 patients died. This difference in mortality rates was significant (p < 0.001).

Table 5. Distribution of skin test responses and mortality rates in patients with or without cancer

Response	Cancer (n = 57)		Non-cancer (n = 206)	
	No.(%)	Mortality(%)	No.(%)	Mortality(%)
Anergic	61.4[a]	68.6[b]	53.4	38.2
Relative anergic	22.8	30.8[b]	23.8	22.4
Normal	15.8[a]	33.3[b]	22.8	14.9

a
Patients with cancer showed a higher rate of anergy (p < 0.001) and a lower rate of normal skin test response (p < 0.001).

b
All patients with cancer showed higher mortality rates (p < 0.001).

Table 5 shows the distribution of skin test responses and percentage of mortality in anergic, relative anergic and normal patients with (n = 57) or without (n = 206) cancer. Patients with cancer showed a significantly higher rate of anergy (61.4%) than patients without cancer (53.4%)(p < 0.001). The percentage of normal skin test responses was significantly lower in the cancer group of patients (15.8%) as compared to 22.8% in patients without cancer (p < 0.001). The percentage of relative anergy differed insignificantly in the two groups of patients. All cancer patients had significantly higher mortality rates (p < 0.001) compared with patients without cancer, regardles of skin test responses.

DISCUSSION

Cell-mediated immunity was assessed in 148 surgical and 115 medical patients, all critically ill, within 48 h after admission to the intensive care unit. The distribution of skin test responses showed approximately 55% anergic, 24% relative anergic responses and 21% normal reaction to four recall antigens. A distribution of skin test responses, approximately that of our study was found by Meakins et al.[6] and Pietsch et al.[7].

They found roughly 63% anergic, 22% relative anergic responses and 15% normal reactions. In preoperative patients Meakins et al.[5] found a majority of normal reactions, about 70%. 21% showed anergic and 9% relative anergic responses. Our data indicate that critically ill patients react to recall antigens in much the same way as patients after major surgery or trauma, although almost half of the patients in our study were medical patients. The skin test responses thus suggest a correlation between severity of disease and immunological competence. Our results thus contrast with the findings of Bancewicz et al.[3], who did not find increased mortality and sepsis rates in patients with abnormal skin test responses. The reason for this is obscure. Sequential skin testing provided even more prognostic information, since the group of anergic patients with unaltered responses showed a mortality rate of almost 90%, whereas patients with improved skin test reactions showed a mortality rate of only 25%.

Sepsis was found in 27.6% of the anergic patients with a mortality rate of 55.0%. In contrast, only 10.7% of the normal responders were septic and the mortality rate was only 33.3% in this group of patients. By examining all 952 patients admitted to the intensive care unit during a three year period, the overall sepsis rate was found to be 15.9% and the mortality rate was 41.1%.

The higher rates of anergy and mortality found in patients with cancer do not essentially alter the results for the total group of patients. Eilber et al.[8] also found higher rates of anergy and mortality in patients with cancer, whereas Johnson et al.[9] and Pietsch et al.[7] found equal distributions of skin test responses between patients with or without cancer. The higher mortality rates in cancer patients could be due to the progression of malignancy per se.

Since the majority of patients received either enteral or parenteral nutrition, it was not possible to relate the skin test responses to the nutritional status of our patients.

In conclusion, our findings show that the responses to skin testing with recall antigens in critically ill patients provide prognostic information on sepsis and mortality rates in good agreement with those found by Meakins et al.[6] and Pietsch et al.[7]. However, our results do not clarify whether changes in immunocompetence precede and cause infectious complications and higher mortality rates or whether they are secondary responses to organ failure in critical illness. Furthermore, the present study does not answer, whether anergy reflects defective T helper cell function, increased T suppressor cell activity[10], or a T cell suppressive factor[11]. Assessment of cellmediated immunity by skin testing with recall antigens cannot be recommended for general use. However, this method can be used in monitoring immunocompetence during future attempts to reverse acquired defects in host-defence mechanisms.

REFERENCES

1. L.D. MacLean. Journal of Trauma, 19: 297 (1979).
2. N.V. Christou, J.L. Meakins and L.D. MacLean. Surgery, Gynecology and Obstetrics, 152: 297 (1981).
3. J. Bancewicz, R. Brown, J. Hamid, C. Ward and M.H. Irving in Clinical Nutrition. (Eds. R.I.C. Wesdrop and P.B. Soeters). Churchill Livingstone (Edinburgh) 265 (1982).
4. F. Janstrup, L. Dragsted, S.H. Johansen and A. Engquist in Shock Research. (Eds. D. Lewis and U. Haglund). Elsevier Science Publishers (Amsterdam) 289 (1983).
5. J.L. Meakins, N.V. Christou, H.M. Shizgal and L.D. MacLean. Annals of Surgery, 190: 286 (1979).
6. J.L. Meakins, J.B. Pietsch, O. Bubenik, R. Kelly, H. Rode, J. Gordon and L.D. MacLean. Annals of Surgery, 186: 241 (1977).
7. J.B. Pietsch, J.L. Meakins and L.D. MacLean. Surgery, 82: 349 (1977).
8. F.R. Eilber and D.L. Morton. Cancer, 24: 362 (1970).
9. W.C. Johnson, F. Ulrich and M.M. Meguid. American Surgeon, 137: 536 (1979).
10. A.M. Munster. Lancet ii: 1329 (1976).
11. A.J. McIrvine and J.A. Mannick. Surgical Clinics of North America, 63: 245 (1983).

NEURAL RELEASE MECHANISMS OF THE INJURY RESPONSE

H. Kehlet, C. Lund and S. Schulze
Department of Surgical Gastroenterology
Hvidovre University Hospital, DK-2650 Hvidovre, Denmark

INTRODUCTION

The body reacts to injury both locally and generally. The inflammatory reaction contributes to the local response, which is considered to
be important for healing and defense against infection. The general response is characterized by hypermetabolism, substrate mobilization, fluid
retention, changes in coagulation and fibrinolysis, and various immuno-
functions.

In contrast to our detailed knowledge on the changes in various hormonal, metabolic, and immunological components of the stress response, information of the exact nature and relative role(s) of the various signals,
which may initiate and potentiate the response is limited and not fully
understood. This chapter is a short updated review of current knowledge
on neural release mechanisms of the injury response. For a detailed information the reader is referred to recent reviews (1,2,3).

ACTIVATION OF AFFERENT NEURAL PATHWAYS TO INJURY

Following an injury, the nociceptive signal(s) to the central nervous
system is transmitted primarily by small myelinated (A𝛿) and unmyelinated
(C) sensory afferent fibers to the substantia gelatinosa in the dorsal
horn, with further rostrad spread to the thalamus (1). The relative role
of various nociceptive stimuli (pressure, vibration, chemical, thermal,
intense mechanical) in arousing the stress response remains to be determined as do the exact nature of the peripheral substrates (transmitters)
involved. However, local synthesis in the traumatized area of histamine,
serotonine, kinins, prostaglandins, and substance P has been shown to initiate or facilitate afferent neural stimuli (1). Concomitantly, noradre-
naline released by efferent neural reflexes potentiates the effect of the
above mentioned factors on firing of afferent nerve fibers.

NEURAL RELEASE OF THE LOCAL INFLAMMATORY REACTION

Swelling and hyperalgesia, two characteristic signs of acute inflammation continue to occur in totally denervated tissues following injury
(4), but the intensity of the response may be modulated by neural influences. Thus, denervation surgically or by local anaesthetics, as well as
capsaicin, and immunosympathectomy reduces swelling, hyperalgesia, as well
as accumulation of leucocytes following an experimental thermal or chemical

injury (5,6). Data from clinical studies are limited, although they suggest a reduced flare response to different stimuli in patients with cord transections (4).

THE ROLE OF AFFERENT NEURAL STIMULI IN RELEASING THE GENERAL RESPONSE TO INJURY

It is well documented that afferent neural stimuli are a major release mechanism of the classical endocrine metabolic response to surgical injury. Thus, neural blockade with local anaesthetics (epidural/spinal) prevents the pituitary response, the adrenocortical and sympathetic response to lower abdominal (gynecological) procedures as well as procedures in the lower extremities (2,3). However, changes in thyroid hormones are unaffected. Likewise, during these minor procedures, neural blockade prevents or reduces the metabolic response since changes in blood glucose, free fatty acids, glycerol are small or negligible (2,3). Hypermetabolism is reduced and protein economy is improved, as assessed by nitrogen balance. Recent experimental studies have failed to demonstrate any effect of epidural analgesia during laparotomy on urinary nitrogen loss, leg glutamine efflux, and decrease in muscle glutamine, although hindquarter nitrogen efflux was diminished (7). In contrast, epidural analgesia in patients undergoing hip replacement prevented the usual postoperative changes in intracellular muscle amino acid concentrations, including the decrease in glutamine, corresponding to an inhibition of the glucose and cortisol response (8).

It is well documented that despite   the blockade of the sympathetic, pituitary, and adrenocortical response to surgery, neural blockade has no influence on the acute phase protein response as well as only a very minor effect on postoperative granulocytosis (2,3).

Neural stimuli may be of some importance in releasing the immunological changes following surgical injury although only few responses are modified by neural blockade. A single dose epidural analgesia has no effect on postoperative leucocyte migration (9), and continuous epidural analgesia is without effect on complement (1o) and immunoglobulin (11) changes postoperatively. During hip surgery epidural analgesia led to an improved monocyte function (spreading and lysis) (12) as well as the usual decrease in blastogenic response of circulating lymphocytes to non-specific and specific mitogens was reduced (13). However, neural blockade had no effect on T-cell subpopulation changes during hip surgery (14). In studies during abdominal surgery where epidural analgesia is less efficient in reducing the stress response (see below) postoperative changes in NK-cell activity (15) as well as postoperative impairment in delayed hypersensitivity (16) was unaltered by neural blockade.

The effect of neural blockade on the stress response to major surgical procedures is well documented to be less pronounced than during minor clean surgery in the lower part of the body (2,3). The explanation hereto is probably dual. Firstly, the currently used epidural techniques do not provide a total and sufficient afferent neural block of the thoracic segments. This has been demonstrated during additional coeliac blockade (2,3) as well as in recent studies using evoked potentials (17). The role of the unmodulated afferent vagal pathway is probably negligible (2,3). Secondly, during major injury other factors than neural stimuli may be of importance.

Selective nociceptive blockade with epidural intrathecal opiates has in several studies been demonstrated to provide only a minor or no alteration of the injury response, despite   sufficient pain relief (2,3).

The influence of nociceptive blockade with drugs inhibiting sensitization of peripheral nerve endings/spinal cord using prostaglandin synthetase inhibitors, substance-P antagonists, antihistamines, etc. has not been completely elucidated. Recent clinical studies suggest that a combination of neural blockade and indometacin may reduce the hyperthermic response to both herniotomy and cholecystectomy, but without influence on postoperative granulocytosis, acute phase protein response, despite total pain alleviation (18,19). Preliminary studies furthermore suggest that even combination of neural blockade with indometacin, histamine 1 and 2 receptor antagonists, serotonine antagonism (ketaserin) as well as inhibition of fibrinolysis (tranexam acid) do not reduce the granulocytosis response to inguinal herniotomy (2o).

CONCLUSION

The available data have demonstrated that minor clean surgical injury activates afferent neural pathways, which secondarily leads to activation of the classical endocrine metabolic response to surgery as well as some immunological changes. Neural-blocking techniques may inhibit the endocrine metabolic response to a major degree during such minor surgical trauma, although only a few components of the immunological response may be modified. The acute phase protein response and granulocytosis are unaffected. The role of the nervous system in releasing the different stress responses to major injury can not be answered at present, since the blocking techniques are insufficient during such procedures. However, humoral factors may probably be of more importance in such situations (21,22, 23,24). Recent studies with combined neural blockade and nociceptive blockade with drugs inhibiting sensitization of peripheral nerve endings leading to total pain relief following surgery have also failed to modulate the acute phase response and granulocytosis.

Although, modulation of the stress response to injury has not been documented to improve survival, the available data from controlled studies on neural blockade in surgical patients (3,25) suggest a benefit on various morbidity parameters. However, a major reservation should be made to consider the stress response a one major response, thereby neglecting possible differences in physiological significance of the various components. Further understanding of the release mechanisms of the components of the injury response is of major importance and may hold potential benefit for the injured and critically ill patient.

REFERENCES

1. Yaksh TL, Hammond DL. Pain, 13:1 (1982)

2. Kehlet H. Clin Anaesthesiol, 2:315 (1984)

3. Kehlet H. in: Neural Blockade (eds: MJ Cousins, PO Bridenbaugh) Lippincott Company (Philadelphia) 1986 (in press)

4. Chapman LF, Goodell H. Ann NY Acad Sci, 116:99o (1964)

5. Helme RD, Andrews PV. J Neurosci Res, 13:453 (1985)

6. Levine JD, Dardick SJ, Basbaum AI, Scipio E. J Neurosci,5:138o(1985)

7. Hulton N, Johnson DJ, Smith RJ, Wilmore DW. J Surg Res, 39:31o (1985)

8. Christensen T, Waaben J, Lindeburg T, Vesterberg K, Vinnars E, Kehlet H. Acta Chir Scand, (1986) in press

9. Edwards AE, Gemmel LW, Mankin PP, Smith CJ, Allen JC, Hunter A. Anaesthesia, 39:lo71 (1984)

lo. Wüst HS, Fiedler H-W, Trobish H, Richter O. Anaesthesist, 31:564(1982)

11. Rem J, Saxtrup Nielsen O, Brandt MR, Kehlet H. Acta Chir Scand, 5o2:51 (198o)

12. Hole A, Unsgaard G, Breivik H. Acta Anaesthesiol Scand, 26:3o1 (1982)

13.   Hole A, Unsgaard G. _Acta Anaesthesiol Scand_, 27:135 (1983)

14.   Hole A, Bakke O. _Acta Anaesthesiol Scand_, 28:296·(1984)

15.   Tønnesen E, Huttel MS, Christensen NJ, Schmitz O. _Acta Anaesthesiol_, 28:654 (1984)

16.   Hjortsø NC, Andersen T, Frøsig F, Neuman P, Rogon E, Kehlet H. _Acta Anaesthesiol Scand_, 28:128 (1984)

17.   Lund C, Selmar P, Hansen OB, Hjortsø NC, Kehlet H. _Anesth Analg_ (1986) in press

18.   Schulze S, Schierbeck J, Sparsø BH, Bisgaard M, Kehlet H.(submitted)

19.   Schulze S, Hasselstrøm L, Jensen NH, Roikjær N, Kehlet H.(submitted)

2o.   Schulze S, Jensen NH, Roikjær N, Kehlet H. (unpublished)

21.   Clowes GHA, George BC, Villee CA, Saravis CA. _N Engl.J Med_, 3o8:545 (1983)

22.   Watters JM, Bessey PQ, Dinarello CA, Wolff SM, Wilmore DW. _Arch Surg_, 121:179 (1986)

23.   Dinarello CA. _J Clin Immunol_, 5:287 (1985)

24.   Wilmore DW. _Clin Nutr_, 5:9 (1986)

SURGERY AND THE IMMUNE RESPONSE

M. Salo

Departments of Anaesthesiology and Medical Microbiology
University of Turku, SF-20520 Turku, Finland

In the last few decades the treatment of surgical patients has continuously advanced. Anaesthetic and surgical techniques have greatly improved.
The treatment of hypovolemia is nowadays better mastered. Great progress has
been made in the maintenance of renal and pulmonary function. The advances
have reduced the numbers of immediate and early deaths after surgery. At
present, many of the patients with a fatal outcome die later from infectious
complications which are still the major hazard during convalescence.
Increasing attention and research is therefore directed to changes during
and after operation in the immune response. As a result, our understanding
of immunological events is constantly improving and we are better equipped
to control immune responses in patients at risk and thus reduce morbidity
and mortality from infections.

IMMUNE RESPONSE TO SURGERY

During surgery a local inflammatory reaction arises at the site of
tissue injury, possibly with systemic effects and the specific immune
response is affected. The distinction of changes in the inflammatory
reaction and the immune response is rather an academic than a practical
question in surgical patients since both mechanisms interact to ensure
survival of the body.
The immune response to surgery is characterized by temporary decreases
in neutrophil functions and in the phagocytic capacity of the mononuclear
phagocyte system (MPS,RES), activation of the complement and overall
decreases in cell-mediated immunity and lesser decreases in humoral immunity.
These changes are progressive, depending on the extent of the operation.

Neutrophil Functions

Neutrophils are the first cells to appear at the site of bacterial
invasion and tissue injury, but their mobilization to skin abrasions is
depressed after surgery[1,2]. This has also been shown *in vitro* tests[3].
Neutrophil adherence is decreased for a day after nephrectomy[3] and open-
heart surgery[4], but increased thereafter[2,4]. Neutrophil motility towards
chemotactic attractants (chemotaxis) and spontaneous migration are depressed
in vitro for some hours to some days after operation[1-3,5]. The extent of the
depression is related to the extent of surgery and it may be due to an

intrinsic defect in neutrophil chemotaxis, serum inhibitors or abnormalities in chemotactic factor generation.

Serum opsonic capacity for facilitation of ingestion of particles to be phagocytosed is slightly depressed after surgery. A decrease in serum opsonic capacity can be observed *in vitro* after cholecystectomy[6] and major abdominal surgery[7] but only at high serum dilutions, which makes the significance of the decrease clinically less important. By contrast, an opsonization defect may be of practical importance in inflamed and poorly perfused tissues and in serum in patients with major postoperative complications, burns or severe infections.

No changes in neutrophil ingestive capacity have been observed after major abdominal surgery[7], nephrectomy[3] or abdominal hysterectomy[8]. The bactericidal capacity of neutrophils is unaltered for *Staph. aureus* after nephrectomy[3] and increased after open-heart surgery[9] but decreased for candida after abdominal hysterectomy[8]. Several studies have measured various microbicidal-related neutrophil variables such as oxygen consumption, nitroblue tetrazolium (NBT) reduction, iodination, chemiluminescence and different enzyme activities. The chemiluminescence responses of granulocytes in phagocytosis of zymosan are depressed for less than one day after nephrectomy and abdominal hysterectomy[5] and are on preoperative levels on the first postoperative day in phagocytosis of zymosan, *Staph. aureus* and *E. coli* after major abdominal surgery[7,10] but depressed for 3-4 days after open-heart surgery[11]. A postoperative decrease has also been observed in NBT reduction[7], neutrophil myeloperoxidase activity[8] and in the activities of the intraneutrophilic granule proteins β-glucuronidase, lysozyme and B[12]-binding protein[12], but no changes were observed in myeloperoxidase-mediated iodination or hexose monophosphate shunt activity after elective general surgery[13].

Anaesthesia in itself produces changes in neutrophil functions. Halothane anaesthesia produces in man a reversible concentration-dependent depression in neutrophil random migration and chemotaxis[14], which is also observed in patients under enflurane and morphine anaesthesia before the start of surgery[15], but inhalation of 60 % v/v nitrous oxide increased neutrophil chemotaxis[16]. Halothane + N_2O + O_2 and thiopentone + N_2O + O_2 anaesthesia decrease ingestion of latex particles and NBT reduction by neutrophils[17]. When separated neutrophils are exposed in tissue culture to increasing concentrations of inhalational, intravenous or local anaesthetics, a dose-dependent inhibition of random and directed motility, ingestion and microbicidal functions of neutrophils is observed[18-21].

Mononuclear Phagocyte System Functions

Both parts of the mononuclear phagocyte system (MPS,RES), monocytes and tissue macrophages, are affected by surgery. The numbers of monocytes in blood circulation increase postoperatively. Their chemotactic motility, ingestive capacity, chemiluminescence responses and lysozyme production have been observed to be increased *in vitro* for about a day after abdominal surgery and decreased thereafter for a week[22,23]. By contrast, the spreading of monocytes on plastic surfaces, which is thought to reflect their phagocytic capacity, is decreased after hip replacement under general anaesthesia and increased under epidural analgesia[24]. Monocyte-mediated cytolysis is decreased after hip replacement under general[24] and combined epidural and general anaesthesia[25], but not after operation under epidural analgesia[24]. This decrease is mediated by serum factors. By contrast, monocyte-mediated haemolytic activity was increased after open-heart surgery in the immediate postoperative period[9].

The number of monocytes which phagocytose latex particles or reduce NBT is decreased after halothane + N_2O + O_2 and thiopentone + N_2O + O_2 anaesthesia[17]. In *in vitro* exposure to 1.7 % v/v enflurane, 0.8 % v/v halothane, 0.2 % v/v methoxyflurane and 70 % v/v nitrous oxide decreases monocyte chemotactic responses[18]. *In vitro* exposure to thiopentone of monocytes depresses their cytolytic capacity in a dose-dependent way but no such depression occurs when monocytes are exposed to a wide range of concentrations of fentanyl, morphine, diazepam, pancuronium or bupivacaine[26].

The total MPS phagocytic capacity, for which the fixed macrophages in the liver (Kupffer cells) and spleen are largely responsible, is depressed after traumas of various origins. This can also be seen after surgery[27]. The depression is only slight even after uncomplicated major surgery[9], it is seen in the immediate postoperative period and it is followed by a period of hyperfunction. Anaesthesia in itself with cyclopropane, diethyl ether, halothane, neurolepts and epidural analgesia may also cause an impairment in MPS phagocytic functions[28]. The impairment may be greater when surgery is performed under halothane instead of combined anaesthesia with the use of neurolepts[29].

The decrease in MPS phagocytic functions is thought to be associated with decreased concentrations of plasma fibronectin which acts as a MPS opsonin. Fibronectin concentrations are decreased for some hours after cholecystectomy[30] and for 1-2 days after major surgery[31]. Other suggested mechanisms for postoperatively decreased MPS functions are decreased blood flow of MPS organs and various inhibitory effects.

Complement

The complement, which is necessary for the inflammatory reaction, neutrophil functions, lysis of cells, bacteria and viruses and participation in many effector systems[32], is activated in a controlled way in connection with various operations[33] including abdominal and cardiovascular surgery[34]. This activation has been shown in decreased haemolytic complement (CH_{50}) values, decreased concentrations of complement components and as the appearance of complement split products into the blood circulation. Many complement components are acute phase reactive and increase unspecifically during the inflammatory reaction. The profiles of complement activation are slightly different in different studies, but the activation of alternate and also of the classic pathways has been suggested to occur during surgery.

A high degree of complement activation can be observed during extracorporeal circulation in open-heart surgery patients[35,36]. The conversion rate of C_3 which in normal health is less than 5 % of total plasma C_3 protein increases during extracorporeal circulation to more than 30 %[37,38] and plasma C_{3a} concentrations may be increased severalfold[35]. C_{5a} is also thought to be released in parallel with C_{3a}, but C_{5a} is readily attached to neutrophil receptors and therefore cannot be in equal concentrations measured in plasma samples. Complement activation during extracorporeal circulation may be somewhat slighter when membrane oxygenators are used instead of bubble ones[36]. Besides extracorporeal circulation, protamine administration during cardiovascular and open-heart surgery may also result in high C_{3a} concentrations[39]. The high levels of activated complement components during open-heart surgery are, however, usually without harmful clinical sequelae.

No differences in complement activation were seen when the patients were anaesthetized with neurolept, halothane or epidural anaesthesia for aortofemoreal surgery[34]. Studies in trauma patients show that the extent of trauma and the presence of nonvital tissue or abscesses affects complement activation. In minor trauma, complement concentrations increase above reference values over a week, but in major trauma they may remain at low levels for a longer time[40].

Cell-mediated Immunity

The effects of an operation on cell-mediated immunity have been well-documented, since cell-mediated immunity responses have a central role in host defences and modern immunological methods were first available for the study of these variables of the immune response. These studies include quantification of effector cell numbers and measurement of their functional variables.

Quantification of Lymphocytes and Their Subsets. Leucocytosis in peripheral blood, an increase in the proportions of neutrophils and monocytes and a decrease in the proportions of eosinophils and lymphocytes occur during and after operation. The numbers of lymphocytes in peripheral blood remain however unaltered during surgery of mild to moderate trauma and decrease after major surgery. The values return to preoperative levels within one or two weeks[41,42].

In minor surgery no changes are observed in the proportions or numbers of T and B lymphocytes, effector cells of cell-mediated and humoral immunity, respectively. As the extent of surgery increases, the numbers of T and B lymphocytes in peripheral blood tend to decrease and their proportions simultaneously change in favour of B cells[42-46]. The numbers of T lymphocyte subpopulations T helper/inducer and T suppressor/cytotoxic cells also decrease, but their proportions either remain unchanged or change in favour of T suppressor cells after operations with increasing trauma severity[42-48].

Changes in Functional *in vitro* Test Values. Most studies of the immune response in surgery deal with *in vitro* measurement of lymphocyte proliferative responses to mitogens, antigens and allogenic cells in peripheral blood samples from patients undergoing an operation. An impairment of T lymphocyte proliferative responses has been observed after operations from minor to major surgery[43,49-53]. Responses to phytohaemagglutinin (PHA) and concanavalin A (Con A), the widely used T lymphocyte mitogens, already start to decrease during operation. They are lowest at the end of surgery and usually return to preoperative values after uncomplicated surgery over a period of 1 to 10-14 days. Similar decreases in lymphocytic responses have been observed to antigens such as purified protein derivative of tuberculin (PPD), streptokinase-streptodornase, mumps and candida and to allogenic cells in a mixed lymphocyte culture. The decreases may be greater in neonates and young children than in older children and adults[54,55]. A decrease in lymphocytic responses can be observed after operations done under general or regional anaesthesia. Some studies show smaller changes in lymphocyte proliferative responses after operations done under regional analgesia than after those done under general anaesthesia[52,53], but this depends on how well afferent neurogenic impulses from the area of surgery can be blocked and perhaps also on how general anaesthesia is performed. However, what is considered more important than anaesthesia in determining the postoperative lymphocyte proliferative responses is the extent of surgical trauma[41,56].

Other functional cell-mediated immunity test values are also depressed postoperatively. Such changes have been found when the effects of lymphokines, effects of lymphocytes on target cells and antibody-dependent cellular cytotoxicity were measured[43]. The sensitivity of lymphocytes to prostaglandin E_2 is increased postoperatively[50] and the amount of serum needed to halve PPD-induced lymphocytic responses is smaller after operation than before it[48]. Natural killer (NK) cell activity is increased during operation but decreased after operation up to a week. This has been observed after hip replacement[57], major abdominal surgery[58] and open-heart surgery[59]. NK cell activity may fall even to undetectable values in patients with postoperative complications[60].

Delayed Hypersensitivity Skin Test. In studies of the effects of an operation on delayed hypersensitivity skin test results, two basically different ways to express the results are used: measurement of reaction diameters/areas induced by recall antigens or grading of patients into reactive, relatively anergic and anergic ones on the basis of to how many antigens they react with an induration of at least 5 mm in diameter. The two types of studies give slightly different results.

292

Those studies measuring the diameters or areas of the indurations show decreased delayed hypersensitivity skin test reactions after nephrectomy[61] and major abdominal surgery[62,63]. By contrast, those studies grading patients into reactive, relatively anergic or anergic ones show no effects of minor or uncomplicated major abdominal surgery on delayed hypersensitivity skin test results[64,65]. Even after major cardiovascular surgery less than half of patients became anergic[65].

Humoral Immunity

A general consensus is that humoral immunity is less affected by surgery than cell-mediated immunity, but, on the other hand, humoral immunity variables have been studied much less. As shown above, the numbers of B lymphocytes decrease in peripheral blood postoperatively as the extent of surgical trauma increases, although less than the numbers of T lymphocytes. Lymphocyte proliferative responses to *Staph. aureus* strain *Cowan I* (StaCw I), which mainly stimulates B lymphocytes to transform, are decreased on the first day after minor orthopaedic surgery and hip replacement and recover by the 3-4th postoperative day (Salo, unpublished). Several studies also show a postoperative decrease in lymphocytic responses to pokeweed mitogen (PWM), which stimulates both T and B lymphocytes[43].

Serum immunoglobulin concentrations decrease postoperatively, especially after major surgery. However, such changes are more likely to be due to haemodilution and loss of protein into extravascular tissues than to disturbances in immunoglobulin production. *In vitro* studies show that the numbers of immunoglobulin synthetizing and secreting cells in peripheral blood[66] decrease after open-heart surgery and there is a simultaneous decrease in the synthesis and secretion of IgG, IgM and IgA into the culture supernatants after PWM stimulation. However, such decreases in immunoglobulin production cannot be observed after hip replacement or minor orthopaedic surgery (Salo, unpublished).

The opinion that humoral immunity is less affected by anaesthesia and surgery is partly based on animal studies measuring specific antibody responses. These studies show that IgG and IgM antibody production is well maintained in response against different types of antigens during anaesthesia and that antibody responses may even be enhanced during surgery[43]. However, recent studies show that the level of IgG anti-tetanus toxoid antibody production is lower in surgical patients than in controls. The impairment is greater in anergic than reactive patients[67]. Depressed antibody production has also been observed in burned patients[68] but no general agreement exists on postoperative antibody responses.

CHANGES IN PATIENTS WITH MALIGNANT DISEASE

Basically similar changes in the immune response are found in patients with malignant and benign disease. Some differences have been found in changes reported below but there is no evidence that the general patterns of postoperative immune response changes differ in these patients. During major abdominal surgery, the patients' nutritional state unlike the malignancy of the disease had an effect on postoperative lymphocytic responses[69]. However, when a tumour has been an immunologic burden to the patient, its removal may improve immune responses[70].

The postoperative increase in monocyte numbers and in their phagocytic capacity in patients with benign disease was not observed in patients with malignant disease[23,71]. Lymphocytic proliferative responses to mitogens and bacterial antigens decrease postoperatively but in patients with malignant disease lymphocyte responses to tumour antigens may also be depressed. Depressed lymphocyte cytotoxicity to tumour cells has been measured postoperatively in patients with mammary carcinoma[72], Wilms' tumour[73] and malignant melanoma[74] and depressed leucocyte migration inhibition in

293

patients with malignant melanoma and breast cancer[75]. These responses
usually return to preoperative values within a week.

Generation of cytotoxic cells is depressed after major abdominal surgery in patients both with malignant and benign disease[76]. No such increases
in NK cell activity was found during surgery in patients with disseminated
malignancy as was observed in patients with benign disease or localized
primary tumours[77] and the postoperative decrease in NK cell activity was
of longer duration in patients with malignant disease[74,78]. However, NK
cells were equally unaffected by 2 % v/v halothane and 66 % v/v nitrous
oxide in patients with benign and malignant disease[79].

Another difference between patients with malignant and benign disease
is found in postoperative antibody-dependent cellular cytotoxicity (ADCC,
killer (K) cell function) reactions. ADCC responses decrease postoperatively
in patients with advanced malignancy and in septic and cachetic patients
but not in patients with benign disease or early cancer[80]. This difference
like that found in NK cell activity may, however, be due to other reasons,
e.g. the nutritional state, than malignancy. Accordingly, in another study
K cell activity was equally depressed after surgery in patients with benign
and malignant disease[81].

Halothane anaesthesia depresses mitogen-induced lymphocytic responses
more in tumour-bearing mice than in controls[82]. However, halothane anaesthesia in patients decreased cytotoxicity as much as balanced anaesthesia
without the use of halogenated anaesthetics[74].

ETIOLOGY OF THE CHANGES

Anaesthesia

In general, anaesthesia is considered to have a smaller role than
surgery in decreasing immune responses during operation. The effects of
anaesthesia on the immune response are twofold. Self-evident is the protective role of anaesthesia as it makes the operation painless and reduces
responses to surgery. The immune depressing effects of anaesthesia may
manifest through its direct effects on host defences but also through its
indirect effects on the endocrine and metabolic balance, oxygenation,
ventilation and tissue perfusion, i.e. through its effects on the internal
milieu of the cells.

Most intravenous, inhalational and local anaesthetics depress mitogen-
induced lymphocytic responses, when separated lymphocytes are exposed *in
vitro* to increasing concentrations of the anaesthetic[43]. Similar observations
have been made on cytotoxicity, NK cell activity and immunoglobulin production by lymphocytes and on granulocyte functions.

Animal studies show decreasing effects of anaesthesia on several
variables of the immune response. Most studies deal with the effects of
halothane but effects of other anaesthetics have also been studied[43,83].
Although most studies are carefully conducted the risk always remains,
especially in handling small animals, that disturbances in homeostasis may
contribute to the extent of changes.

Few studies have been made of the effects of clinical anaesthesia on
man. During balanced anaesthesia with thiopentone + N_2O + analgesics +
muscle relaxants before the start of surgery no changes were observed over
1-2 hrs in the PHA- or Con A-induced lymphocyte proliferative responses[43].
5-7 hrs´ anaesthesia with halothane or enflurane did not affect PHA-induced
lymphocyte responses in volunteers[84] but another study found a decrease in
PHA-induced lymphocytic responses after 3 hrs´ halothane anaesthesia[85].
More changes in the immune response due to anaesthesia may arise in patients
with disturbed homeostasis and in those with marginal immunocompetence.
Anaesthesia may also have other effects on host defences, such as those
on cilial function[86] and on thoracic duct lymph flow and bacterial clearance
when high positive end-expiratory airway pressures (PEEP) are used during
controlled ventilation[87].

The immune response during surgery may be modified to some extent by
blocking afferent impulses from the operation area with regional analgesia,
which is possible at lower abdominal and lower extremity surgery, where
granulocyte and lymphocyte numbers, mitogen- and antigen-induced lymphocyte proliferative responses, monocyte functions and serum suppressive
activity change less by high epidural analgesia than general anaesthesia[24,52,53,88,89]. By contrast, no differences in the immune response
were observed when minor surgery was performed under halothane anaesthesia
with spontaneous ventilation, balanced anaesthesia with mechanical ventilation or regional analgesia[90].

Operative Trauma

The degree and extent of operative trauma are considered to be crucial
in the postoperative changes of the immune response but not exclusively.
Surgical trauma is basically different from accidental trauma in producing
changes in the immune response since during surgery the anaesthetist and
surgeon are able to minimize by prophylactic and therapeutic measures
homeostatic disturbances. Therefore, the changes in the immune response
produced by a surgical trauma can be expected to be slighter than those
after an equal accidental trauma.

Correlations have been observed between the extent of surgical trauma
and the changes in leucocyte and differential counts, numbers of lymphocytes
in peripheral blood and mitogen- and antigen-induced lymphocyte proliferative
responses[41,56]. A similar correlation has been found between the extent of
surgery and T lymphocyte, their subtype and NK cell numbers and the plasma
volume needed to halve PPD-induced lymphocytic responses[48].

Other Factors

The effects of blood transfusion depend on the situation in which blood
is needed and on the blood preparation used. Blood transfusion may sometimes
further deteriorate trauma-decreased immune responses but in some cases it
may improve immune responses, since the maintenance of sufficient oxygenation and tissue perfusion is of primary importance for the patient and
the adequacy of his immune response. The clinically significant deleterious
effects of blood transfusion are suggested by reports on a worse prognosis
in patients with colorectal or some other cancer who received blood transfusion compared to those who did not[91].

Several drugs including many antibiotics, cytostatics and corticosteroids are known to disturb immune responses. Severe malnutrition with
its immunosuppressive effects is deleterious to the patient undergoing
surgery. Psychological factors may be important in postoperative immunosuppression, but immobilization, disturbed diurnal rhythm and preoperative
fasting are less important.

MECHANISMS OF THE CHANGES

The main mechanisms of the postoperative decrease of cell-mediated and
humoral immunity consist of the neuroendocrine response induced by surgery
and anaesthesia, immunosuppressive factors liberated from the surgical
area[65] and pharmacologic effects of drugs given during and after the
operation. Among the mediator mechanisms can be distinguished the effects
of corticosteroids, catecholamines, prostaglandins and the imbalance between
proteases and protease inhibitors[92].

These mechanisms lead to postoperative redistribution of neutrophils
and lymphocytes and to functional changes in effector cells. Intravascular
leucocyte populations are not equal during and after the operation to what
they were before it. Moreover, these cells represent only a small fraction

of cells which participate in the immune response. The effector cells in the tissues might therefore react in other ways than the cells found free in the blood circulation[1].

Activation of suppressor cells is currently under active investigation. Increased suppressor T lymphocyte function has been observed in man after burns[93] and, after surgery, monocytes have been observed to be suppressive in the cytotoxicity reaction[76]. By contrast, decreased Con A-induced suppressor cell function has been found after open-heart surgery[94] and after hysterectomy[95]. The suppressor system thus seems to be very complex and maximum suppression may be observed with a combination of different suppressors[96]. The lymphocytes may moreover be postoperatively increasingly sensitive to endogenous suppressors, such as prostaglandin E, as shown in connection with coronary bypass surgery[50].

The leucocyte transmitters, interleukins 1 and 2, may also have a role in the changes. The ability of leucocyte suspensions *in vitro* to produce interleukin 1 which is an initiator of the immune response, an inducer of the acute phase response and a factor in protein catabolism is at preoperative levels on the fifth day after operation[97]. However, the kinetics of its production during surgery is not known in more detail. By contrast, the capacity of T helper cells to produce interleukin 2, which causes rapid proliferation of effector cells to adequate numbers, is depressed after major surgery for a week but not after minor surgery[98]. Removal of adherent cells abolished the decrease in interleukin 2 production which suggests a monocyte-mediated mechanism in the decrease.

CLINICAL SIGNIFICANCE OF THE ALTERATIONS

The inflammatory reaction and immune response are part of the general physiological response to surgery and occur in interaction with other homeostatic mechanisms. The inflammatory reaction is not only important in defence against microbes but also in wound healing and repair of damaged tissue. The postoperatively depressed immune response is considered to be beneficial to the patient in preventing the body from reacting against its own antigenic structures exposed and released during operation. However, autoantibodies appear in spite of postoperatively depressed immune responses after noncardiac and open-heart surgery[99]. During open-heart surgery, lymphocytes may become sensitized to cardiac mitochondrial antigens[100]. The levels of various anti-heart antibodies and the degree of migration inhibition correlate with the frequency of the postpericardiotomy syndrome suggesting the clinical importance of these autoimmune phenomena.

The changes in the immune balance during surgery may also impair resistance to infections. To what extent the changes in the immune response contribute to the rise of postoperative infections depends on the situation as a whole, since changes also occur in outer host defences and the degree of bacterial contamination varies. Unfavourable location of the surgical trauma, presence of traumatized tissue with haematoma and poor oxygenation and tissue perfusion of the local area may also favour the rise of infection. Surgically broken epithelial barriers and the presence of the intubation tube, intravenous and other canules and catheters all give access to the microbes into the interior. Moreover, changes occur in the mucociliary clearance, secretions and in the natural microbial flora. The functions of complement, granulocytes and MPS, which react first in the blood and tissues against invading organisms, are only slightly affected by uncomplicated surgery causing minor or medium trauma. The complement is capable of mounting an inflammatory reaction and generating chemotactic and opsonic substances, and immunoglobulins are available.

Thus, changes in the immune response are not exclusively responsible for postoperative infections but may contribute to their rise. This is especially true if the disturbances in the immune response are severe and prolonged or the patient is preoperatively immunocompromised. Decreases in the postoperative immune response may thus manifest as increased sensitivity to postoperative infections[101], recrudescence of latent infections[102], reactivation of a virus without causing infection[103] or as a rise of an opportunist infection. The correlation between postoperatively decreased immune responses and frequency of postoperative septic complications and mortality has been best documented by the delayed hypersensitivity skin test. Those patients who are anergic or become anergic postoperatively are at most risk to develop sepsis and die[104].

Another suggested deleterious consequence of postoperatively decreased immune responses is the spread of malignant disease. Although surgery is certainly the most effective method of treating cancer, there are reports of rapid dissemination of cancer shortly after radical surgery and exacerbation of cancer after surgery for an independent or related condition[105]. Although these observations cannot be generalized and the proof of a possible connection is not conclusive, this may point to decreased host defences of clinical importance when large amounts of malignant cells may be free in the blood circulation during surgery. Significantly, there is an abundance of animal studies showing connections between anaesthesia, surgery, immune response and spread of malignancy[106].

Massive complement activation in itself may be harmful to the patient. Massive activation in bacteremia may result in shock during surgery[107], and in endotoxemia it is with other activated mediator systems a major mechanism in the induction of septic shock. C_{5a} in excessive amounts may lead to leucostasis and endothelial damage, and contributes to the rise of adult respiratory distress syndrome (ARDS). Massive stimulation of neutrophils by complement and other chemotactic factors may result in their inability to respond to chemotactic stimuli and to move to the site of microbial invasion.

RECOMMENDATIONS FOR PATIENT CARE

The maintenance of homeostasis is of primary importance for the immune response. The body is able to correct even severe alterations in the immune response if failures in the vital functions can be corrected and the underlying disease can be treated.

High age, malnutrition, certain drugs and diseases are known to affect immune responses preoperatively. These factors should always be taken into account. The best results in the correction of malnutrition have been obtained in severely malnourished patients with preoperative nutrition of at least a week's duration. The use of drugs known to be immunosuppressive must be considered on an individual basis.

In anaesthesia its general performance is more important than the selection of some special anaesthetic agent or method. This means good anaesthetic management and careful maintenance of homeostasis. Halothane with its well-documented immunosuppressive effects cannot be recommended in patients with severe infections. The need to modify operation-induced responses by high analgesic doses or epidural analgesia must primarily be considered from other viewpoints than that of the immune response.

Since the extent of operative trauma is crucial for the postoperative immune response, an atraumatic approach to operation with minimal blood loss is also recommendable from the immunological viewpoint. An open question is what the most suitable time is to make the final reconstructions in the multiple trauma patient. This also concerns patients with postoperatively decreased immune responses before further elective operations. In view of endocrine and metabolic alterations, a stabilization period has been recommended after the primary operations. It might also be beneficial for the immune response. By contrast, in patients with nonvital tissue, abscesses or other toxic processes, immediate operation is a prerequisite for restoration of the immune response although the immune response may transiently be further compromised by the operation.

In experimental work several possibilities already exist in the modifi-cation of the immune response by specific and unspecific means[108]. Ibuprofen, cimetidine[47], levamisole[109] and interleukin 2[110] can prevent changes in postoperative immune responses. In the future, these and other immunomodula-tors may be given prophylactically or per- and postoperatively to high-risk surgical patients to improve their immune responses. This, however, requires a good understanding of the basic immune mechanisms by the operative team and laboratory services available for measurement.

REFERENCES

1. J.H. Wandall, Leucocyte mobilization and function *in vitro* of blood and exudative leucocytes after inguinal herniotomy, Br.J.Surg. 69:669 (1982).
2. J.S. Morris, J.L. Meakins, and N.V. Christou, In vivo neutrophil delivery inflammatory sites in surgical patients, Arch.Surg. 120:205 (1985).
3. T.K. Bowers, J.S. O'Flaherty, R.L. Simmons, and H.S. Jacob, Postsurgical granulocyte dysfunction: studies in healthy kidney donors, J.Lab.Clin. Med. 90:720 (1977).
4. J. Palmblad, Activation of the bactericidal capacity of polymorphonuclear granulocytes after surgery, measured with a new in vitro assay, Scand. J.Haematol. 23:10 (1979).
5. J.S. Solomkin, M.P. Bauman, R.D. Nelson, and R.L. Simmons, Neutrophils dysfunction during the course of intra-abdominal infection, Ann. Surg. 194:9 (1981).
6. J. Perttilä, E.-M. Lilius, and M. Salo, Effects of anaesthesia and surgery on serum opsonic capacity, Acta Anaesthesiol.Scand. 30:173 (1986).
7. W.C. van Dijk, H.A. Verbrugh, R.E.N. van Rijswijk, A. Vos, and J. Verhoef, Neutrophil function, serum opsonic activity, and delayed hypersensitiv-ity in surgical patient, Surgery 92:21 (1982).
8. H. El-Maallem, and J. Fletcher, Effects of surgery on neutrophil granulo-cyte function, Infect.Immun. 32:38 (1981).
9. B. Schildt, L. Berghem, G. Holm, C. Jarstrand, G. Lahnborg, J. Palblad, and K. Rådegran, Influence of cardiopulmonary bypass on some host defence functions in man, Scand.J.Thorac.Cardiovasc.Surg. 14:207 (1980).
10. J. Perttilä, M. Salo, and A. Rajamäki, Granocyte microbicidal function in patients undergoing major abdominal surgery under balanced anaes-thesia, Acta Anaesthesiol.Scand., submitted (1986).
11. J. Perttilä, O.-P. Lehtonen, M. Salo, and R. Tertti, Effects of coronary bypass surgery under high-dose fentanyl anaesthesia on granulocyte chemiluminescence, Br.J.Anaesth. , in press (1986).
12. J.M. Davies, K. Sheppard, and J. Fletcher, The effects of surgery on the activity of neutrophil granule proteins, Br.J.Haematol. 53:5 (1983).
13. L. Bröte, and O. Stendahl, The function of polymorphonuclear leukocytes after surgical trauma, Acta Chir.Scand. 141:565 (1975).
14. G.E. Hill, T.H. Stanley, J.K. Lunn, W.-S. Liu, J.B. English, E.A. Loeser, R. Kawamura, C.R. Sentker, and H.R. Hill, Neutrophil chemotaxis during halothane and halothane-N_2O anesthesia in man, Anesth.Analg. 56:696 (1977).
15. T.H. Stanley, G.E. Hill, M.R. Portas, N.A. Hogan, and H.R. Hill, Neutro-phil chemotaxis during and after general anesthesia and operation, Anesth.Analg 55:668 (1976).
16. G.E. Hill, J.B. English, T.H. Stanley, R. Kawamura, E.A. Loeser, and H.R. Hill, Nitrous oxide and neutrophil chemotaxis in man, Br.J.Anaesth. 50:555 (1978).
17. B.F. Cullen, R.B. Hume, and P.B. Chretien, Phagocytosis during general anesthesia in man, Anesth Analg 54: 501 (1975).
18. G.C. Moudgil, J. Gordon, and J.B. Forrest, Comparative effects of volatile anaesthetic agents and nitrous oxide on human leucocyte chemotaxis *in vitro*, Can.Anaesth.Soc.J. 31:631 (1984).

19. W.D. Welch, Effect of enflurane, isoflurane, and nitrous oxide on the microbicidal activity of human polymorphonuclear leukocytes, Anesthesiology 61:188 (1984).
20. R. Hammer, C. Dahlgren, and O. Stendahl, Inhibition of human leukocyte metabolism and random mobility by local anaesthetics, Acta Anaesthesiol.Scand. 29:520 (1985).
21. M. Nakagawara, K. Takeshige, J. Takamatsu, S. Takahashi, J. Yoshitake, and S. Minakami, Inhibition of superoxide production and Ca^{2+} mobilization in human neutrophils by halothane, enflurane, and isoflurane, Anesthesiology 64:4 (1986).
22. J.P. Neoptolemos, P. Wood, N.W. Everson, and P.R.F. Bell, Monocyte function following surgery in man, Eur.Surg.Res. 17:215 (1985).
23. N.W. Everson, J.P. Neoptolemos, D.J.A. Scott, R.F.M. Wood, and P.R.F. Bell, The effect of surgical operation upon monocytes, Br.J.Surg. 68:257 (1981).
24. A. Hole, G. Unsgaard, and H. Breivik, Monocyte functions are depressed during and after surgery under general anaesthesia but not under epidural anaesthesia, Acta Anaesthesiol.Scand. 26:301 (1982).
25. A. Hole, Per- and postoperative monocyte and lymphocyte functions: Effects of combined epidural and general anaesthesia, Acta Anaesthesiol.Scand. 28:367 (1984).
26. A. Hole, Depression of monocytes and lymphocytes by stress-related humoral factors and anaesthetic-related drugs, Acta Anaesthesiol. Scand. 28:280 (1984).
27. A. Donovan, The effect of surgery on reticuloendothelial function, Arch.Surg. 94:247 (1967).
28. B. Löfström, and B. Schildt, Reticuloendothelial function under general anaesthesia, Acta Anaesthesiol.Scand. 18:34 (1974).
29. A. Doenicke, and W. Kropp, Anaesthesia and the reticulo-endothelial system: comparison of halothane-nitrous oxide and neuroleptanalgesia, Br.J.Anaesth. 48:1191 (1976).
30. A.B. Robbins, J.E. Doran, A.C. Reese, and A.R. Mansberger Jr, Effect of cholecystectomy on cold insoluble globulin, Arch.Surg. 115,1207 (1980).
31. W.O. Richards, W.A. Scovill, and B. Shin, Opsonic fibronectin deficiency in patients with intra-abdominal infection, Surgery 94:210 (1983).
32. R.E. Lewis Jr, J.M. Cruse, and J.V. Richey, Effects of anesthesia and operation on the classical pathway of complement activation, Clin.Immunol.Immunopathol. 23:666 (1982).
33. J. Hahn-Pedersen, H. Sørensen, and H. Kehlet, Complement activation during surgical procedures, Surg.Gynecol.Obstet. 146:66 (1978).
34. H.J. Wüst, H.-W. Fiedler, H. Trobisch, and O. Richter, Fibrinolyse- und Komplementsystem-Profile während aortofemoraler Bypassimplantation, Anaesthesist 31:564 (1982).
35. D.E. Chenoweth, S.W. Cooper, T.E. Hugli, R.W. Stewart, E.H. Blackstone, and J.W. Kirklin, Complement activation during cardiopulmonary bypass, N.Engl.J.Med. 304:497 (1981).
36. N.C. Cavarocchi, J.R. Pluth, H.V. Schaff, T.A. Orszulak, H.A. Homburger, E. Solis, M.P. Kaye, M.S. Clancy, J. Kolff, and G.M. Deeb, Complement activation during cardiopulmonary bypass, J.Thorac.Cardiovasc.Surg. 91:252 (1986).
37. B.J. Collett, N.B. Abdullah, D. Vergani, and R.J. Ware, Complement activation during cardiopulmonary bypass (abstract), Br.J.Anaesth. 55:914P (1983).
38. J. Watkins, and G. Wild, The early diagnosis of impeding coagulopathies following surgery and multiple trauma, Klin.Wochenschr. 63:1019 (1985).
39. P.G. Loubser, Complement activation during surgery with and without cardiopulmonary bypass (abstract), Anesth.Analg. 65:S90 (1986).
40. M. Heideman, C. Saravis, and G.H.A. Clowes, Effect of nonviable tissue and abscesses on complement depletion and the development of bacteremia. J.Trauma 22:527 (1982).

41. M. Salo, Effect of anaesthesia and surgery on the number of and mitogen-induced transformation of T- and B-lymphocytes, Ann.Clin.Res. 10:1 (1978).
42. E. Fosse, H. Opdahl, A. Aakvaag, J.-L. Svennevig, and S. Sunde, White blood cell populations in patients undergoing major vascular surgery, Scand.J.Thorac.Cardiovasc.Surg. 19:247 (1985).
43. M. Salo, Effects of anaesthesia and surgery on the immune response, in: "Trauma, stress and immunity in anaesthesia and surgery", J. Watkins, and M. Salo, ed., Butterworth, London (1982).
44. A. Hole, and O. Bakke, T-lymphocytes and the subpopulations of T-helper and T-suppressor cells measured by monoclonal antibodies (T11, T4, and T8) in relation to surgery under epidural and general anaesthesia, Acta Anaesthesiol.Scand. 28:296 (1984).
45. J.F. Hansbrough, E.M. Bender, R. Zapata-Sirvent, and J. Anderson, Altered helper and suppressor lymphocyte populations in surgical patients, Am.J.Surg. 148:303 (1984).
46. S. Madsbad, K. Buschard, O. Siemssen, and C. Røpke, Changes in T-lymphocyte subsets after elective surgery, Acta Chir.Scand. 152:81 (1986).
47. J.F. Hansbrough, R.L. Zapata-Sirvent, and E.M. Bender, Prevention of alterations in postoperative lymphocyte subpopulations by cimetidine and ibuprofen, Am.J.Surg. 151:249 (1986).
48. T.W.J. Lennard, B.K. Shenton, A. Borzotta, P.K. Donnelly, M. White, L.M. Gerrie, G. Proud, and R.M.R. Taylor, The influence of surgical operations on components of the human immune system, Br.J.Surg. 72:771 (1985).
49. H. Kehlet, M. Thomsen, M. Kjaer, and P. Platz, Postoperative depression of lymphocyte transformation response to microbial antigens, Br.J.Surg. 64:890 (1977).
50. J.S. Goodwin, S. Bromberg, C. Staszak, P.A. Kaszubowski, R.P. Messner, and J.F. Neal, Effect of physical stress on sensitivity of lymphocytes to inhibition by prostaglandin E_2, J.Immunol. 127:518 (1981).
51. R.M. Keane, A.M. Munster, W. Birmingham, R.A. Winchurch, T.R. Gadacz, and C.B. Ernst, Suppression of lymphocyte function after aortic reconstruction, Arch.Surg. 117:1133 (1982).
52. P. Whelan, and P.J. Morris, Immunological responsiveness after transurethral resection of the prostate: general versus spinal anaesthetic, Clin.Exp.Immunol. 48:611 (1982).
53. A. Hole, and G. Unsgaard, The effect of epidural and general anaesthesia on lymphocyte functions during and after major orthopaedic surgery, Acta Anaesthesiol.Scand. 27:135 (1983).
54. P. Ryhänen, Effects of anesthesia and operative surgery on the immune response of patients of different ages, Ann.Clin.Res. 9:suppl 19 (1977).
55. P. Puri, J. Brazil, and D.J. Reen, Immunosuppressive effects of anesthesia and surgery in the newborn: I Short-term effects, J.Pediatr.Surg. 19:823 (1984).
56. B.F. Cullen, and G. van Belle, Lymphocyte transformation and changes in leukocyte count: effects of anesthesia and operation, Anesthesiology 43:563 (1975).
57. E. Tønnesen, H. Mickley, and N. Grunnet, Natural killer cell activity during premedication, anaesthesia and surgery. Acta Anaesthesiol. Scand. 27:238 (1983).
58. E. Tønnesen, M.S. Hüttel, N.J. Christensen, and O. Schmitz, Natural killer cell activity in patients undergoing upper abdominal surgery: relationship to the endocrine stress response, Acta Anaesthesiol. Scand. 28:654 (1984).
59. P. Ryhänen, K. Huttunen, and J. Ilonen, Natural killer cell activity after open-heart surgery, Acta Anaesthesiol.Scand. 28:490 (1984).

60. E. Tønnesen, M.M. Brinkløv, A.S. Olesen, and N.J. Christensen, Natural killer cell activity in a patient undergoing open-heart surgery complicated by an acute myocardial infarction, Acta Pathol.Microbiol. Immunol.Scand.(C) 93:229 (1985).

61. M.S. Slade, R.L. Simmons, E. Yunis, and L.J. Greenberg, Immunodepression after major surgery in normal patients, Surgery 78:363 (1975).

62. W.C. van Dijk, H.A. Verbrugh, R.E.N. van Rijswijk, A. Vos, and J. Verhoef, Neutrophil function, serum opsonic activity, and delayed hypersensitivity in surgical patients, Surgery 92:21 (1982).

63. N.-C. Hjortsø, and H. Kehlet, Influence of surgery, age and serum albumin on delayed hypersensitivity, Acta Chir.Scand. 152:175 (1986).

64. J.B. Pietsch, J.L. Meakins, and L.D. MacLean, The delayed hypersensitivity response: application in clinical surgery, Surgery 82:349 (1977).

65. G.A. McLoughlin, A.V. Wu, I. Saporoschetz, R. Nimberg, and J.A. Mannick, Correlation between anergy and a circulating immunosuppressive factor following major surgical trauma, Ann.Surg.190:297 (1979).

66. J. Eskola, M. Salo, M.K. Viljanen, and O. Ruuskanen, Impaired B lymphocyte function during open-heart surgery, Br.J.Anaesth. 56:333 (1984).

67. C.W. Nohr, N.V. Christou, H. Rode, J. Gordon, and J.L. Meakins, *In vivo* and *in vitro* humoral immunity in surgical patients, Ann.Surg. 200:373 (1984).

68. J.J. Wood, J.B. O'Mahony, M.L. Rodrick, R. Eaton, R.H. Demling, and J.A. Mannick, Abnormalities of antibody production after thermal injury, Arch.Surg. 121:108 (1986).

69. P. Neuvonen, J. Takala, M. Salo, T. Havia, and R. Heinonen, Prolonged impairment of cell-mediated immunity after major abdominal surgery in malnourished patients, Clin.Nutr. 1:283 (1983).

70. J.A. Roth, S.H. Golub, E.A. Grimm, F.R. Eilber, and D.L. Morton, Effects of operation on immune response in cancer patients: sequential evaluation of in vitro lymphocyte function, Surgery 79:46 (1976).

71. I Crzelak, W.L. Olszewski, and E. Engeset, Influence of operative trauma on circulating blood mononuclear cells: analysis using monoclonal antibodies, Eur.Surg.Res. 16:105 (1984).

72. B.M. Vose, and G.C. Moudgil, Effect of surgery on tumour-directed leucocyte responses, Br.Med.J. 1:56 (1975)

73. S. Kumar, and G. Taylor, Effect of surgery on lymphocytotoxicity against tumour cells, Lancet 2:1564 (1974).

74. F. Møller-Larsen, A. Møller-Larsen, and S. Haahr, The influence of general anesthesia and surgery on cell-mediated cytotoxicity and interferon production, J.Clin.Lab.Immunol. 12:69 (1983).

75. A.J. Cochran, W.G.S. Spilg, R.M. Mackie, C.E. Thomas, Postoperative depression of tumour-directed cell-mediated immunity in patients with malignant disease, Br.Med.J. 4:67 (1972).

76. S. Miyazaki, T. Akiyoshi, S. Arigana, F. Koba, T. Wada, and H. Tsuji, Depression of the generation of cell-mediated cytotoxicity by suppressor cells after surgery, Clin.Exp.Immunol. 54:573 (1983).

77. C.D. Griffith, R.C. Rees, A. Platts, A. Jermy, J. Peel, and K. Rogers, The nature of enhanced natural killer lymphocyte cytotoxicity during anesthesia and surgery in patients with benign disease and cancer, Ann.Surg. 200:753 (1984).

78. A. Utchida, R. Kolb, and M. Micksche, Generation of suppressor cells for natural killer activity in cancer patients after surgery, JNCI 68:735 (1982).

79. C.D.M. Griffith, and M.B. Kamath, Effects of halothane and nitrous oxide anaesthesia on natural killer lymphocytes from patients with benign and malignant breast disease. Br.J.Anaesth. 58:540 (1986).

80. J.A. McCredie, H.R. MacDonald, and S.B. Wood, Effect of operation and radiotherapy on antibody-dependent cellular cytotoxicity, Cancer 44:99 (1979).

81. J. Micheels, G. Degiovanni, and A.M. Cayet, Effect of surgery and anaesthesia on cell mediated immunity, Acta Anaesthesiol.Belg.30:Suppl:33 (1979).

82. R.E. Lewis Jr, J.M. Cruse, and J. Hazelwood, Halothane-induced suppression of cell-mediated immunity in normal and tumour-bearing C3H$_f$/He mice, Anesth.Analg. 59:666 (1980).
83. J. Thomas, M. Carver, C. Haisch, F. Thomas, J. Welch, and R. Carchman, Differential effects of intravenous anaesthetic agents on cell-mediated immunity in the Rhesus monkey, Clin.Exp.Immunol. 47:457 (1982).
84. P.G. Duncan, B.F. Cullen, R. Calverly, N.T. Smith, E.I. Eger II, and R. Bone, Failure of enflurane and halothane anesthesia to inhibit lymphocyte transformation in volunteers, Anesthesiology 45:661 (1976).
85. A. Doenicke, B. Grote, H. Suttmann, K.-J. Graf, U.V. Sprecht, H. Ott, B. Sarafoff, and C. Bretz, Effects of halothane on the immunological system in healthy volunteers, Clin.Res.Rev. 1:23 (1981).
86. K.S. Lee, and S.S. Park, Effect of halothane, enflurane, and nitrous oxide on tracheal ciliary activity in vitro, Anesth.Analg. 59:426 (1980).
87. M. Last, L. Kurtz, T.A. Stein, L. Wise, Effect of PEEP on the rate of thoracic duct lymph flow and clearance of bacteria from the peritoneal cavity, Am.J.Surg. 145:126 (1983).
88. J. Rem, M.R. Brandt, and H. Kehlet, Prevention of postoperative lymphopenia and granulocytosis by epidural analgesia. Lancet i:283 (1980).
89. A. Hole, Per- and postoperative monocyte and lymphocyte functions: effects of sera from patients operated under general or epidural anaesthesia, Acta Anaesthesiol.Scand. 28:287 (1984).
90. A.E. Edwards, L.W. Gemmell, P.P. Mankin, C.J. Smith, J.C. Allen, and A. Hunter, The effects of three differing anaesthetics on the immune response, Anaesthesia 39:1071 (1984).
91. R.S. Foster Jr, M.C. Costanza, J.C. Foster, M.C. Wanner, and C.B. Foster, Adverse relationship between blood transfusions and survival after colectomy for colon cancer, Cancer 55:1195 (1985).
92. P.K. Donnelly, B.K. Shenton, A.M. Alomran, D.M.A. Francis, G. Proud, and R.M.R. Taylor, The role of protease in immunoregulation, Br.J.Surg. 70:614 (1983).
93. J.L. Ninnemann, Immunosuppression following thermal injury through B cell activation of suppressor T cells, J. Trauma 20:206 (1980).
94. M. Salo, E. Soppi, O. Lassila, and O. Ruuskanen, Suppressor lymphocytes during open heart surgery, J.Clin.Lab.Immunol. 5:159 (1981).
95. I. Crzelak, W.L. Olszewski, and A. Engeset, Decreased suppressor cell activity after surgery, J.Clin.Lab.Immunol. 16:201 (1985).
96. J. Lundy, and C.M. Ford, Surgery, trauma and imme response. Evolving the mechanism, Ann Surg 197: 434 (1983).
97. J.J. Duncan, L.L. Moldawer, B.R. Bistrian, and G.L. Blackburn, In vitro leukocyte endogenous mediator production is not impaired following surgical stress in moderately malnourished patients, JPEN 8:174 (1984).
98. T. Akiyoshi, F. Koba, S. Arinaga, S. Miyazaki, T. Wada, and H. Tsui, Impaired production of interleukin-2 after surgery. Clin.Exp.Immunol. 59:45 (1985).
99. B. Maisch, P.A. Berg, and K. Kochsiek, Clinical significance of immunopathological findings in patients with post-pericardiotomy syndrome I. Relevance of antibody pattern, Clin.Exp.Immunol. 38:189 (1979).
100.S. Specter, A. Cerdan, C. Cerdan, K. Chang, and H. Friedman, Cell-mediated immune responsiveness to cardiac extracts by peripheral blood leukocytes from patients after myocardial infarction or open-heart surgery, Clin.Immunol.Immunopathol. 30:19 (1984).
101.J.W. Alexander, and R.A. Good, "Immunobiology for surgeons" p. 125, W.B. Saunders Co, Philadelphia (1970).
102.S.J. Eykyn, M.V. Braimbridge, Open heart surgery complicated by postoperative malaria (letter), Lancet ii:411 (1977).
103.C. Porteous, J.A. Bradley, D.N.H. Hamilton, I McA. Ledingham, G.B. Clements, and C.G. Robinson, Herpes simplex virus reactivation in surgical patients, Crit.Care Med. 12:626 (1984).

104. L.D. MacLean, Host resistance in surgical patients, J.Trauma 19:297 (1979).
105. P.H. Lange, K. Hekmat, G. Bosl, B.J. Kennedy, and E.E. Fraley, Accelerated growth of testicular cancer after cytoreductive surgery, Cancer 45:1498 (1980).
106. J. Lundy, E.J. Lovett III, S. Hamilton, and P. Conran, Halothane, surgery, immunosuppression and artificial pulmonary metastases, Cancer 41:827 (1978).
107. I. Brandslund, B. Teisner, P. Hole, J.G. Grudzinskas, and S.-E.Svehag, Complement activation in shock associated with a surgically provoked bacteriaemia, Acta Pathol.Microbiol.Immunol.Scand. (C) 91:51 (1983).
108. M. Salo, Immune response to shock, in: "Treatment of shock. Principles and practice", J. Barrett, and L.M. Nyhus, ed., Lea & Febiger, Philadelphia (1986).
109. H. Miwa, T. Kawai, H. Nakahara, and K. Orita, Decrease in cell-mediated immunity by surgical intervention and its prevention by levamisole, Int.J.Immunopharmacol. 2:31 (1980).
110. T. Akiyoshi, F. Koba, S. Arinaga, T. Wada, and H. Tsuji, In vitro effect of interleukin 2 on depression of cell-mediated immune response after surgery, Jpn.J.Surg. 15:375 (1985).

IMMUNE FUNCTION AND SEVERE INFECTION

V. Andersen
Laboratory of Medical Immunology, Department of Medicine TTA
Rigshospitalet, University Hospital, Copenhagen, Denmark

The general immunosuppression during acute bacterial infection

In patients with an acute bacterial infection such as meningococcal meningitis, lymphocyte proliferative responses in vitro are depressed soon after the onset of symptoms[1]. Responses to microbial antigens are more sensitive than the responses to mitogens such as phytohemagglutinin (PHA) and are as a rule completely suppressed during the acute stage. The response to suboptimal concentrations of PHA may be a more sensitive indicator of depressed lymphocyte function than when the usual dose is employed[2].

In uncomplicated cases, lymphocyte responsiveness normalizes within 14 days. The increasing responses obtained when the offending microorganism is employed for stimulation in vitro are similar to those observed with unrelated microbial antigens. However, if dose-response studies are carried out, a shift in the optimal concentration is seen for the offending microorganism, providing evidence for development of increased lymphocyte sensitivity to the causative microorganism[1].

In case of a complicated clinical course, lymphocyte proliferative responses fluctuate corresponding to the febrile periods. Again, the course of the response to the causative microorganism does not differ from the responses to other microbial preparations.

When the infection has subsided, and recirculating memory cells are established, separation of sensitized and non-sensitized persons in the lymphocyte proliferation assay is best obtained at low antigen concentrations giving submaximal stimulation[3].

The polyclonal stimulation of lymphocytes following infection

Serum antibodies rise fast following acute bacterial infection, whereas a more prolonged increase is seen in total immunoglobulin concentrations[1]. One factor to consider is polyclonal activation of lymphocytes by microbial products[4]. In some cases the factor responsible is well characterized, e.g. the T lymphocyte mitogen LPF (Lymphocytosis Promoting Factor), from Bordetella pertussis[5] and the B lymphocyte mitogen Protein A, from Staphylococcus aureus. However, many polyclonal activators are still incompletely known, and one microbial species may produce several; thus, polyclonal B lymphocyte activation may be induced by a strain of Staphylococcus aureus that is deficient in Protein A[6].

Of importance for the late rise in immunoglobulins following infection is the emergence of immunoregulatory antibodies such as anti-idiotypic antibodies and antibodies against immunoglobulin and complement components.

Detection of pre-existing immune deficiency in patients with infection

The impact of infection on the immune system makes laboratory investigation for underlying immune deficiency difficult. Studies on lymphocytes and other cells must be interpreted with special caution if carried out during or shortly after infection. A careful history should always be obtained as soon as possible, concerning predisposing factors and previous infections. However, studies of lymphocyte and phagocyte function are best postponed until after recovery.

References

1. V Andersen, NE Hansen, H Karle, I Lind, N Høiby, and B Weeke. Clin exp Immunol, 26: 469 (1976).

2. CS Hosking, MG Fitzgerald, and MJ Simons. Clin exp Immunol, 9: 467 (1971).

3. HH Mogensen and V Andersen. Acta path microbiol scand Sect C, 89: 365 (1981).

4. G Banck and A Forsgren. Scand J Immunol, 8: 347 (1978).

5. V Andersen, JB Hertz, SF Sørensen, P Bækgaard, PE Christensen, W Ram-
   høj, GA Hansen, AC Wardlaw, and Y Sato. <u>Acta path microbiol scand</u>
   Sect C, 85: 65 (1977).

6. H Effersøe, F Espersen, and V Andersen. <u>Acta path microbiol scand</u>
   Sect C, 92: 121 (1984).

Chapter 4

Mechanisms of immune response impairment in critically ill patients

INTERLEUKINS : AN OVERVIEW

S. K. Durum

Laboratory of Molecular Immunoregulation, Biological Response
Modifiers Program, Division of Cancer Treatment
National Cancer Institute, NIH, Frederick Cancer Research
Facility, Frederick, MD 21701-1013

Immune responses are in the beginning, specific in that a given lymphocyte is initially triggered via its membrane receptors for a specific
antigen. But following the initial antigen trigger, a specific T lymphocyte elaborates a number of nonspecific soluble mediators with a wide
range of effects on other cells, both proximal and distal, both within
the immune system and outside it. A complex network develops in which
these immune cytokines exert positive and negative feedback effects on
one another. Immune cytokines include: interleukin 1 (produced by macrophages and other cells, with actions on many cell types), interleukin 2
(produced by T cells, with actions on T and B cells), interleukin 3
(produced by T cells, with actions on hematopoietic stem cells), inter-
feron-γ (produced by T cells, with actions on many cell types), B cell
stimulating factor or BSF$_1$ (produced by T cells, with actions on B and T
cells), lymphotoxin (produced by T cells, actions on tumor cells and
various normal cell types) and tumor necrosis factor also termed
cachectin (produced by macrophages and various other cell types, with
actions on fat cells, macrophages and other cells.

INTRODUCTION:

The immune system can be divided into two broad categories of components:
specific elements that engage foreign antigens (lymphocytes and antibodies) and nonspecific elements that mediate and amplify defense mechanisms. Lymphokines are a part of the latter category of non-specific
mediators and amplifiers. The term "lymphokine" is used to encompass
those extracellular products that are elaborated by lymphocytes. Recent
years have witnessed great advances in the characterization of these
molecules and the focus of this brief review will be those lymphokines
that have been cloned and are thereby the best characterized.

Figure 1 illustrates the spectrum of lymphokines that emanate from the
interaction of a helper T lymphocyte engaging its specific antigen on
the surface of an antigen-presenting macrophage (Mϕ). All the lymphokines shown are polypeptides of 15-19 Kd and none are constitutive cell
products; they are produced only following cell activation. They occur
in nanomolar to picomolar concentrations in extracellular fluids. All
probably mediate their effects on target cells via high affinity membrane

receptors. The target cells can be as proximal as the producing cell itself or as distal as the central nervous system. The following discussion will analyze each of the lymphokines in further detail.

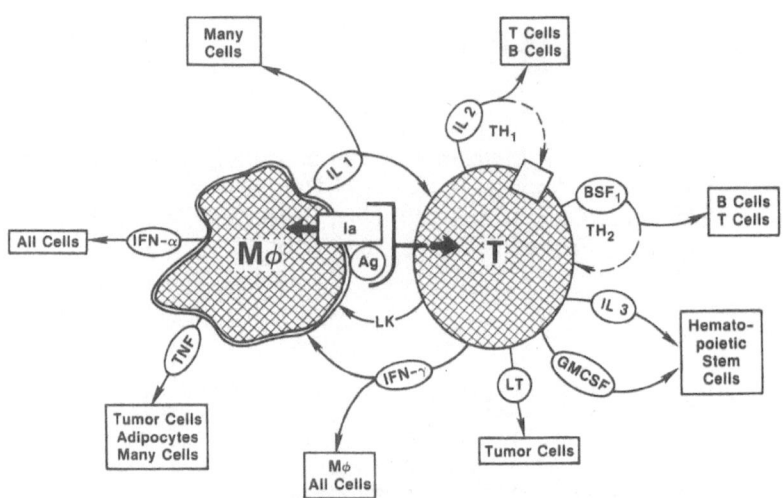

Figure 1. The Lymphokine Spectrum

IL 1 and the T-Mϕ interaction

Figure 2 depicts the earliest events in the T-Mϕ interaction. Antigens are taken up by Mϕ, and in the case of peptide antigens, partly catabolized and then presented on the plasma membrane. T cells employ antigen receptors to recognize the antigen (for which the T cell is specific) and simultaneously engage Ia molecules on the Mϕ. T cells are initially activated through the antigen receptor but require interleukin 1 (IL 1) (1,2) from the Mϕ to become further activated. IL 1 can be induced by T cells via two mechanisms. The first mechanism requires T-Mϕ cell contact and appears to involve Ia molecules which may transduce a signal to release IL 1 (3). A second type of signal can also be delivered from T cells, a lymphokine that has not been fully characterized but is distinct from other known Mϕ-activating lymphokines (such as IFN-γ or CSF1) (4). In the absence of T cells, a number of microbial products can also stimulate Mϕ to produce IL 1. IL 1 so generated then acts on T cells but also has a wide range of biological effects as indicated in Figure 2. The T cell itself is thought to require IL 1 as a stimulus to produce interleukin 2 (IL 2) or in the case of some T cells, to express IL 2 receptors.

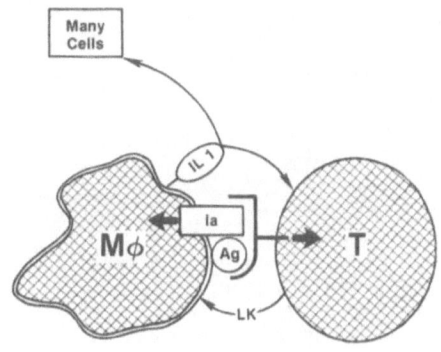

Interleukin 1 (IL 1)

Producer: Mφ, many other cells (e.g., keratinocytes, astrocytes)

Molecule: IL 1α Precursor–33 kd, no signal peptide
                    Mature–17 kd, pI 5
         IL 1β Precursor–33 kd, no signal peptide
                    Mature–17 kd, pI 7

Receptor: 80 kd, 50–3,000/cell, 10^{10}/M

Targets: T cells – IL 2 secretion, IL 2 receptor expression
           B cells – cofactor for proliferation
           Hypothalamus – fever
           Liver – acute phase reactants, many targets and
                   acute inflammatory effects

Figure 2. Interleukin 1

IL 2 and BSF$_1$: lymphocyte growth

Following the initial activation stages involving antigen, Ia and IL 1, T cells produce and utilize growth factors as shown in Figure 3. IL 2 (5) is a growth factor for T cells and can, under some conditions stimulate B cell growth and also promote growth and activation of lymphokine-activated killer cells and natural killer cells. The receptor for IL 2 has also been cloned and is the best characterized lymphokine receptor (6). T cell clones that produce IL 2 have been designated TH$_1$, and appear to be a subset distinct from that designated TH$_2$, which produce B cell stimulating factor (BSF$_1$) (7). BSF$_1$ (also termed "IL 4") (8) activates resting B cells to enlarge, increase Ia expression, and become receptive to other signals, such as cross-linking of surface immunoglobulin. BSF$_1$ also promotes switching from IgM to IgG. Proliferation of B cells requires an additional lymphokine, BCGF$_2$, which has not been cloned as of this writing. BSF$_1$ also supports the growth of some T cell and mast cell lines, and supports the growth of other hematopoietic lineages as well.

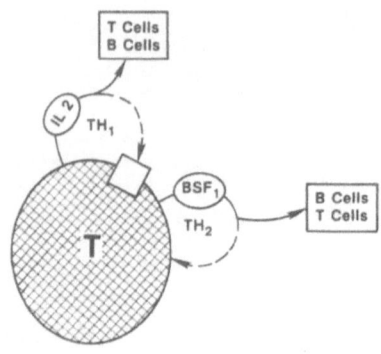

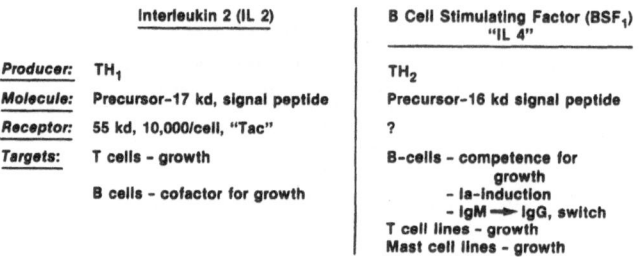

Interleukin 2 (IL 2)		B Cell Stimulating Factor (BSF₁) "IL 4"
Producer:	TH₁	TH₂
Molecule:	Precursor–17 kd, signal peptide	Precursor–16 kd signal peptide
Receptor:	55 kd, 10,000/cell, "Tac"	?
Targets:	T cells - growth	B-cells - competence for growth
	B cells - cofactor for growth	- Ia-induction
		- IgM ➔ IgG, switch
		T cell lines - growth
		Mast cell lines - growth

Figure 3. Interleukin 2 and B Cell Stimulating Factor

IL 3 and GM-CSF: hematopoiesis

In addition to producing the growth factors for differentiated lymphocytes (IL 2 and BSF₁), T cells also produce growth factors (IL 3 and
GM–CSF) for cells at early stages of hematopoiesis, as shown in Figure 4.
Interleukin 3 (IL 3, also termed multi–CSF) (9) induces several lineages
to develop from cultures of bone marow stem cells, mast cells are the
longest lived of the cells arising in these cultures. Granulocyte–monocyte colony stimulating factor (GMCSF) (10) also has growth effects for
multiple lineages, as shown. Although IL 3 and GMCSF have overlapping
activities, they act via different cell receptors.

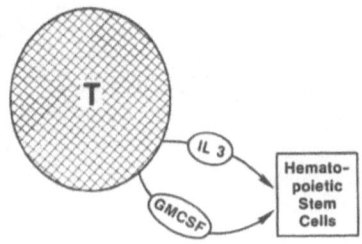

	Interleukin 3 (IL 3, Multi-CSF)	Granulocyte-Monocyte Colony-Stimulating Factor (GMCSF)
Producer:	T cells	T cells, variety of tissues
Molecule:	Precursor-17 kd, signal peptide Mature-15 kd core 28 kd glycosylated	Precursor-17 kd, signal peptide Mature-15 kd core 25 kd glycosylated
Receptor:	60 kd, 3,000-5,000/cell 10^{12}/M	70 kd, 300-1,000/cell 10^9/M
Targets:	Hematopoietic stem cells (granulocyte, monocyte, eosinophil, megakaryocyte, erythroid)	Hematopoietic stem cells (granulocyte, monocyte, eosinophil, megakaryocyte, + erythropoietin → erythroid)

Figure 4.   Interleukin 3 and Granulocyte-Monocyte-Colony-Stimulating Factor

TNF and LT: inflammation and anti-neoplasia

Tumor necrosis factor (TNF) and lymphotoxin (LT), shown in Figure 5, are distinct molecules (28% homology) and are produced by different cells, yet they share a common target cell receptor and thus should have similar activities. Actions of TNF (11) have been studied more extensively than LT (12) and include the dramatic cachexia (wasting) effect which was dis-covered quite independently of the antitumor effect.

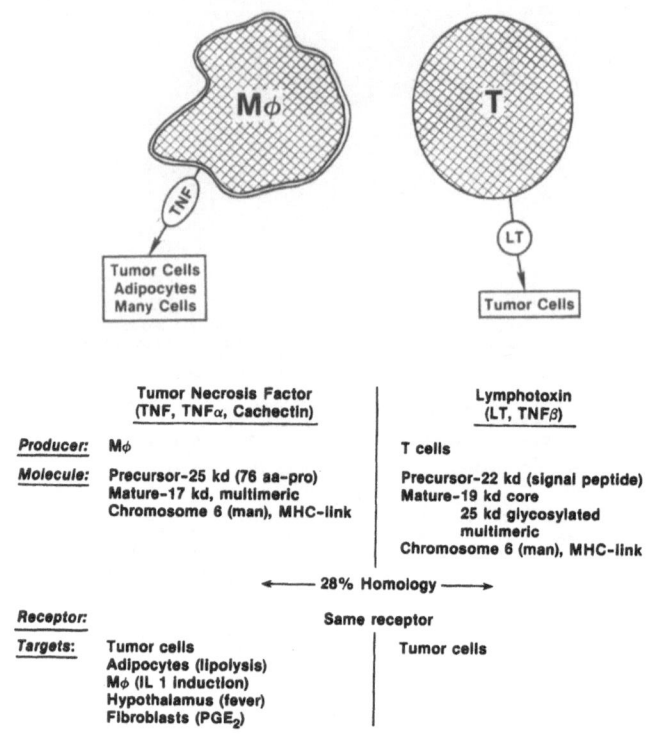

Tumor Necrosis Factor (TNF, TNFα, Cachectin)	Lymphotoxin (LT, TNFβ)
Producer: Mφ	T cells
Molecule: Precursor-25 kd (76 aa-pro) Mature-17 kd, multimeric Chromosome 6 (man), MHC-link	Precursor-22 kd (signal peptide) Mature-19 kd core 25 kd glycosylated multimeric Chromosome 6 (man), MHC-link

← 28% Homology →

Receptor:	Same receptor
Targets: Tumor cells Adipocytes (lipolysis) Mφ (IL 1 induction) Hypothalamus (fever) Fibroblasts (PGE₂)	Tumor cells

Figure 5. Tumor Necrosis Factor and Lymphotoxin

Interferon: anti-virus and immunoregulation

Interferon-α (IFN-α) is produced by activated Mφ (as well as many other cells) (13) whereas IFN-γ is released from activated T cells (14) as shown in Figure 6. Both interferons have antiviral actions and are anti-mitotic. IFN-γ has powerful Mφ-activating properties, arming them to kill intracellular pathogens and tumor cells - a dramatic increase in Mφ Ia expression is also elicited by IFN-γ. Cellular receptors for IFN-γ are known to be distinct from those for IFN-α.

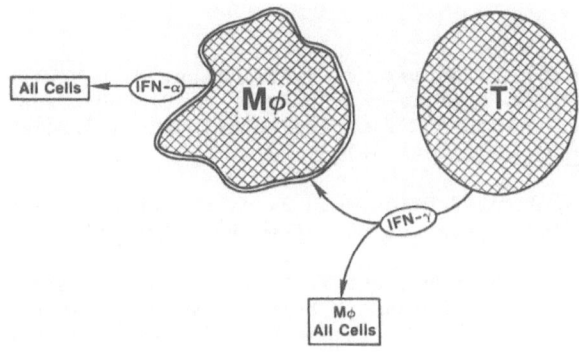

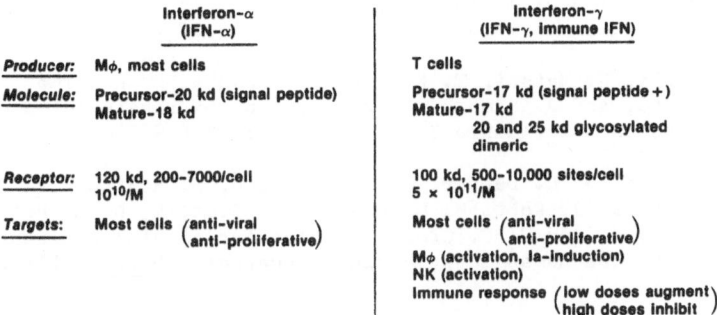

Interferon-α (IFN-α)		Interferon-γ (IFN-γ, immune IFN)	
Producer:	Mφ, most cells	T cells	
Molecule:	Precursor-20 kd (signal peptide) Mature-18 kd	Precursor-17 kd (signal peptide +) Mature-17 kd 20 and 25 kd glycosylated dimeric	
Receptor:	120 kd, 200-7000/cell 10^{10}/M	100 kd, 500-10,000 sites/cell 5×10^{11}/M	
Targets:	Most cells (anti-viral anti-proliferative)	Most cells (anti-viral anti-proliferative) Mφ (activation, Ia-induction) NK (activation) Immune response (low doses augment high doses inhibit)	

Figure 6.   Interferon-α and Interferon-γ

In conclusion, lymphokines are non-antigen specific mediators and amplifiers of the immune response. Although the lymphokines are similar in
certain physical characteristics, such as molecular weight, they are the
products of unrelated genes, rather than a gene family like the immunoglobulin genes. The best studied lymphokines have now been cloned but
we can expect more lymphokines to be cloned, considerable impetus coming
from the potential use of lymphokines as therapeutics. For each lymphokine in its post-cloning phase, the future will bring studies of regulation of its gene, structure-function studies of the biologically active
peptide, determinations of the full spectrum of biological actions and
analyses of the receptor.

317

REFERENCES

1. P. T. Lomedico, U. Gubler, C. P. Hellmann, M. Dukowich, J. G. Giri, Y. C. E. Pan, K. Collier, R. Semanow, A. O. Chua, and S. B. Mizel. Cloning and expression of murine interleukin 1 cDNA in Escherischia coli. Nature 312:458 (1984).

2. S. K. Durum, J. A. Schmidt, and J. J. Oppenheim. Interleukin 1: An immunological perspective. Ann. Rev. Immunol. 3:263 (1985).

3. S. K. Durum, C. Higuchi, and Y. Ron. Accessory cells and T cell activation. The relationship between two components of accessory cell function: Ia and IL 1. Immunobiol. 168:123 (1984).

4. L. Takacs, J. A. Berzofsky, T. Akahoshi, and S. K. Durum. Regulation of IL 1-induction by murine T cell clones. (Abstract) Sixth International Congress of Immunology, p. 172:2.44.2C (1986).

5. T. Taniguchi, H. Matsui, T. Fujita, C. Takaoka, N. Kashima, R. Yoshimoto, and J. Hamuro. Structure and expression of a cloned cDNA for human interleukin 2. Nature 302:305 (1983).

6. W. J. Leonard, J. M. Depper, G. R. Crabtree, S. Rudikoff, J. Pumphrey, R. J. Robb, M. Kronke, P. B. Sretlik, N. J. Peffer, T. A. Waldmann, and W. C. Greene. Molecular cloning and expression of cDNAs for the human interleukin 2 receptor. Nature 311:626 (1984).

7. T. R. Mosmann, H. Cherwinski, M. W. Bond, M. A. Giedlin, and R. L. Coffman. Two types of murine helper T cell clones. I. Definition according to profiles of lymphokine activities and secreted proteins. J. Immunol. 136:2348 (1986).

8. Y. Noma, P. Sideras, T. Naito, S. Bergstedt-Lindquist, C. Azuma, E. Severinson, T. Tanabe, T. Kinashi, F. Matsuda, Y. Yaoita, and T. Honjo. Cloning of cDNA encoding the murine IgG1 induction factor by a novel strategy using SP6 promoter. Nature 319:640 (1986).

9. M. C. Fung, A. J. Hapel, S. Ymer, D. R. Choen, R. M. Johnson, H. D. Campbell, and I. G. Young. Molecular cloning of cDNA for murine interleukin 3. Nature 307:233 (1984).

10. G. W. Wong, J. S. Witek, P. A. Temple, K. M. Wilkens, A. C. Leary, D. P. Luxemberg, S. S. Jones, E. L. Brown, R. M. Kay, E. C. Orr, C. Shoemaker, D. W. Golde, R. J. Kaufman, R. M. Hewick, E. A. Wang, and S. C. Clark. Human GM-CSF: Molecular cloning of the complementary DNA and purification of the natural and recombinant proteins. Science 228:810 (1985).

11. D. Pennica, G. E. Nedwin, J. S. Hayflick, P. H. Seeburg, R. Derynck, M. A. Palladino, W. J. Kohr, B. B. Aggarwal, and D. W. Goeddel. Human tumor necrosis factor: precursor structure, expression and homology to lymphotoxin. Nature 312:724 (1984).

12. P. W. Gray, B. B. Aggarwal, C. V. Benton, T. S. Bringman,
    W. J. Henzel, J. A. Jarrett, D. W. Leung, B. Moffat, P. Ng,
    L. P. Svedersky, M. A. Palladino, and G. E. Nedwin. Cloning and
    expression of cDNA for human lymphotoxin, a lymphokine with
    tumor necrosis activity. Nature 312:721 (1984).

13. D. W. Goeddel, D. W. Leung, T. J. Dull, M. Gross, R. W. Lawn,
    R. McCandliss, P. H. Seeburg, A. Ullrich, E. Yelverton, and
    P. W. Gray. The structure of eight distinct cloned human leukocyte
    interferon cDNAs. Nature 290:20 (1981).

14. P. W. Gray and D. V. Goeddel. Cloning and expression of murine
    IFN-γ cDNA. Proc. Natl. Acad. Sci. USA. 80:5842 (1983).

CELLULAR IMMUNE FUNCTION AND INJURY

J. A. Mannick

Department of Surgery, Brigham & Women's Hospital and Harvard Medical School, Boston, MA.

Defects in cellular immune responsiveness have been repeatedly described in patients after serious injury or major burns[1-10]. The exact relationship between impairment of cellular immunity and septic complications in such injured patients is unknown. Nevertheless, there are several reports which indicate that clinical anergy in patients with major burns, for example, is significantly correlated with mortality, particularly with death from sepsis[4,5,10,11]. A number of years ago, we reported anergy in patients after major burn injury was associated both with failure of circulating lymphocytes from such patients to respond normally to the T cell mitogens phytohemagglutinin (PHA) and Concanavalin A (Con A), and with the presence in the patients' serum of circulating substances suppressive of activation of normal human T lymphocytes[4].

In order to investigate the apparently abnormal T lymphocyte function in injured patients and burn patients, we investigated circulating T lymphocyte subsets with monoclonal antibodies in patients with burns of greater than 30% body surface area and those with lesser degrees of burn injury[12]. We showed that patients after major burn injury had a significant early alteration in the ratio of circulating helper-inducer to cytotoxic-suppressor T cell phenotypes. This reduction tended to return towards normal several weeks after burn injury, and it was again altered in association with the appearance of systemic sepsis. We also found similar changes in patients suffering from major traumatic injury[13]. More recent studies have indicated that patients with severe burns have an early and persistent reduction in the percentage of circulating T cells (Figure 1), and that the reversal of the ratio of T helper-inducer to cytotoxic-suppressor subsets is caused chiefly by an absolute diminution in the number of circulating T helper cells[14]. This diminution was not found consistently in trauma patients. As a correlary to these studies, we also found that patients after serious thermal injury often had circulating lymphocytes of immature phenotypes not normally seen in the circulating blood, including lymphocytes expressing the OKT6 and OKT9 antigens[15].

We and others have examined the question of whether or not increased circulating suppressor cell activity could account for the impairment of T lymphocyte activation seen in patients following major injury[12,16-18]. In patients with major burns, we found that circulating peripheral blood mononuclear cells (PBMC) were on some occasions capable of inhibiting proliferation of lymphocytes from normal human volunteers in response to T cell mitogens or alloantigens in vitro[12]. However, because of

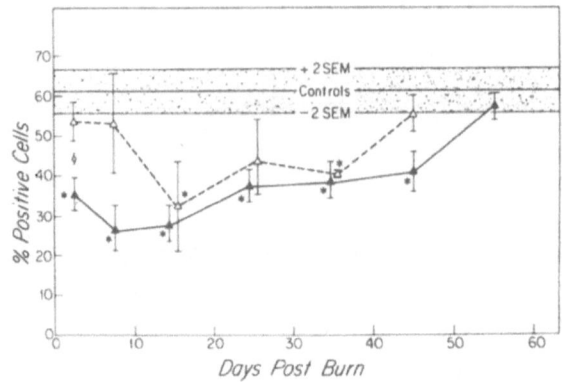

Fig. 1. OKT3 positive cells (total circulating T cells)
following burns. 17 patients with burns, 30%
total body surface area or greater (▲) are
compared to 8 with burns less than 30% total body
surface area (△) and shown as the mean±1SEM. The
control range is shown. (*p<0.05 patients versus
controls,⌀p<0.05 burns greater than 30% versus
burns less than 30%). (Reprinted from Annals of
Surgery)

potentially factitious results inherent in assays which involve the
simultaneous culture of multiple allogeneic cell populations, these
results and similar results reported from other laboratories clearly must
be interpreted with caution. Experiments to demonstrate suppressor cell
activity in autologous culture systems by subtracting and re-adding
appropriate cell types have proved difficult and so far have not produced
sufficiently consistent results to permit firm conclusions.

One assay used in many laboratories for the detection of suppressor
cell activity in an autologous system, namely the inhibition of polyclonal immunoglobulin (Ig) production by T lymphocytes in response to
pokeweed mitogen (PWM) has not yielded positive results in seriously
injured patients (Figure 2). In fact, patients with major burns, for
example, often have background polyclonal Ig production by circulating B
cells that is in excess of PWM stimulated Ig production by similar cells
from normal control individuals[19]. Thus, circulating suppressor T cells
in burn patients, if present, are clearly not effective in inhibiting PWM
stimulated B cell Ig production, a finding consistent with the
observation that there is not an increased percentage of cells bearing
the cytotoxic-suppressor phenotype in the PBMC of these patients.

Because of the persistent impairment of T lymphocyte activation seen
in seriously injured and burn patients, we have more recently studied the
ability of PBMC from such individuals to produce the cytokines necessary
for the initiation of the immune response. Adherent cells from the PBMC
population of seriously injured patients and patients with major burns
were studied for their ability to produce interleukin 1 (IL 1) in
response to endotoxin and the PBMC of the same individuals were studied
for their ability to produce interleukin 2 (IL 2) in response to PHA
stimulation in vitro[20]. We have found that IL 1 production by the

322

Ig Synthesis in 11 patients after thermal injury		
	IgG(ug/mL)	
	Mean ± SEM	P*
Controls	7.57 ± 1.1	...
Days after burn		
0-2	4.01 ± 1.6	NS
3-6	13.7 ± 3.4	0.04
7-10	21.1 ± 6.08	0.017
11-15	35.6 ± 2.2	0.0004
16-20	27.1 ± 3.9	0.0003
21-25	24.7 ± 6.6	0.007
26-30	17.4 ± 5.17	0.048
31-40	17.98 ± 4.1	0.016
41-50	11.2 ± 2.76	NS
51+	8.1 ± 4.1	NS

*Statistics, patients vs. controls; Mann-Whitney U test; NS = P greater than 0.05

Fig. 2.

adherent cells of seriously injured patients is not inhibited at any time in the post-injury or post-burn course. An initial increase in IL 1 production is the customary finding (Figure 3). On the other hand, there

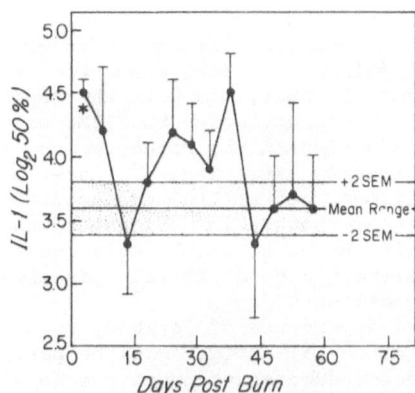

Fig. 3. Serial measurements of IL-1 in 23 burn patients expressed as mean±1SEM. Control values shown as the mean±2SEMs. * indicates p<0.05 patients versus controls. (Reprinted from Annals of Surgery)

is a profound and persistent depression in the ability of circulating lymphocytes from the same seriously injured patients to produce IL 2 (Figure 4). This impairment of IL 2 production, as might be expected, parallels the inhibition of lymphocyte activation seen in the same group

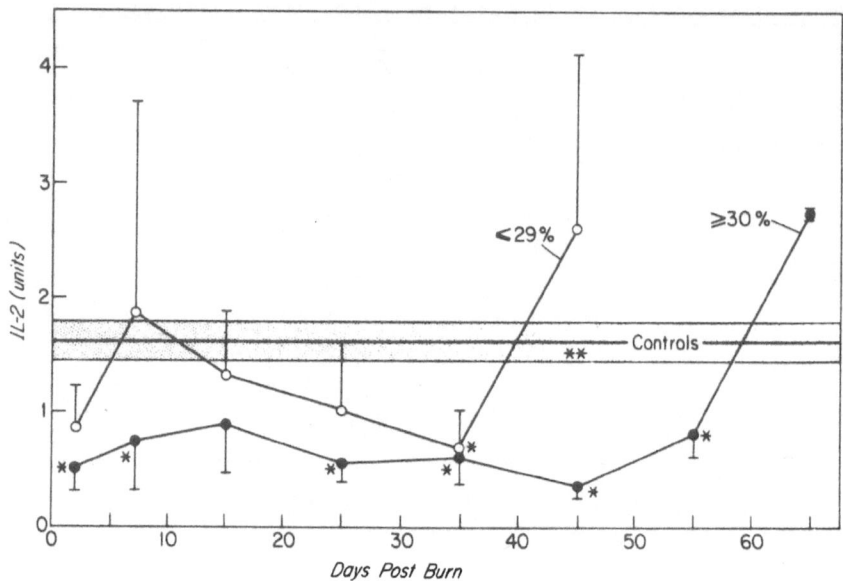

Fig. 4. Serial measurement of IL-2 production in 13
patients with burns, 30% total body surface area
or greater compared to 10 patients with burns, 29%
total body surface area or less. Values are mean±
1SEM. The mean±SEM of IL-2 measurements in the
control group is shown. * indicates p<0.05 compared
to controls, ** indicates p<0.05 when the groups
were compared with one another. (Reprinted from
Annals of Surgery)

of patients and recovers on recovery from the burn wound or traumatic
injury. The degree of impairment of IL 2 production is clearly
associated with the degree of injury and with the appearance of systemic
sepsis. Impairment of IL 2 production by the PBMC of seriously injured
patients parallels the reduction in the percentage of circulating T cells
with the helper phenotype early after injury. Later on in the
post-injury course, impairment of the IL 2 production persists after
recovery of the helper T cell population suggesting that the reduction in
circulating helper T cells, known to be the chief source of IL 2
synthesis, does not completely account for the depression in IL 2
production seen in these patients[20].

Because of the known dependence of antibody formation against a
variety of bacteria and bacterial antigens on the participation of helper
T cells, and the more recent demonstration of the role of IL 2 in
initiating and maintaining the antibody response, it seemed appropriate
to study antibody formation in seriously injured patients despite the
fact that earlier work by other investigators had suggested that antibody
formation was not impaired in such individuals[21]. We therefore studied
the ability of patients with major burns to make a secondary antibody
response to a booster injection of tetanus toxoid[22]. We found that as a
group these patients were impaired in their ability to make an initial
antibody response as compared with normal control individuals (Figure 5),
and that those patients who did make an initial response failed to
maintain antibody production. A subgroup of these patients was
simultaneously studied for the ability of their lymphocytes to produce IL
2. It was found that the failure to make an appropriate anti-tetanus
toxoid antibody response was closely correlated with deficient IL 2
production in the same patients[22].

324

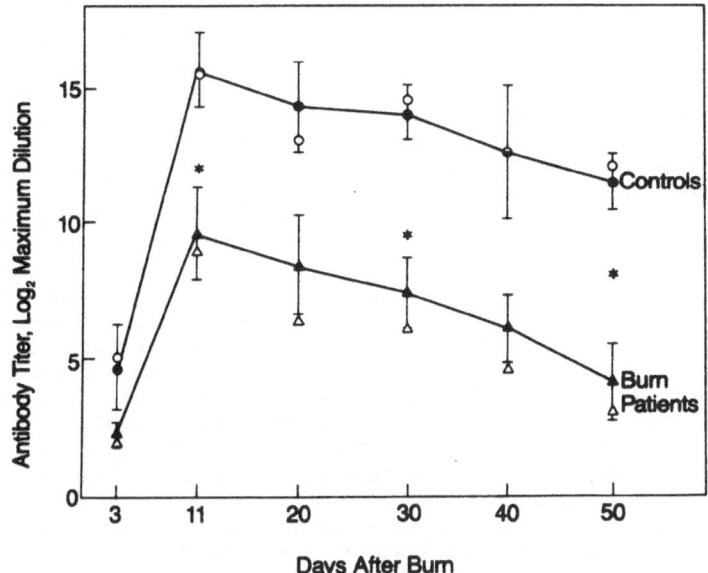

Fig. 5. Propagation of anti-tetanus toxoid antibody
response measured by hemagglutination in 14 burn
patients studied for more than 3 weeks, and in 5
controls. Closed symbols indicate mean±SEM; open
symbols, median; and * p<0.05 patients versus
controls. (Reprinted from Archives of Surgery)

We have also evaluated natural killer (NK) cell activity in
seriously injured patients[23]. We have found that patients with severe
burns had significantly depressed NK activity for a 40-day period
following injury. Patients with lesser burns had reduced NK function for
an initial 10-days post- burn, after which NK activity slowly returned to
the normal range. Traumatically injured patients had depressed NK
function in the 3-6 day range post-injury. All the above changes occured
in the presence of normal numbers of circulating cells being phenotypic
markers for NK effector cells[23].

Suppression of lymphocyte activation by serum or serum fractions
from seriously injured patients has also been repeatedly reported. Our
group was perhaps the first to suggest that serum factors play a role
in the cellular immune deficiency seen in trauma and burn patients[24].
The serum suppressive activity is apparently concentrated in a low
molecular weight polypeptide-containing fraction. This fraction, which
is non-toxic, is inhibitory of normal T lymphocyte activation by
mitogens, and also of T cell-dependent antibody formation in vitro.
These suppressive serum fractions from trauma and burn patients are also
inhibitory of IL 2 production by normal human PBMC, thus suggesting at
least one additional mechanism underlying the marked deficiency of IL 2
production by cells from such individuals[25]. Moreover, suppressive serum
fractions have been found to inhibit the effect of IL 2 on IL 2-dependent
cell lines[25]. The source of the serum suppressive peptide fraction
remains unknown.

Circulating suppressor substances have been described by other
investigators in a variety of disease states. However, the low molecular
weight suppressor peptide material found in burn patient serum by Ozkan and
Ninnemann[26] clearly resembles the material purified from a similar source

in our own laboratory. Ozkan and Ninnemann have attributed the activity
of this peptide fraction to its ability to bind prostaglandins,
particularly prostaglandin E[2] (PGE[2]). Kato and Askenase[27] have also
described an antigen- specific suppressor factor found in the mouse which
requires bound PGE[2] for its activity. The low molecular weight
suppressor factor recovered by Webb and associates[28] from prostaglandin-
treated T lymphocytes also clearly resembles the suppressive material
recovered from trauma patients' serum in our laboratory.

Prostaglandins of the E series have long been known to suppress
lymphocyte activation. It was at first believed that this was entirely
the result of elevation of intracellular cyclic AMP levels, but more
recent studies, including those of Webb et al[28] and Fischer et al[29],
indicate that PGE's also may act through stimulation of a subset of T
suppressor lymphocytes which in turn may release suppressor factors.
Kunkel et al have more recently shown that PGE's act as endogenous
mediators of IL 1 production[30]. The latter findings reinforce the
concept that PGE's may be important in in vivo immunoregulation, since
the relevance of in vitro observed PGE effects for the in vivo situation
has been questioned on the grounds that PGE(s) do not accumulate in vivo
because of diffusion and catabolism.

Recent experiments from our laboratory have shown that PBMC from
seriously injured patients, particularly patients with major burns, are
much more sensitive to inhibition by low concentrations of PGE[2] than are
PBMC from normal individuals[31]. Very low concentrations (10^{-7} or 10^{-8} M)
of PGE[2] added to cultures of patient cells markedly inhibit their
response to a range of dosages of PHA. This finding is particularly true
in patients whose PBMC are inhibited in their PHA response in the absence
of PGE. The suppression of the PHA response to patients' PBMC can be
partially reversed by the addition of Indomethacin to the cell cultures.
These results suggest that one mechanism for the inhibition of T
lymphocyte activation seen in seriously injured patients is PGE[2]
production by the adherent cells in the PBMC population. These
observations are supported by animal experiments which point to PGE
production by adherent cells as an important mechanism inhibiting immune
responses after trauma or burn injury[32,33]

In summary, it appears that after serious injury in man cellular
immune responses are inhibited. The mechanism underlying this inhibition
appears to be chiefly a failure of production of IL 2 by T lymphocytes.
This results not only in an impairment of cellular immunity against
certain micro-organisms, but also in impairment of antibody formation
against bacterial products. In turn, the mechanism underlying the
impairment of IL2 production by T lymphocytes in these patients may be
inhibition by PGE[2] produced by activated cells of the monocyte/macrophage
lineage.

REFERENCES
1. P.R. Riddle and M.C. Berenbaum, Postoperative depression of the
    lymphocyte response to phytohaemagglutinin. Lancet, I:746-748 (1967).
2. S.K. Park, J.I. Brody, H.A. Wallace, and W.S. Blakemore, Immuno-
    suppressive effect of surgery. Lancet, I:53-55 (1971).
3. M.S. Slade, R.L. Simmons, E. Yunis, and L.J. Greenberg, Immuno-
    depression after major surgery in normal patients. Surgery, 78:363-
    372 (1975).
4. J.H.N. Wolfe, A.V.O. Wu, N.E. O'Connor, I. Saporoschetz, and J.A.
    Mannick, Anergy, immunosuppressive serum, and impaired lymphocyte
    blastogenesis in burn patients. Arch. Surg., 117:1266-1271 (1982).
5. F.T. Rapaport, J.M. Converse, L. Horn, D.L. Ballantyne, and J.H.
    Mulholland, Altered reactivity to skin homografts in severe thermal
    injury. Ann. Surg., 159:390-395 (1964).
6. J.R. Batchelor and M. Hackett, HL-A matching in treatment of burned
    patients with skin allografts. Lancet, II:581-582 (1970).

7. L.D. McLean, J.L. Meakins, K. Taguchi, J.P. Duignan, K.S. Dhillon, and J. Gordon, Host resistance in sepsis and trauma. Ann. Surg., 182:207-217 (1975).

8. J.L. Meakins, J.B. Pietsch, O. Bubenick, R. Kelly, H. Rode, J. Gordon, and L.D. McLean, Delayed hypersensitivity: Indicator of acquired failure of host defenses in sepsis and trauma. Ann. Surg., 186:241-250 (1977).

9. N.V. Christou, A.P.H. McLean, and J.L. Meakins, Host defense in blunt trauma: Interrelationships of kinetics of anergy and depressed neutrophil function, nutritional status and sepis. J. Trauma, 20:833-841 (1980).

10. J.H.N. Wolfe, I. Saporoschetz, A.E. Young, N.E. O'Connor and J.A. Mannick, Suppressive serum, suppressor lymphocytes, and death from burns. Ann Surg., 193:513-520 (1981)

11. J.M. Hiebert, M. McGough, G. Rodeheaver, J. Tobiasen, M.T. Edgerton, and R.F. Edlich, The influence of catobolism on immunocompetence in burned patients. Surgery, 86:242-247 (1979).

12. A.J. McIrvine, J.B. O'Mahony, I. Saporoschetz, and J.A. Mannick, Depressed immune response in burn patients: Use of monoclonal antibodies and functional assays to define the role of suppressor cells. Ann. Surg., 148:303-306 (1982)

13. J.B. O'Mahony, S.B. Palder, J.J. Wood, A. McIrvine, M.L. Rodrick, R.H. Demling, and J.A. Mannick. Depression of cellular immunity after multiple trauma in the absence of sepsis, J. Trauma, 24:869-875 (1984)

14. J.B. O'Mahony, J.J. Wood, M.L. Rodrick, and J.A. Mannick, Changes in T-cell subsets following injury: Analysis by flow cytometry and relationship to sepsis. Ann. Surg. 202:580-586 (1985)

15. J.J. Wood, J.B. O'Mahony, M.L. Rodrick, and J.A. Mannick, Immature T lymphocytes following injury characterized by morphology and phenotypic markers, (Submitted for Publication).

16. A.M. Munster, Post-traumatic immunosuppression is due to activation of suppressor T cells. Lancet, I:1329 (1976).

17. C.L. Miller and C.B. Baker, Changes in lymphocyte activity after thermal injury: The role of suppressor cells. J. Clin. Invest. 63:202-210 (1979)

18. J.L. Ninneman, A.E. Stockland and J.T. Condie, Induction of prostaglandin synthesis-dependent suppressor cells with endotoxin: occurrence in patients with thermal injuries. J. Clin. Immunol., 3:142-150 (1983)

19. J.J. Wood, J.B. O'Mahony, M.L. Rodrick, and J.A. Mannick, Immunoglobulin production and suppressor T cells after thermal injury. Surgical Forum, 35:619-621 (1984)

20. J.J. Wood, M.L. Rodrick, J. B. O'Mahony, S.B. Palder, I. Saporoschetz, P. D'Eon, and J.A. Mannick, Inadequate interleukin 2 production: A fundamental immunological deficiency in patients with major burns. Ann. Surg., 200:311-320 (1984)

21. J.W. Alexander, and J.A. Moncrief, Alterations of the immune response following severe thermal injury. Arch. Surg., 93:75-83 (1966)

22. J.J. Wood, J.B. O'Mahony, M.L. Rodrick, R. Eaton, R.H. Demling, and J.A. Mannick, Abnormalities of antibody production after thermal injury. An association with reduced Interleukin 2 production. Arch. Surg., 121:108-115 (1986)

23. B.A. Blazar, M.L. Rodrick, J.B. O'Mahony, J.J. Wood, P.Q. Bessey, D.W. Wilmore and J.A. Mannick, Suppression of natural killer-cell function in humans following thermal and traumatic injury. J. Clin. Immmunol. 6:26-36 (1986)

24. M.B. Constantian, J.O. Menzoian, R.B. Nimberg, K. Schmid, and J.A. Mannick, Association of circulating immunosuppressive polypeptide with operative and accidental trauma. Ann. Surg., 185:73-79 (1977)

25. M.L. Rodrick, I.B. Saporoschetz, J.J. Wood, J.T. Grbic, S.B. Palder and J.A. Mannick, Inhibition of human Interleukin 2 (IL 2) production and action by human serum, Proc. International Lymphokine Workshop (1984) (In Press)

26. A.N. Ozkan and J.L. Ninnemann, Suppression of in vitro lymphocyte and neutrophil responses by a lower molecular weight suppressor active peptide from burn-patient sera. J. Clin. Immunol., 5:172-179 (1985).

27. K. Kato, and P.W. Askenase, Reconstitution of an inactive antigen-specific T cell suppressor factor by incubation of the factor with prostaglandins. J. Immunol., 133:2025-2031 (1984).

28. D.R. Webb, K.J. Wieder, T.J. Rogers, C.T. Healy, and I. Nowowiejski-Wieder, Chemical identification of a prostaglandin-induced T suppressor (PITS), Lymphokine Res., 4:139-149 (1985)

29. A. Fischer, A. Durandy, and G. Griscellio, Role of prostaglandin E_2 at the induction of nonspecific T lymphocyte suppressive activities. J. Immunol., 126:1452 (1981)

30. S.L. Kunkel, S.W. Chensue, and S.H. Phan, Prostaglandins as endogenous mediators of interleukin 1 production. J. Immunol., 136: 186 (1986)

31. J.T. Grbic, J.J. Wood, A. Jordan, M.L. Rodrick, and J.A. Mannick, Lymphocytes from burn patients are more sensitive to suppression by prostaglandin E_2. Surgical Forum 38:108-109 (1985)

32. B.S. Wang, E.H. Heacock, and J.A. Mannick, Generation of suppressor cells in mice after surgical trauma. J. Clin. Invest., 66:200-209 (1980)

33. B.S. Wang, E.H. Heacock, and J.A. Mannick, Characterization of suppressor cells generated in mice after surgical trauma. Clin. Immunol. Immunopathol., 24:161-170 (1982)

METABOLIC MEDIATORS IN INJURY

D. W. Wilmore

Harvard Medical School, Brigham and Women's Hospital, 75 Francis Street, Boston, Massachusetts

Physiologic responses following injury are initiated by signals arising locally from the wound or inflammatory focus. Both neural and circulatory pathways are utilized in the transmission of these signals. The pathways to and integration of the signals in the central nervous system are essential for many of the responses to occur. For this reason, the myriad of mediators and physiologic mechanisms which are activated can be considered as part of a "reflex arc" with mediators and response pathways serving as afferent or efferent "limbs" in relation to the central nervous system. In addition, local or circulatory mediators may bypass the central nervous system and initiate responses directly. (Figure 1) Our understanding of these regulators and pathways is incomplete; some responses, for example synthesis of actue phase proteins, may occur without mediation via the central nervous system.

AFFERENT SIGNALS

Neural Afferents

One of the most important initiators of the injury response, at least in the early period following trauma, is the stimulation of peripheral nerve endings in the damaged tissue. This neural pathway is the most rapid route by which to signal the brain that tissue injury has occurred, and it is primarily manifest as pain. Egdahl and Richards demonstrated that stimulation of the femoral nerve in the anesthetized dog resulted in an immediate rise in 17-hydroxycorticoid output from the adrenal gland[1]. When all tissues of the canine hindlimb had been severed except the femoral nerve, artery, and vein, a similar response was obtained to a burn injury on the limb[2]. This immediate rise in cortisol secretion was prevented or abolished if the nerve was transsected before or following the burn injury, demonstrating that an intact afferent nerve supply from the area of injury was essential for the normal adrenocortical response to occur. In this model, circulating factors did not appear to mediate responses. In further dog studies, the spinal cord or medulla oblongata was transsected, the median eminence of the hypothalamus was ablated or the anterior hypothalamus or pituitary removed[3]. These studies demonstrated the necessity for each of these structures to be intact in order for a normal adrenocortical response to trauma to occur. Removal of the cerebral cortex and thalamus did not alter the ability of the animals to respond appropriately.

These experiments are consistent with the observations in humans that the cortisol response to operative injury is absent in patients with spinal cord transsection above the level of injury. Spinal and epidural anesthesia appear to have similar effects with respect to operative stimuli, although cephalad transmission of nervous signals may not be blocked as completely as with cord transsection, and co-existing afferent pathways, such as splanchnic and vagus nerves, may be present.

Afferent pain pathways are not the sole route for nervous stimulation of the hypothalamus and anterior pituitary gland. Conscious appreciation of pain is not necessary for the adrenocortical response to injury or operation, although it may be a sufficient stimulus. The animal experiments described above were carried out under barbiturate anesthesia, and active cortisol responses were documented when the integrity of the neuroendocrine axis was maintained. In patients underoing abdominal hysterectomy, administration of epidural morphine, which inhibits nociceptive pathways without concomitant sympathetic blockade, had minimal effects on intraoperative elevations of plasma cortisol and glucose, but blocked the usual postoperative rises.[4] These responses were totally prevented by the use of epidural bupivicaine which does interrupt afferent sympathetic pathways.

MAJOR PATHWAYS MEDIATING METABOLIC RESPONSES
TO TRAUMA AND SEPSIS

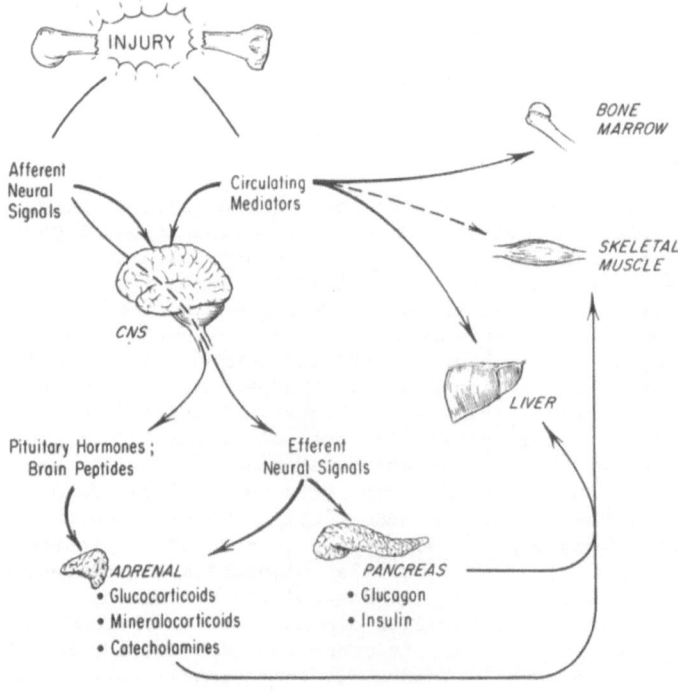

Figure 1

Afferent nervous signals from the area of injury are transmitted rapidly via the spinal cord to the hypothalamus and pituitary. These signals are fundamental in initiating adrenocortical responses following operation and injury. Nonetheless, there is abundant evidence that humoral substances serve, in addition to neural signals, as afferent mediators. Denervation of burn wounds by the application of topical anesthetic or interruption of neural afferents to the brain by spinal anesthesia fails to alter the hypermetabolic response to burn injury.[5] Synthesis of acute phase proteins is unaffected in quadriplegic burn patients. Operative injury in denervated areas of paraplegic patients resulted in altered albumin metabolism, α_2 globulin concentrations, and erythrocyte sedimentation rates comparable to responses observed in non-paralyzed patients.[6] Epidural anesthesia blocks afferent sympathetic impulses and hormonal responses to lower abdominal surgery, but does not affect changes in coagulation, fibrinolysis or acute phase proteins.[7] Regulatory systems distinct from the neuroendocrine axis must also mediate host responses to injury and sepsis.

Interleukin-1 (Endogenous Pyrogen and Other Lymphokines)

Fever is a universally recognized sign of disease processes, whether they are infectious, traumatic, inflammatory or immunologic in nature. In 1948 Beeson and Bennett demonstrated that sterile inflammation in the rabbit peritoneal cavity and canine pleural cavity induced host white blood cells to produce a substance termed "endogenous pyrogen," which caused fever when injected intravenously into rabbits.[8] This pyrogenic activity is only one of numerous biologic effects attributed to the protein molecule (or closely related family of molecules) now referred to as interleukin-1 (IL-1). Interleukin-1 is synthesized by macrophages, both circulating and fixed, in response to a wide variety of stimuli; bacteria and their products appear to be among the most potent stimulators of IL-1 production. Because of its multiple biologic effects, interleukin-1 has been referred to variously as endogenous pyrogen (EP), leukocytic pyrogen (LP), lymphocyte activating factor (LAF), leukocyte endogenous mediator (LEM), and mononuclear cell factor (MCF). Its activities include induction of fever, neutrophilia, and synthesis of acute phase proteins and depression of plasma iron and zinc, all of which are characteristic responses observed following injury and infection. In addition, it has been shown to have a role in activation of B- and T-lymphocytes, augmentation of natural killer cell activity, and stimulation of fibroblast proliferation in vitro. Isolation of interleukin-1 or active fragments from the blood of patients following injury and during sepsis has been achieved only recently, and its precise amino acid structure has only recently been determined. Our knowledge of its biologic effects in humans is incomplete. Nonetheless, IL-1 appears to be a key mediator of physiologic responses with important host defense and wound healing functions.

Interleukin-1 is absent in unstimulated macrophages or present only in very small quantities. It must be synthesized de novo in response to a specific stimulus, requiring a number of hours. The biologic effects of interleukin-1 likely result from nonspecific increases in membrane phospholipase activity.[9] Arachidonic acid is released and serves as substrate for either the cyclo-oxygenase or lipoxygenase pathway, depending on the specific cell type, and yielding prostaglandins, or leukotrienes. The proposed mechanism by which IL-1 induces fever is the stimulation of prostaglandin E2 synthesis in the thermoregulatory center of the hypothalamus, resulting in elevation of body temperature setpoint. Induction of fever by IL-1, but not all of its other activities, can be blocked by inhibitors of the cyclo-oxygenase pathway of prostaglanding metabolism such as aspirin, indomethacin, and ibuprofen. The mechanism(s) by which IL-1 achieves its other effects is unknown, but does not appear to be mediated by cyclo-oxygenase inhibitors. IL-1 may act directly on tissues and cells, for instance activating T- and B-lymphocytes and inducing hepatic acute phase protein synthesis, without the central nervous system playing a role. Other "serum" factors may be required to mediate these changes. These circulating factors, such as tumor necrosis factor (TNF) and the interferons, are elaborated by macrophages following injury. The role of these substances following trauma is yet to be determined.

Exogenous Pyrogens

Microorganisms and their products are powerful inducers of host IL-1 synthesis, but endotoxin, a lipopolysaccharide moiety of gram-negative bacterial cell walls, also acts independently of IL-1 as a circulating factor capable of initiating a variety of host responses.[10,11] Egdahl demonstrated that endotoxin acts directly in the central nervous system, causing fever and accelerated adrenal release of cortisol and epinephrine via ACTH and efferent sympathetic pathways, respectively.[12] Circulating endotoxin has profound systemic effects, particularly on the cardiovascular system, causing hypotension, and may account, along with other mediators such as IL-1, and TNF, for the qualitative differences in host responses to sepsis compared to accidental injury uncomplicated by infection. There has long been speculation about the role of endotoxin in shock states and prolonged critical illness. Although gram-positive bacteria are not thought to yield endotoxin, these organisms still have pyrogenic activity. Enteric organism may translocate across the gastrointestinal tract following injury and thus mediate many of the responses observed.

Other Humoral Mediators

The prostaglandins are oxygenated fatty acids derived from membrane phospholipids, which serve as potent biological regulators. Elevated levels of prostaglandins have been demonstrated in the lymph fluid draining burn wounds, and in the blood of patients during shock. Prostaglandins have also been proposed as regulators of a variety of cell and organ functions, and their local production can be stimulated by bradykinin, thrombin, hypoxia, and hypotension. Thus, local regulation of metabolic pathways may occur. One such effect may be the initiation of skeletal muscle proteolysis via prostaglandin E_2 (PGE_2), which appears to be a locally mediated event causing increased degradation of skeletal muscle protein and release of amino acids.[13] There are undoubtedly many substances, such as the plasma kinins, superoxide radicals, leukotrienes, and others, which are synthesized or released as a result of local tissue injury. Their roles in systemic responses are not clearly defined.

EFFERENT SIGNALS

The Role of the Catabolic Hormone

The catabolic hormones epinephrine, glucagon, and cortisol are all generally elevated following critical illness but their exact role was not established until Sherwin and associates described their synergistic interaction in a series of animal and human experiments.[14] The separate infusion of cortisol, glucagon, and epinephrine each resulted in a transient increase in hepatic glucose production. When the three hormones were infused together, a synergistic effect was noted, and glucose production was consistently elevated during the 6 hour period of infusion.

These studies were continued and expanded by Bessey and colleagues, who administered the three catabolic hormones intravenously, alone or in combination, to normal subjects for 72 hours. A variety of metabolic measurements were performed to determine the metabolic effects of these hormones.[15] Combined hormonal infusion resulted in hyperglycemia, hyperinsulinemia,

Table 1.   Some Metabolic Effects Which Occur with Infusion
of Glucagon, Epinephrine, and Hydrocortisone

	Control	Hormonal Infusion[a]
Metabolic Rate (kcal/hr·m^2)	32.24±1.05	38.42±1.31
Urinary Nitrogen (g/day)	11.8±0.7	14.8±0.9
Nitrogen Balance (g/day)	-0.2±0.4	-3.5±0.4
Sodium Excretion (mEq/day)	237±17	132±14
Potassium Excretion (mEq/day)	93±2	122±5
Basal Glucose (mg/dl)	94±2	133±4
Basal Insulin (µU/ml)	8±1	22±3

[a]All values different, $p < 0.05$.

diminished whole body insulin-mediated glucose disposal, and reduced forearm glucose uptake. At the end of the 72 hour infusion, hepatic glucose production was significantly elevated in comparison with controls who received only normal saline. When endogenous glucose and/or insulin was given, endogenous glucose production was not completely suppressed, in contrast to total suppression observed during the control study. In addition, hypermetabolism occurred throughout the period of hormonal infusion, and metabolic rates increased approximately 20% (Table 1). The patients maintained nitrogen equilibrium during the control arm of the study, but increased their urinary excretion of nitrogen during hormonal infusion and excreted approximately 17 g of nitrogen per day. Because food intake was fixed at approximately 14 g protein per day, negative nitrogen balance of approximately 3 g per day occurred in all subjects during hormonal infusion.

Isotopic studies demonstrated that the hormonal environment increased protein turnover and was particularly associated with increased protein catabolism.

The Hormonal Environment and Inflammatory Mediators

The infusion of catabolic hormones into healthy, normal subjects, in part, mimics the endocrine environment of critically ill patients. As a result of hormonal infusion, there is an increase in metabolic rate, hyperglycemia, hyperinsulinemia, skeletal muscle insulin resistance, accelerated nitrogen turnover, and negative nitrogen balance. However, a variety of alterations commonly observed in critically ill patients was not seen; in particular, hormonal infusion did not induce fever, marked leukocytosis, hypoferremia or alterations in circulating concentrations of acute phase proteins. To induce tissue inflammation and evaluate the systemic responses which occur following inflammation, we injected etiocholanolone, a naturally occurring pyrogenic steroid, intramuscularly in normal volunteers. Some subjects received only etiocholanolone injections[16], while others received the injections along with an infusion of catabolic hormones[17]. Etiocholanolone injection resulted in fever, leukocytosis, elevation in C-reactive protein, and hypoferremia. However, concentrations of the catabolic hormones remained at normal levels and blood glucose and protein metabolism

Table 2. The Induction of Interleukin-1 by the
Intramuscular Injectionof Etiocholanolone
and its Metabolic Effects   (Mean±SEM)

	Control	Etiocholanolone Injection
Peak Rectal Temperature (°F)	99.3±0.1	100.9±0.3[a]
WBC (cells x $10^3/mm^3$)	6.9±0.4	11.2±0.5[a]
C-Reactive protein (mg/dl)	0.6	3.6±0.7[a]
Serum Iron (µg/dl)	72±13	30±4[a]
Urinary Nitrogen Excretion (g/day)	11.9±0.4	12.1±0.6
Nitrogen Balance (g/day)	-0.5±0.3	-0.2±0.5
Plasma Glucose (mg/dl)	94±2	96±2
Serum Insulin (µU/ml)	8±1	4±1[a]

[a]$p < 0.05$, when compared to controls

was no different than in control studies. The fed subjects remained in
nitrogen balance (Table 2).

When etiocholanolone was given in conjunction with hormonal infusion
the entire spectrum of responses commonly observed in critically ill
patients occurred. Some of these responses could be ascribed solely to the
hormonal infusion, others to the inflammatory mediators; and finally, some
responses were a result of interaction between the two stimuli. Hypermetabolism and hyperglycemia, insulin resistance, and negative nitrogen
balance were predominantly mediated by the infusion of the counter-regula-
tory hormones. Etiocholanolone injection resulted in fever, acute phase
protein synthesis, and hypoferremia. Leukocytosis, temperature, and
C-reactive protein response were reflected in interaction between the two
stimuli. Because of this interaction, we hypothesize that both inflammatory
and endocrine mediators are important stimuli of the metabolic response to
critical illness. Both sets of mediators are necessary for a complete
manifestation of the host response, although in our model system the
predominant catabolic features of the response were attributable to the
hormonal mediators, while 'acute phase' responses were mediated primarily
by the presence of the inflammatory mediators.

REFERENCES

1. Egdahl R. and Richards J. Surg. Forum, 7:142 (1957).
2. Egdahl R. J. Clin. Invest., 38:1120 (1959a).
3. Hume D. and Egdahl R. Ann. Surg., 150:697 (1959).
4. Christensen P., Brandt M., Rem J., and Kehlet H. Br. J. Anaesth., 54:
    23 (1982).
5. Wilmore D.: The Metabolic Management of the Critically Ill. Plenum
    Medical Book Company (New York and London) p (85) (1977).
6. Davies J., Liljedahl S-O. and Reizenstein P. Injury, 1:271 (1970).
7. Kehlet H., Brandt M., and Rem J. JPEN, 4:152 (1980).

8. Beeson P. _J. Clin. Invest._, 27:524 (1948).
9. Dinarello C. _Rev. Inf. Dis._, 6:51 (1984).
10. Berry L. _CRC Crit. Rev. Toxicol._, 5:239 (1977).
11. Wolff S. _J. Inf. Dis._, 128 (Suppl.):S259 (1973).
12. Egdahl R. _Surgery_, 46:9 (1959).
13. Baracos V., Rodeman H., Dinarello C., and Goldberg A. _N.E.J.M._, 308:553 (1983).
14. Shamoran H., Hendler R., and Sherwin R. _J. Clin. Endo. Metab._, 52:1235 (1981).
15. Bessey P., Watters J., Aoki T., and Wilmore D. _Ann. Surg._, 200:264 (1984).
16. Watters J., Bessey P., Dinarello C., Wolff S., and Wilmore D. _Surgery_, 98:298 (1985).
17. Watters J., Bessey P., Dinarello C., Wolff S., and Wilmore D. _Arch. Surg._ 121:179 (1986).

MEDIATORS AFFECTING IL2 FUNCTION IN BURN IMMUNOSUPPRESSION

B.G. Sparkes, J.A. Teodorczyk-Injeyan[1], W.J. Peters[2], R.E. Falk[1]
The Defence & Civil Institute of Environmental Medicine
Downsview, Ontario, Canada
[1] Department of Surgery, University of Toronto, [2] The Ross Tilley
Burn Centre, Wellesley Hospital, Toronto, Canada

INTRODUCTION

Thermal injury induces the most complex array of physiological dysfunction known. Extensive research effort has led to an understanding of some of the changes occurring following a major burn, yet early pathological changes have been documented to be different from those of later phases and great difficulties have arisen in determining cause-and-effect relationships in these changes. For example, intensive first treatment using specific, systemic and local therapy successfully delays mortality in the early phase of very severe burns but has little relationship to the outcome. This was observed where burn victims, treated early, survived initially in greater numbers than those devoid of early treatment (48% greater at day 4, 32% at day 12), whereas the difference was only 7.3% at 2 months [1]. The fatal outcome of the late phase is ascribed to multiple organ system failure (MOSF) but not always to concomitant sepsis, for often bacteria cannot be detected in up to half the patients who die [2]. Where sepsis had been well controlled and was not confirmed at death, it had been surmised that infection was not the primary cause of death. In this situation, devitalized tissue and/or circulating endotoxin have been thought to perpetuate some mediator-induced response which lead to MOSF[2]. Evidence in animal models suggests, however, that endotoxic shock and burn shock are two different pathologies. Whereas endotoxin mediates its effects through lipid peroxidation via the production of oxygen free radicals, and catalase and superoxide dismutase can reverse these effects, burn shock is not protected by these enzymes [3]. Thus endotoxin, when present, may not be the main mediator of late death in burns.

One significant system which can undergo failure is the immune system, specifically the T cell system (anergy). It was shown that the degree of T cell failure correlated well with mortality rates [4], but in the debate on whether anergy causes sepsis or is caused by sepsis, arguments are offered for both interpretations. Early studies on T cell function in burn injury [5,6] ascribed the failure to an activation of non-specific suppressor cells, as cocultures of normal lymphocytes with burn patients' lymphocytes gave lower than normal responses. Again bacterial endotoxin was claimed as the mediator which activated suppressor cells [7], as it was known that polyclonal B cell activators could lead to generation of suppressor lymphocytes [8]. In particular, suppression was claimed to involve the macrophage which releases prostaglandin E (PGE) and interleukin-1 (IL1) upon stimulation with endotoxin. PGE would then induce the suppressor T cell [9]. However in burn patients' circulating cells IL1 levels proved to be more or less in the normal range up to 50 days post burn, and PGE levels in plasma did not correlate with immune suppression [10]. Thus, the question of endotoxin-mediated suppressor cell activation in burn trauma may be factitious, especially as a result of direct measurements of the suppressor lymphocyte marker OKT8. After the first week, numbers of post burn OKT8-positive cells were shown to be normal, and the low T4/T8 ratio was due to a decrease in number of T4 helper lymphocytes [11,12].

Our analysis of burn patient's humoral and cellular immune status revealed an abnormally high immunoglobulin (Ig) production *in vitro* in the first few weeks, followed by a depressed response which was either transient in survivors, or permanent in non-survivors. These changes did not correlate well with the proliferative responses to conventional mitogens [13] but were well correlated with the allogenic responses in the mixed lymphocyte reaction (MLR) [14,15] suggesting T helper cell functional failure.

We observed further, that production of both IL2 and its receptor (IL2R) in burn patient's lymphocytes was markedly suppressed during or shortly before the functionally observed immunosuppression [16,17]. IL2R levels decreased over the post burn period, but if they remained above 50% of the first day level they increased again leading to survival. Once below 50% of the first day level, expression of IL2R continued to drop until death. Addition of IL2 to survivors' cell cultures increased both the number of IL2R and the MLR response. In non-survivors' cells, IL2 also raised the IL2R levels, but did not reverse the low MLR functional response. Thus, terminal burn injury appears to affect the expression of IL2 functional receptors [18].

This report deals with our preliminary examination of IL2 function and factors implicated in post burn immunosuppression, such as post burn serum, PGE_2, and a very well characterized lipid-protein complex induced in skin by burning (Cutaneous Burn Toxin, CBT). This toxin is known to depress the *in vivo* immune response to *Pseudomonas* [19] and the *in vitro* lymphocyte proliferative response [20].

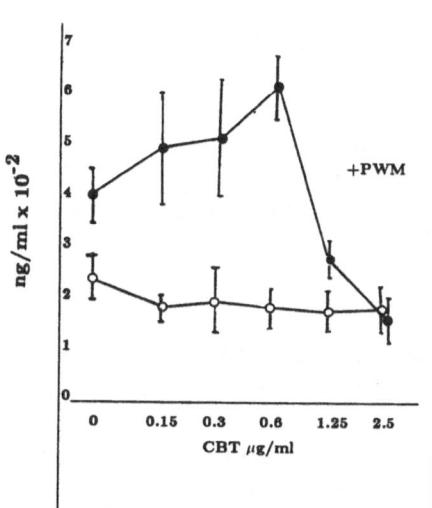

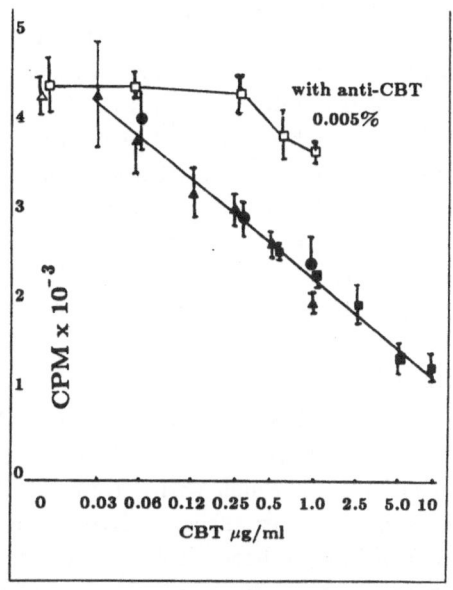

Fig. 1. Immunoglobulin G (ng/ml) in supernatants of 8 day cultures of 5 x 10^5 PBMC from a normal subject, containing CBT with (•) or without (o) PWM. With PBMC from two other subjects the PWM-induced IgG concentrations with $1.25\mu g/ml$ CBT, were 76% and 60% of their respective control values without CBT.

Fig. 2. Proliferative response of IL2-dependent NK cells in the presence of IL2, assayed in three separate experiments (▲, •, ■) with CBT. Growth of these cells in the presence of CBT and antiserum to CBT at 0.005% (□). Cells were cultured at 10^5/ml for 64 hours with a tritiated thymidine pulse for the last 16 hours.

MATERIALS & METHODS

Preparation of peripheral blood mononuclear cells (PBMC), culture conditions and enumeration of IL2R-expressing cells by flow cytometry, were performed as described in detail elsewhere [17]. Assessment of Ig production in pokeweed mitogen stimulated PBMC cultures was performed as described before [15]. Post burn sera were tested at a concentration of 10% (v/v), IL2 (Biogen Corp.) was used at 20U/ml, PGE_2 (Sigma Co.) at 10^{-6} to 10^{-8}M and indomethacin (Sigma Co.) at a concentration of $1\mu g/ml$. CNBr-activated Sepharose 4B (Pharmacia) and monospecific anti-PGE_2 serum (Sigma Co.) were utilized for the affinity chromatography adsorption of post-burn sera. CBT was isolated and purified from burned human skin following the procedures of Dr. G. Schoenenberger [21] in his laboratory. Antiserum to CBT was produced in sheep and the immune globulin fraction was the gift of Dr. G. Schoenenberger. The human IL2-dependent cell line [22] was generously provided by Dr. R. M. Gorczynski of the Ontario Cancer Institute, Princess Margaret Hospital, Toronto. These cells were cultured at 10^5/ml in microwell plates in volumes of 0.1 ml with tritiated thymidine ($1\mu Ci$ in $50\mu l$) added at the 48th hour. Plates were harvested at the 64th hour. Standard IL2, when added in $50\mu l$, was in a dilution calculated to promote proliferation registering about 4000 cpm.

RESULTS

IL2-bearing cells were enumerated in ConA-stimulated PBMC from normal controls and from burn patients at day 1 and at the time of maximal immunosuppression which varied from day 10 to day 40. IL2 added to these cultures increased the percentage of IL2R-positive cells in all cultures (Table 1). The addition of indomethacin (IM) to cultures had no effect on normal receptor levels or on receptor expression at the time of immunosuppression; but it did increase the expression on cells isolated on the first day of the injury.

When PGE_2 was added to cultures of normal PBMC in the presence of normal human serum (NHS), there was a drop in percentage of IL2R-positive cells. IL2 in these cultures reversed the inhibition caused by PGE_2 (Table 2). Post burn (PB) serum isolated on day 1 was not suppressive of IL2R expression, but serum isolated at the time of immunosuppression (days 10 to 40) was. Removal of PGE_2 from immunosuppressive serum, using antibody to PGE_2 did not completely reverse the inhibition. These results suggested that immunosuppressive sera contain factors other than PGE_2 which affect IL2 receptor expression.

The toxin from burned skin is known to appear in patient plasma increasingly after thermal injury [23]. To study the viability of PBMC in the presence of CBT, cultures of 10^6 cells/ml were established with CBT ranging from 0.25 to 10 $\mu g/ml$. The percent viable cells after 48 hours was enumerated by trypan blue staining and the numbers of cells per ml were counted using the Coulter Counter. CBT had no effect on the viability of PBMC (Table 3). As a test for its effect on immunoglobulin production, CBT was incubated *in vitro* with 5 x 10^5 PBMC/ml and pokeweed mitogen at 10μg/ml. From 0.15 to 0.6 μg/ml, CBT did not inhibit IgG production but 1.25 and 2.5 μg/ml concentrations were suppressive (Fig 1). Supernatants from cultures of 10^6 cells/ml incubated in the presence of 5μg/ml phytohemagglutinin, and with CBT ranging from 0.12 to 10μg/ml, produced decreasing amounts of IL2. The IL2 was assayed by the growth of an IL2-dependent human T cell line (data not shown). The direct effect of CBT on proliferation of this IL2-dependent cell line, in the presence of a standard amount of IL2, was to inhibit the growth increasingly as the concentration of CBT increased (Fig 2). Fifty percent of the optimal growth occurred in the presence of about 1.0μg/ml CBT in this system.

DISCUSSION

Some laboratories have recently focused on the failure of IL-2 function in T lymphocytes of severely burned patients [10,11,17]. Since it is known that PGE inhibits the production of human IL2 [9], we tested the sensitivity of lymphocytes from the burn patients, to induction of receptors for IL2, in the presence of indomethacin. At the time of the patients' immunosuppression (days 10 to 40), the decreased percentage of IL2R-bearing cells was not augmented by IM. On the first day, however, IM did augment the percentage of IL2R-positive cells. This indicated that endogenous PG production in PBMC was active on the first day but not at the time of immunosuppression. A biphasic release of eicosanoid metabolites in burn patients was reported [24] to be very high on day 1 and was then exhausted. Our findings are consistent with this pattern. Thus, the immunosuppression of the later period, characterized by reduction in helper cell numbers and in IL2 function, was not dependent on endogenous eicosanoid metabolism.

When PGE_2 was added to cell cultures of normal subjects, in concentrations known to be immunosuppressive *in vitro,* it reduced the percentage of IL2R-positive cells. However, this reduction was reversed in the presence of IL2. Serum from patients on the first post-burn day did not suppress levels of IL2R expression, but at the time of immunosuppression burn serum did lower the percentage of IL2R-positive cells. However added IL2, and the removal of PGE_2 both brought the percentages of IL2R-positive lymphocytes up to similar levels, but not completely to the normal control level. This suggested that factors affecting IL2R, other than PGE_2, might exist in the immunosuppressive serum.

In recent years, immunosuppressive serum factors have been gaining attention and are reported to be of a variety of molecular sizes [10, 24, 25, 26]. The lipid-protein complex from burned skin, CBT, was known to have immunosuppressive properties [19,20]. It appears in the plasma of burn patients at concentrations ranging from 10 to 100 μg/ml [23], with a peak at about the sixth day post-burn [27]. When it was incubated with PBMC, to observe its effect on cell viability, no deleterious effects were noted *in vitro* in 48 hours. However in assays of immune function, CBT at 1.25 and 2.5 μg/ml caused a marked

Table 1. Effect of IL2 (20U/ml) and indomethacin (1 μg/ml) on IL2R expressed by ConA-stimulated PBMC from normal donors and burn patients.

Culture Conditions	% IL2R-positive cells at 72 hr			
	Burn Patients			Controls
	Day 1	Day 10-40		
		Surviving	Non-surviving	
PBMC	26.1±6.3	16.6±8.1	<5	31.4±12.5
PBMC + IL2	37.9±8.2	34.2±11.2	15.5±3.1	54.1±5.3
PBMC + indomethacin	41.7±7.9	14.3±7.0	<5	33.4±8.5

Table 2. Effect of post-burn (PB) sera (10%) and IL2 (20U/ml) on expression of IL2R in ConA-stimulated cultures of normal PBMC*

Serum	Expression of IL2R after 72 hr [% of control level]§	
	without IL2	with IL2
NHS, pooled	100	145.9
PB, day1	117.2	135.7
PB, day 10-40	26.0	74.8
PB, day 10-40/anti-PGE$_2$	76.0	108.9
NHS + PGE$_2$ 10^{-6}M	50.4	77.2
NHS + PGE$_2$ 10^{-7}M	46.3	101.6
NHS + PGE$_2$ 10^{-8}M	52.8	105.7

*Table 2 presents the means of results from 7 separate experiments in which 69 individual serum samples collected from burn patients 1-40 days post-burn were tested.
§ Results are expressed as a percentage of the IL2R-bearing cells enumerated in ConA-activated cultures incubated in the presence of normal human serum (NHS, pooled).

Table 3. Effect of CBT on growth and viability of PBMC cultured at 37 °C for 48 Hrs.

	CBT concentration (μg/ml)							
	0	0.25	0.5	1.0	2.5	5	10	
Cell concentration ($\times 10^{-6}$/ml)								Mean
0 Hrs:	1.0	1.0	1.0	1.0	1.0	1.0	1.0	
48 hrs:	1.73	1.76	1.63	1.86	1.68	1.87	1.80	1.77±0.10
Viability %								
0 Hrs:	94.8	94.8	94.8	94.8	94.8	94.8	94.8	
48 Hrs:	93.2	91.5	86.9	90.5	90.4	86.3	87.0	88.8±2.3

inhibiton of PWM-induced IgG production by PBMC at 5 x 10^5/ml. As well, PHA-induced IL2 production was arrested. The complex, at about 1.0 μg/ml, directly inhibited 50% of the proliferative response of a human IL2-dependent cell line cultured initially at 10^5 cells/ml in the presence of IL2. It is yet unknown whether CBT exerted its effect by binding IL2, or had a direct effect on the membrane IL2R.

Proper IL2 function now appears to be critical to the late burn mortality, as non-survivors' lymphocytes cannot functionally respond to exogenous IL2 [18]. CBT had earlier been implicated in late burn mortality [28], because CBT plasma levels correlated with patient death, to a highly significant degree, by probit analysis[19]. It is significant therefore, that CBT interferes with a mechanism so critical to recovery from thermal injury. Curiously, early eschar excision and wound closure, involving extensive surgery and multiple anaesthesics, are procedures which themselves should be immunosuppressive. Yet they are known to improve the immunologic depression profoundly [29]. This is supported in animal models [30,31]. Thus, evidence exists which links thermal injury of skin to an immune failure in severe burns. Skin has an immunological function [32] and CBT may be a complex of some immunoreactive material in skin perhaps reacting with IL2 or its receptor. However, immunosuppression mediated by CBT will have to be weighed in consideration with other immunosuppressive factors.

REFERENCES

[1] G. Arturson. The Los Alfaques disaster: a boiling liquid, expanding-vapour explosion. *Burns* **7**: 233 - 251 (1980).

[2] R. H. Demling. Burns. *New Eng. J. Med*. **313**: 1389 - 1398 (1985).

[3] T. Yoshikawa, N. Yoshida, H. Miyagawa, T. Takemura, T. Tanigawa, M. Murakami and M. Kondo. Role of free radical lipid peroxidation in burn and endotoxin shock. *These Proceedings*.

[4] J. M. Heibert, M. McGough, G. Rodeheaver. The influence of catabolism on immunocompetence in burned patients. *Surgery*: **86**: 242 - 247 (1979).

[5] C. L. Miller, C. C. Baker. Changes in lymphocyte activity after thermal injury. *J. Clin. Invest*. **63**: 202 - 210 (1979).

[6] J. H. N. Wolfe, I. Saporoschetz, A. E. Young, N. E. O'Connor and J. A. Mannick. Suppressive serum, suppressor lymphocytes and death from burns. *Ann. Surg*. **193**: 513 - 520 (1981).

[7] J. L. Ninneman, A. E. Stockland, J. T. Condie. Induction of prostaglandin synthesis-dependent suppressor cells with endotoxin: Occurrence in patients with thermal injuries. *J. Clin. Immunol*. **3**: 142 - 150 (1983).

[8] J. L. Ninneman. Immunosuppression following thermal injury through B cell activation of suppressor T cells. *J. Trauma* **20**: 206 - 213 (1980).

[9]   S. Chouaib, L. Chatenoud, D. Klatzmann and D. Fradelizi. The mechanisms of inhibition of human IL 2 production. II PGE$_2$ induction of suppressor T lymphocytes. *J. Immunol.* **132**: 1851 - 1857 (1984).

[10] J. A. Mannick, M. L. Rodrick. Cellular immune function and injury. *These Proceedings*.

[11] A. C. Antonacci, S. E. Calvano, L. E. Reaves, A. Prajapati, R. Bockman, K. Welte, R. Mertelsmann, S. Gupta, R. A. Good and G. T. Shires. Autologous and allogeneic mixed-lymphocyte responses following thermal injury in man: The immunomodulatory effects of interleukin 1, interleukin 2 and a prostaglandin inhibitor, WY-18251. *Clin. Immunol. & Immunopath.* **30**: 304 - 320 (1984).

[12] M. Braquet, P. Lavaud, D. Dormont, R. Garay, R. Ducousso, J. Guilbaud, M. Chignard, P. Borgeat and P. Braquet. Leukocytic functions in burn-injured patients. *Prostaglandins* **29**: 747 - 764 (1985).

[13] J. A. Teodorczyk-Injeyan, B. G. Sparkes, W. J. Peters and R. E. Falk. Lymphoproliferative response to phytohemagglutinin in the burn patient - no paradigm in vitro. *J. Burn Care Rehab.* **7**: 112 - 116 (1986).

[14] B. G. Sparkes, J. A. Teodorczyk-Injeyan, W. J. Peters and R. E. Falk. Polyclonal antibody production in the burn patient - kinetics and correlation with T cell activity. *Proc. Can. Fed. Bio. Soc.* **28**: 131 (1985).

[15] J. A. Teodorczyk-Injeyan, B. G. Sparkes, W. J. Peters and R. E. Falk. Polyclonal immunoglobulin production in the burned patient - kinetics and correlation with T cell activity. *J. Trauma* **26** (9): - (1986).

[16] J. A. Teodorczyk-Injeyan, B. G. Sparkes, W. J. Peters and R. E. Falk. Inhibiton of T cell activation in burn patients during post-burn immunosuppression. *Eur. Soc. Surg. Res.* **17** (S1): 27 (1985).

[17] J. A. Teodorczyk-Injeyan, B. G. Sparkes, G. B. Mills, W. J. Peters and R. E. Falk. Impairment of T cell Activation in burn patients: a possible mechanism of thermal injury-induced immunosuppression. *Clin. Exp. Immunol.* **65** (2): - (1986).

[18] J. A. Teodorczyk-Injeyan, B. G. Sparkes, G. B. Mills, W. J. Peters and R. E. Falk. Impaired expression of interleukin-2 receptor (IL 2R) in the immunosuppressed burn patient: Reversal by exogenous IL 2. *J. Trauma* **27** (1): - (1987).

[19] G. A. Schoenenberger. Klinik und Forschung bei schweren Verbrennungen. *Schweiz. Rdsch. Med. PRAXIS* **67** (32): 1163 - 1174 (1978).

[20] J. A. Ninneman and M. D. Stein. Suppression of in vitro lymphocyte response by "burn toxin"-isolates from thermally injured skin. *Immunol. Lett.* **2**: 339 - 342 (1981).

[21] G. A. Schoenenberger. Burn toxins isolated from mouse and human skin. Their characterization and immunotherapy effects. *Monogr. Allergy,* **9**: 72 -

139 (1975).

[22] R. M. Gorczynski, J. Rogers and J. J. Harris (1985). Inhibition of human natural killer (NK) mediated lysis of K562 with purified defined carbohydrates and monoclonal antibodies. *In* Genetic control of host resistance to infection and malignancy. p769 - 774. E. Skamene & P. Kongshaven *Eds.* Alan R. Liss Inc., Montreal.

[23] P. H. Hasler, M. Allgower and G. A. Schoenenberger. Immunoradiometric assay (IRMA) for quantitative determination of a cutaneous human burn toxin in plasma of severely burnt patients. *Eur. Soc. Surg. Res.* **16** (S1):107 - 108 (1984).

[24] M. Braquet. Impaired host defence mechanisms in burn and sepsis. *These proceedings*.

[25] J. L. Ninneman. Immmune depression and sepsis following injury: the role of circulating mediators. *These Proceedings*.

[26] C. R. Baxter. Polymorphonuclear leukocyte function in the immune response in burns. *These Proceedings*.

[27] B. G. Sparkes, P. H. Hasler, M. Allgower and G. A. Schoenenberger. Unpublished data.

[28] M. Allgower, M. Graf, P. Hasler, B. Kistler, B. Kremer and G. A. Schoenenberger. Evidence of possible involvement of a cutaneous burn toxin in the late burn disease/mortality. *Bull. & Clin. Rev. Burn Inj.* **1** (3):27 - 29 (1984).

[29] J. L. Ninneman. Frontiers in understanding burn injury. *J. Trauma* **24** (9): S 81 - S 83 (1984).

[30] C. Echinard, D. Bernard, C. Vescovali and J. P. Jouglard. Kinetics of immune depression in burned rats: the sixth day phenomenon. *J. Burn Care & Rehab.* **6** (3):256 - 260 (1985)

[31] J. F. Hansbrough, V. Peterson, E. Kortz and J. Piacentine. Post burn immunosuppression in an animal model: monocyte dysfunction induced by burned tissue. *Surgery* **93** (3):415 - 423 (1983).

[32] R. L. Edelson and J. M. Fink. The immunologic function of skin. *Scientific American* **252** (6):46 - 53 (1985).

T CELL SUBPOPULATIONS FOLLOWING THERMAL INJURY : EFFECT OF THE ACUTE* STRESS RESPONSE

S.E. Calvano, J.D. Albert, A. Legaspi, B. Organ, K.J.Tracey, S.F. Lowry, G.T. Shires, A.C. Antonacci
Department of Surgery, New York Hospital-Cornell University Medical Center, New York, New York 10021, U.S.A.

INTRODUCTION

In his now famous treatise of 1897, "Adaptation in Pathological Processes," Doctor William Welch discussed the dynamic physiologic and evolutionary mechanisms at work in an organism's adaptive response to various life-threatening pathologic processes. In his paper, Doctor Welch aptly suggested:

> that the adaptability of this mechanism to bring about useful adjustments has been in large part determined by the factors of organic evolution, but that in only relatively few cases can we suppose these evolutionary factors to have intervened on behalf of morbid states. For the most part, the agencies employed are such as exist primarily for physiologic uses, and while these may be all that are required to secure a good pathologic adjustment, often they have no special "fitness for this purpose"

In any discussion of the adaptive responses to the pathologic processes induced by trauma, the alterations observed in host defense mechanisms are among the most important to the survival of the organism in the morbid state. Whether a good pathologic adjustment is secured depends, in large part, on a successful "series" of pathologic adjustments to the recurrent threats of immune deficiency and septic infection. However, these threats often arise as the final "coup de grace" in an organism's frequently unsuccessful struggle to adapt to the complex phenomenon of trauma. As such, trauma must be viewed as a dynamic insult, extending and increasing in complexity well beyond the initial encounter with host as a series of interactions among adaptive responses, and including the pathophysiologic sequelae of adaptation and the introduction of new influences. Indeed, host defense must not be viewed as an isolated entity, but rather as an integral element of the clinical didactic, sensitive to the effects of organ system change and capable of exerting its own beneficial and deleterious effects on other organ systems. Successful immunologic adaptation, therefore, is more accurately determined by how significantly immunologic homeostatic balance is affected by the complex phenomenology of trauma, i.e., the immunologic effects of (a) the acute stress (adrenocortical response), (b) neuroendocrine influences, (c) nutritional deficiency, (d) the infection itself, or (e) medical intervention, and others.

This paper will concentrate on the immunologic effects of the acute stress response to burn induced immunodeficiency.

BACKGROUND

Severe injury be it from surgery, trauma, or burn, represents a monumental challenge to homeostasis. Indeed, it has long been assumed that during the response to injury, homeostasis, in the strict sense, is foregone as regulatory systems coordinate to maintain vital functions at the expense of less vital ones. However, following the most severe injuries it is clear that physiologic regulation often does not occur in a coordinated and adaptive fashion. This should not be suprising since prior to modern clinical intervention, such severe injuries were invariably fatal, with survival past the shock phase a rare occurrence. Thus, from an evolutionary perspective, adaptive mechanisms to cope with post-shock injury could not have been selected and, as a result, disordered and inappropriate regulatory system responses following recovery from shock would be the rule rather than the exception.

Thermal injury represents such a condition of severe stress, both physiologically and psychologically. Intuitively, one would therefore expect that stress-associated hormones such as corticosteroids and catecholamines would be elevated following injury. Indeed, this is the case with many of the deleterious physiologic sequelae of injury being attributed to chronic activation of "stress hormone" systems.

Corticosteroids have long been employed as effective immunosuppressive agents (Gabrielsen and Good, 1967[1]; Makinodan et al., 1970[2]). Although species differences exist, administration of corticosteroids or their derivatives generally produces a consistent pattern of hematologic and immunologic change. This includes rapid lymphopenia, monocytopenia, eosinopenia and granulocytosis (Dougherty and White, 1944[3]). These cellular changes have been considered to be due either to lysis (Dougherty and White, 1945[4]; Claman, 1972[5]), or redistribution of cells to an extravascular compartment (Claman, 1972[5]; Cohen, 1972[6]; Berney, 1974[7]; Fauci, 1974[8].) Immunologic changes include thymic involution (Dougherty and White, 1945[4]; Claesson and Ropke, 1969[9]), depressed cell-mediated immune responses reflected by decreases in: T killer and natural killer functions (Fernandes et al., 1975[10]; Parrillo and Fauci, 1978[11]; Oshimi et al., 1980[12]), T cell blastogenesis (Webel et al., 1974[13]; Blomgren and Andersson, 1976[14]; Neifeld and Tormey, 1979[15]), mixed lymphocyte responsiveness (Ilfeld et al., 1977[16]; Katx and Fauci, 1979[17]), graft versus host reactions (Medawar and Sparrow, 1956[18]), and delayed hypersensitivity responses (Derbes et al., 1950[19]; Gabrielsen and Good, 1967[1]). Humoral immunity has been reportedly enhanced (Fauci et al., 1978[20]) or attenuated (Elliott and St. C. Sinclair, 1968[21]) by corticosteroids while phagocytic cell functions are uniformly inhibited (Rhinehart et al., 1974; 1975[22,23]).

When administered in pharmacological doses to normal human subjects, corticosteroids induce a profound lymphopenia (Dale et al., 1974[24]; Fauci and Dale, 1974[25]; Yu et al., 1974[26].) However, this lymphopenia does not reflect a generalized reduction in numbers of all lymphocytes. Only T cells are affected and, specifically, only T cells with helper/inducer function (T cells bearing the T4 or Leu3 cell-surface markers). Numbers of cells with suppressor/cytotoxic function (T cells bearing the T8 or Leu2 cell-surface markers) remain constant or decrease only slightly. This effect of corticosteroids has been documented by Fauci and co-workers (Haynes and Fauci, 1978[27]; Cupps et al., 1984[28]) using both rossetting and immunofluorescence analyses. In the rossetting studies, numbers of circulating T_{mu} cells (T cells with helper activity) decreased by approximately $1000/mm^3$ with only slight reduction in T_{gamma} cells (T cells with suppressor activity). When similar studies were performed using immunofluorescence techniques, identical results were obtained. T4 positive cells were specifically depleted from the circulation following corticosteroid administration while T8 positive cells were affected only slightly. It should be noted that these changes in numbers of T cells belonging to the two subsets reversed within 24 hours following termination of corticosteroid treatment. Further, it seems likely that this quantitative decrease in helper/inducer cells and the corresponding relative increase in suppressor/cytotoxic cells in the blood is responsible, at least in part, for the immunosuppressive effects of corticosteroids.

On the basis of these studies, the question that immediately arises is, where do the T cells disappear to during corticosteroid treatment and where do they return from following termination of treatment? Animal studies (Levine and Claman, 1970[29]; Cohen, 1972[6]; Fauci, 1974[8]) have indicated that corticosteroids induce sequestration of T cells in the bone marrow where they are not normally found in significant numbers. Data reported by Antonacci et al. (1982[30]; 1984b[31]) have demonstrated that, following burn injury, an identical pattern of change occurs in peripheral blood lymphocytes as observed following corticosteroid administration. Thus, thermal injury was shown to induce lymphopenia that was limited to T cells bearing receptors for the F_c fragment if IgM (T_{mu} cells) or cells bearing the T4 marker. As was the case following corticosteroid treatment, numbers of peripheral blood T_{gamma} or T8 positive cells were little affected by burn injury. Normalization of these cell numbers occurred during recovery from the injury.

In contrast to the approximate 3-fold increase in total cortisol observed in most studies, Calvano et al. (1984a[32]) demonstrated a 10-fold increase in free cortisol relative to normal controls. This was due to the increase in total cortisol along with significant decreases in the cortisol-binding proteins, transcortin and albumin. Since endocrine dogma asserts that only unbound or free cortisol exhibits biological activity, this 10-fold increase reflects the amount of bioactive cortisol actually present. It would seem that this dramatic increase in free cortisol following burn injury represents a large enough increment to account for the same peripheral T cell subset changes as observed following exogenous administration of corticosteroids.

Corticosteroids also have direct suppressive effects on lymphoid cell activties, as opposed to modulating numbers of cells available to participate in and regulate immune responses. T lymphocytes are much more strongly affected than are B lymphocytes (Fauci et al., 1974[25]). Inhibition of T cell proliferation is the most commonly observed effect when corticosteroids are administered in vivo (Fauci and Dale, 1974[25]; Webel et al., 1974[13]) or added to mitogen- or antigen-stimulated mononuclear cells in vitro (Blomgren and Andersson, 1976[14]; Crabtree et al., 1979[33]; Neifeld and Tormey, 1979[15].) However, under proper conditions, attenuation of cytolytic T cell activity has also been reported (Gillis et al., 1979b[34]; Schleimer et al., 1984[35]). It has now been established that inhibition of T cell proliferation by corticosteroids is, in turn, due to inhibition of interleukin-2 (IL-2) or T cell growth factor production (Crabtree et al., 1979[33]; Gillis et al., 1979a[36]; Larsson, 1980[37]). Exogenous administration of IL-2 to cultures treated with corticosteroids restores proliferation to normal indicating that the T cells are primed to respond (i.e., express receptors for IL-2) but simply lack a sufficient amount of IL-2 in order to do so (Crabtree et al., 1979[33]; Gillis et al., 1979b[34]; Larsson, 1980[37]). It is not known if inhibition of IL-2 secretion by corticosteroids is in turn mediated by increased leukocyte cyclic AMP levels (Coffey and Hadden, 1985[38]).

Most assay systems for quantifying cortisol (as opposed to synthetic corticosteroid) -induced inhibition of T cell proliferation require substantial ($>5 \times 10^{-7}$M) concentrations of steroid for an effect to be observed (Ilfeld et al., 1977[16]; Katz and Fauci, 1979[17]; Neifeld and Tormey, 1979[15]). However, one assay that is approximately an order of magnitude more sensitive is the autologous mixed lymphocyte reaction (Ilfeld et al., 1977[16]; Katz and Fauci, 1979[17]). In this system, an individual's purified T cells are induced to proliferate when co-cultured with their own irradiated (or mitomycin C treated) non-T cells. Data reported by Antonacci et al. (1984a[39]) demonstrated marked inhibition of proliferation in autologous mixed lymphocyte culture following burn injury. Further, and perhaps most important, addition of exogenous IL-2 to burn patient autologous mixed lymphocyte cultures restored proliferation to levels observed in normal control cultures. As pointed out, corticosteroids inhibit T cell proliferation in vitro by decreasing lymphocyte production of IL-2, and addition of exogenous IL-2 to corticosteroid-treated cultures can restore proliferation to normal. These results suggest that injury-induced elevations of corticosteroids may, in addition to influencing numbers of circulating leukocytes, directly affect T cell activity by inhibiting the production of IL-2. Further support for this hypothesis is provided by the work of Wood et al. (1984[40]), who directly measured IL-2 production in cultures of mitogen-stimulated mononuclear cells from burn patients and normal controls. IL-2, but not IL-1, production was found to be significantly reduced in the cultures of cells from thermally-injured subjects.

The obvious similarity between corticosteroid administration and burn injury on changes in numbers of peripheral blood T cells belonging to the helper/inducer and suppressor/cytotoxic subsets strongly suggests that increased output of endogenous corticosteroids following thermal injury is responsible for the specific T4 lymphopenia in the blood of patients sustaining this type of injury. Whether these T cells are sequestered in the bone marrow as happens during corticosteroid treatment remains to be demonstrated definitively, but preliminary experiments in a rat model of thermal injury indicate a significant increase in helper/inducer T cells in the bone marrow on day two post-injury (Calvano et al., 1984b[41]).

RESULTS

In the present study, the relationship between increased corticosteroids and immunologic changes following thermal injury was elucidated further by administration of a constant infusion of hydrocortisone (cortisol) to normal human subjects at a rate designed to produce steady-state plasma concentrations which mimic those observed after burn.

Infusion of hydrocortisone was carried out over a 6 hour period. Prior to infusion, during infusion at 4, 4.5, 5, 5.5 and 6 hours, and at 16 hours following termination of infusion, blood samples were obtained and the plasma assayed for hydrocortisone by radioimmunoassay. Pre- and post-infusion plasma concentrations of hydrocortisone were within the expected normal range, averaging 9.3 and 11.1 ug/dl, respectively. However, during infusion, plasma levels of hydrocortisone reached a steady-state of 35-40 ug/dl. These plasma concentrations were pathophysiological, not pharmacological. A group of 24 thermally-injured human subjects had a mean hydrocortisone concentration of 47.5 ug/dl at 24-48 hours post-injury, a level only slightly higher than that achieved during hydrocortisone infusion. These data indicate that the infusion rate employed was able to increase plasma hydrocortisone in normal subjects to an extent that mimics levels observed during severe physiologic stress.

Prior to infusion, subjects were slightly lymphopenic as evidenced by an average lymphocyte count of $1139/mm^3$ (Normal Range = 1500-$3500/mm^3$). The reason for this is not known, but may be related to prior manipulation of the subjects for placement of central venous and arterial lines. Nonetheless, during infusion, significant lymphopenia ensued with the lowest number ($534/mm^3$) of circulating lymphocytes observed after 6 hours of infusion. However, by 16 hours post infusion, lymphocyte counts had normalized to $2171/mm^3$ and the pre-existing slight lymphopenia was no longer present. Hydrocortisone infusion also induced a significant monocytopenia with a time course very similar to that of the lymphopenia.

A group of thermally-injured subjects (N = 26) showed only slight lymphopenia at 24 hours post-injury, and completely unlike the response to hydrocortisone infusion, these patients manifested dramatic monocytosis. These differences between hydrocortisone infusion subjects and burn patients may be due to the high rate of catecholamine secretion known to occur after severe injury.

Corticosteroids cause granulocytosis, and during hydrocortisone infusion, numbers of granulocytes were significantly elevated by 4 hours and had approximately doubled by 6 hours of infusion. Unlike lymphocyte counts, numbers of granulocytes were completely normal prior to infusion, and there was no difference between pre-infusion and 16 hour post-infusion numbers.

Thermally-injured patients also manifested granulocytosis, but to a greater degree than did the infusion subjects. Again, this may have been due to a burn-induced catecholamine response since catecholamines, like corticosteroids, cause granulocytosis. These data suggest that the two hormones may be additive in their effect on granulocyte numbers.

T cells, expressed as the percent T3+ or T11+ of total lymphocytes, declined gradually during the course of hydrocortisone infusion with statistically significant decreases observed at 5.5 and 6 hours of infusion. Sixteen hours after infusion, the percentage of T cells had returned to pre-infusion levels.

The decline in the percentage of T cells could be accounted for almost entirely by a similar decline in the percentage of cells comprising the T4 subset. Thus, the percentage of T4+ lymphocytes went from 45% prior to infusion to 31% by 6 hours of infusion. Following infusion, the percentage of T4+ cells rose, reaching 48% at 16 hours. Since in peripheral blood T4 and T8 are mutually exclusive T cell subsets, a decrease in the percentage of T4+ cells should be associated with an increase in the percentage of T8+ cells if, as suggested above, T4+ rather than T8+ cells left the blood during hydrocortisone infusion. The percentage of T8+ lymphocytes did increase during infusion, but at no time were these changes statistically significant relative to the pre-infusion percentage of T8+ cells.

At 24-48 hours post-injury, thermally-injured subjects showed a pattern of change in the percentage of T cells and T cell subsets similar to that of normal individuals undergoing hydrocortisone infusion. Thus, following burn, percentages of T3+ and T4+ cells were dramatically decreased while the percentage of T8+ lymphocytes increased slightly. It is to be noted that the percentage of T3+ cells decreased to a greater extent in the burn patients than in the infusion subjects, but the percentages of T4+ and T8+ cells were quite similar in burn and infusion subjects. Although burn patients actually manifested an inversion in the T4:T8 ratio, this ratio also was close to 1 after 6 hours of hydrocortisone infusion. In the thermally-traumatized subjects percentages of T3+ and T4+ lymphocytes normalized during recovery from injury.

Following burn injury, the percentage of B cells (identified by monoclonal anti-Ia) in burn patients was remarkably similar to that found for HLA-DR+ lymphocytes during hydrocortisone infusion.

Numerical changes in lymphocyte subsets during hydrocortisone infusion were dominated by the profound lymphopenia. The T cell lymphopenia brought about by hydrocortisone infusion was mostly limited to the T4 subset. Thus, while numbers of T4+ cells decreased to an extent similar to T3+ cells, numbers of T8+ cells declined only slightly. The increase in all T cells following infusion was most likely due to the previously mentioned compensation for the slight lymphopenia that was present prior to initiating infusion.

Burn patients showed decreased numbers of T3+ and T4+ cells, but this decrease was not as great as that during infusion. This was because lymphopenia in the burn subjects at 24 hours post-injury was not nearly as great in magnitude as was the lymphopenia in the infusion subjects. In contrast, numbers of T8+ lymphocytes in thermally-injured subjects were not different from normal. However, the data does indicate that during hydrocortisone infusion the numbers of T8+ cells did drop lower than observed in the burn patients. Again, this can be attributed to the more profound lymphopenia during hydrocortisone infusion.

Numbers of B cells (HLA-DR+) and LGLs (Leu7+) did not change during hydrocortisone infusion. However, at 16 hours post-infusion, there was a significant increase in B cells but not LGLs. This suggests that the slight lymphopenia observed prior to infusion can be attributed to a decrease in both B and T cells, but not LGLs, since these cells did not increase in number by 16 hours post-infusion.

In thermally-traumatized subjects, numbers of B cells (Ia+) were higher than seen during hydrocortisone infusion, almost identical to pre-infusion numbers, and lower than those observed following infusion. The conclusion therefore is that B cells are decreased only slightly or not at all following burn injury, a pattern similar to that observed during hydrocortisone infusion.

DISCUSSION

Previous reports have shown that exogenous administration of corticosteroids produces the effects documented here: lymphopenia, monocytopenia, granulocytosis, a specific decrease in T4+ lymphocytes, and inhibition of lymphocyte proliferation to mitogens. However, these previous studies employed either oral dosing or bolus intravenous injection of glucocorticoids. Such routes of hormone administration prevented precise control of plasma steroid concentrations. Thus, peak concentrations of corticosteroids attained were well above the normal range and must therefore be considered "pharmacologic", not "physiologic."

The present study differs from the above-described reports in that plasma hydrocortisone concentrations were controlled by means of a constant infusion and were never allowed to rise above levels which the adrenal glands normally are capable of achieving. This was confirmed by comparing the steady-state infusion plasma hydrocortisone concentrations to those observed for a group of 24 thermally-injured patients and noting that both groups had mean hydrocortisone levels of approximately 40-50 ug/dl. Interestingly, however, the hematologic/immunologic effects of generating these pathophysiological hydrocortisone concentrations in normal subjects were similar to those documented in previous studies in which pharmacologic levels of corticosteroids were employed. This suggests that the threshold for a corticosteroid effect is actually quite low and may be relevant in less severe states of stress than thermal injury. Alternatively, it may be that the total mass of corticosteroid administered is a relevant variable since, in the present study, a 60 kg subject would have received 65 mg of hydrocortisone over the 6 hour infusion period.

Considering these injury-induced changes, one at a time, as resulting from the combined effects of elevations in corticosteroids and catecholamines, the following can be predicted. First, injury-associated lymphopenia produced by elevated corticosteroids may be attenuated by the lymphocytosis induced by catecholamines. It is to be noted that this is supported by the fact that burn patients had hydrocortisone concentrations similar to those of the infusion subjects, but only a slight/moderate lymphopenia compared to that observed during infusion. Second, monocytosis is observed following burn injury suggesting a predominant catecholamine influence on these cells since hydrocortisone infusion alone produces monocytopenia. However, at this time it is not known if the magnitude of thermal injury-induced monocytosis would be much greater if there were not concomitant elevation of corticosteroids. Finally, it appears that granulocytosis following burn injury might be the result of an additive effect of corticosteroids and catecholamines since both hormones act to increase plasma granulocyte numbers, and granulocytosis after burn injury is greater than that produced by hydrocortisone infusion alone.

As has been suggested previously, the specific decrease in T4+, as well as T_{mu}, lymphocytes may be related to immune dysfunction commonly observed following burn injury. However, it has become clear recently that the T4 subset further can be resolved into two sub-subsets, helper-effector and suppressor-inducer. Two-color immunofluorescence analyses employing anti-T4 along with either anti-Leu8, anti-2H4, or anti-4B4 can be employed to identify these subsets of T4+ cells. Such analyses need to be performed following thermal injury as well as after corticosteroid administration. But, based on an immunodeficiency hypothesis of thermal injury-induced immune dysfunction, the prediction is that the decline in T4+ cells in both injury and corticosteroid conditions will be largely limited to the helper-effector subset of T4. In addition, it appears that elevations of plasma hydrocortisone to levels that mimic those observed following burn injury are sufficient to produce a functional immunologic deficit.

357

The data presented here as well as previous data from this laboratory and others thus lend support to the hypothesis that elevated corticosteroids are intimately involved in the immunologic changes occurring in the early period following thermal injury. Further, since immune regulation is a temporal process orchestrated by complex interactions among the individual elements comprising the immune system, it is suggested that these corticosteroid-induced early events may "set the stage" for others, including active immunosuppression, occurring later in the post-burn course.

* Data presented in this paper has been previously presented in the following publications :

1. Calvano, S.E., et al.
   Surgery, Gynaecology and Obstetrics , January, 1987, in press.

2. Calvano, P.E.
   Advances in Host Defence Mechanism.
   Edited by G.T. Shires and J.M. Davis, Raven Press: New York, Chapter 6, 1986.

REFERENCES

1) Gabrielsen, A.E., and Good, R.A. Adv. Immunology 6:91 (1967).

2) Makinodan, T., Santos, G.W., and Quinn, R.P. Pharmacological Reviews 22:189 (1970).

3) Dougherty, T.R., and White, A. Endocrinology, 35:1 (1944).

4) Dougherty, T.R., and White, A. American Journal of Anatomy, 77:81 (1945).

5) Claman, H.N. New England Journal of Medicine, 287:388 (1972).

6) Cohen, J.J. Journal of Immunology, 107:841 (1972).

7) Berney, S. Clinical Research, 22:414A (1974).

8) Fauci, A.S. Federation Proceedings, 33:750 (1974).

9) Claesson, M.H., and Ropke, C. Acta. Pathol. Microbiol. Scand., 76:376 (1969).

10) Fernandes, G., Yunis, E.J., and Good, R.A. Clinical Immunology and Immunopathology, 4:304 (1975).

11) Parrillo, J.E., and Fauci, A.S. Scandinavian Journal of Immunology, 8:99 (1978).

12) Oshimi, K., Gonda, N., Sumiya, M., and Kano, S. Clinical Exp. Immunology, 40:83 (1978).

13) Webel, M.L., Ritts, R.E., Jr., Taswell, H.F., Donadio, J.V., Jr., and Woods, J.E. Journal Lab. of Clinical Medicine, 83:383 (1974).

14) Blomgren, H., and Andersson, B. Experimental Cell Research, 97:233 (1976).

20) Fauci, A.S., Pratt, K.R., and Whalen, G. Journal of Immunology, 119:598 (1978).

21) Elliott, E.V., and St. C. Sinclair, N.R. Immunology, 15:643 (1968).

22) Rhinehart, J.J., Balcerzak, S.P., Sagone, A.L., and LoBuglio, A.F. Journal of Clinical Investigation, 54:1337 (1974).

23) Rhinehart, J.J., Sagone, A.L., Balcerzak, S.P., Ackerman, G.A., and LoBuglio, A.F. New England Journal of Medicine, 292:263 (1975).

24) Dale, D.C., Fauci, A.S., and Wolf, S.M. New England Journal of Medicine, 291:1154 (1974).

25) Fauci, A.S., and Dale, D.C. Journal of Clinical Investigation, 53:240 (1974).

26) Yu, D.T.Y., Clements, P.J., Paulus, H.E., Peter, J.B., Levy, J., and Barnett, E.V. Journal of Clinical Investigation, 53:565 (1974).

27) Haynes, B.F., and Fauci, A.S. Journal of Clinical Investigation, 61:703 (1978).

28) Cupps, T.R., Edgar, L.C., Thomas, C.A., and Fauci, A.S. Journal of Immunology, 132:170 (1984).

29) Levine, M.A., and Claman, H.N. Science, 167:1515 (1970).

30) Antonacci, A.C., Good, R.A., and Gupta, S. Surgery, Gynecology & Obstetrics, 155:1 (1982).

31) Antonacci, A.C., Reaves, L.E., Calvano, S.E., Amand, R., deRiesthal, H.F., and Shires, G.T. Surgery, Gynecology & Obstetrics, 159:1 (1984b).

32) Calvano, S.E., Chiao, J., Reaves, L.E., Antonacci, A.C., and Shires, G.T. Journal of Burn Care and Rehabilitation, 5:143 (1984a).

33) Crabtree, G.R., Gillis, S., Smith, K.A., and Munck, A. Arthrit. Rheum., 22:1246 (1979).

34) Gillis, S., Crabtree, G.R., and Smith, K.A. Journal of Immunology, 123:1632 (1979b).

35) Schleimer, R.P., Jacques, A., Shin, H.S., Lichtenstein, L.M., and Plaut, M. Journal of Immunology, 132:266 (1984).

36) Gillis, S., Crabtree, G.R., and Smith, K.A. Journal of Immunology, 123:1624 (1979a).

37) Larsson, E.L. Journal of Immunology, 124:2828 (1980).

38) Coffey, R.G., and Hadden, J.W. Federation Proceedings, 44:112 (1985).

39) Antonacci, A.C., Calvano, S.E., Reaves, L.E., Prajapati, A., Bockman, R., Welte, K., Mertelsmann, R., Gupta, S., Good, R.A., and Shires, G.T. Clinical Immunology and Immunopathology, 30:304 (1984a).

40) Wood, J.J., Rodrick, M.L., O'Mahony, J.B., Palder, S.B., Saporoschetz, I., D'Eon, P., and Mannick, J.A. Annals of Surgery, 200:311 (1984).

41) Calvano, S.E., Chiao, J., Antonacci, A.C., and Shires, G.T. Surgical Forum, 35:153 (1984b).

NATURAL KILLER CELL ACTIVITY AND LYMPHOCYTE FUNCTION DURING AND AFTER OPEN-HEART SURGERY IN RELATION TO THE ENDOCRINE STRESS RESPONSE

E. Tonnesen, M. M. Brinklov and N. J. Christensen
Department of Anaesthesia, Odense University Hospital, DK-5000
Odense C, Denmark

The effect of elective open-heart surgery on natural killer (NK) cell activity was studied in 20 patients allocated to two different anaesthetic techniques. To clarify the mechanisms behind changes in NK cell activity, the distribution of lymphocyte subpopulations and lymphocyte blastogenesis to PHA were determined. The endocrine response to surgery was measured as serum cortisol and plasma catecholamines.

ANAESTHETIC DATA

Premedication consisted of diazepam 0.25 mg/kg b.w., morphine 0.25 mg/kg b.w. and scopolamine 0.003 mg/kg b.w.

Group I received high dose fentanyl(75-125 μg/kg b.w.) maintained with etomidate-fentanyl infusion. Ventilation: oxygen/air.

Group II received a balanced anaesthesia consisting of fentanyl ($\leq$ 20 μg/kg b.w.) maintained with midazolam infusion. Ventilation: oxygen/nitrous oxide/halothane.

METHODS

NK cell cytotoxicity was measured in a 6 hours ^{51}Cr-release assay against K562 target cells (1).

Circulating lymphocyte subpopulations were determined using mono-clonal antibodies against the T-cell population (OKT3), the T-helper cell (OKT4), the T-suppressor cell (OKT8), the B-cell population (B1) and NK cells (Leu11). Conventional fluorescence microscopy was used.

Plasma epinephrine and norepinephrine were measured by a single isotope-derivative assay (2) and serum cortisol was measured by a competitive protein binding technique (3).

SAMPLE TIMES

From each patient 12-14 consecutive blood samples were drawn from an arterial line:

Sample 1:   The day before operation
Sample 2:   Before induction (after cannulation of the radial artery)
Sample 3:   10 min after intubation
Sample 4:   10 min after sternotomy
Sample 5:   Before cardiopulmonary by-pass

Sample 6a: After 10 min on by-pass
Sample 6b: After 40 min on by-pass
Sample 6c: After 70 min on by-pass or 2 min after initiation of partial by-pass
Sample 6d: After 100 min on by-pass or 2 min after initiation of partial by-pass
Sample 7:   15 min after termination of by-pass
Sample 8:   During skin suture
Sample 9:   First postoperative day
Sample 10: Third postoperative day
Sample 11: Sixth postoperative day

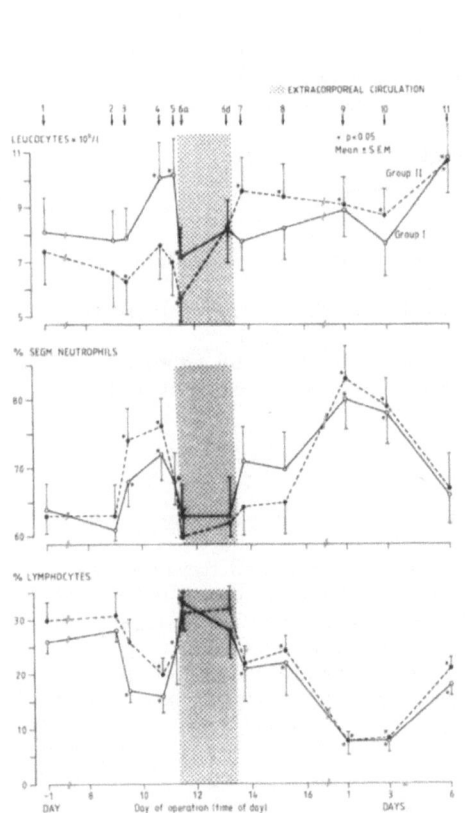

Changes in leucocyte count, the percentage of segmented neutrophils and lymphocytes before, during and after open-heart surgery. Asterix indicates significant differences (P < 0.05) between preoperative value (Sample 1) and the indicated value.

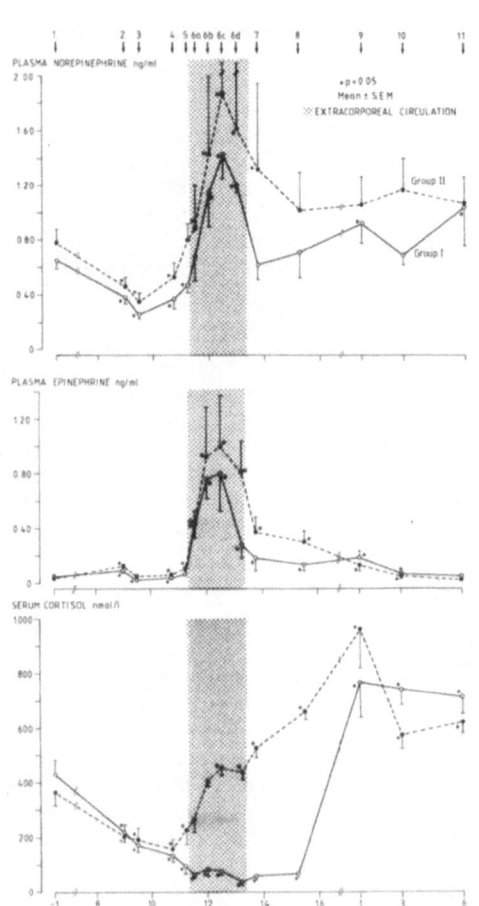

Plasma norepinephrine, plasma epinephrine, and serum cortisol concentrations before, during and after open-heart surgery.

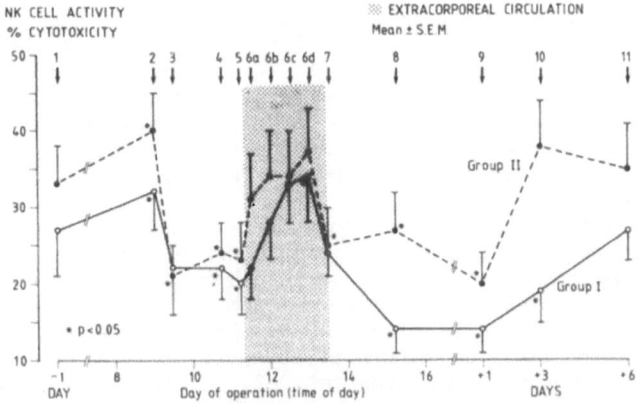

NK cell activity expressed as %
cytotoxicity before, during and
after open-heart surgery.

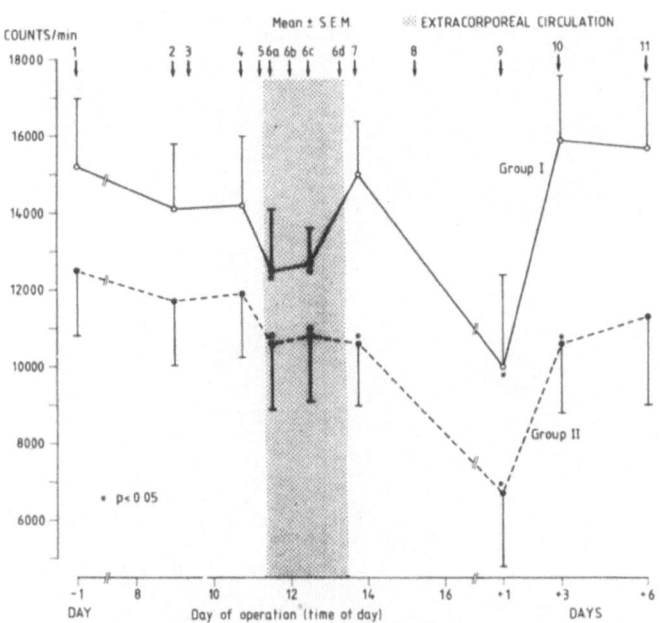

PHA response of lymphocytes expressed as counts per minute before,
during and after open-heart surgery.

RESULTS AND DISCUSSION

With respect to sex, age, weight and duration of surgery the two groups were comparable.

NK cell activity fluctuated in the same way in the two groups. After a transient increase just before induction of anaesthesia the activity, decreased significantly until start of extracorporeal circulation (ECC) which was accompanied by a gradual increase. Postoperatively, NK cell activity and the lymphocyte response to PHA were significantly depressed for 3-6 days. These changes were accompanied with severe lymphopenia affecting the T-lymphocytes (OKT3) with a selective loss of the helper cell (OKT4) subset (data not shown).

The immunological parameters showed only minor differences between the groups, whereas the endocrine response differed with a delayed cortisol response in Group I. A steep rise in plasma catecholamines occurred in both groups during EEC and plasma epinephrine remained significantly elevated, although at a lower level, until the sixth postoperative day.

The results suggest that coronary by-pass surgery is associated with two periods, pre-bypass and postoperatively, of marked immunosuppression, probably mediated by different mechanisms. Finally, the study suggests a correlation between NK cell activity and plasma epinephrines during ECC.

REFERENCES

1. E. Tønnesen, H. Mickley, and N. Grunnet. Acta Anaesthesiol. Scand. 27: 238 (1983).
2. N. J. Christensen, J. P. Vestergaard, T. Sørensen, and O. J. Rafaelsen. Acta Psychiatr. Scand., 14: 178 (1980).
3. H. Kehlet, C. Binder, and C. Engbæk. Acta Endocrinol. (Copenhagen), 75: 119 (1974).

NATURAL KILLER CELL ACTIVITY DURING EPINEPHRINE AND CORTISOL INFUSION IN HEALTHY VOLUNTEERS

E. Tønnesen, M.M. Brinkløv, N.J. Christensen
Department of Anaesthesia, Odense University Hospital
DK-5000 Odense C, Denmark

Using a continuous intravenous infusion pattern to simulate some of the hormonal changes induced by major surgery the effect of epinephrine and cortisol on natural killer (NK) cell activity and circulating lymphocyte subpopulations were investigated.

MATERIAL

Twenty healthy fasting volunteers, aged 25-40 years, were allocated to receive either a continuous intravenous infusion of cortisol 5 µg/kg b.w./min for five hours (Group I), epinephrine 0.05 µg/kg b.w./min for one hour (Group II), cortisol for five hours with simultaneous infusion of epinephrine during the last hour (Group III) or saline for five hours with simultaneous infusion of saline during the last hour (Group IV).

BLOOD SAMPLING

The first sample was taken immediately before start of infusion (0 hour) after a rest period of 30 min. Subsequently, in Group I, III, and IV, blood was drawn every 60 min during the first four hours of infusion followed by sampling every 15 min in the fifth hour. The last blood sample was drawn 15 min after completing the infusions.
In Group II blood was drawn every 15 min during epinephrine infusion and finally 15 min after infusion.

METHODS

NK cell cytotoxicity was measured in a 6 hours ^{51}Cr-release assay against K562 target cells.
Circulating lymphocyte subpopulations were determined using monoclonal antibodies against the entire T-cell population (OKT3), the T-helper cell (OKT4), the T-suppressor cell (OKT8), the B-cell population (B1) and NK cells (Leu 11). Conventional flouroscence microscopy was used.
Total leucocyte count was determined by a Coulter Counter S and the differential count was performed automatically.
Serum cortisol was measured by a competitive protein binding technique. Plasma epinephrine and norepinephrine were measured by a single isotope-derivative assay.

RESULTS

Compared with preinfusion levels cortisol did not induce changes either in activity or fraction of NK cells (Leu 11).

Increasing leucocytosis, neutrophilia and lymphopenia could be demonstrated after two hours infusion. The lymphopenia was accompanied by a significant reduction of the T-cell fraction (OKT3 and OKT4) (Data not shown).

During epinephrine infusion the fraction of NK cells (Leu 11) increased significantly (Data not shown). The increase in number of NK cells was accompanied by a simultaneous increase in NK cell activity followed by a return to pre-infusion levels after termination of infusion.

Significant leucocytosis occurred without changes in the percentages of segmented neutrophils and lymphocytes.

The T-cell (OKT3) fraction decreased significantly with a selective loss of T-helper (OKT4) cells (Data not shown).

During simultaneous infusion of cortisol and epinephrine NK cell activity as well as the number of NK cells (Leu 11) in peripheral blood increased to the same level as during epinephrine infusion alone.

Leucocytosis and neutrophilia equal to the additive response to the two hormones seperately could be demonstrated, whereas the cortisol-induced lymphopenia was abolished by the addition of epinephrine.

Infusion of saline 0.9% for five hours did not induce any changes in the resting level or activity of any cell type analyzed (Data not shown).

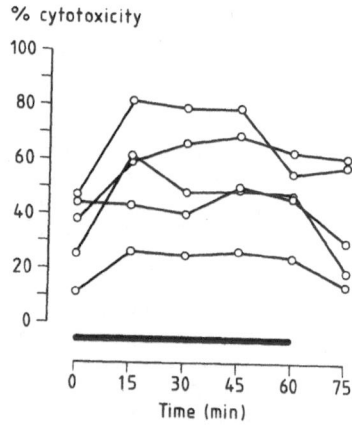

CORTISOL INFUSION   EPINEPHRINE INFUSION

NK cell activity expressed as % cytotoxicity. Effector to target ratio 50:1.

CORTISOL INFUSION

n=5 (median, range)
* p< 0.05

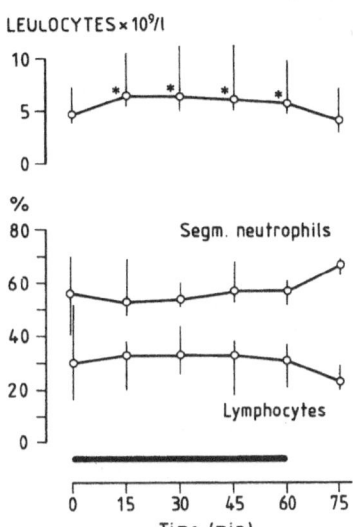

LEUCOCYTES × 10⁹/l

Segm. neutrophils

Lymphocytes

Time (hours)

EPINEPHRINE INFUSION

n=5 (median, range)
* p<0.05

LEUCOCYTES × 10⁹/l

Segm. neutrophils

Lymphocytes

Time (min)

Leucocyte and differential count.

CORTISOL AND EPINEPHRINE INFUSION

n=5 (median, range)
* p < 0.05

LEUCOCYTES × 10⁹/l

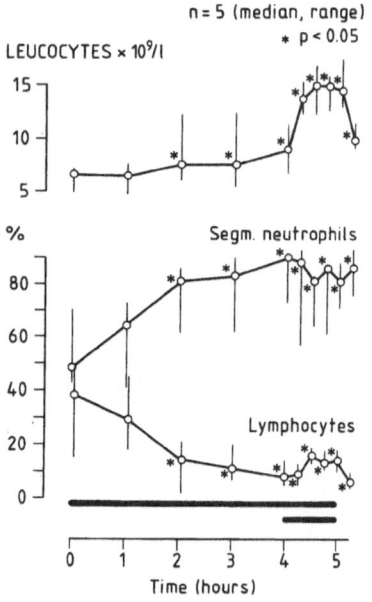

Segm. neutrophils

Lymphocytes

Time (hours)

Leucocyte and differential count.

CORTISOL AND EPINEPHRINE INFUSION

% cytotoxicity

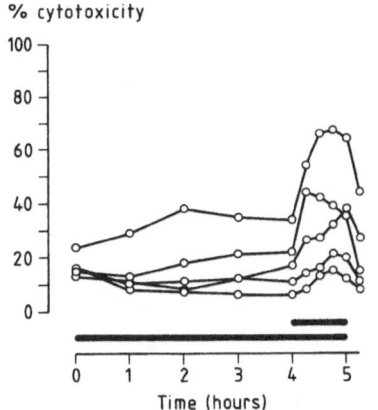

Time (hours)

NK cell activity expressed as %
cytotoxicity. Effector to target
ratio 50:1.

CONCLUSION

Our results confirm the role of epinephrine as a potent inducer of NK cell activity in vivo, whereas no effect of cortisol on NK cell activity could be demonstrated in the actual infusion design.

A LONGITUDINAL STUDY OF THE INFLUENCE OF ACUTE MYOCARDIAL INFARCTION ON NATURAL KILLER CELLS.

K. Klarlung, B. K. Pedersen, T. G. Theander and V. Andersen

Department of Cardiology, Copenhagen County Hospital, Gentofte
Copenhagen, Denmark. Laboratory of Medical Immunology
Rigshospitalet, University Hospital, Denmark
Lymphocyte Laboratory, Department of Infectious Diseases
Rigshospitalet, University Hospital, Copenhagen, Denmark

INTRODUCTION

Natural killer (NK) cells are thought to play an important role in immune surveillance against cancer and certain infections (1). NK cells are augmented by interferon (IF) and interleukin 2 (Il-2) and inhibited by certain prostaglandins (PG), e.g. PGE1, PGE2, PGA1 and PGA2, by activated PMNs and by physical stress. The acute myocardial infarction represents a condition with the involvement of cellular necrosis and severe physical stress. In this study we measured baseline, IF- and Il-2-enhanced NK cell activity within 24 hours after clinical symptoms of AMI and regularly thereafter for 6 weeks. Furthermore NK cells were quantitated by counting large granular lymphocytes (LGL).

PATIENTS AND METHODS

Ten patients with AMI, (8 males, 2 females), mean age 58 years, and 62 sex and age matched healthy controls were studied. Blood samples were collected within 24 hours after the AMI and then on days 3,7 and 14 and at 6 weeks. NK cell activity of mononuclear cells, isolated by Ficoll Isopaque gradient centrifugation was measured against K 562 target cells in a Cr-release assay, as previously described in details (2). NK cell activity was measured in triplicates at effector/target (E/T) cell ratios=80/1, 40/1 and 20/1, only E/T cell ratio=80/1 is given below. In vitro incubation of mononuclear cells with IFN-α and Il-2 was performed as described (2). LGLs were enumerated by identification of medium to large-sized lymphocytes with a relatively high ratio of cytoplasma to nucleus, azurophilic cytoplasmic granules and often an indented nucleus in cytocentrifuged preparations, stained with May-Grünwald (2).

RESULTS.

Baseline NK cell activity on days 1,3,7,14 and at 6 weeks was

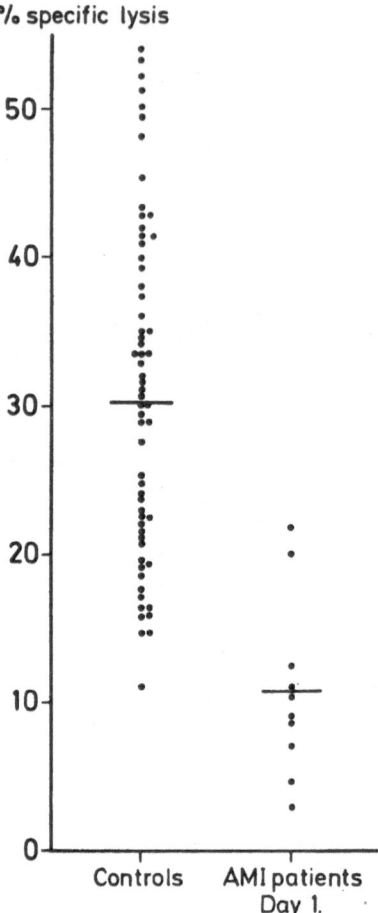

% specific lysis

Figure 1.　Baseline NK Activity

significantly lower than that of controls (p<0.01 on day 1,3,7, p<0.05 on
day 14, p=0.05 at 6 weeks). The NK cell activity of the patients on day 1
was 10.9±6.1% versus 31.1±11.3% in the control group (mean+SD), Fig.1. IF
as well as Il-2 significantly enhanced the NK cell activity (p<0.01) on
all five days tested, Fig.2, but neither IF nor Il-2 restored the
suppressed NK cell activity, except when measured at six weeks. The NK
cell activity (mean+SEM) of the patients measured at 6 weeks was after
exposure to IF 47.9±2.9% versus 53.2±2.2% in controls (NS), and after
exposure to Il-2 36.7±2.2% versus 41.3±4.8% (NS). NK cells were
quantitated by counting LGLs. The proportion and concentration of LGLs on
days 1,3,14 and at 6 weeks did not differ from the values observed in
controls and LGLs did not differ significantly between the days measured,
except on day 7, where LGLs were increased when compared to day 1 and to
6 weeks, P<0.01. Similarly a slight, but statistically not significant
increase in Leu 11 positive cells on day 7 was found. Serum-cortisol
concentration (mean+SEM) measured day 1 was 0.62±0.10 umol/l versus
0.44+0.4 umol/l at 6 weeks (NS)(normal range 0.0-0.83). We were not able
to demonstrate any correlation between maximum ASAT levels, maximum LDH
levels, leukocyte count,body temperature, plasma cortisol concentration
and NK cell activity. Furthermore we did not find any correlation between
medical treatment and NK cell activity.

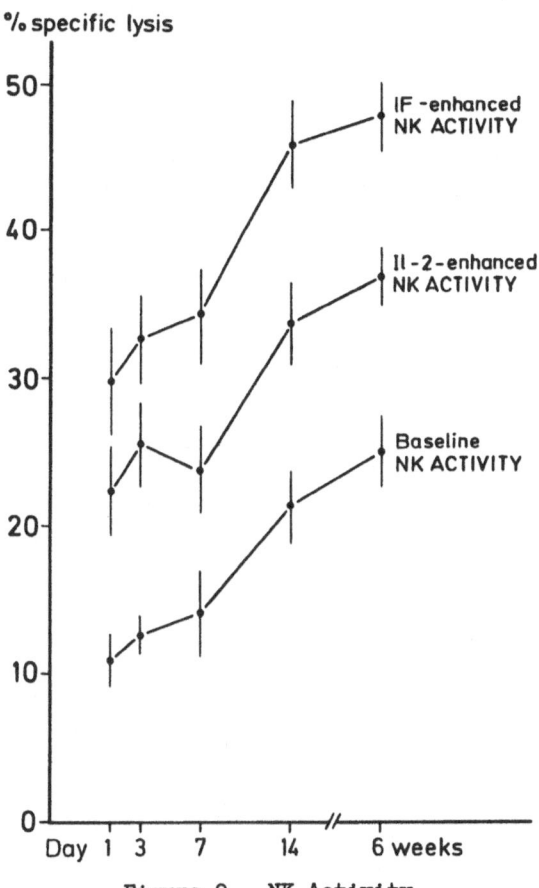

Figure 2. NK Activity

DISCUSSION

This report demonstrates significantly reduced NK cell activity after AMI, which could not be restored by addition of IF or Il-2, indicating that the NK cell activity was not suppressed because of low production of these compounds. We were not able to demonstrate any correlation between serum-cortisol and NK cell activity, and the low NK cell activity is thus not likely to be due to a corticosteroid effect. The proportion and concentration of LGLs and Leu 11 positive cells were normal in the patients. The NK cell defect in patients with AMI is thus functional.

REFERENCES

1. B.K. Pedersen. Allergy, 40, 547 (1985).
2. B.K. Pedersen, P.Oxholm, R.Manthorpe and V.Andersen. Clin.Exp.Immunol.63, 1 (1986)

IMMUNE DEPRESSION AND SEPSIS FOLLOWING INJURY : THE ROLE AND IDENTITY OF CIRCULATING MEDIATORS

J. L. Ninnemann
Department of Surgery, University of California, San Diego
La Jolla, CA 92093

INTRODUCTION

It is well known that profound immunological disturbances often occur as a result of thermal or traumatic injuries (1,2). These changes, which include serum protein alterations (e.g. complement activation, fibronectin depletion, and the generation of immune complexes) as well as compromise of cellular functions (e.g. depressed neutrophil chemotaxis and intracellular killing, suppressor T-cell activation and depression of lymphocyte blastogenesis), have been summarized elsewhere (3). Immune problems in patients are so far reaching that it has been cynically suggested that the study of any immunologic parameter following injury is sure to reveal defects of potential clinical importance. Unfortunately, this statement is basically true. In general, the greater the extent of the injury, the greater the likelihood that clinically significant immunological depression will occur. In burn patients, this means that all patients with a greater than 40% body surface area injury can be expected to have serious, multiple immunological problems and, therefore, be at high risk to sepsis.

Two clinical manipulations seem critical to reversing the immune changes observed in burn patients: removal of the dead or injured tissue and restoration of the surface barrier to wound colonization through wound closure. Immunological perturbations will persist in burn patients as long as these two conditions have not been met. With closure of the burn wound often comes rapid restoration of immunological competence and, frequently, full recovery of the patient. The earlier each of these clinical manipulations can be accomplished, the earlier immunological restoration will be achieved, thereby decreasing the septic threat and increasing patient chances for survival. This has become the dominant surgical philosophy for treating major thermal injuries (4).

The question of how immunologic depression in the burn patient is mediated, has received a great deal of attention. Experimental studies have shown that the general circulation of patients after injury, a major operation, or the onset of disease often contains factors affecting vascular permeability, causing hemolytic changes in red blood cells, depressing cardiac output and renal function, causing profound immunosuppression, and, in some cases, precipitating multiple organ failure. Especially interesting have been cross-perfusion experiments in which it has been found that the transfer of blood from one animal after thermal injury,

can produce these effects in another, normal animal (5). Our own in vitro
experiments to study lymphocyte competence following injury have shown that
the response of normal cells placed in burn patient serum is often profoundly impaired (6). On the other hand, the response of burn patient
lymphocytes removed from the "burn environment" and placed in normal serum,
returns toward normal following a period of incubation (7). Warden et al.
have made similar observations using neutrophil chemotaxis as the test
system (8). It is clear, therefore, that post-injury serum contains mediators which dictate the behavior of, at the very least, lymphocytes and
neutrophils. In addition, cell behavior appears to be controlled by mediator induced, nonspecific suppressor T cells. The activity of these cells
is usually mitomycin-resistant, and appears to be short lived since it can
be abrogated by a short period of cell culture, which may explain the return to normal of cellular immune response upon removal from the "burn
environment".

Typical suppressive activity of sequential serum samples drawn from
patients with major thermal injuries is summarized in Figure 1, which plots
post-injury day vs the suppressive activity of patient serum as measured in
mixed lymphocyte culture. Such studies reveal the presence of immunosuppressive substances in the serum of almost all patients with major thermal
injuries. These substances appear very quickly following injury, and in
fact can often be detected within 4-6 hours. Suppressive serum activity
can persist for weeks to months, depending on the severity of the injury.
As suggested earlier, patient serum loses its suppressive nature as the
wound is surgically closed and the patient approaches hospital discharge.
Patients who do not survive have suppressive serum to the very end, and
suppressive activity often reaches a dramatic peak in terminally injured
patients shortly before death (10).

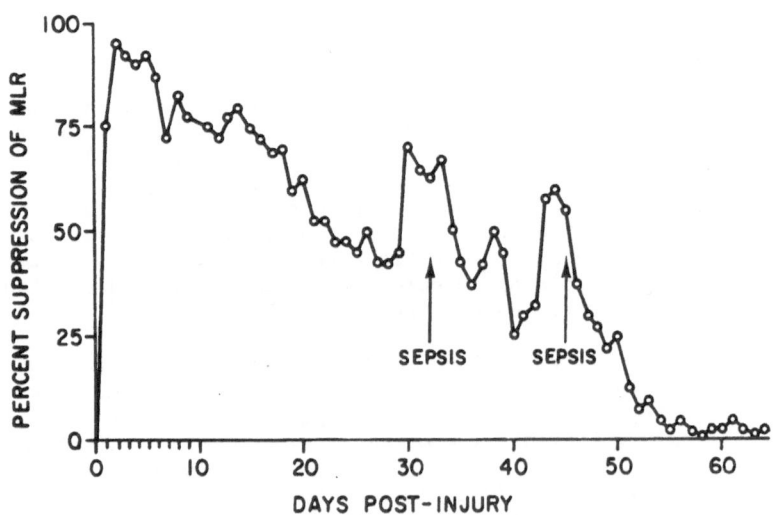

Fig. 1 : SUPPRESSIVE ACTIVITY OF PATIENT SERA

Four potential sources have been the focus of investigations concerning the mediators of injury-induced immune depression: (a) the products of the injury itself, which are released into the circulation from the wound; (b) endogenous immunoregulatory molecules such as the prostaglandins; (c) immunologically active substances from exogenous sources, such as bacterial endotoxin; and (d) iatrogenic suppressors, such as derivatives of topical agents used for burn therapy. It seems likely that all four categories of mediators play clinically significant roles.

Suppressive Products of Injury: SAP

Various unique products of burn injury have been proposed to have immunologic activity. Perhaps the best known of these is the controversial burn-toxin molecule of Allgower and Schoenenberger (11). These researchers reported the isolation and biochemical characterization of a high molecular weight lipid-protein complex with cell membrane affinity and a nonspecific, noncytotoxic immunosuppressive activity. The formation and release of this complex appeared to be blocked by the topical application of cerium nitrate to the burn wound. In 1977, Hakim reported the isolation and partial purification of an immunosuppressive peptide from fractionated burn serum, which had a molecular weight of 10,000, an ability to inhibit guinea pig peritoneal macrophage migration, and suppressive activity vs mitogen-induced blastogenesis of normal human peripheral blood lymphocytes (12). A similar factor was reported by Garner et al. (13) who isolated a 5-10 kilodalton peptide which suppressed the transformation of mitogen-stimulated T and B lymphocytes. We have reported the isolation and characterization of a low molecular weight glycopeptide in patients with 40% body surface area burns. This isolate, given the acronym SAP (suppressor active peptide) has been found to suppress T lymphocyte blastogenesis in mixed lymphocyte cultures through the generation of a suppressor T-cell population, to inhibit neutrophil chemotaxis in a receptor mediated yet nonspecific manner, and to hemolyze human erythrocytes by increasing membrane fragility (14-16). Biochemical analysis of SAP has revealed a complex structure, comprised of a 4000 dalton peptide component rich in glycine and serine and containing hydroxylysine and hydroxyproline, and a carbohydrate component containing sialic acid (hexose to protein ratio of 1:9 w/w). That SAP can be isolated within hours of injury indicates an etiology other than drug, topical or treatment related, and suggests an origin other than bacterial. We feel, instead, that SAP may be generated as a result of the inflammatory process, and our results suggest that the proteolytic degradation of a normal tissue component may produce this immunologically active byproduct. The preliminary amino acid analysis of the peptide portion of SAP reveals a characteristic structure related to collagen. It is interesting to note that burn toxin isolates of Schoenenberger contained collagen-like material, and that attempts to remove this "contaminant" also eliminated biological activity (GA Schoenenberger, unpublished observation). Such collagen-related material could be released by the proteolytic degradation of tissue collagen which is known to occur during inflammation (17), or by the breakdown of the collagen-like portion of serum Clq. Since Clq participates in the inflammatory response and contains a collagen-like sequence, we have tested in vitro, the putative generation and immunological activity of Clq peptide fragments under physical conditions present in the burn patient (18). Nanogram quantities of heat or enzyme generated fragments of Clq were clearly shown to be suppressive in vitro to neutrophil chemotaxis and mixed lymphocyte response (MLR). The addition of lymphocytes pre-treated with Clq fragments suppressed ongoing MLR indicating the activation of suppressor cells by the peptides. Finally, anti-Clq globulin was found to reduce the suppressive activity of collagen-like SAP, isolated from human burn sera (18).

The products of the arachidonic acid cascade are among the "normal" mediators which maintain immunological homeostasis in the healthy host. It is now clear that prostaglandins are potent lymphokines (22-24) which along with the leukotrienes, have immunoregulatory properties (25,26). They have also received increasing attention as potentially important "abnormal" mediators following injury. Heggers et al. described the direct release of large quantities of prostaglandins from skin cells as a result of thermal injuries (19), and prostaglandins appear to be involved in the generation of many burn injury-related vascular and tissue viability problems (20,21). In our own work, we could readily demonstrate the suppressive activity of prostaglandins (particularly PGE_2) using an in vitro, one way mixed lymphocyte response as the test assay system. However, we found that the quantity of PGE demonstrated in burn patient whole sera by RIA, while generally quite elevated, was not correlated directly with the suppressive activity of the sera in lymphocyte cultures. While cortisol quantities were also generally quite elevated, the additional consideration of cortisol and its possible synergistic activity with PGE (27) did not improve the correlation. We feel that the lack of correlation between serum PGE levels and patient immunologic depression was due to a) the presence of additional suppressive substances in patient sera (indeed, the net effect of many of these substances upon lymphocyte response appears to be the same); and b) the fact that suppressor cell activity will continue to be expressed even after the disappearance of PGE (and possibly other suppressors) from the system. We have been able to demonstrate this fact in our culture system by the preincubation of responder lymphocytes in PGE_2 (or burned patient serum) for 16 hr, followed by washing, and the addition of stimulator cells. Therefore, clinically, the activity of suppressor cells may well be present long after the decline in concentration of serum-borne suppressors.

That PGE is indeed responsible for at least some of the immunological disturbances observed in burn patients is indicated by our early work with crude fractions of burn sera, and by our more recent work with SAP as reviewed above. Both delipidation (the removal of fatty acids, including the prostaglandins) and the addition of rabbit anti-PGE significantly reduced the suppressive activity of low molecular weight serum fractions. The suppressive activity of PGE_2, added directly to cultures, could also be blocked by the addition of anti-PGE_2. The antiserum was not suppressive to the stimulation levels in control cultures (28). Higher molecular weight, prostaglandin-containing serum fractions were incapable of suppressing the in vitro lymphocyte response. The suppressive activity of PGE therefore appears to depend upon its lack of association with albumin, a common prostinoid carrier in the serum, and an association with a specific low molecular weight serum component. The immunologic activity of this complex is abrogated by delipidation, or blocked by the addition of antiserum.

When tritium-labeled PGE_2 (in buffer) was applied to our Sephadex G-200 column, radioactivity was not detected in any of the protein-containing fractions collected. This is consistent with the low molecular weight of the prostaglandin (352 daltons), and its probable retention on the column. When tritium-labeled PGE_2 was applied to our Sephadex G-200 column after overnight preincubation in normal serum, however, radioactivity eluted clearly in two regions which corresponded to the presence of PGE_2 in a mw 68,000 fraction (consisting primarily of albumin), and a mw 5,000 fraction (consisting primarily of low molecular weight peptide material). Since fractions from each of the two regions eluted in identical positions upon rechromatography, we felt that we were indeed observing stable, PGE-containing molecular complexes of 68,000 daltons and 5,000 daltons, respectively.

When we first reported this work in 1984, we hypothesized that it might be possible that the carrier allows persistence of PGE in the circulation beyond its usual clearance in the lungs (29). The function of the lungs in clearing prostaglandins may be further compromised by respiratory injuries which often accompany major thermal injuries. The participation of carrier molecules in the physiologic effects of the prostaglandins is not a new concept. It is generally believed that the transport of prostaglandin across biologic membranes (such as is required for the generation of suppressor cells) is carrier mediated, since prostaglandin cannot freely diffuse through membrane structures (39). The peptide fraction we have designated SAP therefore, may confer suppressive activity to PGE by facilitating the transport of this compound across the lymphocyte membrane.

Exogenous Mediators: Bacterial Endotoxin

Certain bacterial products are clearly recognized to possess immunologic activity. For example, bacterial lipopolysaccharide (LPS) has a profound effect upon the induction and expression of cell-mediated immunity. Lagrange et al. have described a depression in T cell activity, measured by relative levels of delayed type hypersensitivity in normal and LPS treated mice, at the same time that antibody production was enhanced (31). They speculated that the opposing effects of LPS on T and B cell responses were due to suppressor T cell generation. Immunosuppressive properties of endotoxin have been observed by others as well (32,33), including Garrido, who studied the suppressive properties of products of Pseudomonas aeruginosa, a common burn wound pathogen (33).

Our own work led us to consider bacterial endotoxin as a contributor to serum-mediated post-injury immunosuppression. We recognized very early that suppression is closely tied to the occurrence of bacterial invasion of the burn wound. Organisms involved are most often Gram negative, and, we have concluded, opportunistic of an impaired cell-mediated immune response (34-36). Production of endotoxin is associated with this same group of organisms. In addition, the physiochemical behavior of some of the suppressive sera suggested endotoxin involvement. Immunosuppressive serum obtained from burn patients slowly lost its activity when stored in the cold (even at -80 degrees C), yet does not lose its activity when heated to 100 degrees C for 30 min. All patient sera, demonstrated to contain bacterial endotoxin by the Limulus test, were profoundly immunosuppressive (37). However less than one third of immunosuppressive sera could be demonstrated to contain endotoxin. While this observation may simply reflect the well known limitations of the Limulus test, it is also likely that endotoxin may initiate the suppressive sequence without circulation in the blood stream. Two recent studies support these findings and suggest that prostaglandin E_2, released by macrophages in response to exposure to endotoxin, may be the agent which circulates and acts directly in the suppression of lymphocyte response. Ellner and Spagnuolo report that the addition of 2.0 to 20.0 ug/ml of LPS suppressed normal lymphocyte response to both PHA and SKSD (38). When they depleted their cultures of adherent cells, the lymphocytes were no longer suppressible. The mechanism they have proposed to explain these observations includes the activation of monocytes and subsequent release of prostaglandin E_2, which is the molecule which suppresses the in vitro lymphocyte response. Indeed, the suppression they observed could be blocked by the addition of indomethacin to the cultures, a drug known to block PGE_2 production by macrophages. They also found that culture supernatants produced by macrophages exposed to LPS were suppressive, implicating PGE_2 in the suppressive sequence. These findings are particularly relevant to our observation of B cell involvement in the suppressive sequence, and in view of a recent study by Wilton et al. (39) which outlines clearcut evidence that B lymphocytes are required both for

macrophage activation and for the release of mediators including PGE_2.
Thus endotoxin containing organisms need not leave the burn wound to exert
a profound influence on patient immune reactivity.

We have demonstrated that additions of as little as 1.0 ng of chroma-
tographically purified endotoxin from Escherichia coli 055:B5,E.coli
0111:B4, Pseudomonas aeruginosa (Fisher-Devlin immunotype 1), Serratia
marcescens, or Salmonella minnesota to human mixed lymphocyte or to mitogen-
stimulated cultures produced statistically significant suppression (40).
In each case, endotoxin was most suppressive when present in the culture
system prior to the introduction of the alloantigen or mitogen. Suppressive effects were dependent upon the participation of peripheral blood
monocytes and could be blocked by the addition of the prostaglandin synthetase inhibitor indomethacin or meclofenamate sodium. Prostaglandin production by monocytes appeared to induce a population of "short lived" suppressor cells, identified by the immediate and delayed addition of lymphocyte cocultures to endotoxin-preincubated cells. The suppressive behavior
of endotoxin-primed lymphocytes was identical to the behavior of burn
patient serum-primed lymphocytes or to lymphocyte populations derived from
a subpopulation of burn patients whose serum was Limulus positive. We,
therefore, feel that endotoxin plays a significant immunologic role in some
patients (40).

Iatrogenic suppressors: povidone-iodine

Finally, it should be mentioned that not all post-burn immunosuppression results from natural causes. Some is inadvertently produced as a result of the clinical treatment the patients must necessarily receive. It
is now known that many antibiotics (41), multiple anesthesia (42), the use
of topical agents (43), multiple surgeries with the administration of blood
products (44,45) can all be non-specifically immunosuppressive.

We have found that the addition of povidone (PVP)-iodine, marketed in
the United States as the topical agent Betadine by Purdue-Frederick Inc,
to in vitro cultures of normal human lymphocytes is profoundly suppressive
to both MLC-and PHA-induced blastogenesis (40). Evidence suggests that it
is the iodine component and not the PVP component of these additions that
produces the suppressive effect. Concentrations of PVP-iodine, equivalent
to iodine concentrations of 10^{-4} M are capable of suppressing the response
of normal lymphocytes when 1) in vitro cultures are supplemented directly,
or 2) when lymphocytes are preincubated in the presence of PVP-iodine for
24 h, washed, then stimulated with PHA or allogeneic lymphocytes.

Our data show that this suppression of normal lymphocyte response is
mediated by the generation of suppressor cells as the result of exposure
to PVP-iodine. These suppressor cells exert their influence even after
treatment with mitomycin-C, and thus closely resemble recently reported
suppressor T cells, generated as the result of lymphocyte exposure to
sodium periodate (46). The activity of such suppressor cells can be abrogated by their brief treatment with sodium borohydride, a powerful reducing agent acting upon aldehyde groups at the cell surface (47).

Finally, the clinical significance of our data is clear. Betadine
is in common use as a topical antibacterial in the treatment of burn
patients. We have shown that the sera of patients with large body surface
injuries often contain high levels of iodine, absorbed as the result of
Betadine treatment. These same sera are profoundly suppressive to the
in vitro responsiveness of normal lymphocytes. Iodine appears to be
selectively absorbed to serum albumin; this serum fraction was found to
contain the suppressive activity.

CONCLUSION

The treatment of thermal injuries requires correction of profound biomechanical, biochemical, physiological, and immunological abnormalities, all of which put the patient at very high risk to sepsis. Indeed, it is recognized that sepsis is probably the most significant remaining barrier to the successful treatment of major thermal injuries, and septic risk appears to be directly related to the degree of immunological depression of the patient. The immune deficiencies acquired as a consequence of injury are complex, and, it is now clear, are due to the activation of a variety of mediator systems. That these mediators have systemic activity can be demonstrated by in vitro treatment of normal immune cells with burn serum or isolated burn serum components. Alternately, at least partial correction of burn cellular defects can be accomplished by in vitro patient cell incubation in a normal serum environment.

The possible systemic mediators which may account for these observations are multiple, and it is likely that more than one mediator system is important in a single patient at various post-burn times. Many of these mediator systems are interrelated, introducing the hope that initial, key mediators can be distinguished from secondary sequelae, and manipulated to patient advantage. Ability to finally identify the key mediators will lie in the study of early post-burn events, and in manipulating the immune system of injured subjects with monospecific antibodies, and such pharmacologic agents as polymyxin B, nonsteroidal antiinflammatory agents, and cymetidine.

REFERENCES

1. J.L. Ninnemann (ed), "Traumatic Injury: Infection and Other Immunologic Sequelae", University Park Press, Baltimore (1983).
2. J.L. Ninnemann (ed), "The Immune Consequences of Thermal Injury", Williams and Wilkins, Baltimore (1981).
3. J.L. Ninnemann, Immunologic defenses against infection: alterations following thermal injuries. J. Burn Care Rehabil. 3:355 (1982).
4. J.F. Burke, W.C. Quinby, C.C. Bondoc, Early excision and prompt grafting as routine therapy for the treatment of thermal burns in children. Surg. Clin. N. Am. 56:477 (1976).
5. C.R. Baxter, J.A. Moncrief, M.D. Prager, A circulating myocardial depressant factor in burn shock, in: "Research in Burns", P. Matter, T.L. Barclay, Z. Konickova, eds., Hans-Huber Verlag, Bern (1971).
6. J.L. Ninnemann, Suppression of lymphocyte response following thermal injury, in: "The Immune Consequences of Thermal Injury", J.L. Ninnemann, ed., Williams and Wilkins, Baltimore (1981).
7. J.L. Ninnemann, M.D. Stein, J.T. Condie, Lymphocyte response following thermal injury: the effect of circulating immunosuppressive substances. J. Burn Care Rehabil. 2:196 (1981).
8. G.D. Warden, A.D. Mason, B.A. Pruitt, Suppression of leukocyte chemotaxis in vitro by chemotherapeutic agents used in the management of thermal injuries. Ann. Surg. 181:363 (1975).
9. C.L. Miller, B.J. Claudy, Suppressor T cell activity induced as a result of thermal injury, Cell. Immunol. 44:201 (1979).
10. J.L. Ninnemann, Immunologic complications associated with thermal injury, in: "Pathophysiology of Combined Injury and Trauma", R.I. Walker, D.F. Gruber, T.J. MacVittie, J.J. Conklin, eds., Kamen-Tempo, Santa Barbara (1984).
11. G.A. Schoenenberger, Burn toxins isolated from mouse and human skin, Monogr. Allergy 9:72 (1975).
12. A.A. Hakim, An immunosuppressive factor from serum of thermally traumatized patients, J. Trauma 17:908 (1977).

13. W.D. Garner, M.D. Prager, C.R. Baxter, Multiple inhibitors of lympho-
    cyte transformation in serum from burn patients, J. Burn Care Re-
    habil. 2:97 (1981).

14. J.L. Ninnemann, A.N. Ozkan, Definition of a burn injury-induced
    immunosuppressive serum component. J. Trauma 25:113 (1985).

15. A.N. Ozkan, J.L. Ninnemann, Suppression of in vitro lymphocyte and
    neutrophil responses by a low molecular weight suppressor active
    peptide from burn-patient sera. J. Clin. Immunol. 5:172 (1985).

16. J.L. Ninnemann, A.N. Ozkan, J.J. Sullivan, Hemolysis and suppression
    of neutrophil chemotaxis by a low molecular weight component of
    human burn patient sera. Immunol. Lett. 10:63 (1985).

17. A. Bertelli, The role of proteases in burns, in: "International Sym-
    posium on Pharmacological Treatment in Burns", A. Bertelli, J.
    Donati, eds., Exerpta Medica Foundation, Amsterdam (1968).

18. J.L. Ninnemann, A.N. Ozkan, The immunosuppressive activity of Clq
    degradation peptides. J. Trauma (in press) (1986).

19. J.P. Heggers, G.L. Loy, M.C. Robson, Histological demonstration of
    prostaglandins and thromboxanes in burned tissue. J. Surg. Res.
    28:110 (1980).

20. M.G. Arturson, Prostaglandins in human burn-wound secretion. Burns
    3:112 (1977).

21. M.G. Arturson, The role of prostaglandins in thermal injury, in:
    "Burn Injuries", L. Koslowski, K. Schmidt, R. Hettich, eds., F.K.
    Schattauer, Stuttgart (1979).

22. M.E. Goldyne, Prostaglandins and the modulation of immunological re-
    sponses. Int. J. Dermatol. 16:701 (1977).

23. B. Sammuelsson, E. Granstrom, K. Green, Prostaglandins. Ann. Rev.
    Biochem. 44:669 (1975).

24. D.R. Webb, T.J. Rogers, I. Nowowieski, Endogenous prostaglandin syn-
    thesis and the control of lymphocyte function. Ann. N.Y. Acad.
    Sci. 332:262 (1979).

25. B. Sammuelsson, S. Hammarstrom, Slow reacting substances and leuko-
    trienes. Immunol. Today 2:3 (1981).

26. D.R. Webb, K.J. Wieder, I. Nowowieski. Prostaglandins in lymphocyte
    suppressor mechanisms, in: "Advances in Immunopharmacology", J.
    Hadden, L. Chedid, P. Mullen, eds., Pergamon Press, New York (1981).

27. M.C. Berenbaum, W.A. Cope, R.V. Bundick, Synergistic effect of cor-
    tisol and prostaglandin E on the PHA response. Clin. Exp. Immunol.
    26:534 (1976).

28. J.L. Ninnemann, A.E. Stockland, Participation of prostaglandin E in
    immunosuppression following thermal injury. J. Trauma 24:201
    (1984).

29. J.B. Lee, Are the prostaglandins hormones? in: "Perspectives on the
    Prostaglandins", J.B. Lee, ed., Medcom Press, New York (1973).

30. L.Z. Bito, M. Wallenstein, R. Barody, The role of transport processes
    in the distribution and disposition of prostaglandins. Adv.
    Prostagl. Thrombox. Res. 1:297 (1976).

31. P.H. Lagrange, G.B. Mackaness, Effects of bacterial lipopolysaccharide
    on the induction and expression of cell-mediated immunity: II.
    Stimulation of the efferent arc. J. Immunol. 114:447 (1975).

32. J.M. Cruse, J.T. Forbes, B.R. Shrivers, G.Y. Gillespie, G.K. Lewis,
    R.W. Scales, J.F. Fields, R.B. Hester, E.S. Watson, H.D. Whitten,
    A synergistic immunosuppressive effect of endotoxin and PHA-M on
    immunologic enhancement in mice. Immun. Forsch. 143:31 (1972).

33. M.J. Garrido, Immunosuppressive effect of the Pseudomonas aeruginosa
    endotoxin on lymphocyte transformation in vitro. Zbl. Bakt. Hvg.
    I. Abt. Orig. A 237:274 (1977).

34. J.L. Ninnemann, J.C. Fisher, H.A. Frank. Prolonged survival of human
    skin allografts following thermal injury, Transplantation 25:69
    (1978).

35. J.L. Ninnemann, Immunosuppression following thermal injury through B cell activation of suppressor T cells. J. Trauma 20:206 (1980).

36. J.L. Ninnemann, J.C. Fisher, T.L. Wachtel. Thermal injury associated immunosuppression: occurrence and in vitro blocking effect of post recovery serum. J. Immunol. 122:1736 (1979).

37. J.L. Ninnemann, M.D. Stein, Bacterial endotoxin and the generation of suppressor T cells following thermal injury. J. Trauma 20:959 (1980).

38. J.J. Ellner, P.J. Spagnuolo, Suppression of antigen and mitogen induced human T lymphocyte DNA synthesis by bacterial lipopolysaccharide: mediation by monocyte activation and production of prostaglandins. J. Immunol. 123:2689 (1979).

39. J.M. Wilton, D.L. Rosenstreich, J.J. Oppenheim, Activation of guinea pig macrophages by bacterial lipopolysaccharide requires bone marrow derived lymphocytes. J. Immunol. 114:388 (1975).

40. J.L. Ninnemann, A.E. Stockland, J.T. Condie, Induction of prostaglandin synthesis dependent suppressor cells with endotoxin: occurrence in patients with thermal injuries. J. Clin. Immunol. 3:142 (1983).

41. A.M. Munster, B. Loadholt, A.G. Leary, M.A. Barnes, The effect of antibiotics on cell-mediated immunity. Surgery 81:692 (1977).

42. W.D. Welch, Effect of enflurane, isoflurane, and nitrous oxide on the microbicidal activity of human polymorphoneuclear leukocytes. Anesthesiology 61:188 (1984).

43. G.D. Warden, R.D. Mason, B.A. Pruitt, Suppression of leukocyte chemotaxis in vitro by chemotherapeutic agents used in the management of thermal injuries. Ann. Surg. 181:363 (1975).

44. N.V. Christou, R. Superina, M. Broadhead, Postoperative depression of host resistance: determinants and effect of peripheral protein-sparing therapy. Surgery 92:786 (1982).

45. M.S. Slade, R.L. Simmons, E. Yunis, Immunodepression after major surgery in normal patients. Surgery 78:363 (1975).

46. J.L. Ninnemann, M.D. Stein, Suppressor cell induction by povidone-iodine in vitro demonstration of a consequence of clinical burn treatment with betadine. J. Immunol. 126:1905 (1981).

47. J.L. Ninnemann, L. Morrison, Reversal of povidone-iodine-induced suppressor cell activity by sodium borohydride. Immunol. Lett. 4:189 (1982).

ELECTIVE IMMUNOSUPPRESSION

M.L. Foegh[1], M.R. Alijani[1], G.B. Helfrich[1], B.S. Khirabadi[2],
K. Lim[1], P.W. Ramwell[2]
Georgetown University Medical Center, [1] Department of Surgery,
Division of Transplantation, [2] Department of Physiology &
Biophysics, Washington D.C.

INTRODUCTION

The anergy which is associated with trauma, burn injury and
sepsis shares similarities with the pharmacologically induced
immunosuppression of organ transplant patients. These patients have
the same high risk of infections which also causes mortality as in the
trauma and burn patients. The increased risk of infection in
transplant patients is mainly due to prednisone, although both
azathioprine, an inhibitor of DNA replication, and cyclosporin A, an
inhibitor of interleukin 2 (IL-2) formation, may play some role in the
increased incidence of infection. The immunodeficiency characteristic
of burn patients has been suggested to be mediated by a deficiency in
IL-2 production[1]. This may however only be a partial explanation since
patients treated with CsA do not seem to have a substantial increased
risk of infections although they are immunosuppressed.

The corticosteroids are the first drug of choice for treating
acute allograft rejection. The corticosteroids exert part of the
immunosuppression by inhibiting interleukin 1 (IL-1) and IL-2
formation[2]. CsA inhibits IL-2 formation through inhibiting messenger
RNA for IL-2[3]. CsA does not seem to increase the incidence of
bacterial infections. Thus the increased risk of infections associated
with corticosteroids is likely to be through the inhibition of IL-1.
This monokine is a chemo-attractant for polymorphonucleocytes (PMN's),
therefore the use of corticosteroids result in decreased chemotaxis of
PMN's. Leukotriene B_4 (LTB_4) is another chemo-attractant factor which
is derived from PMN's. Decreased release of LTB_4 is seen in burn
patients[4], and this too may relate to increased risk of infection. A
further aspect of corticosteroid effects which is different from those
expressed by CsA is its effect in inhibiting macrophage function[5]. The
response of macrophages to IL-1 includes chemo-attraction, activation
of tumor cell killing and release of arachidonate metabolites.

In the absence of convincing data on peripheral blood T_4/T_8
ratios the only measurement of too little or too much
immunosuppression is graft rejection and infection, respectively.
However in patients the measurements of urine immunoreactive
thromboxane B_2 ($i-TXB_2$) can be used as an indicator of allograft
rejection[6,7]. Our longitudinal studies of kidney transplant patients
suggest that urine $i-TXB_2$ may be a suitable indicator of the degree of
immunosupression in the transplant patient.

In an experimental cardiac rat transplant model, we have investigated the possibility of replacing corticosteroids with drugs interfering with eicosanoids and platelet activating factor [8]. These drugs are either cyclooxygenase inhibitors, 5-lipoxygenase inhibitors[9], PAF antagonists[10], thromboxane synthase inhibitors, and thromboxane antagonists[11].

EFFECT OF IMMUNOSUPPRESSIVE DRUGS ON ARACHIDONATE METABOLISM

The immunosuppressive drugs, azathioprine and corticosteroids may affect arachidonic acid metabolism as illustrated in figure 1. Corticosteroids inhibit phospholipase activity through formation of polypeptides, lipocortins,[12] which block arachidonate release and subsequent formation of leukotrienes, prostaglandins, and thromboxane. Azathioprine is 6-mercaptopurine coupled to an imidazole. The latter permits oral administration of 6-mercaptopurine. Following ingestion, azathioprine is cleaved to 6-mercaptopurine and imidazole. Interestingly, imidazole is a thromboxane synthase inhibitor although not very potent and rather non-specific. Inhibition of thromboxane synthesis may relate to the finding that intravenous azathioprine improves kidney allograft survival in dogs better than 6-mercaptopurine[13].

The effect of CsA on arachidonate metabolism is not well documented. Inhibition of arachidonate metabolism has been demonstrated as a decrease in prostacyclin formation in rat blood vessels[14], and as inhibition of prostaglandin synthesis in rat macrophages[15]. However such inhibition may relate to CsA toxicity.

MODULATION OF LYMPHOCYTE PROLIFERATION BY LIPID MEDIATORS

The initiation of the clonal expansion of T-lymphocytes requires two signals from the macrophage; one signal is IL-1 and the other is foreign antigen presentation. The eicosanoids affect this interaction by modulating both formation and action of monokines and lymphokines as well as antigen expression (fig. 2). Prostaglandin E_2 (PGE_2) and corticosteroids inhibit IL1 and IL2 formation whereas leukotrienes have been shown both directly and indirectly to promote IL1 formation[16-19]. The effect of LTB_4 on lymphocyte proliferation has been explored by Goodwin[20] and by Rola-Pleszczinski[19] who found LTB_4 to promote lymphocyte proliferation. Initially, LTB_4 was found to inhibit both mitogen and antigen-induced T4 lymphocyte proliferation[21]. However, further studies indicated that LTB_4 also facilitates proliferation of T_4 lymphocytes when endogenous PGE_2 synthesis is blocked by indomethacin[19]. Recent studies introduced a new concept namely that T4 lymphocytes undergo phenotypic conversion to T8 without de novo protein synthesis[22] and moreover that this conversion is enhanced by LTB_4. Another aspect of immune enhancement is promotion of lymphocyte migration which is also reported to be facilitated by LTB_4[23].

Thromboxane A_2 (TXA_2) synthase inhibitors have been shown to inhibit lymphocyte proliferation which infers that TXA_2 promotes lymphocyte proliferation.[24] This effect could however also be explained by shunting of endoperoxides into PGE_2 and PGI_2[25]. More recently a TXA_2 agonist was shown to promote mitogen induced lymphocyte proliferation, and to reverse the inhibition of lymphocyte proliferation obtained with a TXA_2 synthase inhibitor[26]. In order for the lymphocyte to undergo clonal expansion in response to foreign antigen and IL-1, the lymphocyte needs to recognize DR antigen on the

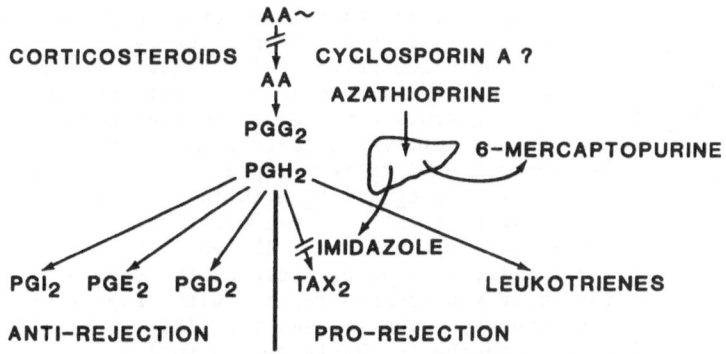

Figure 1. Arachidonic acid cascade showing the anti-rejection and pro-rejection metabolites. The possible sites of interference of the cascade by immunosuppressive drugs are indicated.

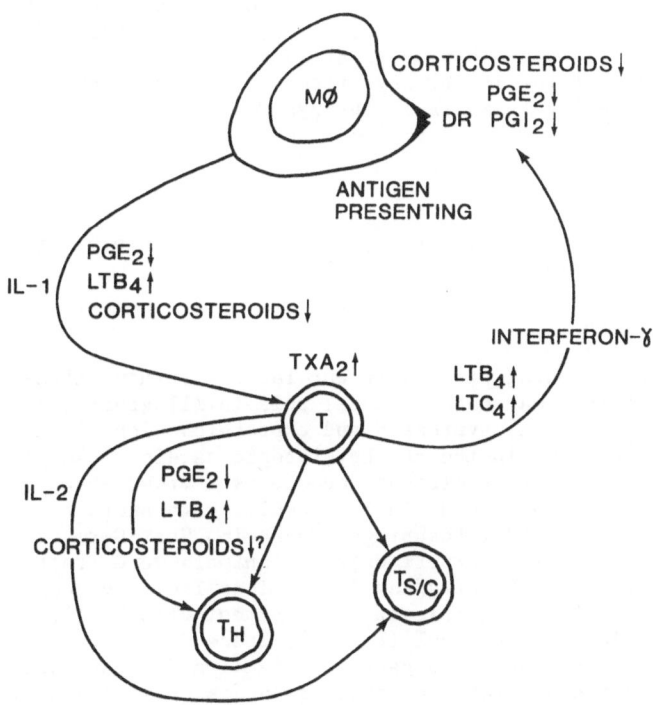

Figure 2. Modulation of macrophage-T-lymphocyte interaction by corticosteroids and eicosanoids.

antigen presenting macrophage. The DR antigen expression can be modulated by the cyclo-oxygenase products and also by gamma-interferon[27,28]. Unanue and coworkers[2,29] showed in vitro that PGE_2, PGI_2 and corticosteroids decrease Ia or DR antigen expression on macrophages. Stable TXA_2 agonists may promote antigen expression but this has not been described yet. Gamma-interferon, enhances the immune response by increasing DR antigen expression by macrophages and endothelial cells. Leukotriene C_4 and D_4 both increase the release of gamma-interferon from lymphocytes[27]. Since corticosteroids and CsA exert their effects through inhibition of IL1 and IL2 it might be possible to replace corticosteroids or decrease the dose of CsA or azathioprine in the immunosuppressive regimen with drugs that prevent the formation of the pro-rejection eicosanoids (TXA_2, LTB_4, LTC_4, LTD_4, and LTE_4) or promote the synthesis of the anti-rejection compounds (PGE_2, PGD_2, 6-keto PGE_1 and PGI_2 or their analogues[30,31].

In addition to the effect of the dihydroxyeicosanoids such as LTB_4, further studies have demonstrated synthesis of trihydroxy eicosanoids, the lipoxins, which are derived from the 15 as well as the 5 lipoxygenase pathway. These lipoxins are reported to inhibit the action of natural killer cells[32], and in lower doses to promote natural killer cell activity[33]. A biological role for the lipoxins is disputed since meaurable amounts of lipoxins can only be formed by adding 15-HPETE to granulocytes. The eosinophils are the most likely cell to produce biological important amounts of lipoxins due to the activity of its 15-lipoxygenase but lipoxins have not yet been demonstrated to be released from eosinophils.

Recently PAF has been implicated as one of the mediators of allograft rejection[10]. The mechanisms are unknown. PAF has direct effects on cells and tissues but it also stimulates arachidonic acid release. Depending on the cell type this latter effect of PAF leads to leukotrienes, prostaglandins or thromboxane synthesis. A stimulatory effect of PAF on lymphocyte proliferation would be expected. This is not the case with in vitro experiments, where inhibition has been described[34]. This inhibition is attenuated in vitro with indomathacin indicating that the inhibitory effect of PAF on lymphocyte proliferation may be indirect through PGE_2.

IN VIVO EXPERIMENTAL STUDIES OF INVOLVEMENT OF LIPID MEDIATORS IN CELL MEDIATED REJECTION

Thromboxane A_2

Experimental studies in our laboratory have already revealed that thromboxane may have a causal role in allograft rejection. Interdiction of TXA_2 synthesis and expression significantly improved allograft survival in the rat heterotopic cardiac transplant model where Lewis rats are recipients and Lewis x Brown-Norway F_1 are donors[9]. We have obtained similar results with another thromboxane antagonist (BM 13,505, Boehringer-Mannheim, West-Germany, 15 mg/kg/day i.m.). As in previous experiments the animals were treated with low-dose of CsA (0.5 mg/kg/day i.m.). This low dose does not prolong allograft survival significantly. Treatment with BM 13,505 prolonged cardiac allograft survival (fig. 3). These data imply that thromboxane contributes in part to rejection. Furthermore the results support the proposition that thromboxane is an indicator with a causal role.

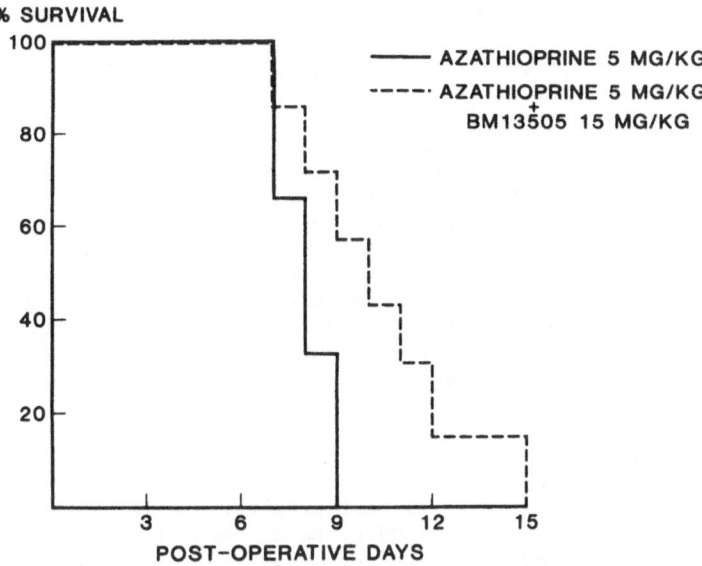

Figure 3.   The effect of azathioprine and azathioprine + a thromboxane A_2 antagonist (BM 13505) on cardiac allograft survival in rats

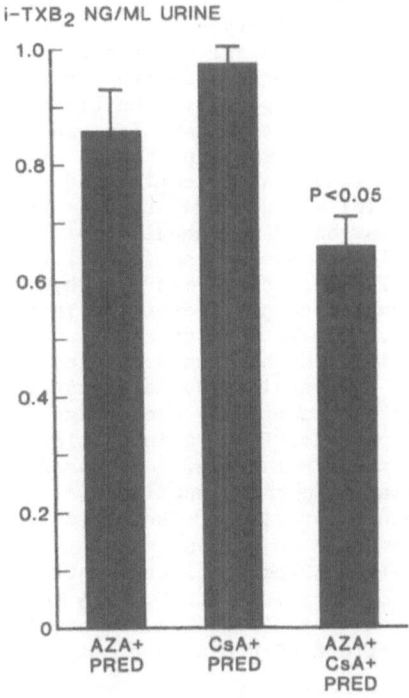

Figure 4.   Urine i-TXB$_2$ values in three groups of kidney transplant patients treated with either azathioprine and prednisone (AZA+PRED) or cyclosporin A and prednisone (CsA+PRED) or azathioprine, cyclosporin A and prednisone (AZA+CsA+PRED). S.E.M. is indicated on the top of the bars.

Leukotrienes

The in vitro studies showing that leukotrienes promote IL1 and IL2 formation as well as IL2 receptor expression, support the thesis that drugs which prevent the formation of leukotrienes (5-lipoxygenase inhibitors) might have the same effect as corticosteroids. The vascular properties of the leukotrienes, namely increased vascular permeability suggest that these products may play a role in the edema linked to rejection. We have found that three different 5-lipoxygenase inhibitors in combination with either low dose CsA or azathioprine will prolong allograft survival to the same extent as the combination with prednisone[35]. In animal models corticosteroid administered alone does not prolong organ allograft survival, nor does non-steroidal anti-inflammatory drugs[36] with the exception of aspirin, which probably acts by other mechanisms.

Prostaglandins

PGE_2, PGD_2 and PGI_2 all have properties which attenuate cell mediated rejection. Numerous studies have shown attenuation of immune responses. A common mechanism whereby prostaglandins may exert their effect is by increasing intracellular cyclic adenosine monophosphate (cAMP) and thereby lowering intracellular calcium.

One of the first descriptions of the use of eicosanoids in transplant rejection was in 1972 by Quagliata et al[37] who showed that PGE treatment prolonged skin allograft survival in mice. This finding was confirmed four years later by Anderson et al.[38]. Then the first studies emerged suggesting the possibility of manipulating transplant survival with polyunsaturated fatty acids[39]. These early experimental studies were performed with skin all ografts which are different to organ allografts in that the former requires growth of a capillary network before the tissue is exposed to the blood elements of the host whereas in organ allografts a large vascular bed of the foreign tissue is perfused immediately. The first implication of eicosanoid association with organ transplant rejection was published 15 years ago and since than only a limited number of studies have been performed.

The first suggestion of a mechanism for the effect of prostaglandins on the immune system was in 1971 when Franks et al.[40] showed that the PGE's increase cAMP in thymic lymphocytes and thereby inhibit lymphocyte proliferation. This was confirmed by other investigators. PGI_2 is more potent than PGE_2, but PGE_2 may be more relevant since the human macrophages produce much more PGE_2 than PGI_2[41]. Few studies have been carried out with PGD_2, which is however synthesized by macrophages in substantial amounts. A further property of PGE_2 is inhibition of lymphocyte migration[42]. The non-immune related effect, vasodilation may be important in maintaining allograft function during rejection. The platelet anti-aggregatory activity of these prostaglandins may also become important during rejection, when platelets accumulate in the organ. Prolongation of skin allograft survival by PGE_1[38] has been extended to organ allografts[43]. PGI_2 reverses rejection in kidney transplant patients. Further a PGI_2 analogue, iloprost, exerted a synergistic effect with CsA in a rat cardiac transplant model[44].

Platelet Activating Factor

The implication of PAF in transplant rejection is a recent discovery. The first association was found in hyperacute rejection in rabbits[45]. Hyperacute rejection is a humoral—mediated event caused by preformed antibodies and involves complement activation. Histologically, hyperacute rejection is dominated by intravascular platelet aggregation. This type of rejection is now rarely seen in patients receiving organ transplants. Most patients experience acute cell mediated rejection.

This hyperacute study was followed by experiments in cell mediated rejection where a PAF antagonist (BN52021, Beaufour Institute, Plessis-Robinson, France) acted synergistically with CsA and azathioprine in prolonging cardiac allograft survival in a rat transplant model of acute rejection[10]. PAF antagonists, L 652,731 (Merck, Sharp and Dohme, Rahway, New Jersey, USA), and RP 48740 (Rhone-Poulenc, France) also exerted synergistic effects with a low dose of CsA, but not with azathioprine. The mechanisms are unknown but likely to be similar to CsA due to the consistant finding of a synergistic effect with this compound. These latter experiments tend to confirm the conclusion that BN 52021 was prolonging graft survival by blocking PAF expression.

Urinary i-TXB$_2$ as an Indicator of Rejection

The primary clinical thrust of our work has been directed to the use of urinary immunoreactive TXB$_2$ as an indicator of renal and cardiac allograft rejection.[6,7,11,46,47]. Urinary i-TXB$_2$ is an early indicator of rejection in renal transplant patients[6,7]. The false positive urine i-TXB$_2$ values observed in kidney transplant patients are due to gross hematuria, surgery, deep vein thrombosis[6] and myocardial infarction[48]. Pneumonia and urinary tract infection do not cause an increase in urine i-TXB$_2$.

URINE I-TXB$_2$ MAY INDICATE DEGREE OF IMMUNOSUPPRESSION IN TRANSPLANT PATIENTS

The effect of immunosuppressants on urine i-TXB$_2$ of kidney transplant patients was evaluated in a recent study [49] by following daily urine i-TXB$_2$ values from the second post-transplant day. Urinary i-TXB$_2$ was determined in 23 kidney transplant patients treated with either CsA and prednisone (N=10), or with azathioprine and prednisone (N=15) or therapy with all three drugs (triple therapy) (N=9). The patients were all within 6 weeks following transplantation and were not undergoing rejection. No difference was seen in urine i-TXB$_2$ values between the group treated with azathioprine and prednisone and the group treated with CsA and prednisone (fig 4). However the patients receiving triple therapy exhibited a significantly (p < 0.05) lower urine i-TXB$_2$. The lower i-TXB$_2$ value was not caused by corticosteroids since all three groups received the same amount of corticosteroids. Furthermore in the triple therapy group the doses of azathioprine and CsA were less than the doses used in the groups receiving either azathiprine and prednisone or CsA and prednisone.

The lower urine i-TXB$_2$ of the patients on triple therapy may reflect the degree of immunosupression and offer an objective measurement of the state of immunosuppression in each individual patient.

Studies in rats revealed that even moderate doses of CsA increase the excretion of urinary i-TXB$_2$[50]. This is not the case in kidney transplant patients where steroids may be suppressing the increased i-TXB$_2$ excretion seen in in the rat.

CONCLUSION

The elective immunosuppression of transplant patients and the side effects occuring from such a treatment relate to some of the findings in trauma and burn patients. The experimental models for transplantation have been used to explore The immunomodulatory role of the lipid mediators in the rejection process. These findings as well as in vitro data support the suggestions that prostaglandins which elevate cyclic AMP attenuate rejection by inhibiting the immune response. In contrast, TXA$_2$ and the leukotrienes act as pro-rejection compounds in promoting both mediator release and initiating lymphocyte proliferation by increasing cytosolic calcium and preventing an increase in cAMP. In addition these products have vascular effects which are similar to rejection related events such as edema and decreased blood flow to the allograft.

Several strategies may be used to decrease the immune response of transplant recipients and thereby promote graft survival. Firstly PGE$_2$, PGD$_2$ & PGI$_2$ synthesis or expression may be increased, secondly, TXA$_2$ synthesis or expression may be inhibited, thirdly, inhibition of leukotriene synthesis and expression may also be inhibited. Finally, attenuation of thromboxane and leukotrienes may be obtained with use of PAF antagonists since PAF releases arachidonate. Our studies in both experimental cardiac allograft models and renal and cardiac transplant patients are in accord with this hypothesis.

Acknowledgement: This work was supported by grants from American Heart Association 84-1147, National Institute of Health, HL32319, and HL31241.

REFERENCES

1.   Antonacci, A.C., Good, R.A., and Gupta, S.: T-cell subpopulations following thermal injury. Surgery, Gynecology and Obstetrics 155:1 (1982).

2.   Snyder, D.S., and Unanue, E.R.: Corticosteroids inhibit murine macrophage Ia expression and interleukin 1 production. J. Immunol.129:1803 (1982).

3.   Kronke, M., Leonard, W.J., Depper, J. et al.: Cyclosporin A inhibits T-cell growth factor gene expression at the level mRNA transcription. Proc. Nat. Acad. Sci. 81:5214 (1984).

4.   Braquet, M, Lavaud, P., Dormont, D. et al.: Leukocytic function in burn-injured patients. Prostaglandins 29:747 (1985).

5.   Drath, D.B., and Kahan B.D.: Phagocytic cell function in response to immunosuppressive therapy. Arch. Surg. 119:156 (1984).

6.   Foegh, M.L., Alijani, M.R., Helfrich, G.B., and Ramwell, P.W.: Eicosanoids and organ transplantation. Ann. Clin. Res. 16:318 (1984).

7.   Foegh, M.L., Khirabadi, B.S., Rowles, J.R., Braquet, P., and Ramwell, P.W.: Inhibition of PAF and leukotriene expression in acute cardiac allograft rejection in rats. Pharmacol. Res. Commun.

8.   Foegh, M.L., Khirabadi, B.S., Braquet, P., and Ramwell,

P.W.: Prolongation of cardiac allograft survival with BN 52021, a specific antagonist of platelet-activating factor. Transplantation 42:86 (1986).

9.   Foegh ML, Khirabadi BS, and Ramwell PW: Prolongation of experimental cardiac allograft survival with thromboxane related drugs. Transplantation 40:124 (1985).

10.   Foegh, M.L., Cooley, C., Helfrich, G.B. e t al.: Urine i-TXB$_2$ in renal allograft rejection. Lancet 2:431 (1981).

11.   Steinhauer, H.B., Wilms, H, Ruther, M., and Schollmeyer, P.: Clinical experience with urine Thromboxane B$_2$ in acute renal allograft rejection. Transpl. Proc. 18 (suppl. 4):98 (1986).

12.   Hirata, F.,Schiffman, E, Venkatasubramanian, K., Solomon, D., and Axelrod, J.: A phosphplopase A$_2$ inhibitory protein in rabbit neutrophils induced by glucocorticoids.Proc. Nat. Acad. Sci. 77:2533 (1980).

13.   Calne, R.Y.: Inhibition of the rejection of renal homografts in dogs by purine analogues. Transplant. Bull. 28:65 (1961).

14.   Neild, G., Rocchi, G., Imberti, L., et al.: Effect of cyclosporine on prostacyclin synthesis by vascular tissue. Thromb. Res. 32:373 (1983).

15.   Fan, T.-P.D., and Lewis, G.P.: Mechanism of cyclosporin A induced inhibition of prostacyclin synthesis by macrophages. Prostaglandins 30:735 (1985).

16.   Farrar, W.L., and Humes, J.L.: The role of arachidonic acid metabolism in the activities of interleukin 1 and 2. J. Immunol. 135:1153 (1985).

17.   Rapaport, R.S., and Dodge, G.R.: Prostaglandin inhibits the production of human interleukin 2. J. Exp. Med. 155:943 (1982).

18.   Dinarello, C.A., Monnoy, S.O., and Rosenwasser, L.J.: Role of arachidonate metabolism in the immunoregulatory function of human leukocytic pyrogen/lymphocyte activating factor/interleukin 1. J. Immunol. 130:890 (1983).

19.   Rola-Pleszczynski, M., and Gagnon, L.: Natural killer cell function modulated by leukotriene B$_4$: mechanism of action. Transplant. Proc. 18 (suppl. 4):44 (1986).

20.   Goodwin, J.S.: Role of leukotriene B$_4$ in T-cell activation. Transpl. Proc. 18 (suppl. 4):49 (1986).

21.   Payan, D.G., and Goetzl, E.J.: Specific supppression of human T-lymphocyte function by leukotriene B$_4$. J. Immunnol. 131:551 (1983).

22.   Goodwinn, J.S., Atluru, D., Sierakowski, S., and Lianos, E.A.: Mechanism of action of glucocorticoids. Inhibition of T-cell proliferation and interleukin 2 production by hydrocortisone is reversed by leukotriene B$_4$. J. Clin. Invest. 77:1244 (1986).

23.   Payan, D.G., and Goetzl, E.J.: The dependence of human T-lymphocyte migration of the 5-lipoxygenation of endogenous arachidonic acid. J. Clin. Immunol. 1:266 (1981).

24.   Kelly, J.P., Johnson, M.C., Parker,C.W., : Effect of inhibitors of arachidonic acid metabolism on mitogenesis in human lymphocytes: possible role of thromboxanes and products of the lipoxygeenase pathway. J. Immunol. 122:1572 (1979).

25.   Foegh, M.L., Maddox, Y.T., and Ramwell, P.W.: Human peritoneal eosiniphils and formation of arachidonate cyclooxygenase products. Scand. J. Immunol. 23:599 (1986).

26.   Ceuppens, J.L., Vertessen, S., Deckmyn, H., and Vermylen, J.: Effect of thromboxane A$_2$ on lymphocyte proliferation. Cell. Immunol. 90:458 (1985).

27.   Johnson, H.M., and Torres, B.A.: Leukotrienes: positive signals for regulation of gamma-interferon production. J. Immunol. 132:413 (1984).

28.   Snyder, D.S., Beller, D.I., and Unanue, E.R.: Prostaglandins modulate macrophage Ia expression. Nature 299:163 (1982).

29.   Snyder, D.S., and Unanue, E.R.: Corticosteroids inhibit murine macrophage Ia expression and interleukin 1 production. J. Immunol. 129:1803 (1982).

30.   Basham, T.Y., and Merigan, T.C.: Recombinant interferon-gamma increases HLA-DR synthesis and expression. J. Immunol. 130:1492 (1983).

31.   Foegh, M.L., Khirabadi, B.S., and Ramwell, P.W.: Improved ratr cardiac allograft survival with non-steroidal pharmacologic agents related to eicosanoids. Transpl. Proc. In press.

32.   Serhan, C.N., Hamberg, M., Ramstedt, U., and Samuelsson, B.: Lipoxins: stereochemistry, biosynthesis and biological activities. Adv. Prostagl. Thrombox. Leukotr. Res. 16:83 (1986).

33.   Rokash, J., and Fitzsimmons, B.: Lipoxins do they have a biological role? Transpl Proc 18 (suppl. 4):7 (1986).

34.   Foegh, M.L., Hartmann, D.-P., Rowles, J.R. et al.: Leukotrienes, thromboxane and PAF in organ transplantation. Adv. Prostagl. Thrombox. Leukotr. Res. 17:144 (1987).

35.   Foegh, M.L., Braquet, P, Khirabadi, B.S., Rowles, J.R., and Ramwell, P.W.: Inhibition of PAF and leukotriene expression in acute cardiac allograft rejection in rats. Biochem. Res. Comm. 18:127 (1986).

36.   Foegh, M.L., Alijani, M.R., Helfrich, G.B., Khirabadi, B.S., Schreiner, G.E., and Ramwell, P.W.: In: Prostaglandins Leukotrienes Lipoxins, Bailey M (ed) New York, Plenum Press, 1985.

37.   Quagliata, F., Lawrence, V.J.W., and Philips-Quagliata, J.M.: Prostaglandin E_1 as a regulator of lymphocyte function. Cell. Immunol. 6:457 (1973).

38.   Anderson C.B., Jaffe, B.M., and Graff, R.J.: Prolongation of murine skin allografts by prostaglandin E. Transplantation 23:244 (1977).

39.   Mertin, J., and Hunt, R.: Influence of polyunsaturated fatty acids on survival of skin allografts and tumor incidence in mice. Proc.Natl. Acad. Sci. 73:928 (1976).

40.   Franks, D.J., MacManus,J.P., and Whitfield, J.F.: The effect of prostaglandins on cyclic AMP production and cell proliferation in thymic lymphocytes. Biochem. Biophys. Res. Commun. 44:1177 (1971).

41. Maddox, Y.T., Foegh, M.L., Zeligs, B., Bellanti, J. Zmudka, M., Ramwell, P.W.: A routine source of human peritoneal macrophages. Scand. J. Immunol. 19:23 (1984).

42.   Van Epps, D.: Suppression of human lymphocyte migration by PGE_2 Inflammation 5:81 (1981).

43.   Strom, T.B., Carpenter, C.B.: Prostaglandin as an effective antirejection therapy in rat renal allograft recipients. Transplantation 35:279 (1983).

44.   Rowles, J.R., Foegh, M.L., Khirabadi, B.S., and Ramwell, P.W.:The synergistic effect of cyclosporine and iloprost on survival of rat cardiac allografts. Transplantation 42:94 (1986).

45.   Ito, S., Camussi, G., Tetta, C., Mílgrom, F., and Andres, G.: Hyperacute renal allograft rejection in the rabbit. Lab. Invest. 51:148 (1984).

46.   Khirabadi, B.S., Foegh, M.L., and Ramwell, P.W.: Urine immunoreactive thromboxane B_2 in rat cardiac allograft rejection.

Transplantation 39:6 (1985).
47.   Foegh, M.L., Alijani, M.R., Helfrich, G.B., et al.:
Thromboxane and leukotrienes in clinical and experimental
transplant rejection. Adv. Prostagl. Thrombox. Leukotr. Res.
13:209 (1985).
48.   Foegh, M.L., Eliasen, K., Johansen, S., Helfrich, B.G.,
and Ramwell, P.W.: Coronary artery thrombosis and elevated urine
immunoreactive thromboxane B_2. Prostaglandins 32:781 (1986).
49.   Foegh, M.L., Alijani, M.R., Helfrich, G.B., Khirabadi,
B.S., Lim, K., and Ramwell, P.W.: Lipid mediators in organ
transplantation. Transplant. Proc. 18 (suppl. 4):20 (1986).
50.   Kawagushi, A., Goldman, M.H., Shapiro, R., et al.:
Increase in urinary thromboxane B_2 in rats caused by
cyclosporine. Transplantation 40:214 (1985).

CHARACTERIZATION OF THE INDUCTION OF HUMAN INTERLEUKIN 1 BY ENDOTOXINS

J-M. Cavaillon and N. Haeffner-Cavaillon
Unité d'Immuno-Allergie, Institut Pasteur, 28 rue du Dr. Roux
75624 Paris Cedex 15, France. INSERM U28, Hopital Broussais
96 rue Didot, 75674 Paris Cedex 14, France

INTRODUCTION

Endotoxins isolated from Gram negative bacteria can induce profound physiological modifications and trigger the immune system. Endotoxins are lipopolysaccharides (LPS) which consist of a polysaccharide chain (PS) co-valently linked to a hydrophobic moiety (Lipid A) via a 2-keto-3-deoxy-octonic acid residue(s) (KDO). The classical endotoxic activities (e.g. local Shwartzman reaction, pyrogenicity, lethality...) as well as the immunological effects of endotoxins (adjuvanticity, mitogenicity, polyclonal activation of immunoglobulin secreting cells...) are mediated by the Lipid A (1,2). However, it has also been shown that the polysaccharide moiety of some LPS act as an adjuvant (3,4); protect mice against lethal irradiation (5,6); induce the production of colony stimulating factors (5-8); induce a "transfer tumor resistance" in serum of BCG-infected mice (9); and mediate B-cell proliferation (10). LPS can also stimulate complex positive and negative immunoregulatory circuits in which monocytes/macrophages play a central role (10-14).

We have previously shown (15) that Bordetella pertussis endotoxin, as well as its isolated polysaccharide chain (PS), induces a polyclonal activation of rabbit immunoglobulin-secreting cells and we have established that such polyclonal activation was dependent upon the presence of macrophages. We have also reported that the same endotoxin binds to rabbit macrophages (16). The binding was saturable, specific, dose-dependent and reversible on peritoneal macrophages, whereas only non-specific binding was detected on rabbit alveolar macrophages. The specific binding was mediated by the polysaccharide moiety. Extension of this work to human monocytes and mouse macrophages using radiolabeled B. pertussis and Neisseria meningitidis LPS led to similar conclusions (17,18). Since interleukin 1 (IL 1), originally described in 1972 by Gery and Waksman (19) as Lymphocyte Activating Factor (LAF), is involved in many of the activities mediated by endotoxin-stimulated macrophages (20-22), we have investigated the interactions between LPS and monocytes which lead to IL 1 release.

PARAMETERS OF INTERLEUKIN 1 INDUCTION BY ENDOTOXINS

The presence of IL 1 was assessed in the supernatants of human peripheral blood mononuclear adherent cells which were purified, cultured and stimulated with various LPS in the absence of serum. These conditions permit

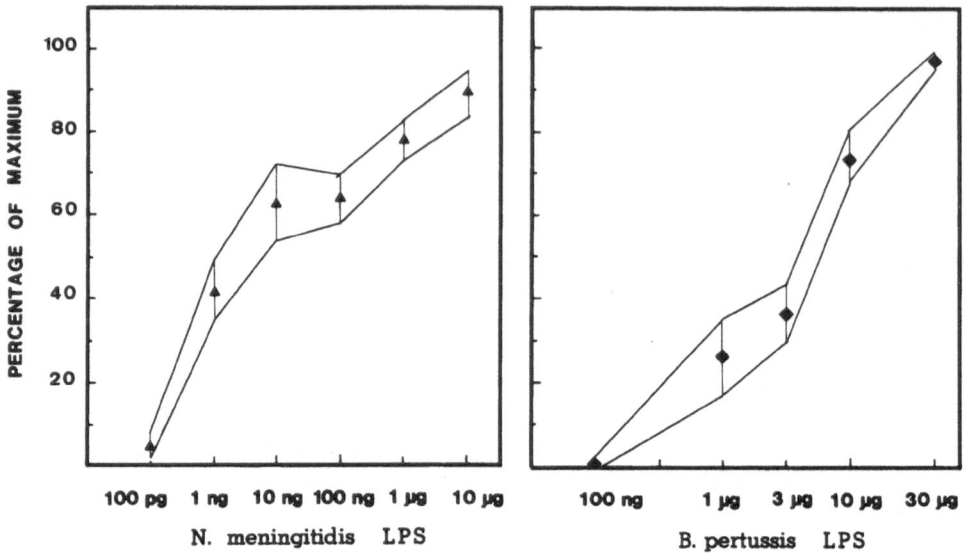

FIGURE 1. *Dose-dependent effect of Neisseria meningitidis and Bordetella pertussis endotoxins on the induction of IL 1 secretion by human monocytes. Each experimental point represents the mean of values obtained from 10 to 22 differents donors ± SE. A suspension of peripheral blood mononuclear cells (0.5ml) containing 10^6 NSE+ cells/ml were led to adhere in 24 wells plastic dishes in the absence of serum for 1 h at 37°C. After washing, the adherent cells were cultured in 0.5 ml RPMI for 24 h in the absence of serum, and in the absence or in the presence of IL 1 inducers. Cell supernatants were harvested, centrifuged and tested for their capacity to induce C3H/HeJ thymocytes proliferation in the presence of suboptimal doses of Con A ; 1:10 diluted supernatants were added to 0.75 x 10^6 thymocytes, in the presence of 4% FCS, 5 x 10^{-5} M 2-mercaptoethanol and 0.075 μg Con A per well. The cultures were maintained 72 h at 37°C and (^{3}H)-thymidine was added 7 h before the end of the culture.*

one to avoid the secretion of "spontaneous" or "constitutive" IL 1 which has been reported by various authors (23,24). The IL 1 activity was detected by the conventional co-mitogenic assay on C3H/HeJ mice thymocytes in the presence of suboptimal doses of concanavalin A (25). We noticed that the capacity of endotoxins to induce IL 1 secretion by human monocytes was dependent upon the origin of the LPS. The most characteristic example can be observed when comparing the dose-response relationship of IL 1 secretion induced by N. meningitidis LPS and B. pertussis LPS (Figure 1). Similar observations were reported by Newton (26) with different LPS. Furthemore, individual variations exist in monocyte sensitivity to LPS depending upon the donors.

Maximal IL 1 secretion was detected in supernatants of adherent cells maintained 12 to 24 h in culture in the presence of endotoxin (Figure 2). However, the presence of lipopolysaccharide throughout the culture period is not required since IL 1 activity was present in 24 h-supernatants when LPS was removed after two hours of culture (Table 1). Human monocytes lose their capacity to secrete IL 1 when maintained in culture more than 24 h prior to LPS stimulation (Table 1). It was shown by several authors (27-29) that the capacity of 24 h-cultured monocytes to secrete IL 1 upon further LPS stimulation can be restored by the addition of interferon. Interferon acts also by itself (30) or synergistically with LPS (31,32) in the induction of IL 1 secretion by human monocytes and alveolar macrophages.

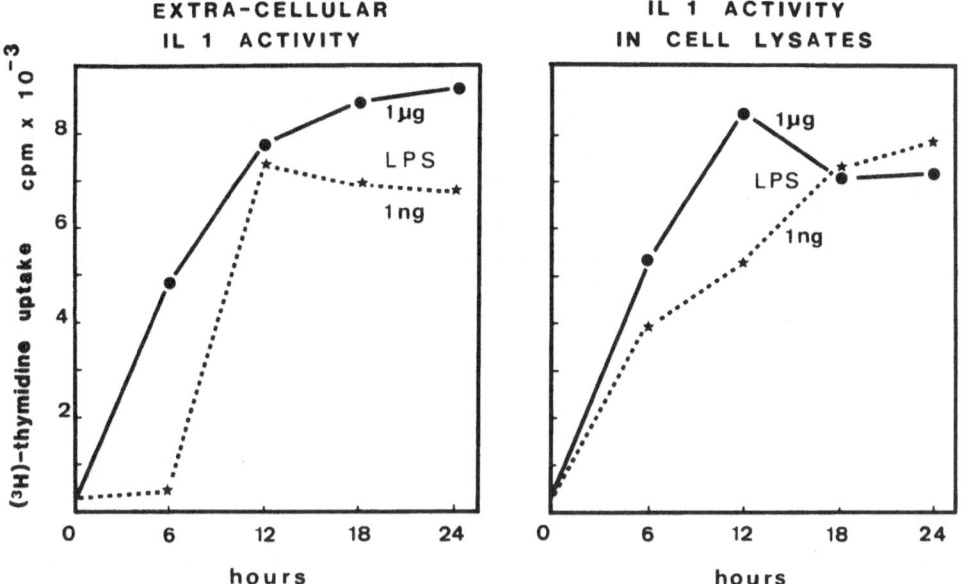

FIGURE 2 . *Kinetics of the induction of IL 1 secretion (extra-cellular IL 1 activity) and IL 1 synthesis (IL 1 activity in cell lysates) by human mono-cytes stimulated by N. meningitidis LPS. Cells were cultured in the presence of 1 ng to 1 µg LPS from 6 to 24 h; after each period of time, supernatants were harvested and 0.5 ml of fresh medium was added to the adherent cells and the cultures were frozen; after freezing/thawing (3x), the supernatants were collected and centrifuged. The supernatants were assayed for IL 1 acti-vity as described in figure 1.*

A dissociation between the production of cytoplasmic IL 1 and extracellular release of IL 1 by murine macrophages and human monocytes has been reported for various IL 1 inducers (24,33). In agreement with Newton's observations (26), we report that low doses of N. meningitidis LPS (10 - 100 pg/assay), which are unable to induce the release of IL 1, have the capacity to induce the synthesis of cytoplasmic IL 1, as judged by the IL 1 activity found in

TABLE 1. Interleukin 1 activity in the supernatants of human monocytes stimulated by LPS during different period of time

PRESENCE OF LPS	TIME OF SUPERNATANT HARVESTING	IL 1 SECRETION INDUCED BY	
		N. meningitidis LPS (1 µg)	B. pertussis LPS (10 µg)
0 — 2 h	2 h	3015 ± 497 [a]	2038 ± 263
0 — 2 h	24 h	14953 ± 2306	19359 ± 1540
0 — 24 h	24 h	13782 ± 1295	18899 ± 340
0 — 48 h	48 h	14442 ± 1171	22330 ± 1425
24 — 48 h	48 h	3023 ± 418	6262 ± 1536

[a] The results are expressed as the mean of triplicate cultures ± SD; the background value (supernatants of unstimulated adherent cells) was 2421 ± 999

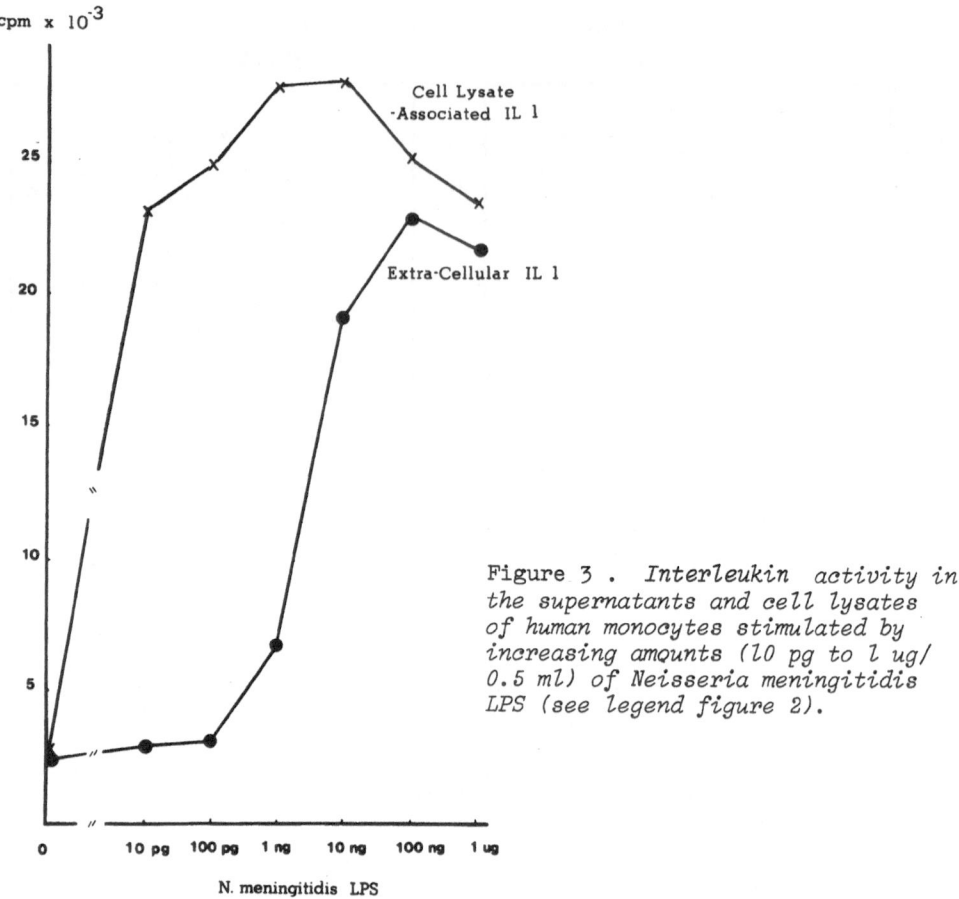

cpm x 10^{-3}

Cell Lysate
-Associated IL 1

Extra-Cellular IL 1

Figure 3 . *Interleukin activity in
the supernatants and cell lysates
of human monocytes stimulated by
increasing amounts (10 pg to 1 ug/
0.5 ml) of Neisseria meningitidis
LPS (see legend figure 2).*

0   10 pg   100 pg   1 ng   10 ng   100 ng   1 ug

N. meningitidis LPS

cell lysates (Figure 3). With high doses of LPS, the kinetics of appearance
of extracellular IL 1 and cell-lysate associated IL 1 are quite similar,
whereas with low doses of LPS, no IL 1 activity was detected in the cell
supernatant after 6 h of culture, but an IL 1 activity was already detectable in cell lysate (Figure 2). Additional information has been provided
by Bayne et al. (34) who showed that 93% of LPS-stimulated human monocytes
contain cytoplasmic IL 1, and by Matsushima et al. (35) who described that
the intracellular IL 1 was mainly associated with the cytosol. Because of
the tremendous sensitivity of the monocytes to LPS with respect to cytoplasmic associated IL 1 (IL 1 can be detected, depending upon the donors, in
cell lysates of cultured monocytes stimulated by 0.1 pg of LPS) , it should
be stressed that this assay is meaningless when used to assess the IL 1 inducing capacity of natural components which may be contaminated by such low
LPS contaminations which are impossible to detect by conventional assays.

INVOLVEMENT OF THE POLYSACCHARIDE MOIETY OF ENDOTOXIN IN IL 1 INDUCTION

As mentioned previously, we have shown that the polysaccharide (PS)
moiety of B. pertussis LPS was able to induce a macrophage-dependent polyclonal activation of immunoglobulin-secreting cells (15) and to mediate the
specific binding of lipopolysaccharide to mononuclear phagocyte plasma membrane (16-18). As discussed in the introduction, IL 1 is thought to play a
role in the LPS-induced polyclonal activation, therefore, we investigated
if the isolated PS was able to induce IL 1 secretion. Indeed, we were able

to show that <u>B. pertussis</u>, as well as <u>N. meningitidis</u> PS had the capacity
to induce the release of IL 1 by human monocytes (36,37). In this context,
it is noteworthy to remember that isolated PS derived from LPS are involved
in various biological effects (3-10) including : adjuvanticity; protection
against irradiation; B-cell proliferation and differentiation; and protection against tumors. These different activities correlate with the capacity
of the PS to induce IL 1 secretion, since it has been shown that interleukin 1 is involved in adjuvanticity (38), radioprotection (39), participates
by itself, or synergistically with B-cell growth factors and B-cell differentiation factors in the maturation and the activation of B-lymphocytes (40-
43) and is able to inhibit the growth of certain tumor cell lines (44-47).

In preliminary studies, we observed that the activity of isolated PS
was dependent upon the hydrolysis used to release it from endotoxin. Thus
PS isolated after hydrolysis of the LPS in acetate buffer (48) was active
(Table 2), whereas those obtained after hydrolysis in acetic acid (1%) or
mineral acid (HCl 0.25N) were devoid of IL 1-inducing capacity. The lack of
activity correlated with a difference in response to the thiobarbituric acid
assay. This assay detects the presence of the KDO molecule at the reducing
end of the polysaccharide. Neither synthetic KDO residues alone (49), nor
synthetic KDO residues substituted in position 5 (the position which is
substituted by a heptose residue in almost all LPS from Gram negative bacteria) by a methyl group or a benzyl group were able to induce IL 1 secretion.

To further investigate the determinant (s) required for IL 1 induction,
we examined the role of the residues present at the reducing end of different LPS-derived polysaccharides. Accordingly, Dr. Martine Caroff (UA 1116,
CNRS, Institut de Biochimie, Orsay) isolated a fragment containing as the
terminal reducing sugar a KDO residue after hydrolysis of "deaminated" <u>B.</u>
<u>pertussis</u> endotoxin in acetate buffer (fragment B). After acidic methanoly-
sis, a second fragment was isolated containing as the reducing sugar the
methyl-ketoside-methylester-KDO residue (see schematic pathway on figure 4).
Fragment B was the only active fragment which induced in a dose dependent
fashion, a level of IL 1 secretion similar to that reached with the entire
PS, suggesting that the carboxyl group of the KDO residue was necessary for
the induction of IL 1. This hypothesis was corroborated by the fact that
after alkaline treatment of the fragment obtained after methanolysis [treat-
ment which restores the carboxyl group], the oligosaccharide obtained was
as active as the fragment B. Similar findings were obtained with a trisaccharide (Hep-Hep-KDO) isolated from <u>Salmonella minnesota</u> Rd1 LPS by Dr.

TABLE 2. IL 1 INDUCTION BY B. PERTUSSIS LPS AND ITS DERIVED
         ENTIRE POLYSACCHARIDE MOIETY AND PENTASACCHARIDE
         REFERRED TO AS FRAGMENT B

Experiment I		Experiment II	
Inducers	cpm ± SD	Inducers	cpm ± SD
none	2691 ± 611	none	2110 ± 361
LPS 10 µg	16656 ± 1215	LPS 30 µg	19513 ± 2751
30 µg	22047 ± 2828	PS 30 µg	12848 ± 1394
PS 10 µg	6615 ± 1817	Fragment B	14762 ± 4264
30 µg	19715 ± 3482	30 µg	

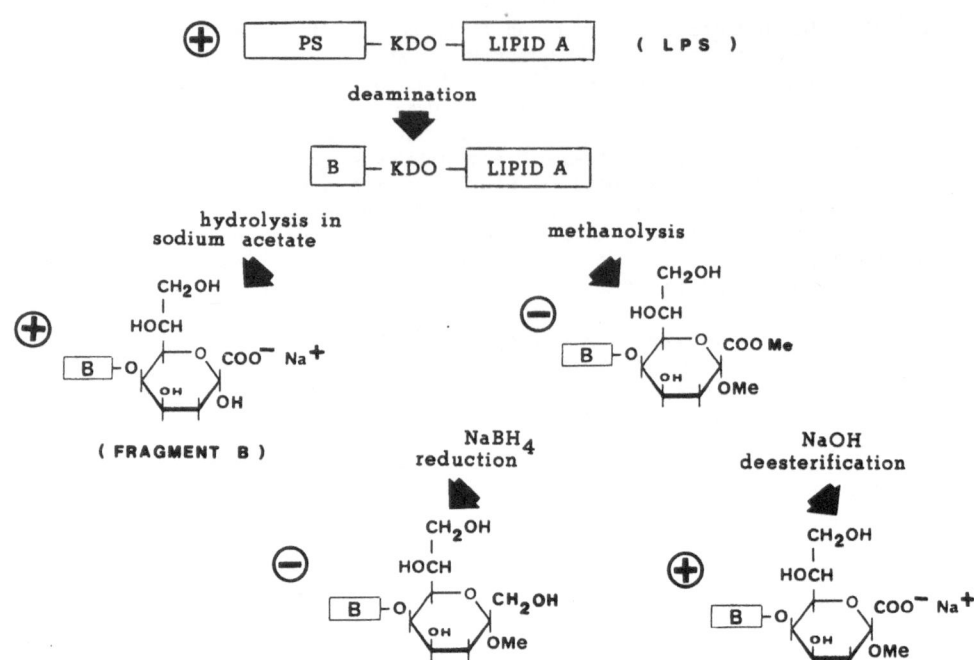

FIGURE 4. *Schematic representation of the fragmentation of the B. pertussis LPS. ⊕ and ⊖ symbols illustrate the capacity of the different fragments to induce comparable level of IL 1 secretion to that obtained with the entire polysaccharide, or the incapacity, respectively.*

Helmut Brade (Forshungsinstitut Borstel, Germany)(49) and representative of the inner core region of many enterobacterial LPS. Isolated Hep-Hep-KDO was able to induce IL 1 secretion in a dose dependent fashion (Figure 5). When the trisaccharide was treated with $NaBH_4$, the carbonyl-reduced trisaccharide was inactive, indicating a conformational requirement in the KDO molecule. Finally we reported that synthetic KDO disaccharides were able to induce IL 1 secretion, confirming univocally that determinants present in KDO residues are involved in IL 1 secretion induced by endotoxins.

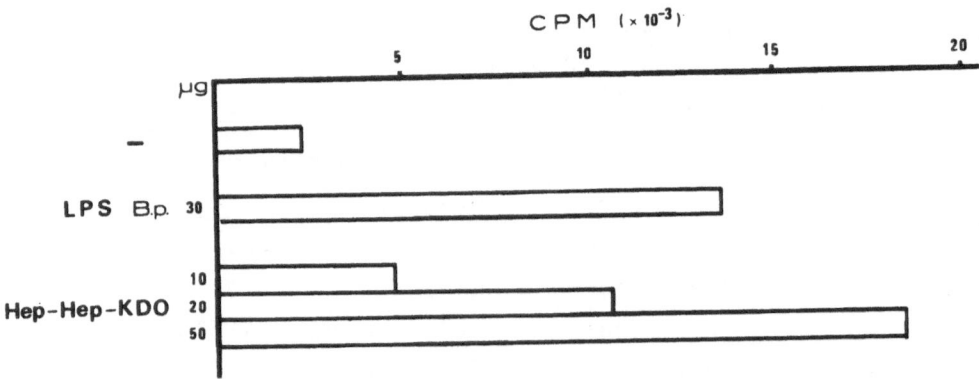

FIGURE 5. *Interleukin 1 secretion induced by the trisaccharide Hep-Hep-KDO isolated from Salmonella minnesota Rd1 mutant LPS.*

INABILITY OF ISOLATED B. PERTUSSIS AND N. MENINGITIDIS PURIFIED LIPID A
TO INDUCE INTERLEUKIN 1 SECRETION

   To further investigate the determinants of the LPS molecule involved
in the induction of IL 1 secretion, we tested the capacity of isolated
Lipid A to stimulate the release of IL 1 by human monocytes. Classical procedures used to isolate and purify Lipid A lead to Lipid A preparations
which are heterogeneous(50) and contaminated with significant amounts (1-
5%) of native and variously degraded LPS, as assessed by thin layer chromatography, gas liquid chromatography, and SDS PAGE analysis (N. Haeffner-
Cavaillon, unpublished observations). Estimation of KDO by the thiobarbituric acid assay or estimation of neutral sugars by colorimetric methods are
not sensitive enough for assessing the purity of Lipid A preparations.
Escherichia coli F515, N. meningitidis, and B. pertussis Lipid A preparations containing trace amounts of LPS, induced IL 1 secretion (Caroff et al.
submitted). However, it should be noted that in the case of B. pertussis
Lipid A, the amount of LPS detected (1-2%) did not induce IL 1 secretion
by itself. When homogeneous and purified B. pertussis Lipid A preparations
were obtained after hydrolysis in acetate sodium buffer (M. Caroff et al.
manuscript in preparation), the Lipid A contained 2 glucosamine residues,
2 phosphate groups, and 5 fatty acids; a composition which is similar to
that reported for enterobacterial derived Lipid A. When tested in biological assays, this preparation was pyrogenic, mitogenic, and induced a Shwartzman reaction, but was unable to stimulate the release of significant amounts
of IL 1 by human monocytes, at doses up to 30 µg (Figure 6). Similar results
were obtained with N. meningitidis Lipid A devoid of any LPS contamination.
One to 10 ng / assay of native N. meningitidis LPS induced significant levels
of IL 1 secretion, whereas 10 to 30 µg / assay of pure Lipid A were inactive
or had low activity depending upon the donors.

   When B. pertussis Lipid A preparations were isolated after hydrolysis
of LPS in 0.25 N. HCl, the preparations were heterogeneous and contained a

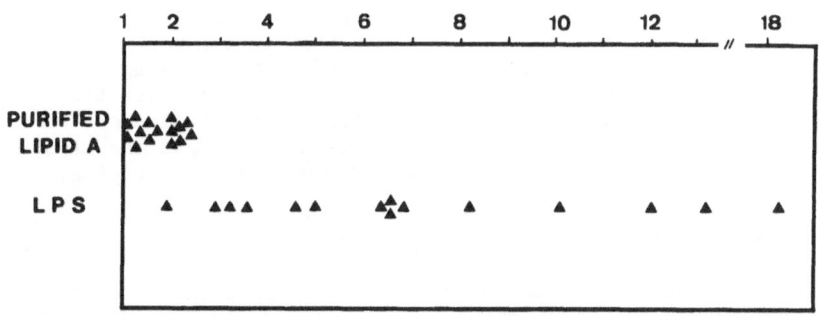

FIGURE 6. *Comparison of interleukin 1 secretion induced by 10 µg of native
Bordetella pertussis LPS and 10 µg purified Lipid A. Each symbol represents
one individual donor among the 15 tested. The results are expressed as Stimulation Index (S.I.) :*

$$S.I. = \frac{cpm \text{ in the presence of supernatants (1:10) from LPS or Lipid A stimulated human adherent cells}}{cpm \text{ in the presence of supernatants from unstimulated adherent cells}}$$

TABLE 3 . PYROGENICITY (EP) AND CAPACITY TO INDUCE LYMPHOCYTE
ACTIVATING FACTOR (LAF) OF B. PERTUSSIS LPS AND
DIFFERENT LIPID A PREPARATIONS.

Preparations	Pyrogenicity [a]	Lymphocyte Activating Factor [b]
LPS	EP ++	LAF ++
Crude Lipid A (acetate buffer)	EP ++	LAF ++
Crude Lipid A (HCl, 0.25 N)	EP ±	LAF ++
Purified Lipid A (Acetate Buffer)	EP ++	LAF -
Purified Lipid A (HCl, 0.25 N)	EP ±	LAF -

[a] pyrogenicity was measured by the standard procedure using New Zealand rabbits (49)

[b] LAF activity was measured by the standard co-mitogenic assay on C3H/HeJ thymocytes. Human monocytes were stimulated with 20 µg/ml of LPS and Lipid A

glucosamine to phosphate group ratio of 2:1 (some fatty acids were also lost during the hydrolysis). The Lipid A preparation containing trace amounts of LPS was not pyrogenic but able to induce IL 1 secretion (Table 3)(M. Caroff et al. submitted for publication). According to our results it can be proposed that different determinants of the LPS molecule are involved in the induction of endogenous pyrogen and IL 1. Similar dissociation between the induction of LAF activity and pyrogenicity was reported with muramyl dipeptide analogs (52). Recently it was demonstrated that IL 1, which was known as "endogenous pyrogen" (53-55) is not indeed the only mediator of fever, and that other factors such as alpha interferon (56,57) and Tumor Necrosis Factor (TNF)(58,59) which can be induced by endotoxins, are also involved in fever induction (60,61). Recently, it was reported by Loppnow et al. (62) that synthetic E. coli Lipid A containing 6 fatty acids, two phosphate groups and two glucosamine residues, was able to induce similar levels of IL 1 to those obtained with S. minnesota Re mutants, whereas synthetic Lipid A containing 4 fatty acids instead of 6 was a very weak IL 1 inducer. Thus the amount of fatty acids within the Lipid A might influence its activity. Moreover, it was reported that the 4'-dephosphorylated synthetic Lipid A was 10 times more active than the di-phosphorylated compound (62). Considering all the results, it can be proposed that the capacity to induce IL 1 secretion by isolated Lipid A depends upon the presence of different substituents on the β-D-glucosaminyl-(1-6)α-D-glucosamine disaccharide, and that the Lipid A moiety may not be equally involved in native LPS from various origins.

INHIBITION OF LPS-INDUCED INTERLEUKIN 1 SECRETION BY POLYMYXIN B

Polymyxin B (PMB) has been widely used to demonstrate the role of Lipid A in many biological assays. Since the lack of IL 1 inducing capacity of isolated Lipid A may be due to the loss of a component present in the native endotoxin during the hydrolysis procedure of the LPS, we further investigated the role of Lipid A in the induction of IL 1 secretion by studying the effect of PMB on native LPS-induced IL 1 secretion (63). Among the different LPS tested, PMB could only inhibit IL 1 induction by E. coli LPS and Acinetobacter calcoaceticus LPS, indicating that for these two endotoxins, the Lipid A moiety may play a role in IL 1 induction. For several other LPS (N. meningitidis, Salmonella enteritidis, S. friedenau)(100 ng/assay), PMB (2 µg/assay) had no inhibitory activity. PMB had only an inhibitory effect when the N. meningitidis was used at low concentrations (PMB ratio to LPS 1000:1). Furthermore, at a concentration of 10 µg/assay PMB acted synergistically with all LPS and at higher doses, PMB itself, probably due to its toxicity, induced IL 1 release (63). Since polymyxin B is known to react with Lipid A (64), these results in conjunction with those obtained with isolated Lipid A indicate that the involvement of the Lipid A moiety is dependent upon the origin of the endotoxins. It is noteworthy that Apte et al. (65) showed that PMB did not inhibit all LPS-induced biological activities.

ENHANCING EFFECT OF SERUM

As mentioned previously, we purified, cultured, and stimulated monocytes in culture medium without serum to avoid the so-called spontaneous release of IL 1. Since we have shown that the specific binding of radiolabeled LPS to human monocytes required, (or was increased by) the presence of serum component(s)(17), we investigated whether serum could also amplify IL 1 induction by LPS. A potentiating effect was observed with low concentrations of serum from different origins (37). It is noteworthy that an enhancing effect of serum has been reported for a wide variety of biological activities induced by LPS (66-68). The addition of very low amounts (0.1%) of human autologous serum which had no significant influence on the background values, led to the secretion of IL 1 when added at doses of LPS which by themselves were unable to induce the secretion of IL 1 (Figure 7). According to the kinetic data, the synergistic effect did not require the simultaneous presence of LPS and serum since sequential addition of serum and LPS led to IL 1 secretion similar to that of simultaneous addition (37). Recently, it was reported by Matsushima et al. (35) that plasminogen is involved in the synergistic effect of serum on LPS-induced IL 1. When tested in LPS specific binding experiments, plasminogen can not be substituted for complete serum (J-M. Cavaillon, unpublished observation). Thus, plasminogen seems to be involved in the regulation of the release of IL 1 rather than in mechanisms related to the triggering signal mediated by the specific binding of LPS to monocytes.

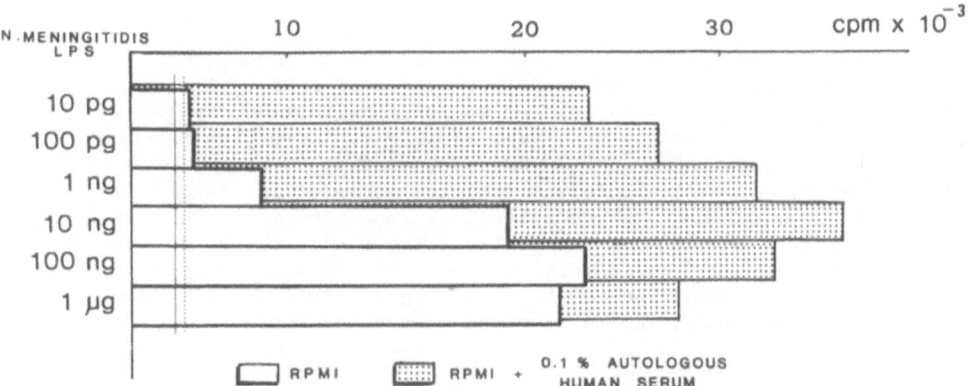

FIGURE 7. *Enhancing effect of 0.1% human serum on interleukin 1 secretion induced by increasing amounts of N. meningitidis LPS (—— and ·····represent the background values of supernatants from cultured monocytes without LPS in the absence and in the presence of 0.1% serum, respectively).*

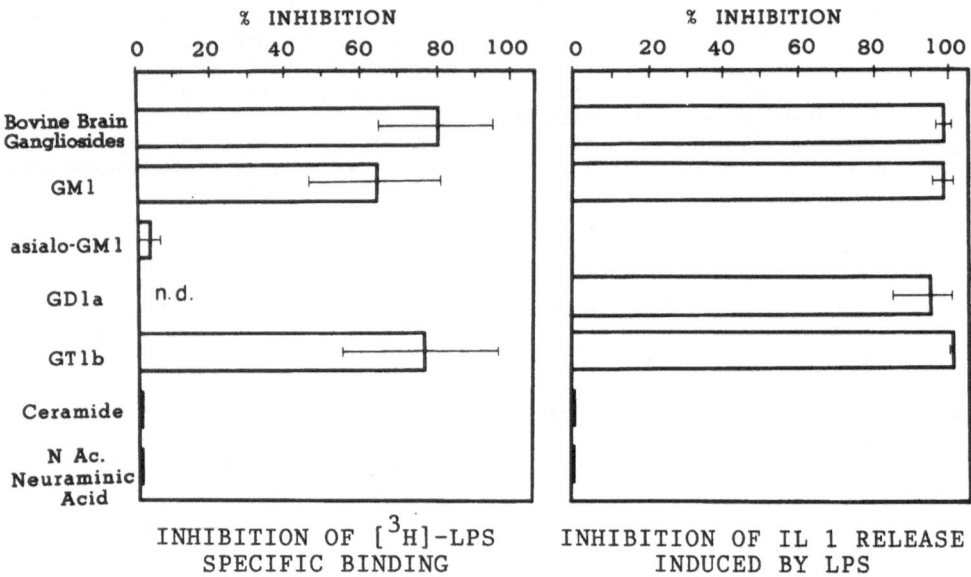

% INHIBITION   % INHIBITION

INHIBITION OF [^{3}H]-LPS   INHIBITION OF IL 1 RELEASE
SPECIFIC BINDING   INDUCED BY LPS

FIGURE 8 . *Inhibition of the N. meningitidis LPS specific binding (0.2 μg*
[^{3}H]-LPS) to human monocytes by gangliosides (5 μg), ceramide (5 μg), and
N. acetyl neuraminic acid (5 μg); and inhibiton of the N. meningitidis LPS
(10 ng) -induced IL 1 secretion by gangliosides (1 μg), ceramide (1 μg) and
N. acetyl neuraminic acid (3 μg). The presented data are the mean of the
results obtained with 3 to 7 different donors.

INHIBITION OF LPS-INDUCED INTERLEUKIN 1 SECRETION BY GANGLIOSIDES

Gangliosides which are glycosphingolipids present in the membranes of
most eukaryotic cells, are potent immunosuppressors(69-72). They have been
shown to inhibit some of the LPS-induced activation of the immune system
(73,74).We have investigated the capacity of such glycosphingolipids to
inhibit LPS-induced interleukin 1 release (75). Crude preparations, as well
as purified individual mono, di and trisialogangliosides (GM1, GD1a and
GT1b respectively) were able to abbrogate the secretion of IL 1 induced by
LPS as well as the synthesis of intracellular IL 1. Interleukin 1 induction
by other inducers, such as MDP and C3a, was unaffected by the addition of
gangliosides. Ceramide, the lipid moiety of gangliosides, failed to inhibit
the IL 1 release induced by LPS. Similar lack of inhibition was obtained
with an asialoganglioside (asialo-GM1), suggesting a crucial role of N.
acetyl neuraminic acid within the ganglioside molecule. Parallel results
were obtained when studying the ability of gangliosides to inhibit the specific binding of LPS to human monocytes (Figure 8). Altogether, these results demonstrate that the specific binding of LPS via its polysaccharide
moiety, is the prerequisite event in the signal provided by endotoxins for
the synthesis of interleukin 1 by human monocytes.

CONCLUSIONS

According to our data, we conclude that at least for B. pertussis and
N. meningitidis endotoxins, the initial event for IL 1 secretion by human
monocytes is a binding of specific determinants in the inner core region of
the endotoxin molecule to a lectin-like receptor present on the monocytes

plasma membrane. This interaction of LPS with monocytes may be inhibited by gangliosides, leading to an inhibition of IL 1 synthesis; this demonstrates the causual relationship between the specific binding to the monocytes recep- tor and the induction of IL 1 secretion. A similar correlation between LPS binding and an LPS-induced function has been reported by Akagawa and Tokunaga (76), studying the capacity of mouse macrophages, activated by LPS, to kill tumor cells. These authors observed the failure of LPS to activate alveolar macrophages , results which correlate with the lack or very low expression of LPS-binding sites. This observation is consistent with our report demons- trating a lack of specific binding of LPS to rabbit alveolar macrophages (16). Interestingly, treatment of murine cells by interferon (IFN) led to a LPS-indu- cible cytotoxicity and to the expression of LPS receptors. Since it was re- ported by Eden and Turino ((32) that IFN allowed the IL 1 secretion by human alveolar macrophages stimulated by LPS, and by Wewers et al. (77) that normal human alveolar macrophages do not secrete IL 1 when stimulated by LPS, it can be proposed that as for mouse (76) and rabbit (16) alveolar macrophages, human alveolar macrophages do not express LPS receptors in the normal state. Regulation of expression of LPS-receptors by IFN will be of physiological relevance since infectious diseases may generate IFN (78).

Induction of IL 1 secretion was promoted by the B. pertussis pentasaccha- ride containing at the reducing end a KDO residue with a free carboxyl group, and by the Salmonella minnesota trisaccharide Hep-Hep-KDO at doses of 10 µg/ assay. The involvement of the inner core region among different endotoxins in IL 1 induction is not astonishing since only 2 regions of LPS have been highly conserved during evolution : the Lipid A and the inner core region. The inner core region constitutes a domain of very limited structural diver- sity, consisting of neutral sugars mainly heptose residues, and KDO residues. There are no natural LPS, nor mutants devoid of KDO molecule(s), and it should be noted that KDO molecules are negatively charged. They may, there- fore, be the first determinants involved in ionic bridges, which are a fea- ture of ligand-receptor interactions. The surrounding residues may promote attachment due to hydrophobic interactions. Isolated Lipid A from B. pertus- sis and N. meningitidis LPS were impotent IL 1 inducers, and polymyxin B, known to inhibit Lipid A-mediated activities, did not or weakly inhibited IL1 secretion induced by native B. pertussis and N. meningitidis LPSs. However, synthetic E. coli Lipid A was able to induce IL 1 secretion (62). The IL 1 induction was dependent upon the amount of fatty acids and phosphate in syn- thetic Lipid A. Interestingly, we observed that PMB inhibited completely IL 1 secretion induced by E. coli LPS. Therefore, according to our results and those reported by others, we suggest that the Lipid A region is not the only active principle for IL 1 induction by all endotoxins, and that both moieties of the endotoxin are required to induce IL 1, the PS mediating the specific binding to the receptor, the Lipid A enhancing the attachment. It can be also proposed that Lipid A is required to obtain a more adequate conformation of the residues present in the inner core region involved in the induction of IL 1. Finally, the exact structure of LPS capable of inducing IL 1 secretion may be different from the native LPS added to the culture, since it has been reported that LPS may be processed by macrophage enzymes leading to a more active molecule (79,80).

REFERENCES

1. Morrison D.C. and Ryan D.L. Adv. Immunol. 23 : 394 (1979).
2. Lüderitz O., Freudenberg M.A., Galanos C., Lehmann V., Rietschel E.T. and Shaw D.H. in "Microbial Membrane Lipids" Current Topics in Membranes and Transport (S. Razin and S. Rottem Eds) Academic Press New York, 17: 79 (1982).
3. Nowotny A. Microbiology pp 247-252 (1977)

4. Scibienski R.J. Mol. Immunol. 17 : 21 (1980).
5. Nowotny A., Behling U.H. and Chang H.L. J. Immunol. 115 : 199 (1975).
6. Urbaschek R. and Urbaschek B. Ann. NY Acad. Sci. 459 : 97 (1985).
7. Nowotny A., Behling U.H., Madani F., Nowotny A.M., Pham P.H., Hertogs C.F., and Pluznik D.H. J. Immunopharmaco. 5 : 93 (1983).
8. Urbaschek R. Adv. Exp. Med. Biol. 121 B : 51 (1980).
9. Butler R.C., Abdelnoor A.M. and Nowotny A. Proc. Natl. Acad. Sci. USA 75 : 2893 (1978).
10. Williamson S.I., Wannemuehler M.J., Jirillo E., Pritchard D.G., Michalek S.M., and McGhee J. J. Immunol. 133 : 2294 (1984).
11. Rosenstreich D.L. and Vogel S.N. Microbiology pp 11-15 (1980).
12. Zabala C. and Lipski P.E. J. Immunol. 129 : 2496 (1982).
13. Sager D.S. and Jasin H.E. Clin. Exp. Immunol. 47 : 645 (1983).
14. Corbel C. and Melchers F. Eur. J. Immunol. 13 : 528 (1983).
15. Haeffner-Cavaillon N., Cavaillon J-M. and Szabó L. Cell. Immunol. 74 : 1 (1982).
16. Haeffner-Cavaillon N., Chaby R., Cavaillon J-M. and Szabó L. J. Immunol. 128 : 1950 (1982).
17. Haeffner-Cavaillon N., Cavaillon J-M., Etievant M., Lebbar S. and Szabó L. Cell. Immunol. 91 : 119 (1985).
18. Haeffner-Cavaillon N., Cavaillon J-M. and Szabó L. in "Handbook of Endotoxin" Vol. 3 : Cellular Biology of Endotoxin (L.J. Berry Ed.) Elsevier Science Publisher pp 1-24 (1985).
19. Gery I. and Waksman B.H. J. Exp. Med. 136 : 143 (1972).
20. Dinarello C.A. Rev. Infect. Diseases 6 : 51 (1984).
21. Durum S.K., Schmidt J.A. and Oppenheim J.J. Annual Rev. Immunol. 3 : 263 (1985).
22. Unanue E.R., Kiely J-M. and Calderon J. J. Exp. Med. 144 : 155 (1976).
23. Treve A.J., Barak V., Tal T. and Fuks Z. Eur. J. Immunol. 13 : 647 (198).
24. Lepe-Zuniga J.L. and Gery I. Clin. Immunol. Immunopathol 31 : 222 (1984).
25. Gearing A.J.H., Johnstone A.P. and Thorpe R. J. Immunol. Methods 83 : 1 (1985).
26. Newton R.C. J. Leuk. Biol. 39 : 299 (1986).
27. Arenzana-Seisdedos F., Virelizier J.L. and Fiers W. J. Immunol. 134 : 2444 (1985).
28. Haq A.U., Rinehart J.J. and Maca R.D. J. Leuk. Biol. 38 : 735 (1985).
29. Newton R.C. Immunology 56 : 441 (1985).
30. Boraschi D., Censini S. and Tagliabue A. J. Immunol. 133 : 764 (1984).
31. Arenzana-Seisdedos F. and Virelizier J.L. Eur. J. Immunol. 13 : 437 (1983).
32. Eden E. and Turino G.M. Am. Rev. Respir. Dis. 133 : 455 (1986).
33. Gery I. Davies P., Derr J., Krett N. and Barranger J.A. Cell. Immunol. 64 : 293 (1981).
34. Bayne E.K., Rupp E.A., Limjuco G., Chin J. and Schmitd J.A. J. Exp. Med. 163 : 1267 (1986).
35. Matsushima K., Taguchi M., Kovacs E.J., Young H.A. and Oppenheim J.J. J. Immunol. 136 : 2883 (1986).
36. Haeffner-Cavaillon N., Cavaillon J-M., Moreau M., and Szabó L. Mol. Immunol. 21 : 389 (1984).
37. Cavaillon J-M. and Haeffner-Cavaillon N. Immunol. Lett. 10 : 35 (1985).
38. Staruch M.J. and Wood D.D. J. Immunol. 130 : 2191 (1983).
39. Neta R., Douches S. and Oppenheim J.J. J. Immunol. 136 : 2483 (1986).
40. Falkoff R.J.M., Muraguchi A., Hong J.X., Butler J.L., Dinarello C.A. and Fauci A.S. J. Immunol. 131 : 801 (1983).
41. Falkoff R.J.M., Butler J.L., Dinarello C.A. and Fauci A.S. J. Immunol. 133 : 692 (1984).
42. Giri J.G., Kincade P.W., Mizel S.B., J. Immunol. 132 : 223 (1984).
43. Pike B.L. and Nossal G.J.V. Proc. Natl. Acad. Sci. USA 82 : 8153 (1985).
44. Onozaki K., Matsushima K., Aggarwal B.B. and Oppenheim J.J. J. Immunol. 135 : 3962 (1985).

45. Lovett D., Kozan B., Hadam M., Resch K. and Gemsa D. _J. Immunol._ 136 : 314 (1986).
46. Tsai S.C. and Gaffney E. _Cancer Res._ 46 : 1471 (1986).
47. Lachman L.B., Dinarello C.A., Llansa N.D. and Fidler I.J. _J. Immunol._ 136 : 3098 (1986).
48. Rosner M.R., Tang J.Y., Borzi Lai J. and Khorana M.G. _J. Biol. Chem._ 254 : 5906 (1979).
49. Lebbar S., Cavaillon J-M., Caroff M., Ledur A., Brade H., Sarfati R. and Haeffner-Cavaillon N. _Eur. J. Immunol._ 16 : 87 (1986).
50. Nowotny A. _in_ "Handbook of endotoxin" Vol. 1 : Chemistry of endotoxin (E. Th. Rietschel Ed.) Elsevier Scince Publisher pp 308-338 (1985).
51. Ayme G., Caroff M., Chaby R., Haeffner-Cavaillon N., Ledur A., Moreau M., Muset M., Mynard M.C., Roumiantzeff M., Schulz D. and Szabó L. _Infect. Immunity_ 27 : 739 (1980).
52. Damais C., Riveau G., Parant M., Gerota J. and Chedid L. _Int. J. Immuno-pharm._ 4 : 451 (1982).
53. Murphy P.A., Simon P.L. and Willoughby W.F. _J. Immunol._ 124 : 2498 (1980).
54. Rosenwasser L.J. and Dinarello C.A. _Cell. Immunol._ 63 : 134 (1981).
55. Hanson D.F. and Murphy P.A. _Infect. Immunity_ 45 : 483 (1984).
56. Matisová E., Bùtorová E., Lackovic and Borecky L. _Acta Virol._ 14 : 1 (1970).
57. Maehara N. and Ho M. _Infect. Immunity_ 15 : 78 (1977).
58. Carswell E.A., Old L.J., Kassel R.L., Green S. Firoe N. and Williamson B. _Proc. Natl. Acad. Sci. USA_ 72 : 3666 (1975).
59. Beutler B.A., Milsark I.W. and Cerami A. _J. Immunol._ 135 : 3972 (1985)
60. Dinarello C.A., Bernheim H.A., Duff G.W., Le H.V., Nagabhusham T.L., Hamilton N.C. and Coceani F. _J. Clin. Invest._ 74 : 906 (1984).
61. Dinarello C.A., Cannon J.G., Wolff S.M., Bernheim H.A., Beutler B., Cerami A., Figari I.S., Palladino Jr. M.A. and O'Connor J.V. _J. Exp. Med._ 163 : 1433 (1986).
62. Loppnow H., Brade L., Brade H., Rietschel E.T., Kusumoto S., Shiba T. and Flad H.D. _E.O.S. Riv. Immun. Immunofarmac._ VI (S3): 203 (1986).
63. Cavaillon J-M. and Haeffner-Cavaillon N. _Mol. Immunol._ 23 : 965 (1986).
64. Morrison D.C. and Jacobs D.M. _Immunochem._ 13 : 813 (1976).
65. Apte R.N., Hertogs C.F. and Pluznik D.H. _J. Immunol._ 118 : 1435 (1977).
66. Graber S.E. and Clancey M.A. _J. Cyclic Nucl. Prot. Phosphor. Res._ 9 : 155 (1983).
67. Miller R.A., Gartner S. and Kaplan H.S. _J. Immunol._ 121 : 2160 (1978).
68. Leslie C.C., Musson R.A., Henson P.M. _J. Leuk. Biol._ 36 : 143 (1984).
69. Ryan J.L. and Shinitzky M. _Eur. J. Immunol._ 9 : 171 (1979).
70. Whisler R.L. and Yates A.J. _J. Immunol._ 125 : 2106 (1980).
71. Ladish S., Ulsh L., Gillard B. and Wong C. _J. Clin. Invest._ 74 : 2074 (1984).
72. Robb R.J., _J. Immunol._ 136 : 971 (1986).
73. Morrison D.C., Brown D.E., Vukajlovich S.W. and Ryan J.L. _Mol. Immunol._ 22 : 1169 (1985).
74. Coleman D.L., Morrison D.C. and Ryan J.L. _Cell. Immunol._ 100 : 288 (1986)
75. Cavaillon J-M. and Fitting C. _Eur. J. Immunol._ 16 : 1009 (1986).
76. Akagawa K.S. and Tokunaga T. _J. Exp. Med._ 162 : 1444 (1985).
77. Wewers M.D., Rennard S.I., Hance A.J., Bitterman P.B. and Crystal R.G. _J. Clin. Invest._ 74 : 2208 (1984).
78. Stewart II W.E. _in_ "The interferon System" Springer Verlag (Wien; New York) pp 38-44 (1979).
79. Duncan Jr. R.L., Hoffman J., Tesh V.L. and Morrison D.C. _J. Immunol._ 136 : 2924 (1986).
80. Munford R.S. and Hall C.L. _Infect. Immunity_ 48 : 464 (1985).

HUMAN ENDOTHELIAL CELLS PRODUCE PLATELET ACTIVATING FACTOR IN RESPONSE TO INTERLEUKIN 1

F. Bussolino, A. Bosia, F. Breviario, M. Aglietta, G. Garbarino,
D. Ghigo, F. Sanavio, G. Camussi, G. P. Pescarmona and
E. Dejana

University of Turin, Cattedra di Chimica Propedeutica Biochimica
V. Santena 5bis 10126 Torino, Italy
Institute Mario Negri, Milan, Italy

Introduction

Earlier consideration of the endothelium has tended to picture it as a passive cells between the blood and the interstitium. There is now a growing awareness that vessel walls participate actively in the development of the immune and inflammatory response by a two-way interaction between endothelial cells (EC) and immunocompetent cells. In the presence of cell-mediated immune response, vasodilation and proliferation of capillary EC has been documented (1) as well as leukocyte adhesion to and passage through endothelial linings in order to localize at inflammatory sites (2). It is therefore of interest to establish which molecules and pathways are involved in this cell-to-cell interaction.

The mediators produced by the immunocompetent cells and involved in this system of comunication are the autocoid substances (prostaglandins, leukotrienes, acetylated alkyl phosphoglycerides, oxygen species) and the classical lymphokines (Table 1).

Table 1. Products of immunocompetent cells acting on EC.

AUTOCOIDS	LYMPHOKINES
−Prostaglandins	−Interferon α
−Leukotrienes	−Interferon γ
−Acetylated alkyl ether	−Interleukin-1
	−Tumor necrosis factor
OTHERS	−Others not well identified
−Antibodies	
−Reactive oxygen species	
−Proteases	

Autocoids modulate rapidly and for limited periods of time the vascular responses and the migration of leukocytes, and they act either on EC or leukocytes themselves (2-6). Lymphokines are also potent activator of the vascular cells (proliferation, migration, production of colony stimulating factors) (7-10). Their effects are slow, and persist for longer periods of time. Furthermore, EC are involved in the host defense as cells able to present the antigen to T lymphocyte in Ia (class II major histocompatibility complex antigen)-restricted manner (11). Interferonγinduces the expression of Ia antigens on the surface of EC (12), thus permitting EC to present the antigen to T cells, perhaps serving to recruit antigen specific-T cells at the site of developing immune response.

It has been recently show that interleukin-1 (IL-1), a pleiotropic multifunctional mediator produced by activated monocytes/macrophages (13-14), is produced by (15-16) and active on EC (17-27). As summarized in Table 2, IL-1 induces a wide range of biological activities on EC. Firstly, IL-1 changes the non -thrombogenic surface of EC into an actively thrombogenic by inducing the synthesis of a tissue factor-like procoagulant (17-18) and of a inhibitor of the plasminogen activator (22), and by inhibithing the protein C anticoagulant pathway (20).Secondly, IL-1 induces EC adhesiveness (18,21,23,26) for blood leukocytes: this effect is partially related to a de novo synthesis of the procoagulant-like activity (18) and of a protein recognized by a monoclonal antibody raised against cultured EC prestimulated by IL-1 (21). Thirdly, this lymphokine induces the synthesis of two autocoids, namely prostacyclin (PGI2) (27) and platelet-activating factor (PAF) (19), that both act on EC by an autocrine mechanism (4,28).

Table 2. Activities of IL-1 on EC.

PRODUCTION OF:

-Tissue procoagulant activity	(Bevilacqua et al.,1984)
-Plasminogen activator inhibitor	(Nachman et al.,1986)
-Platelet activating factor	(Bussolino et al., 1986)
$-PGI_2$	(Rossi et al.,1985)
$-PGE_2$	(Albrightson et al.,1985)
-von Willebrand factor	(Shorer et al.,1985)

INHIBITION OF:

-Protein C anticoagulant pathway	(Nawroth et al.,1986)

CHANGE IN:

-Shape	(Montesano et al.,1985)
-Matrix composition	(Montesano et al.,1985)

INCREASE IN ADHESION OF:

-Neutrophils, monocytes,HL60 and U937 cell lines	(Bevilacqua et al.,1986)
-T lymphocytes	(Cavender et al.,1986)

These EC activities may play an important role in the vascular regulation of thrombus formation, vascular tone and cell interaction with the vessel wall.
Here, we summarize the regulation of the production of PAF, an acetylated alkyl phosphoglycerides involved in the cell-to-cell interaction (29-31), by molecules with IL-1 activity.

IL-1 induces EC production of PAF (19)

Crude and purified IL-1 from stimulated monocytes promotes PAF production in a dose-dependent manner (1-50 U/ml). Other lymphokines (interferons, IL 2) are inactive; endotoxin, at the concentrations that could contaminate IL-1 preparations, has no effect. Similarly to the other activities mediated by IL-1 (13-14), PAF synthesis requires a long interaction with the cells (2 hours), lasts the same time (up to 18 hours) and is blocked by protein synthesis inhibitor. In contrast, other stimuli (A23187, angiotensin II, vasopressin, antibodies-anti factor VIII, ATP, histamine, thrombin, bradykinin) induce a rapid PAF production, their effects last only for a few minutes and they do not require protein synthesis to be active (32-33). Leukotrienes C4 and D4 also induce a prolonged accumulation of PAF persisting up to 2 hours (34). The different time course in PAF production related to the different stimuli, suggests a more complex regulation of PAF metabolic pathways and/or differences in the transmembrane signals coupling the action of the stimulus with the synthesis of PAF. Most of the PAF produced (7-10 ng/ $5x10^5$ cells) remains associated to the cells and only 25% is released in the culture medium.

Modulation of EC PAF production by different molecular species of IL-1 (35)

We tested the ability of five IL-1 preparations, all active in the thymocyte costimulator assay, in inducing PAF and PGI2 production from EC: human natural IL-1 obtained from Staphylococcus albus-stimulated human monocytes and purified by affinity chromatography (Genzyme); "22 K factor", a pure preparation from mitogen-stimulated human mononuclear cells showing extensive homology with the derived aminoacid IL-1β sequence; murine recombinant IL-1 α, having a high homology to the humanα species (murine IL-1 α) (a gift of Dr. LoMedico); human recombinant IL-1β ;human recombinat IL-1 α(a gift of Dr. Gillis).
Human natural IL 1, that contains equal amounts of the different species (αor pI 5 and βor pI 7) of IL-1, elicits PAF and PGI2 production. In contrast, when αand β IL-1 molecules are examined, different responses are elicited by the two molecular species. Murine IL-1 αinduces the synthesis of PAF, but is inactive in inducing PGI2 production. However, in contrast to the murine αspecies, human αIL-1 promotes PAF and PGI2 generation. IL-1 βpreparations (22 K factor and humanβrecombinant IL-1) elicit PGI2 production, but are unable to activate PAF synthesis.

The different response of EC to αand βIL-1 species for PAF production, is the first evidence of a functional dissociation between these two IL-1 forms, and suggests that a slight difference in the molecular structure might greatly affect functional properties.

IL-1 activates 1-O-alkyl-2-lyso-GPC: acetyl CoA acetyltransferase (19,36)

Two specific enzymatic reactions have been documented to be involved in PAF biosynthesis. 1-O-alkyl-2-lyso-GPC: acetyl CoA acetyltransferase (acetyltransferase) catalyzes acetylation of inactive lyso-PAF into the bioactive PAF, whereas 1-O-alkylacetylglycerol: CDP-choline cholinephosphotransferase (cholinephosphotransferase) transfers the phosphobase to 1-O-alkylacetylglycerol (30).
In EC stimulated by A23187, Camussi et al. (32) suggested the importance of an acetylation process in the synthesis of PAF as inferred from the incorporation of labeled acetate into PAF molecule. Acetyl CoA, a substrate for the acetyltransferase, amplifies the response induced by IL-1 confirming the importance of the acetylation step.
The basal activity of acetyltransferase in resting EC is 3.68 nmol/min/mg protein and the Km for acetyl CoA is 92 M. After IL-1 stimulation the activity of the enzyme is increased severalfold in a transient and time-dependent fashion similar to that of PAF production. The Vmax of the acetyltransferase is about 3-5 times the Vmax of unstimulated cells, whereas the Km is not affected. Similarly, Ninio et al. (37) showed that A23187 stimulation of PAF production in rat peritoneal cells is accompanied by a rapid increase in the Vmax of the acetyltransferase with no change in the Km. Furthermore, a mononuclear cell factor, probably related to IL-1 enhances the V max of cyclooxygenase in stimulated fibroblasts producing PGE2, without affecting the Km of the enzyme (38).
Other authors have shown that acetyltransferase activity might be enhanced by reversible stimulation of the enzyme by a phosphorylation / dephosphorylation process (39), and that cAMP-dependent protein kinase increases the Vmax of the enzyme (40).
At present, further data are required to elucidate the mechanism of IL-1 -mediated stimulation of acetyltransferase and either activation of the enzyme or de novo synthesis can be suggested.
The cholinephosphotransferase is present in EC (1.7 nmol/min/mg protein), but its activity is not increased after IL-1 challenge.

IL-1 induces PAF synthesis by a deacylation/reacetylation process (36)

IL-1 causes the incorporation of [^{3}H]acetate into the PAF molecule that parallels with the time course of acetyltransferase activation, suggesting that EC incorporate the exogenous precursor into the PAF as shown in other cell types (30). When EC are preincubated with [^{3}H] lyso-PAF, they incorporate the precursor into PAF molecule.

In non-stimulated EC a greater amount of [3H]lyso-PAF is acylated to 1-O-[3H] alkyl-2-acyl-GPC, in a time dependent manner (plateau after 2 hours labeling). IL-1 induces PAF production that parallels with the fall of 1-O-[3H] alkyl-2-acyl-GPC, while [3H]lyso-PAF is not directly transformed into [3H] PAF. These experiments suggest that PAF is produced by a deacylation of 1-O-alkyl-2-acyl-GPC into lyso-PAF, which is then acetylated by the acetyltransferase.

1-O- [3H] alkylacetylglycerol, the substrate of the cholinephosphotransferase is not significantly incorporated in PAF molecule.

After 12-18 hours, the generated PAF is catabolized. In experiments performed with labeled precursors, this event correlates with an increase of 1-O-alkyl-2-acyl-GPC. It is still not possible to decide whether PAF is inactivated by a direct transacylation or whether lyso-PAF is an essential intermediate in PAF catabolism (30).

IL-1 induces transient intracellular calcium mobilization (41)

The model of IL-1 as a calcium ionophore agrees with most of the evidence presently available and provides for a probable mechanism of action. Indirect evidence derives from the fact that many of the biological activities of IL-1 are mimicked by the calcium ionophore A23187: prostaglandin production in different tissues, neutrophil degranulation, T-cell mitogenesis. Moreover, calcium channel blockers inhibit the fever due to IL-1 (13-14). This lymphokine activates phospholipase A2, a calcium-dependent enzyme, in rabbit chondrocytes, so eliciting arachidonic acid release from intracellular stores and its conversion to PGE2 (42).

Recently, we have observed that IL-1 induces a clearcut, dose-dependent Ca++ influx in quin-2 loaded EC. Murine and human α recombinant IL-1 showed a more powerful effect than β recombinant IL-1 . With Ca++ labeling studies we obtained a dose-dependent Ca++ efflux elicited by α and β IL-1; the major efficacy of IL-1 α was confirmed. Differing from other agonists inducing Ca++ influx in EC, such as thrombin and arachidonic acid (41), IL-1 evokes a prolonged (20-30 minutes) rise of intracellular Ca++ levels.

A hypothesis

The model presented in figure 1 outlines possible pathways for IL-1-mediated PAF production in EC. IL-1 interacts with a binding site on EC surface (43) inducing a rise in Ca++ in the cytosol. It is possible that IL-1 activates the protein kinase C, as suggested by the study of Levine and Xiao (44) showing a synergic effect between IL-1 and activators of protein kinase C. Furthermore, IL-1 is able to phosphorylate a 41 K protein in K562 cell line (45). The increased Ca++ in the cytosol and the possibly activated protein kinase C are intracellular signals for de novo synthesis of the acetyltransferase or of a kinase that activates this enzyme. IL-1 could activate a phospholipase A2 or increase its cellular concentration, which deacylates 1-O-alkyl-2-acyl-GPC to lyso-PAF (30).

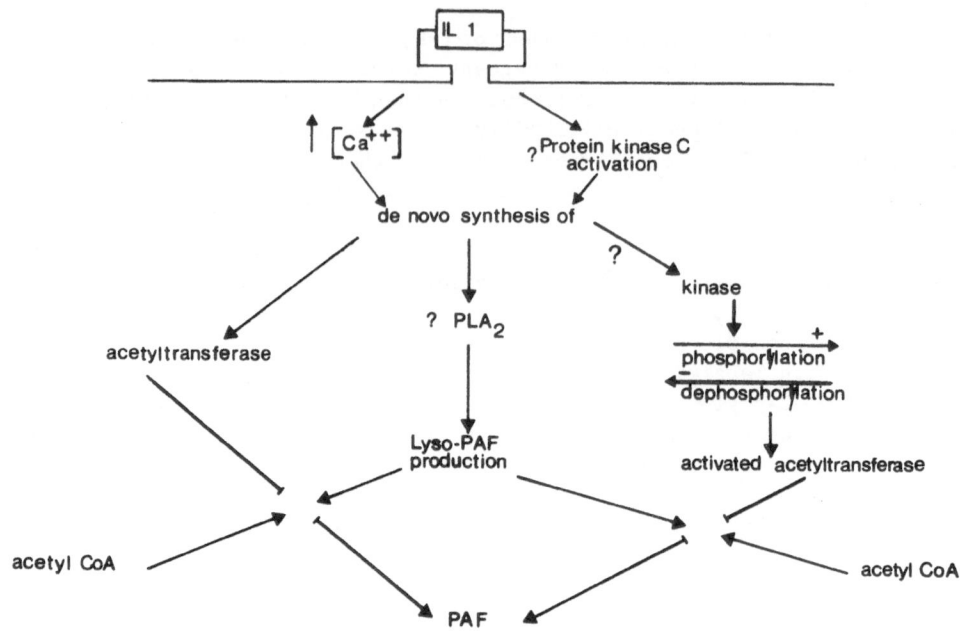

Figure 1. Hypothesis on the mechanisms inducing PAF production from EC after IL-1 activation. PLA2= phospholipase A2.

Implications of IL-1-induced PAF production from EC

The biological role of IL-1 mediated EC production of PAF remains to be characterized. The range of biologically active PAF concentrations is very close to those reached in EC after IL-1 challenge. Most PAF produced is cell-bound, but can be extracted from the cells. Unfortunately, proof is not given that PAF associated to EC is exposed in outer membrane, as hypothyzed by McIntyre et al.(33). However, PAF associated to EC could modulate the adhesion of blood circulating cells. IL-1 has been shown to promote the adhesion of neutrophils to EC (18,21). This effect is only partially mediated by the generation of procoagulant like activity or of a protein recognized by a specific monoclonal antibody, that partially inhibits the adhesion induced by IL-1 (21). Our preliminary experiments shown that neutrophils preincubated with PAF-specific antagonists or desensitized to PAF itself, adhere to IL-1 treated EC to a lesser extent than non-treated neutrophils. This suggests that PAF can partially mediate the adhesion of neutrophils caused by IL-1.
Furthermore, PAF released from IL-1-stimulated EC is rapidly inactivated by binding to target cells and humoral inhibitors, suggesting its action only in the microvascular network. This released PAF can activate platelets, neutrophils and monocytes at the site of immunologically-mediated processes, or act through an autocrine mechanism on EC influencing the vascular permeability.

References

1) Polverini P.J. , Cotran R.S., and Sbolley M.M. J. Immunol., 118:529 (1977).

2) Harlam, J.M. Blood, 65:513 (1985).

3) Pologe L.G., Cranar E.B., Pawlosky N.A., Abraham E., Cohn Z.A., and Scott W.A. J.Exp. Med., 160:1043 (1984).

4) Bussolino F., Aglietta M., Sanavio F., Stacchini A., Lauri D., and Camussi G. J. Immun.,135: 2748 (1985).

5) Bult H., Herman A.G. and Rampart E. Br.J.Pharmacol. 84:329 (1985)

6) Hoover R.L., Karnovsky M.J, Austen K.F., Corey E.J., and Lewis R.A. Proc.Natl. Acad. Svi.USA 81:2191 (1984).

7) Martin B.M., Jr.Gimbrone M.A., Unanue E.R. and Cotran R.S.. J.Immunol., 126:1510 (1981).

8) Polverini P.J., Cotran R.S., Jr.Gimbrone M.A., and Unanue E.R. Nature 269:804 (1977).

9) Cohen M.C., Picciano P.T., Douglas W.J., Yoshida T., Krentzer D.L. and Cohen S. Science 215:301 (1982).

10) Bagby C.G., McCall E., Bergstrom K.A., and Burger D. Blood, 62:663 (1983).

11) Nunez G., Ball E.J. and Stastny P. J. Immunol., 131:666 (1983).

12) Pober J.S., Gimbrone Jr. M.A., Cotran R.S., Reiss C.S., Burakoff S.J., Fiers W. and Ault K.A. J.Exp.Med. 157:1339 (1983).

13) Oppenheim J.J., Kovacs E.J., Matsushima K. and Durum S.K. Immuol. Today, 7:45 (1985).

14) Dinarello C.A. New Engl.J.Med., 311:1413 (1984).

15) Nawroth P.P., Bank I., Handley D., Cassimeris J., Chess L. and Stern D. J.Exp.Med., 163,1363 (1986).

16) Miossec P., Cavender D., and Ziff M. J.Immunol., 136:2486 (1986).

17) Bevilacqua M.P., Pober J.S., Majeau G.R., Cotran R.S. and Jr.Gimbrone M.A. J.Exp.Med. 160:618 (1984).

18) Bevilacqua M.P., Pober J.S., Wheeler M.E., Cotran R.S. and Gimbrone M.A. Jr. Am.J.Pathol., 121:393 (1985).

19) Bussolino F., Breviario F., Tetta C., Aglietta M., Mantovani M. and Dejana E. J. Clin. Invest., 77:2027 (1986)

20) Nawroth P.A., Handley D.A., Esmon C.T. and Stern D.M. Proc. Natl. Acad. Sci. USA 83:3460 (1986).

21) J.S.Pober J.S., Bevilacqua M.P., Mendrick D.L., Lapierre L.A., Fiers W., and Gimbrone M.A. Jr. J.Immunol., 136:1680 (1986).

22) Nachman R.L., Hajjar K.A., Silverstein R.L. and Dinarello C.A. J.Exp.Med. 163: 1595 (1986).

23) Cavender D.E., Haskard D.O., Joseph B. and Ziff. MJ. Immunol. 136:203 (1986).

24) Schorer A.E., Moldow C.F. and Rich M.E.. Blood suppl.1:358a (1985).

25) Montesano R., Orci L., and Vassalli P. J.Cel.Physiol. 122:424 (1985).

26) Bevilacqua M.P., Pober J.S., Wheeler M.E., Cotran R.S. and Gimbrone M.A.Jr. J.Clin.Invest., 76:2003 (1985).

27) Rossi V., Breviario F., Ghezzi P., Dejana E. and Mantovani A. Science 229:174 (1985).

28) Gordon J.L. and Pearson J.D. in: Pathobiology of Endothelial Cells (H.L. Nossel and H.S. Vogel), Academic Press, New York, p.433 (1982).

29) Benveniste J. and Vargaftig B.B. in: Ether Lipids: Biochemical and Biomedical Aspects (H.K. Mangold and F. Paltauf), Academic Press, New York, p.355 (1983).

30) Snyder F. Med.Res.Rev. 5: 104 (1985).

31) Camussi G. Kidney Int., 29: 469 (1986).

32) Camussi G., Aglietta A. ,Malavasi F., Tetta C., Sanavio F., Piacibello W. and Bussolino F. J. Immunol 131: 2397 (1983).

33) McIntyre T.M., Zimmerman G.A., Satoh K., and Prescott S.M. J. Clin. Invest. 76: 271 (1985).

34) McIntyre T.M., Zimmerman G.A. and Prescott S.M. Proc. Natl. Acad. Sci. USA. 83:2204 (1986).

35) Dejana E., Breviario F., Erroi A., Bussolino F., Mussoni L., Gramse M., Pintucci G., Casali B., Dinarello C.A., Van Damme J. and Mantovani A., submitted.

36) Bussolino F., Breviario F., Aglietta M., Sanavio F., Bosia A. and Dejana E. submitted.

37) Ninio E., Mencia-Huerta J.M. and Benveniste J.. Biochim.Biophys.Acta., 751:298 (1983).

38) Whiteley P.J. and Needleman P. J.Clin.Invest. 74:2249 (1985).

39) Leinhan D.J., and Lee T.C. Biochem. Biophys.Res.Comm., 120: 834 (1984).

40) Gòmez-Cambronero J., Velasco S., Mato J.M., and Sanchez-Crespo M. Biochim. Biophys. Acta 845, 516 (1985).

41) Ghigo D., Bussolino F., Aglietta M., Dejana E., Garbarino G., Pescarmona G.P. and Bosia A. in: International symposium in honor of Luigi Sabbatani on calcium ion: membrane transport and cellular regulation. Padova-Venezia, Italy,July 1-4, Abstract book p.135 (1986).

42) Chang J, Gilman S.C. and Lewis A.J. J.Immunol.136:1283 (1986).

43) Dower S.K., Kronheim S.R., March J.C., Conlon P.J., Hopp T.P.,Gillis S. and Urdal D.L. J.Exp.Med. 162:501 (1985).

44) Levine L. and Xiao D.M. J. Immunol. 135:3430 (1985).

45) Martin M. Immunobiology, 171:165 (1986).

46) Albrightson C.R., Baenziger N.L. and Needleman P. J.Immunol. 135:1872 (1985).

INTERLEUKIN-1 STIMULATION OF VASCULAR ENDOTHELIAL CELL ADHESIVENESS FOR LEUCOCYTES : *IN VITRO* AND *IN VIVO* STUDIES

C. J. Dunn, W. E. Fleming, R. G. Schaub, A. J. Gibbons,
J.W. Paslay and D. E. Tracey
Department of Hypersensitivity Diseases Research and
Department of Atherosclerosis and Thrombosis
The Upjohn Company, Kalamazoo, MI 49001, U.S.A.

INTRODUCTION

Under normal conditions, circulating leukocytes adhere to the vascular bed through a reversible adhesive interaction with endothelial cells (ECs).[1] However, a dramatic increase in leukocyte adherence is observed following tissue injury which is restricted to vessels in the affected area.[2] During the early stages neutrophils predominate.[1] Although this is a fundamental aspect of inflammation, little is known of the actual mechanisms which initiate leukocyte adhesion. It has been known for some time that divalent cations are essential and removal of cell surface sialic acid residues augments EC-leukocyte adhesion.[1,3] Development of EC culture techniques has permitted more extensive studies on the adhesive interactions between these cells. Recent observations suggest a role for altered fibronectin,[4] chemotactic factors[5,6] and arachidonic acid metabolites,[7,8] indicating the polymorphonuclear leukocyte (PMN) as the target cell. Enhanced EC-PMN adhesion following preincubation of ECs with C'5a[3], or C'5a desarg,[9] suggested that these cells could be stimulated to become hyperadhesive for PMNs, although these observations are controversial.[5] Leukotriene B$_4$ (LtB$_4$) has been shown to initiate adhesion of human PMNs to bovine ECs immediately after contact with endothelium in vitro, although the adhesiveness of PMNs was also directly affected by LtB$_4$.[7,10]

In this communication we provide evidence that interleukin (IL-1) acts as an endogenous mediator of leukocyte adhesion to vessel walls via a selective effect on cultured endothelium which is independent of leukocyte stimulation. In addition to these in vitro findings IL-1 was also shown to stimulate vascular-leukocyte interactions in vivo indicating a potential role for this cytokine in thrombotic and inflammatory diseases.

MATERIALS AND METHODS

Cell Culture, Separation and Labeling

Endothelial cells (ECs) were cultured from human umbilical veins as previously described.[11] ECs were grown to confluency and subcultured at 37ºC in RPMI 1640 culture medium (KC Biological, Kansas) containing 20% fetal bovine serum (Sterile Systems Inc., Logan, Utah), 1mM glutamine, 10 µg/ml streptomycin and 2.5 µg/ml fungizone (Gibco, Grand Island, New York) in 24-well tissue culture plates (Falcon, no. 3008). After 3-4 days the sub-confluent EC monolayers were used for adhesion studies. ECs were not used beyond the third subculture passage. The purity of EC culture was confirmed by morphology and anti-factor VIII antigen fluorescent staining as previously described.[12]

A human fibroblast cell line (from the American Type Culture collection, - CRL 1445 Rockville, MD) was cultured under the same conditions and used for adhesion assays where appropriate.

Human PMNs were isolated by centrifugation of citrated human blood (300xg, 4°C for 20 min). Platelet poor plasma (PPP) was obtained by centrifugation of plasma at 2,500 x g for 15 min. The erythrocyte-leukocyte suspension was made up to 50 ml with 2.5% gelatin in sterile saline and allowed to settle for 30 min (37°C). The cells were centrifuged for 15 min at 250 x g (room temperature) and resuspended in 4 ml saline. Erythrocytes were removed by hypotonic lysis and the remaining leukocytes resuspended in PPP ($20x10^6$ cells/ml). This preparation yielded $\geq$ 90% PMNs, of which the remaining cells were monocytes.

The human mononuclear cell line, U-937 (ATCC-CRL 1593) was maintained in culture medium for use in adhesion assays. Human peripheral blood lymphocytes (PBLs) were separated according to previously described techniques.[13] Leukocytes were labeled with ^{51}Cr as described.[14] Cell viability was $\geq$ 90%.

Endothelial Cell-Leukocyte Adhesion **in vitro**

^{51}Cr-labeled leukocytes were added to EC cultures at a final concentration of 2 x 10^6 leukocytes/ml. After incubation at 37°C (30 min for PMNs; 60 min for U-937 cells; 120 min for PBLs) non-adherent leukocytes were removed by gentle washing. The remaining adherent leukocytes were removed by trypsin-EDTA treatment and their radioactivity determined using a gamma counter. Leukocyte adhesion was expressed as the percent labeled leukocytes remaining after adhesion relative to the total labeled leukocytes added.

Highly purified human monocyte-derived interleukin-1 (IL-1) was prepared by C.A. Dinarello (Tufts Univ.) and human recombinant IL-1α and IL-1β were obtained from Genzyme Corp., Boston, Mass., in phosphate buffered saline, pH 7.4, containing 0.5% bovine serum albumin, and diluted in culture medium as required. The activity of IL-1 was established in the mouse thymocyte proliferation assay[16] and expressed as half maximal thymocyte stimulation units (U). IL-1 was endotoxin-free as assessed by the Limulus amebocyte assay.[15]

Freshly washed subconfluent EC cultures were incubated with IL-1 for varying time periods at 37°C. Excess IL-1 was removed by washing in culture medium (x3), ^{51}Cr-labeled leukocytes added and adhesion assessed. In some experiments, ECs were preincubated with IL-1 for 5-60 min, washed, incubated in culture medium for varying periods of time, washed again and tested for leukocyte adherence.

Similar experiments were performed with lipopolysaccharide (LPS- E. Coli derived, 0.55:B5 phenol extract, Sigma). The direct effects of IL-1 and LPS on EC-leukocyte adhesion were determined by co-incubation of either substance with ECs and PMNs or U-937 cells for 30 min or 60 min, respectively. U-937 cells were also preincubated with IL-1 or LPS for 3.5 hr, washed and tested for adhesion to untreated ECs. Serum was omitted from all LPS experiments to avoid the non-specific effects of LPS-activated serum proteins.

Phytohemagglutinin (PHA) and phorbol myristate acetate (PMA) were obtained from Sigma and also tested for stimulation of EC-PMN adhesion.

Actinomycin D and cycloheximide (Sigma) were added to EC cultures in the absence and presence of IL-1 or LPS for 3.5 hr. Cultures were washed (x3) and leukocytes added to assess adhesion.

All experiments were carried out twice with 6 wells per group.

Stimulation Of Vascular-Leukocyte Interactions By IL-1 **In Vivo**

Adult male New Zealand White rabbits (4.5-5 kg) were anesthetized with Nembutal (30 mg/kg iv). The neck was shaved and prepared for a sterile cut-down procedure. Both jugular veins were carefully exposed to avoid unnecessary trauma or occlusion. Ethylene vinyl acetate (EVA) disks containing either BSA (250 mg/disk) or human interleukin 1 (Genzyme, Boston, MA, 50 U/disk) were prepared according to previously described techniques,[17] carefully positioned around each vein and sutured in place with 4.0 Dexon. Each EVA disk was positioned to avoid obstruction of blood flow. The incisions were closed and the animals recovered and were maintained for 24 hours in individual holding cages. After 24 hours, the rabbits were anti-coagulated with heparin (3000 Units iv) and overdosed with intravenous Nembutal. The chest was opened and an 18 gauge needle was positioned into the left ventricle for whole body perfusion at 90-110 mmHg pressure. Perfusion fluid was removed from an incision in the right ventricle. Two liters of Tyrode's buffer (pH 7.4) was followed by 1 liter of 1% glutaraldehyde/4% formaldehyde fixative prepared in Tyrode's buffer. Each jugular vein was dissected, removed, and further processed for microscopic analysis.

RESULTS

Stimulation of EC-Leukocyte Adhesion **In Vitro**

Incubation of ECs with IL-1 for 3.5 hr. resulted in increased EC adhesiveness for PMN leukocytes compared with basal adhesion exhibited by untreated ECs (Table 1). The effect was dose related, maximal adhesion (+89%) being observed around 2.5×10^{-1} U/ml IL-1. Minimal effective doses occurred between 2.5×10^{-4} - 2.5×10^{-5} U/ml IL-1. Similar results were found with LPS, peak activity (+53.6%) occurring between 10-100 ng/ml (Table 1).

Table 1. Dose-related augmentation of PMN adhesion to endothelial cells pretreated with IL-1 or LPS.

	Group	PMN adhesion (±sem)	% change PMN adhesion
	Control	26.7 (±0.65)	--
	2.5×10^{-1}	50.6 (±1.11)	+ 89.5 *
	2.5×10^{-2}	42.3 (±1.77)	+ 58.4 *
IL-1 (U/ml)	2.5×10^{-3}	37.0 (±0.92)	+ 38.6 *
	2.5×10^{-4}	37.1 (±0.78)	+ 39.0 *
	2.5×10^{-5}	32.0 (±0.74)	+ 19.9
	Control	23.3 (±0.54)	--
	100	36.8 (±0.81)	+ 53.6 *
	10	35.3 (±0.92)	+ 51.5 *
LPS (ng/ml)	1	32.1 (±0.95)	+ 37.8 *
	1×10^{-1}	26.2 (±0.47)	+ 12.4
	1×10^{-2}	25.5 (±0.44)	+ 9.4
	1×10^{-3}	22.3 (±0.40)	- 4.3

* $P < 0.001$ treated cultures compared with untreated controls.

PMN adherence to fibroblasts (13.9% ± 0.44) was not increased by either IL-1 (1-10 U/ml) or LPS (1-100 ng/ml) pretreatment, indicating specificity for ECs.

Supernatants transferred from IL-1 or LPS-treated ECs did not stimulate adherence of leukocytes to fresh untreated EC cultures.

Treatment of ECs with IL-1 or LPS for various time periods revealed the requirement for a minimum of 1-2 hr following treatment with these stimuli for expression of hyperadhesiveness. EC-PMN adhesion increased progressively as the preincubation time was extended (up to 3.5 hr).

Variation of the "exposure time" of ECs to IL-1 or LPS stimulus showed that periods as short as 5 min were sufficient for enhanced EC-PMN adhesion provided at least 2 hr was allowed to elapse between removal of stimulus and addition of PMNs. Maximal activity occurred when ECs were exposed to IL-1 or LPS for 30-60 min followed by a 2 hr incubation period.

PMN adhesion to IL-1 stimulated ECs diminished to basal levels 24 hr after treatment. However, an additional IL-1 treatment at this time resulted in augmented EC-PMN adhesion resembling that following initial IL-1 stimulation (fig. 1).

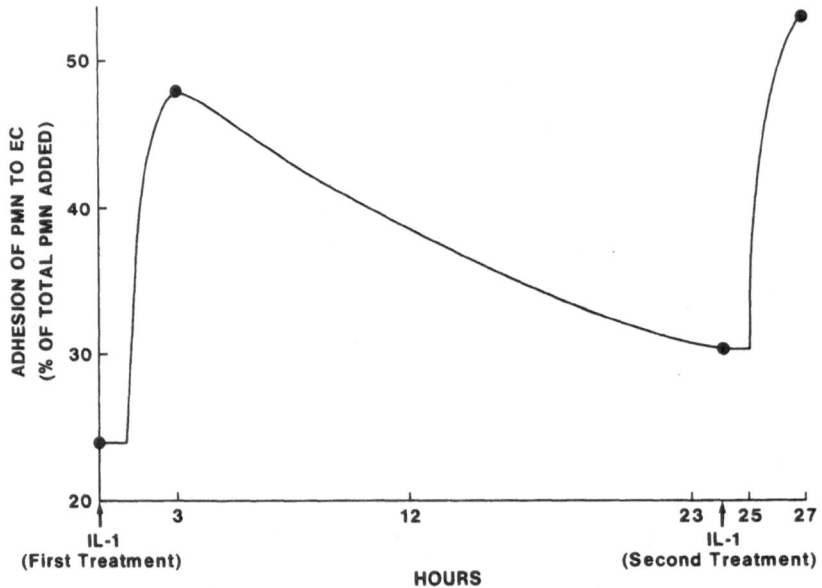

Figure 1. The effects of re-exposure of ECs to IL-1 on EC-PMN adhesion 24 hr following initial stimulation (IL-1 10 U/ml).

Pretreatment of ECs with actinomycin D (5 µg/ml) or cycloheximide (5 µg/ml) completely blocked IL-1 and LPS-stimulated EC-PMN adhesion, whereas unstimulated basal adhesion was unaffected (fig. 2).

Adhesion of U-937 cells to ECs pretreated with IL-1 (10 U/ml) or LPS (100 ng/ml) for 3.5 hr was enhanced by 98.4% and 106%, respectively (fig. 3). PBLs depleted of adherent leukocytes also showed increased adhesion (+85%) to ECs similarly pretreated with IL-1.

Coincubation of ECs and PMNs, or ECs and U-937 cells (30 min and 60 min, respectively), with IL-1 (10 U/ml) or LPS (100 ng/ml) during adhesion did not cause increased adhesion compared with untreated cells. Furthermore, prolonged pretreatment (3.5 hr) of U-937 cells with IL-1 or LPS failed to alter their subsequent adherence to untreated ECs (fig. 3).

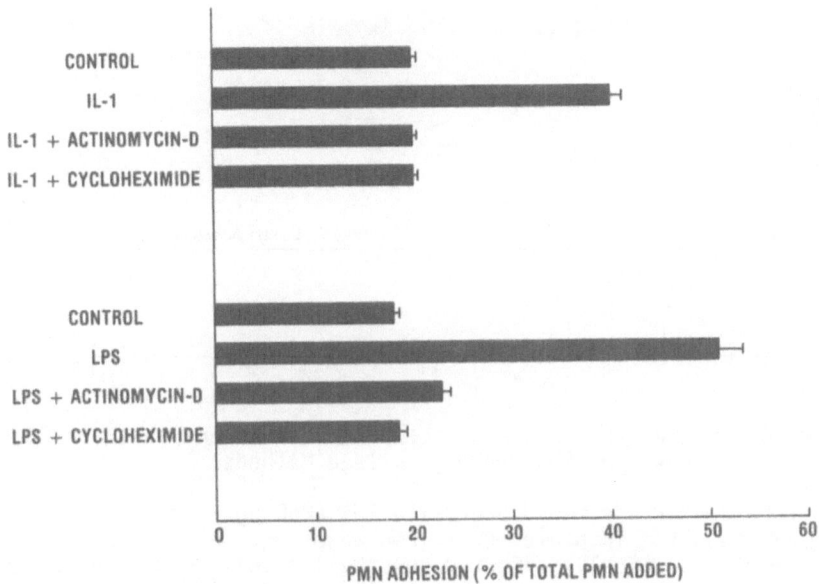

Figure 2. The inhibitory effects of protein synthesis inhibitors on IL-1 and LPS-induced EC-PMN adhesion (see Methods).

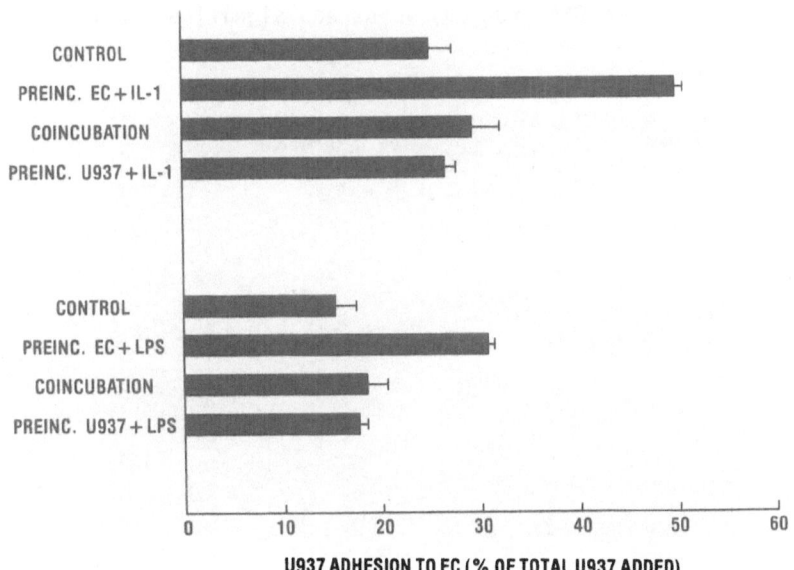

Figure 3. IL-1 stimulated EC hyperadhesiveness for U-937 cells (see Methods).

Studies with purified human recombinant IL-1 (rIL-1) indicated that rIL-1β was 6-7 times more active than rIL-1α with respect to stimulation of EC hyperadhesiveness for PMNs (fig. 4).

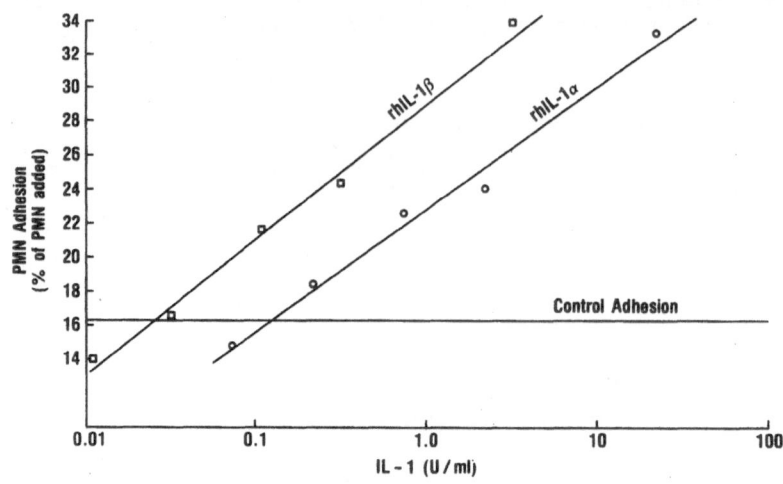

Figure 4. Comparison of human recombinant IL-1α and IL-1β on EC hyperadhesiveness for PMN leukocytes (see Methods).

Preincubation of ECs for 3.5 hr with PHA (250 ng - 25 μg/ml) or PMA (1 ng - 100 ng/ml) resulted in augmented EC-PMN adhesion. As for IL-1 and LPS, such treatment specifically affected endothelial cell adhesiveness and did not alter EC morphology as determined by light and scanning/transmission electron-microscopy (data not shown).

Stimulation Of Vascular-Leukocyte Interactions By IL-1 **In Vivo**

Morphological examination by scanning electron microscopy (SEM) showed that 4 of 8 control (EVA/BSA treated) veins had significant alterations of

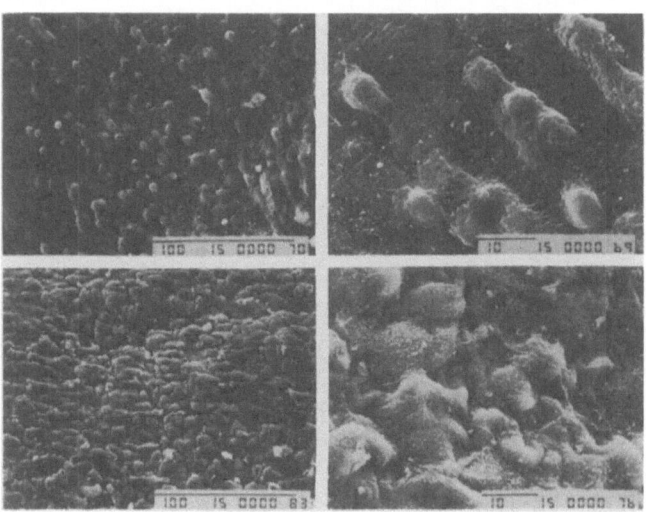

Figure 5. Scanning electron micrographs of veins exposed to control EVA disks demonstrate the basal injury produced by surgical intervention and disk placement. Endothelial cells exhibited rounding, separation of cell junctions and filopodia. Some subendothelial migration of PMNs was also observed.

the endothelial cell layer. Two of these veins had rounding of the endothelial cells, separation of intercellular junctions and partial detachment of endothelium from the basement membrane. Few adherent leukocytes were observed on these vessels (fig. 5). The remaining 2 veins with significant vascular alterations had leukocyte migration under the endothelium (fig. 5) and the associated endothelial cell detachment which results from leukocyte migration. The remaining 4 control veins exhibited less severe morphologic changes which included formation of endothelial cell filopodia, separation of intercellular junctions and scattered leukocyte adhesion and migration.

The responses of veins to IL-1 (EVA/IL-1 treated) could be distinguished from those observed in the control veins. Four of the 11 IL-1 treated veins examined exhibited evidence of injury resembling that of the most severely affected control veins (i.e. separation of endothelium, formation of endothelial cell filopodia and scattered migration of leukocytes under the endothelium). The remaining 7 IL-1 treated veins had leukocyte adhesion and vascular injury which was significantly greater than in control veins (fig. 6).

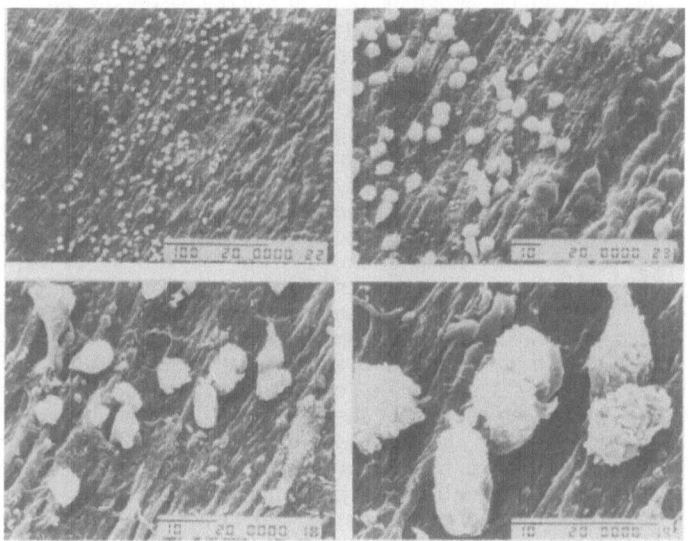

Figure 6. Veins exposed to IL-1 EVA disks had significantly more injury and PMN migration. Large areas exhibited deendothelialization. Many of these areas had both PMNs and platelets adherent to the exposed basement membrane. The subendothelial migration of PMNs in areas adjacent to those with exposed basement membrane suggests that PMN migration was the cause of endothelial cell loss in these veins.

In each of these 7 veins large areas of endothelial cell denudation with exposure of basement membrane and platelet accumulation was observed. Leukocyte migration was the most likely cause for the loss of endothelium. Intact endothelium, adjacent to deendothelialized zones, had large numbers of migrating leukocytes in the process of detaching the cells from the basememt membrane (fig. 6). Other conspicuous features of these 7 responder IL-1 treated veins included multiple leukocyte aggregates and the association of platelets with adherent leukocytes (fig. 6). One of the responder veins had a large mural thrombus on the lumenal surface.

DISCUSSION

Several studies have shown that a wide range of biological substances interact directly with leukocytes, causing increased adhesiveness.[3,4,5,6,7,8,9] However, historical evidence derived from in vivo[1,2] observations strongly suggests that it is the EC which becomes hyperadhesive following a variety of inflammatory stimuli. Our experiments were initially designed to determine the nature of such a response using the in vitro endothelial cell-leukocyte adhesion assay system. We pursued the hypothesis that EC damage results in increased adhesiveness of these cells for circulating leukocytes. Direct physical trauma to ECs (heat, pressure) failed to alter EC hyperadhesiveness or to induce secretory products possessing this activity (Dunn and Fleming, unpublished data). Since endotoxin causes injury to bovine ECs in vitro,[18] we decided to determine whether this effect could lead to enhanced EC adhesiveness for leukocytes. Surprisingly, LPS caused a delayed increase in EC adhesiveness for leukocytes which was not associated with EC toxicity. On the contrary, this proadhesive action of LPS was dependent on EC protein synthesis. Similar results were obtained with human ECs which are not susceptible to LPS-induced injury.[17] Such a novel phenomenon had not been previously described. At this time reports appeared on the stimulatory effects of LPS and IL-1 on EC procoagulant activity,[19,20,21] an event which was also protein-synthesis dependent. We reasoned, therefore, that IL-1 may represent a potential endogenous mediator of EC-leukocyte adhesion, via direct activation of ECs. As the in vitro data show, IL-1 indeed proved to be a potent stimulator of EC adhesiveness for a variety of leukocyte types (PMNs; U-937 mononuclear cells; peripheral blood lymphocytes). As with LPS, IL-1 caused a dose-dependent delayed increase in EC-PMN adhesion without EC injury, was specific for ECs and did not involve leukocyte stimulation. Exposure of ECs to stimulus for as little as 5 min resulted in EC hyperadhesiveness provided a 2 hr latent period was allowed to elapse prior to addition of leukocytes. Thus, adhesion could be induced by brief exposure of ECs to IL-1 (or LPS) which was not reversible upon removal of stimulus. Preincubation of ECs with actinomycin D or cycloheximide resulted in complete inhibition of IL-1 and LPS-induced EC-PMN adhesion, suggesting a requirement for protein synthesis. This, coupled with the fact that IL-1 or LPS-treated EC culture supernatants could not transfer activity, suggest that adhesion is probably mediated via EC surface membrane changes. These observations have been confirmed and extended by the work of other investigators[20] who have shown that IL-1 induces the expression of a specific EC surface protein which is thought to be the proadhesive molecule. Such a protein, to which a monoclonal antibody, H4/18, has been raised, correlated with the development of delayed-type hypersensitivity vascular lesions in man.[22] That the IL-1 induced EC procoagulant activity may be related to the observed hyperadhesive response seems unlikely since addition of tissue thromboplastin or various anti-thrombin agents failed to modify adhesion (Dunn & Fleming, unpublished observations) and the EC procoagulant response is refractory to repeated IL-1 stimulation[20] whereas adhesion is not.

Other alterations in EC surface membrane properties have been reported, including increased expression of Fc (IgG) and C'3 receptors[23] following viral infection, which paralleled increased adhesion of PMNs to the infected endothelium. IL-1 has also been shown to stimulate formation of an EC pericellular glycosaminoglycan matrix.[24] Further studies are required to determine whether there is a relationship between these events and IL-1 induced EC hyperadhesiveness.

In addition to its proadhesive effects, IL-1 induces EC synthesis of procoagulant activity,[20,21] tissue plasminogen activator inhibitor[25] and delayed release of prostacyclin[26] in vitro. Accordingly, we attempted to determine the effects of IL-1 on leukocyte-vascular interactions in vivo. A model was constructed whereby either bovine serum albumin (BSA) or IL-1 was applied locally from a slow-release polymer (EVA) positioned around the adventitial side of the jugular vein. Control EVA-BSA treated veins showed evidence of variable pathological changes, probably due to a combination of the irritant effects of surgery and the presence of the EVA-BSA disk. It is also possible that the perivascular disk caused some intermittent turbulence or obstruction of blood flow during the 24 hr incubation period. Such changes, combined with surgical trauma, are important requirements for vascular injury.[27] In spite of the increased baseline injury observed in control veins, the IL-1 treated veins showed a significantly greater degree of injury, apparently mediated by leukocyte adhesion and migration. Supernantants from IL-1 disks, but not BSA disks incubated in culture medium, enhanced EC-PMN adhesion in vitro indicating that IL-1 was released from EVA polymer disks in vivo (Dunn and Fleming, data not shown). It would appear that IL-1 diffused from the polymer to the vein lumen, activating ECs to become hyperadhesive for leukocytes. In most IL-1 treated veins (7 of 11), larger numbers of individual aggregated leukocytes were observed than in control veins. This would be consistent with the in vitro observations reported herein. The subsequent migration and activation of accumulated leukocytes would induce EC loss, basement membrane exposure and platelet accumulation. That the EC changes were mediated by an inflammatory reaction generated by IL-1 around the vein seems unlikely since 1) subcutaneous implantation of EVA disks containing IL-1 produces minimal inflammation which does not differ from control BSA or phosphate buffered saline (Dunn and Gibbons, unpublished observations), 2) the inflammatory response to IL-1 is brief, peaking around 3 hr and requires a relatively high local concentration of IL-1,[28] and 3) the inflammatory response to surgery would probably mask any IL-1 induced inflammation. These results are in accord with recent studies which also demonstrate IL-1 stimulation of EC procoagulant activity in vivo.[29]

Collectively, these observations indicate that IL-1 and LPS augment EC hyperadhesiveness for leukocytes in vitro via mechanisms dependent upon EC protein synthesis. The significant pathological changes observed in the blood vessel lumen following perivascular application of IL-1 in vivo suggests a potential role for this cytokine as an endogenous mediator of inflammatory and thrombotic diseases. In contrast, the stimulatory effects of LPS on EC adhesiveness may be a major contributory factor to the extensive adhesion of leukocytes observed throughout the vascular bed in septic shock.[30] The relevance of PMA and PHA-stimulated EC-leukocyte adhesion to vascular diseases has yet to be determined although these stimuli may represent useful tools in gaining a better understanding of the mechanism(s) of induced vascular hyperadhesiveness at the molecular level.

REFERENCES

1. A. Atherton and G.V.R. Born, Quantitative investigations of the adhesiveness of circulating polymorphonuclear leucocytes in blood vessel walls, J. Physiol., 222:447(1982).
2. F. Allison, M. R. Smith and W. B. Wood, Studies on the pathogenesis of acute inflammation. I. The inflammatory reaction to thermal injury as observed in the rabbit ear chamber, J. exp. Med., 102:655 (1955).
3. R. L. Hoover, R. T. Briggs and M. J. Karnovsky, The adhesive interaction between polymorphonuclear leukocytes and endothelial cells in vitro, Cell, 14:423 (1978).
4. G. M. Vercellotti, J. McCarthy, L. T. Furcht, H. S. Jacob and C. F. Moldow, Inflamed fibronectin: an altered fibronectin enhances neutrophil adhesion, Blood, 62:1063 (1983).
5. M. G. Tonnesen, L. Smedly, A. Goins and P. M. Henson, Interaction between neutrophils and vascular endothelial cells, in "Cologne Atherosclerosis Conference No. 1: Inflammatory Aspects," p. 25, M. J.

Parnham and J. Winkelmann, ed., Birkhauser Verlag, Basel-Boston-Stuttgart (1982).

6. H. S. Jacob, P. R. Craddock, D. E. Hammerschmidt and C. F. Moldow, Complement-induced granulocyte aggregation: an unsuspected mechanism of disease, The New Engl. J. Med., 302:789 (1980).

7. J. Palmblad, C. L. Malmsten, A-M. Uden, O. Radmark, L. Engstedt and B. Samuelsson, Leukotriene B$_4$ is a potent and stereospecific stimulator of neutrophil chemotaxis and adherence, Blood, 58:658 (1981).

8. E. J. Goetzl, M. L. Brindley and D. W. Goodman, Enhancement of human neutrophil adherence by synthetic leukotriene constituents of the slow-reacting substance of anaphylaxis, Immunol., 50:35 (1983).

9. I. F. Charo, C. Yuen, H. D. Perez and I. M. Goldstein, Chemotactic peptides modulate adherence of human polymorphonuclear leukocytes to monolayers of cultured endothelial cells, J. Immunol., 136:3412 (1986).

10. R. L. Hoover, M. J. Karnovsky, K. F. Austen, E. J. Corey and R. A. Lewis, Leukotriene B$_4$ action on endothelium mediates augmented neutrophil-endothelial adhesion, Proc. natl. Acad. Sci., USA, 81:2191 (1984).

11. E. A. Jaffe, R. L. Nachman, C. G. Becker and C. R. Minick, Culture of human endothelial cells derived from umbilical veins, J. Clin. Invest., 52:2745 (1973).

12. E. A. Jaffe, L. W. Hoyer and R. L. Nachman, Synthesis of antihemophilic factor antigen by cultured human endothelial cells, J. Clin. Invest., 52:2757 (1973).

13. A. Boyum, Separation of leucocytes from blood and bone marrow, Scand. J. Clin. Lab. Invest., 21:Suppl. 97 (1968).

14. J. Gallin, R. C. Clark and H. P. Kimball, Ganulocyte chemotaxis: an improved in vitro assay employing ^{51}Cr-labeled granulocytes, J. Immunol., 110:233 (1973).

15. J. Fine, Limulus assay for endotoxin, N. Engl. J. Med., 289:484 (1973).

16. I. Gery, R. Gershon and B. H. Waksman, Potentiation of the T-lymphocyte response to mitogens. I. The responding cell, J. exp. Med., 136:128 (1972).

17. W. D. Rhine, D. S. T. Hsieh and R. Langer, Polymers for sustained macromolecule release: procedures to fabricate reproducible delivery systems and control release kinetics. J. Pharm. Sci., 69:265 (1980).

18. J. M. Harlan, L. A. Harker, M. A. Reidy, C. M. Gajdusek, S. M. Schwartz and G. E. Striker, Lipopolysaccharide-mediated bovine endothelial cell injury in vitro, Lab. Invest., 48:269 (1983).

19. T. Lyberg, K. S. Galdal, S. A. Evensen and H. Prydz, Cellular cooperation in endothelial cell tissue thromboplasin synthesis, Br. J. Haematol., 53:85 (1983).

20. M. P. Bevilacqua, J. S. Pober, M. Elyse Wheeler, R. S. Cotran and M. A. Gimbrone, Interleukin-1 activation of vascular endothelium. Effects on procoagulant activity and leukocyte adhesion, Am. J. Pathol., 121:393 (1985).

21. A. E. Shorer, C. F. Moldow, G. H. R. Rao, M. Oken and M. E. Kaplan, Interleukin 1 (IL-1) promotes tissue factor expression in human endothelial cells, Blood, 64:(Nov. Suppl.) 267a (1984).

22. R. S. Cotran, M. A. Gimbrone, M. P. Bevilacqua, D. L. Mendrick and J. S. Pober, In situ induction and detection of a human endothelial cell activation antigen in immunologic inflammation, Fed. Proc., 45:379, Abstr. no. 1309 (1986).

23. D. B. Cines, A. P. Lyss and M. Bina, Fc and C'3b receptors induced by herpes simplex virus on cultured endothelial cells, J. Clin. Invest. 69:123 (1982).

24. R. Montesano, A. Mossaz, J-E. Ryser, L. Orci and P. Vassalli, Leukocyte interleukins induce cultured endothelial cells to produce a highly-organized glycosaminoglycan-rich pericellular matrix, J. Cell. Biol., 99:1706 (1984).

25. J. J. Emeis and T. Kooistra, Interleukin 1 and lipopolysaccharide induce an inhibitor of tissue-type plasminogen activator in vivo and in cultured endothelial cells, J. exp. Med., 163:1260 (1986).
26. E. Dejana, F. Breviario, V. Rossi, P. Ghezzi and A. Mantovani, Stimulation of vascular cell prostacyclin production by interleukin-1, in: "The Physiologic, Metabolic and Immunologic Actions of Interleukin-1," p. 55, M. J. Kluger, J. J. Oppenheim and M. C. Powanda ed., A. R. Liss, New York (1985).
27. R. G. Schaub, C. A. Simmons, M. H. Koets, P. J. Romano II and G.J. Stewart, Early events in the formation of a venous thrombus following local trauma and stasis, Lab. Invest., 51:218 (1984).
28. G. Beck, G. S. Habicht, J. L. Benach and F. Miller, Interleukin-1. A common endogenous mediator of inflammation and the local Schwartzman reaction, J. Immunol., 136:3025 (1986).
29. P. P. Nawroth, D. A. Handley, C. T. Esmon and D. M. Stern, Interleukin 1 induces endothelial cell procoagulant while suppressing cell-surface anticoagulant activity, Proc. natl. Acad. Sci. USA., 83:3460 (1986).
30. D. C. Morrison and R. J. Ulevitch, The effects of bacterial endotoxins on host-mediation systems-A review, Am. J. Path., 93:527 (1978).

ACKNOWLEDGEMENTS

   We would like to thank Mrs. Diane Ulrich and Mr. Robert Simmons for their expert help in the design and development of figures and photographs. We are also grateful to Mrs. Irena Hrabowy who organized and typed the manuscript.

SYSTEMIC ALTERATION OF ENDOTHELIAL CELLS IN SHOCK

C. Mittermayer, R.P. Franke, G. Kalff, U.N. Riede
Department of Pathology and Department of Anaesthesiology
Aachen University of Technology, D-5100 Aachen
Klinikum der RWTH, West Germany

1. PATHOLOGY OF EARLY CHANGES IN SEPSIS

From the standpoint of pathology the number of patients dying from sepsis has been constantly increasing in the last 10 - 20 years. Diagnosis of this complex condition has become more difficult recently since it is not always possible to detect bacteria, fungi or viruses. This can be related to invasive medicine and intensive care following surgery and extensive burn.

Moreover, the spectrum of pathologic characteristics seems to have changed. The organ weights of patients suffering and dying from shock following trauma and sepsis differed in the 1950s, 1960s and 1970s: At autopsy the weight of the kidneys was high in the 1950 - 1960 period while the weight of the lungs was low in this period. In the 1970s the lung weights increased 2 - 3 times at autopsy compared to that in the 1950s indicating a shift in the shock syndrome symptoms (Mittermayer et al., 1973).

Complications arising from the lung function then came into the foreground and are still dominating the intensive care scene. By analyzing lung probes from deceased patients and lung biopsies of patients with the aid of electron microscopy and morphometric methods the following was shown:

> Within seconds after trauma the weight of the lungs increases from average 230 g to values up to 1000 g (Joachim at al., 1976) due to fluid assembly in the interstitial space (Riede et al., (1981). Endothelial cells are not damaged at this point. This stage is reversible.
>
> Between the 6th and 10th day after onset of shock a morphological transformation of the lung architecture takes place. Lung fibrosis is the final result. This kind of fibrosis cannot be discerned from other forms of lung fibrosis of different etiology on a morphological basis alone. Taking into correlation specific diffusion capacity of the alveolar wall and PaO2 (Riede et al.,

1984) the differential diagnosis can be made. Apparently shock-induced fibrosis of the lung is brought about by disturbances of diffusion and perfusion.

In any case of shock-induced lung damage the first morphological sign at the cellular level is damage of endothelial cells.

2. ENDOTHELIAL DAMAGE - INITIAL FEATURE OF SHOCK

The mean barrier thickness of the alveolar wall in human lungs decreases during the first 6 days in shock (Riede et al., 1978). This is due to blebbing and shedding of endothelial cells. Finally capillaries without any endothelial lining can be observed (Figure 1). The replacement of the lost endothelial cells then depends on the efficiency of the regenerative forces. About 50 % of these patients die; the remainder survives and shows severe morphological residues in the lung (Mittermayer, 1978).

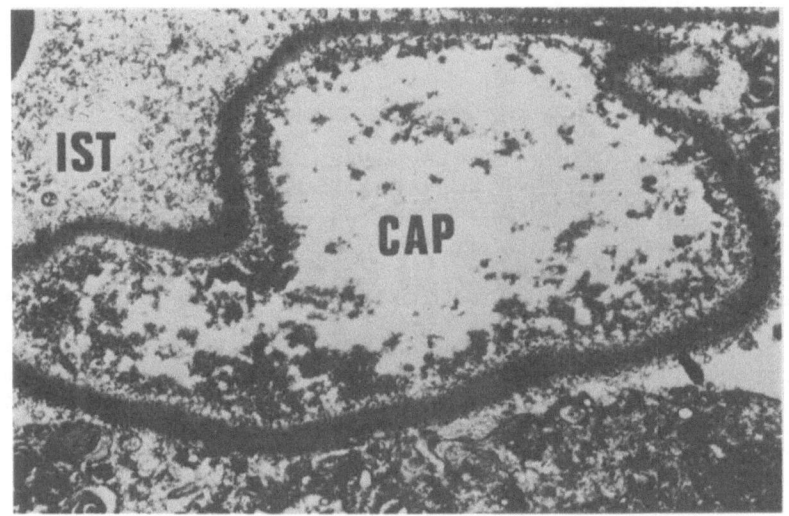

Figure 1

Human lung in septic shock. Capillary (CAP) is devoid of endothelial layer; basement membrane thickened. The interstitium (IST) is broadened by edema.

Damage of endothelial cells during shock seems to be a universal phenomenon in man. In the heart "shock-endocarditis" occurs (Mittermayer et al., 1971a). This is an aggregation of platelets along the closing zone of the valves (Figure 2). In animal experiments it was demonstrated that endotoxin induced shock produced endothelial damage after 30 minutes followed by endocarditis thrombotica. This occurred when endothelial cell death exceeded reparative forces at the valves. Endotoxin or Lipid A in combination with the mechanical alteration occurring at the closure of the valves evokes endothelial damage, which is followed in the porcine heart by a 1.200-fold regeneration burst compared to normal cell turnover there (Mittermayer, 1971b).

3. REDUCED REGENERATION OF HUMAN ENDOTHELIAL CELL CULTURES

At present there is still only scarce knowledge of how endothelial cell damage occurs during shock in various regions of the human body. At the Department of Pathology, Aachen, Technical University, an in vitro system of human endothelial cells was used to study proliferation under different conditions (Scharffetter et al., 1986). These cells were derived from umbilical veins or veins from vein-stripping surgery in adults. Grown on a polymer-base and

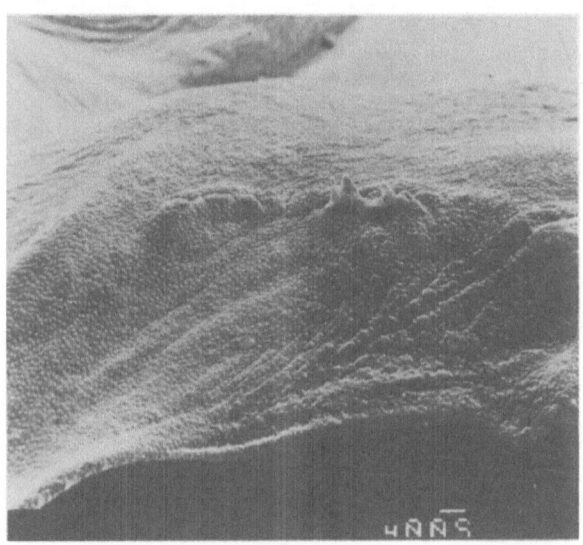

Figure 2

Scanning Electronmicroscopy (marker = 500 um)
Endocarditis thrombotica in pig aortic valve 1 hour after
i.v. endotoxin injection (20 µg/kg body weight)

coated with extracellular matrix, they exhibited Factor VIII related antigen, prostaglandin production and the appearance of silver-stained cell borders upon confluency (Figure 3). If non-confluent cultures are used and fed with serum from a patient with septic shock it can be seen that marked decrease in proliferation activity occurs (Schöffel et al., 1982a). Moreover, the prognosis of patients correlates grossly with the degree of this inhibition (Schöffel et al., 1982b). Thus the septical patient may not be able to repair endothelial damage in vivo and develops the clinical signs of increased vascular permeability and microthrombosis as seen in the lung.

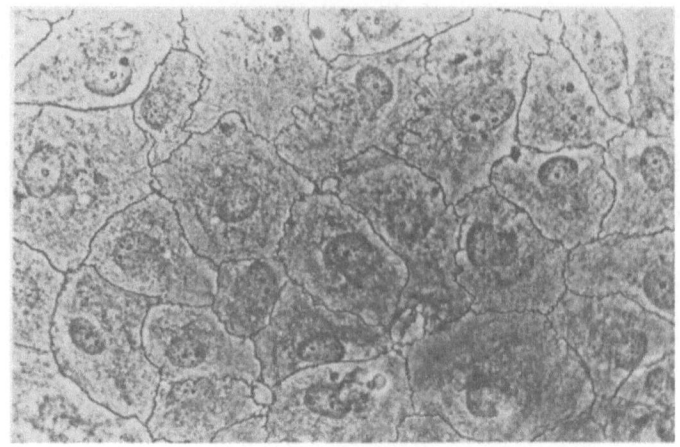

Figure 3

Human endothelial cells in confluent state. Silver-staining
of the cell borders. The cells are polygonal and cell borders
are indented (280 x).

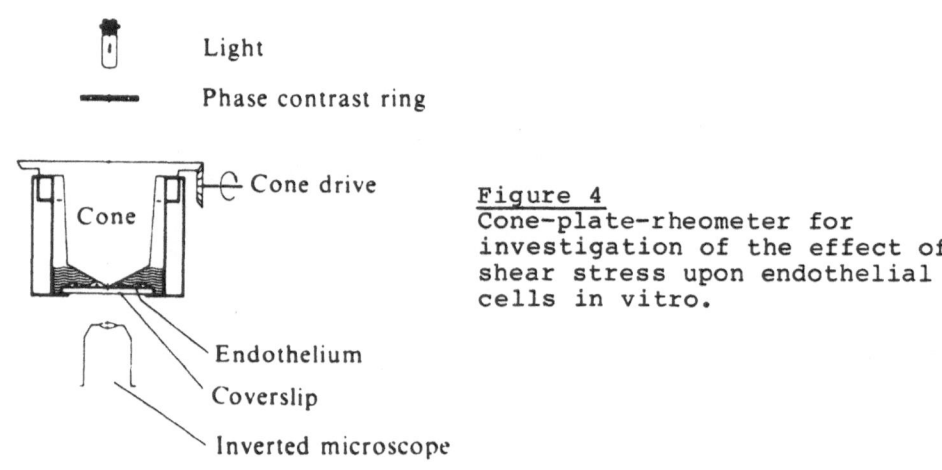

Light

Phase contrast ring

Cone drive

Cone

Figure 4
Cone-plate-rheometer for
investigation of the effect of
shear stress upon endothelial
cells in vitro.

Endothelium

Coverslip

Inverted microscope

4. DISTURBANCE OF ENDOTHELIAL CELL ADHERENCE IN SEPTIC SHOCK

Human endothelial cells are grown on a transparent base in a
cone-plate rheometer (Figure 4) and can be observed
continously by microscope and camera (Franke et al., 1984).
Confluent human endothelial cells in culture show a circum-
ferencial microfilament network. No stress fibres are seen
at this stage crossing the cytoplasm (Schnittler et al.,
1986) Shear stress of 2,5 dyn/cm2 over 3 hours is able to
induce large amounts of stress fibres crossing the cytoplasm. (Figure 5).

The range between 2 and 6 dyn/cm2 shear stress is equivalent
to the venous and the arterial shear stress in man. In this
dynamic type cell culture, cells must be highly confluent
(5th day after confluency) in order to be fully non-
thrombogenic in the fluid dynamic experiments with blood. If
lipid A, the effective constituent of endotoxin, is added in
growth medium without blood (200 ug/ml), rupture of the cell
layer is encountered. The same holds true for supernatants of
highly pathogenic E. coli and clostridium strains (Franke et
al., 1985).

Figure 6 demonstrates the effect of sera from 8 patients
suffering from hemorrhagic shock and 12 patients suffering
from septic shock, two of them after burns. It is seen that
endothelial cell layer is lost to the artificial stream of
culture medium during shear stress. If whole blood of healthy
donors is then added to the chamber thrombi growing at the
site of endothelial damage can be seen. In control
experiments with sera and blood from healthy donors no
thrombi are formed during the period of the experiment.

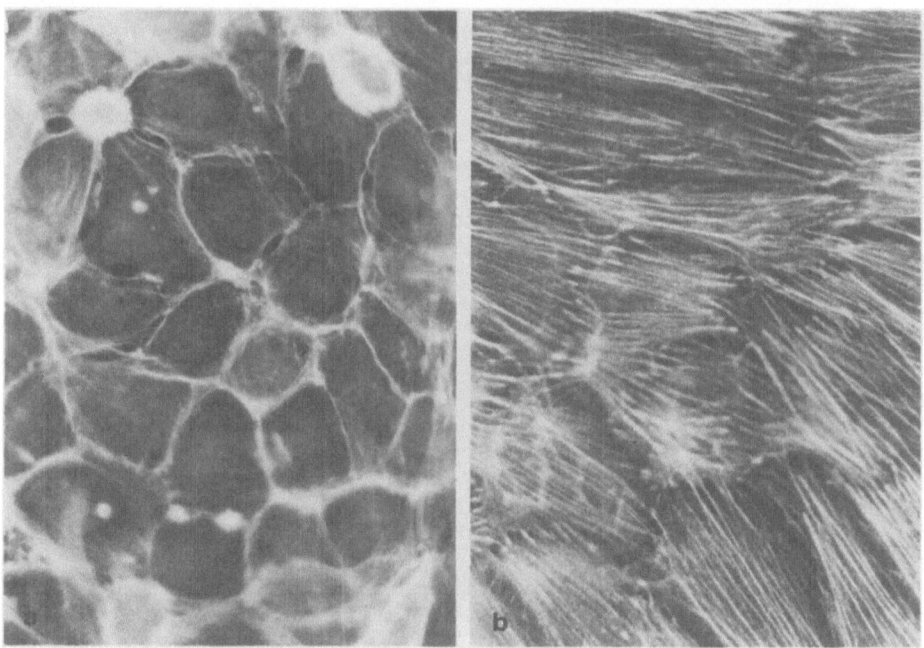

Figure 5

          Pattern of actin filament in human endothelial
cell culture visualized by phaloidin-rhodamin.
a) Human endothelial cells in confluent state show only
   a marginal microfilament network. No stress fibres cross
   the cytoplasm.
b) Human endothelial cells after 3 hours of exposure to low
   levels of fluid shear stress (2,5 dyn/cm2); large amount
   of stress fibres pass through the cells.

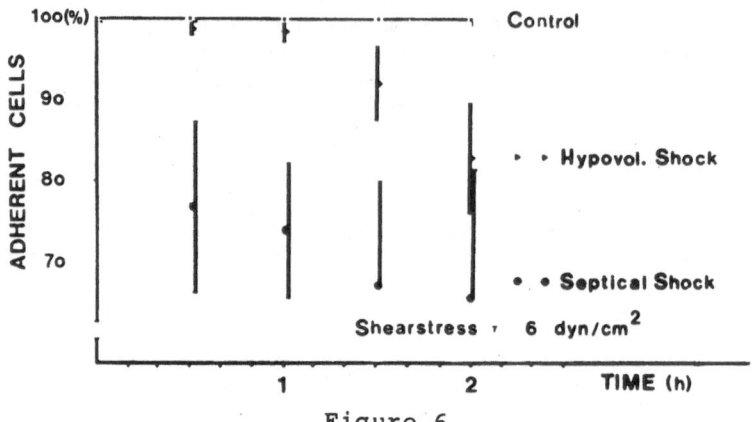

ADHERENCE of Human Endothelial Cells
in Cone-Plate-Rheometer

Figure 6

Confluent human endothelial cells are incubated (48 hours) with sera from shock patients. Cells were then subjected to fluid shear stress (6 dyn/cm2).

It is hoped, that with this dynamic cell system problems of cell adherence and cell regeneration seen in shock patients can be elucidated by standardized cell models.

Acknowledgement

This study is supported by SFB 109 and SFB 106 of the Deutsche Forschungsgemeinschaft.

REFERENCES

1. Franke R.-P., Gräfe M., Schnittler H., Seiffge D., Drenkhahn D. and Mittermayer C.
   Nature 307 : 648 (1984).

2. Franke R.P., Dauer U., Sobczak H. and Mittermayer C.
   Fortschr. antimikrob. u. antineopl. Chemother. 4 : 873 (1985).

3. Joachim H., Vogel W. and Mittermayer C.
   Z. Rechtsmedizin 78 : 13 (1976).

4. Mittermayer C., Waldthaler A., Vogel W. and Sandritter W.
   Beitr. Path. Anat. 143 : 29 (1971a).

5. Mittermayer C., Madreiter H., Schindera F. and Huth K.
   Verh. dtsch. Ges. Path. 55 : 350 (1971b).

6. Mittermayer Ch., Thomas C., Rengholt R., Schäfer H., Martinez G. and Sandritter W.
   Klin. Wschr. 51 : 37 (1973).

7. Mittermayer C., Hassenstein H. and Riede U.N.
   Path. Res. Pract. 162 : 73 (1978).

8. Riede U.N., Joachim H., Hassenstein J., Costabel U.,
   Sandritter W., Augustin P. and Mittermayer C.
   Path. Res. Pract. 162 : 41 (1978).

9. Riede U.N., Mittermayer C., Horn R. and
   Sandritter W.
   Pathobiology of the Alveolar Wall in Human Shock Lung
   in: Pathophysiology of Shock, Anoxia, and Ischemia
   Editors: R. Adams Cowley, B.F. Trump
   Williams & Wilkins Baltimore/London p. 358-371 (1981).

10. Riede U.N. and Mittermayer C.
    Anaesthesiol. u. Reanimat. 9 : 143 (1984).

11. Scharffetter K., Blasius S., Franke R.-P., Köhler M.,
    and Mittermayer C.
    VASA 15 : 33 (1986).

12. Schnittler H.-J., Franke R.P., Fuhrmann R.,
    Mittermayer C. and Drenckhahn D.
    Europ. J. of Cell Biol. (in press) (1986).

13. Schöffel U., Kopp K.H., Männer H., Vogel F. and
    Mittermayer C.
    European Journal of Clinical Investigation 12 : 165
    (1982a).

14. Schöffel U., Shiga J. and Mittermayer C.
    Circulatory Shock 9 : 499 (1982b).

ARTERIAL WALL REACTION FOLLOWING ELECTRICAL LOCAL DE-ENDOTHELIALIZATION

R. H. Bourgain, R. Andries, C. Bourgain and L. Maes
Department of Physiology, V.U.B., Laarbeeklaan 103
B-1090 Brussel, Belgium

INTRODUCTION

Endothelial continuity is essential for vascular homeostasis as its maintenance prevents local thromboformation and atherogenesis (1). If it is well known that small, restricted endothelial denudation zones may occur as the result of hemodynamics (2), homocystinemia (3), generation of free oxygen species (4), and angiotensin challenge (5), in these conditions, only a few endothelial cells are affected and no myointimal thickening occurs. More recently, Hansson and Schwartz (6), described focal spontaneous cell death in rat aortic endothelium following a long phase of gradual detachment of the affected endothelial cells, but without significant exposure of the subendothelial structures because migration of the normal surrounding cells rapidly covered the area of denudation.

In a previous paper by Potvliege and Bourgain (7), evidence was adduced that local endothelial cell loss induced by the application of an electrical current, is invariably followed by a typical reaction pattern involving the surrounding endothelium and the media: indeed both smooth muscle cell migration and accumulation of lipids and collagen within the vessel wall is observed and appears to be directly related to direct exposure of subintimal tissue to the blood stream. Guilbaud et al. (8), observed major modifications of the vascular and particularly the arterial endothelium in patients suffering from electrocution complicated by burn, while Braquet et al. (9), convincingly demonstrated the involvement of the arachidonate cascade and liberation of free oxygen species in severely burned patients.

In view of these observations, we felt the need for a systematic approach to the study of arterial wall reaction by performing local electrically induced de-endothelialization in small vessels using a method developed and standardized by Bourgain et al. (10).

Inbred male white Wistar rats weighing between 250 and 300 g as well as guinea-pigs (500 to 600 g) were investigated in order to study ultrastructure of and the effect of PAF-acether antagonists on electrically induced thromboformation. All experiments were performed under general anesthesia by thalamonal (3 ml/kg body weight IM) and the animals were kept under artificial ventilation. Following a longitudinal incision in the abdominal wall, a loop of small intestine was gently extracted from the abdominal cavity and a branch of the mesenteric artery (250 to 300 um in diameter) was dissected free from the surrounding mesenteric tissue over a length of 2 to 3 mm and kept under continuous superfusion by isotonic Ringer solution at 37 °C. The induction of local de-endothelialization was performed using a micromanipulator for the application of a platinum electrode onto the vessel wall in the middle of the arterial segment. This electrode measured 100 um in diameter and was insulated except at the tip, where the conductive diameter was restricted to 25 um. A direct current was then applied for 60 seconds; its polarity was inversed every 5 seconds while the intensity of the current was varied according to the diameter of the vessel, ranging from 30 to 50 uA. The inversion of the polarity avoided vasospastic phenomena. The electrical current application resulted in a retraction followed by desquamation of approximately 20 adjoining endothelial cells and exposure of the subintimal structures to the blood stream, but without interference with the normal blood flow pattern.

For ultrastructural analysis, at well indicated time intervals, the animals were perfusion-fixed at physiological pressure through a catheter in the left common carotid (1.5 % glutaraldehyde in 0.1 M cacodylate buffer with 2 % sucrose and 2 mM $CaCl_2$, pH 7.4, 37 °C). After overnight immersion fixation of the whole mesenterium in the glutaraldehyde fixative (4°C), the arterial segment was isolated from the surrounding structures, postfixed in 1 % osmiumtetroxide for 1 h, dehydrated in ascending ethanol concentrations and liquid CO_2 critical point-dried. The specimens were sputtercoated with 15 nm gold and examined with a Philips SEM 505 scanning electron microscope at 30 kV. For transmission electron microscopy the arterial segment was seized in a forked 5 mm wide forceps, excised and immediately plunged in 2.5 % glutaraldehyde in 0.075 M cacodylate buffered to pH 7.2 and left to fix for 2 hours at room temperature. After overnight rinsing, they are postfixed for 1 h in unbuffered 1 % osmiumtetroxide. The specimens were stained in bulk for 2 h at 37 °C by 0.5 % uranyl acetate buffered to pH 5.3 before being embedded in Spurr's resin. Transverse sections were cut over at least half the length of each specimen. Semi-thin sections, taken at regular intervals, were stained with alkaline methylene blue and examined by light microscopy. Thin sections, taken in uninvolved as well as in de-endothelialized portions were contrasted by lead citrate and examined with a Zeiss EM-9 electron microscope.

When in the course of the current application or immediately afterwards, a local thrombus appeared, it was

measured by means of the optoelectronic device we developed
accordingly. The PAF acether antagonist used was the gink-
golide BN52021 (Ipsen Lab).

RESULTS

Ultrastructural observations

     Normal endothelium. SEM investigation revealed the
typical ultrastructure of normal endothelium in the rat.
The endothelial monolayer formed a thin carpet of closely
adjoining, flattened, polygonal and elongated cells, their
nuclei bulging slightly into the vascular lumen. Except at
bifurcations where the orientation of the cells is more
randomlike; the long axis of the endothelial cells always
paralleled the direction of the blood stream as described
by Maes et al. (11).

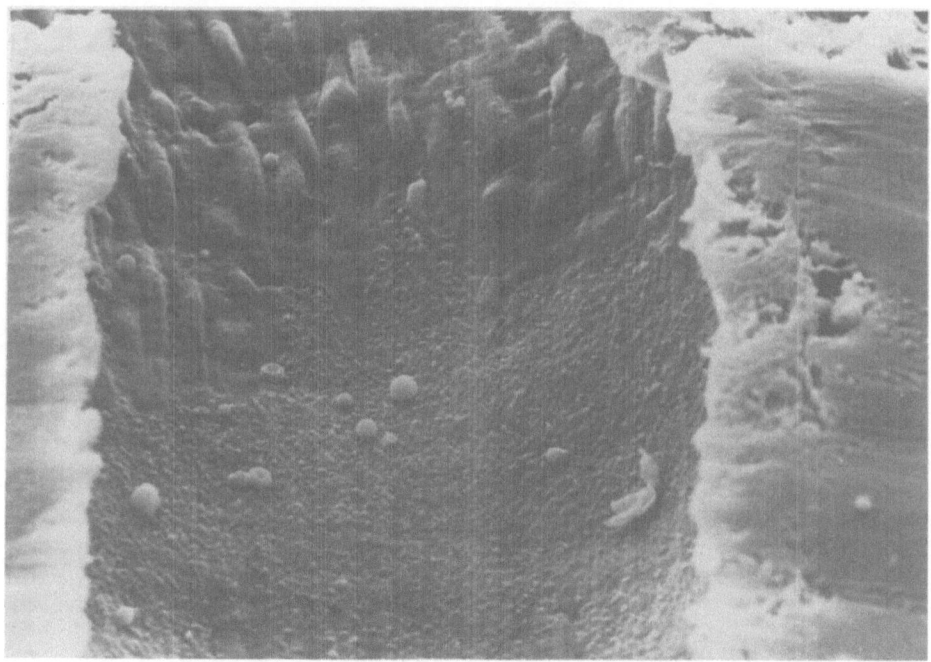

Figure 1. SEM-picture of an arterial segment excised 3 days
after the induction of the lesion, showing the
activated endotelial cells and a large de-endo-
thelilialized area, covered by platelets.

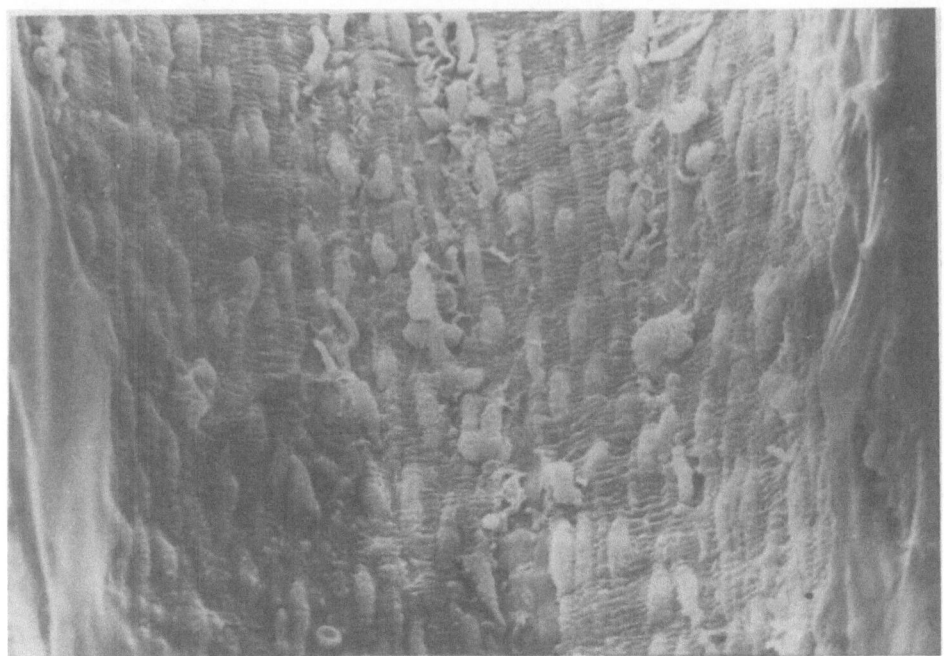

Figure 2. SEM-picture of an arterial segment excised 7 days after the induction of the lesion, showing activated endothelial cells.

Intimal lesion. The application of the electrical current resulted in an immediate, localized, slightly oval lesion which consisted firstly in a retraction of the endothelial cells at the site of the application of the current followed by the detachment of 15 to 20 adjoining endothelial cells. Sometimes, small localized crevices became evident in the underlying subintima. Very few ultrastructural alterations were present at this stage in the media. If present, they consisted of some membrane irregularities, eventually pycnotic nuclei and some degree of disrupture of the organelles. The lamina elastica was never ruptured and maintained a normal aspect. Few if any platelets adhered onto the denuded area but a few hours later more adhesion occurred. When observed 3 min following the application of the electrical current, the edge between the area of the lesion and the nondenuded endothelium consisted of flaps of partially detached endothelial cell remnants. Over time the area of the lesion extended. Some endothelial cells surrounding the affected area demonstrated detachment from the subendothelium and manifested bulging into the vascular lumen (11).

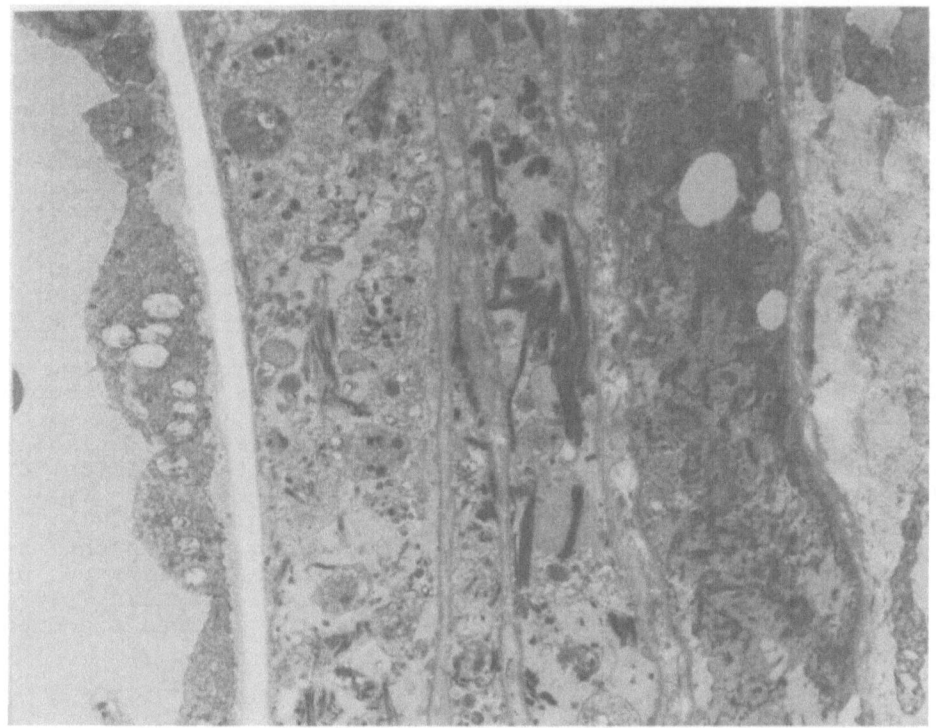

Figure 3. TEM-picture of a 3 days old lesion, clearly demonstrating the presence of platelets in the
medial structures (*4000).

These cells subsequently then completely desquamated,
delimiting an area of 30 to 35 cells. This phenomenon was
observed within 24 to 48 h. The surrounding endothelial
cells showed a very marked bulging of the nuclei and elongation of their structure revealing the characteristic
feature of activated cells. Transmission electron microscopy demonstrated in the cytoplasm of these activated
cells extensive enlargement of the endoplasmic reticulum
and important hyperplasia. Observations made on days 3 and
7 indicated that the activated endothelial cells both at
the upstream and downstream locations of the affected zone,
were moving in order to cover the de-endothelialized area;
all of these cells demonstrated extreme bulging and at
cetain sites they clearly overlapped the edges of their
adjoining neighbours. Flaps of these cells apparently
floating freely into the vascular lumen as a result of this
phenomenon. In all specimens examined, these findings were
constantly observed. At day 7, the de-endothelialized area
was completely covered by the endothelial cells, very occasionally a small crevice could still be detected between
some cells, but this was rather exeptional. The general
aspect of the cells was still one of some degree of activation (figures 1 and 2).

Medial lesions. The modifications which appear immediately after application of the electrical current were
described in the preceding section. Within 24 h, the media
underlying the site of application of the current, demon-

strated extensive invasion by platelets. Eventually, a few
erythrocytes were seen together with the invading platelets
located deep in the vessel wall in between the smooth
muscle cells. This type of recruitment of thrombocytes by
the vessel wall was observed in all specimens; surprisingly, no leukocyte diapedesis was evident. These phenomena
were observed to a higher degree at day 3; by this time,
the platelets had degranulated and the few erythrocytes
became deformed and were clearly hemolyzing (figure 3). At
day 7 the phenomena had abated to some extent, but the
evidence of platelet recruitment and degranulation was
still very much conspicuous. The smooth muscle cells presented very marked vacuolization but this remained restricted to the area underlying the site of application of
the electrical current. The plasma membranes of the smooth
muscle cells did not show any disruption, but at some
sites, the cytoplasm presented small vacuoles, tightly
packed together and higly suggesting fatty infiltration
(figure 4).

Interestingly, in animals subjected to a high cholesterol and high fat diet for at least 2 months, this type of
lesion was also observed by Potvliege and Bourgain (12).
Fibrin deposition within the vessel wall media was a common

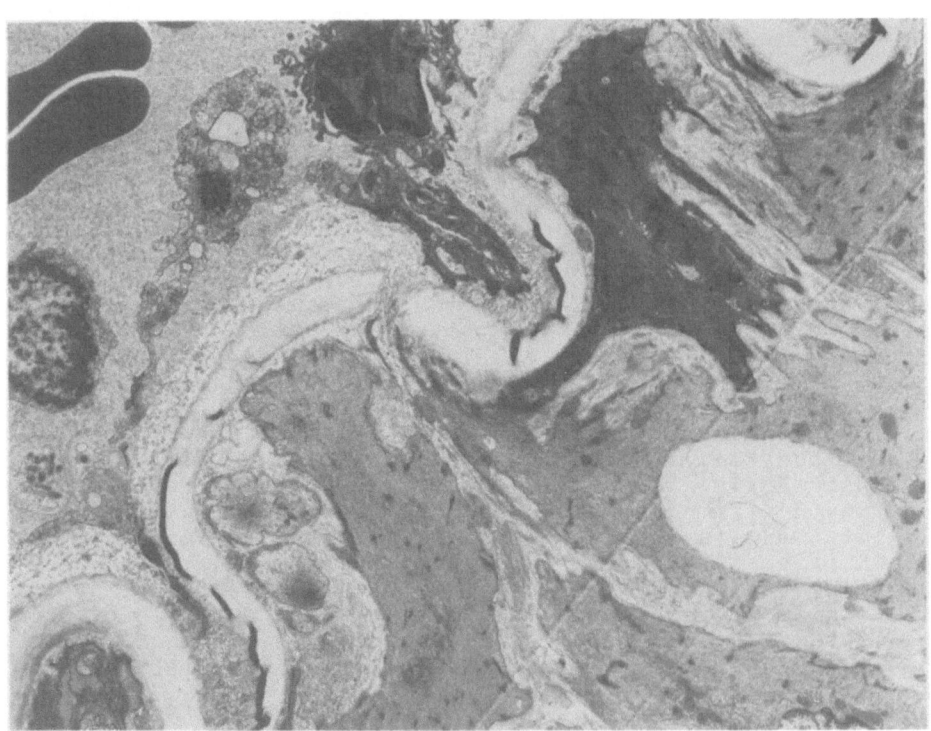

Figure 4. TEM-picture of a 7 days old lesion showing the
deposition of fibrin (*4000).

finding in 48 h to 3 to 4 days specimens. When compared to
normal arteries, the alignment of the smooth muscle cells

was drastically modified and no longer presented a well defined circular appearance. Marked irregularity of their contours was seen, admittedly restricted to the current application site but not exceeding this zone to any major extent. Important collagen and elastin deposition was present in the medial stuctures near the adventitia.

Thrombotic Complications

In the described experimental conditions, electrical current application onto the arterial wall, performed in the guinea-pig in identical conditions as in the Wistar rat, frequently resulted in the appearance of a thrombus at the site of endothelial cell lesion. The thrombus consisted mainly of platelets although at a later stage, between 4 to 6 minutes, some leukocytes were recruited, surrounding and penetrating the thrombus. Topical superfusion by BN52021 at 10-3 M or intravenous administration (4 mg/kg) of the drug, invariably resulted in the disappearance of the thrombotic phenomenon. This observation convincingly demonstrated, that at least in these experimental conditions, the thrombotic phenomenon following desquamation of the endothelial cells was related to generation of PAF-acether by the vessel wall, possibly by the endothelial cells surrounding the desquamated site. In a previous paper, Bourgain et al. (12) documented these findings in detail.

DISCUSSION

Our data clearly demonstrate that the local application of an electrical current in well standardized conditions, induces specific alterations at the site of the endothelium and the media of the arterial wall which do not extend into the adventitia. Within a few minutes following current application, the endothelial cells demonstrate marked retraction, rapidly followed by desquamation from the underlying intimal structures. Immediately following the electrical challenge, few if any platelets adhere onto the exposed intimal structures; however, within minutes, more platelets adhere onto the denuded areas and within a few hours, invasion of the vascular wall by platelets occurs complying with the observation that the local application of an electrical current results ultimately in platelet adhesion onto the de-endothelialized area. The presence of fibrin within the vessel wall also clearly demonstrates that plasma proteins infiltrate the vessel wall and participate in a process of inflammatory reactions comprising the intima and media of the arterial wall. The appearance of fatty vacuoles in the smooth muscle cells as well as the deposition of fats, collagen and elastin in the interstitial spaces separating these cells is indicative of a process specifically affecting the arterial media. The reaction of the endothelial cells surrounding the affected area is identical in all arterial specimens examined. The bulging phenomenon and the marked elongation of the cells as well as the superposition of their edges, are indicative of a mobilisation to cover the denuded area; this activity is present upstream as well as downstream of the de-endothelialized area. Transmission electron microscopic data clearly indicate that the endothelial cells present a very high degree of metabolic activity as demonstrated by the increase of the endoplasmic reticulum and cell size and conspicuous signs of hyperplasia. The observation that thromboformation can be inhibited by specific PAF-acether antagonists is highly suggestive that this mediator is indeed intimately involved in the accompanying processes of platelet-vessel wall interaction and vascular challenge.

REFERENCES

1. Ross, R., Glomset, J. and Harker, L. Am. J. Path. 86: 675 (1977)
2. Fry D. Circulation Res. 22: 165 (1968)
3. Harker L., Ross R., Slichter S. and C. Scott. J. Clin.Invest. 58: 731 (1976)
4. Sachs, T., Moldow, C.F., Craddock, P.R. and Bowers, T.K. J. Clin. Invest. 61: 1161 (1978)
5. Reidy, M.A. and Schwartz, S.M. Microvasc. Res. 24: 158 (1982)
6. Hansson, G.K. and Schwartz, S.M. Am. J. Path. 112: 278 (1983)
Maes L., Andries R. and R. Bourgain, Endothelial injury and platelet thrombosis in mesenteric arteries of rats: A scanning electron microscopy study, Blood Vessels, 23: 1 (1986)
7. Potvliege. P. and Bourgain. R. Br.J. Exp Path, 63: 116 (1982)
8. Guilbaud, J. Unpublished data (1986)
9. Braquet, M., Ducousso, R., Garay, R., Guilbaud, J., Carsin, H. and Braquet P. The Lancet 8409: 976 (1984)
10. Bourgain, R., Vermarien, H., Andries, R., Vereecke, F., Jaqueloot, J., Rennies, J., Blockeel, E. and Six, F. Adv. Exp. Med. Biol. 180: 635 (1984)
11. Maes, 1., Andries, R. and Bourgain, R. Blood Vessels 23: 1 (1986).
12. Bourgain, R., Maes, L., Braquet, P., Andries, R., Touqui, L. and Braquet, M. Prostaglandins 30: 185 (1986).

Investigation supported by grants No 3.0067.83 and 3.0053.84 FGWO.
We thank Mrs. H. De Backer-De Zutter for skilled technical assistance.

Chapter 5

The role of
Lipid Mediators
in the
immune response

THE ROLE OF PLATELET ACTIVATING FACTOR
IN THE IMMUNE RESPONSE

P. Braquet[1], M. Paubert-Braquet[2] and M. Rola-Pleszczynski[3]

[1] IHB, 17 avenue Descartes, F-92350 Le Plessis Robinson, France
[2] Unité de Recherche Clinique, Centre de traitement des Brûlés, Hôpital d'instruction des Armées, Percy, F-92140 Clamart, France
[3] Faculté de Médecine, University of Sherbrooke, Sherbrooke (Canada)

INTRODUCTION

It has been demonstrated that histamine is released into rabbit plasma during the acute allergic reaction following IgE-dependent activation of basophils (1-3) and this was ascribed to a factor actively released from leukocytes by a calcium - and temperature - dependent process (4,5). From 1972 onwards, Benveniste et al. (6-7), described the method used to obtain this substance which they named platelet-activating factor (PAF), and they started its characterization, demonstrating that it was released from rabbit basophils by an IgE-dependent process. It was subsequently evidenced by structural elucidation that PAF is a phospholipase A_2 sensitive phospholipid (8-10), identified as 1-alkyl-2(R)-acetyl-glycero-3-phosphorylcholine (10,11). The two main molecular species of PAF produced by human neutrophils or rabbit basophils have a hexadecyl or octadecyl alkyl moiety, but more advanced separation techniques have shown that PAF comprises multiple (at least 16) molecular species containing saturated or unsaturated alkyl or acyl chains, or even a polar head group other than choline (12,13).

PAF has been termed as both PAF-acether (ace for acetate and ether for the alkyl fond), and AGEPC (Acetyl Glyceryl Ether Phosphorylcholine) in the literature. In this paper, 1-0-octadecyl (or hexadecyl)-2(R)-acetyl-glycero-3-phosphorylcholine, is simply termed PAF.

As a potent mediator of anaphylaxis and hyperreactivity, PAF may be involved in shock, graft rejection, post-ischemic disorders, ovoimplantation, and certain central nervous system (CNS) disorders (rev. in 14-18). This review presents some data indicating that PAF is also a potent modulator of the immune response.

CELLULAR SOURCES OF PAF

As shown in Fig. 2, there are many cellular origins of PAF (6, 19-26). PAF is released by sensitized basophils and mast cells following antigen or anti-IgE challenge. Monocytes and macrophages respond to phagocytic stimuli such as bacteria, opsonized zymosan or immune complexes, or to the calcium ionophore A 23187, whereas neutrophils respond in addition to phorbol esters and the chemotactic agents C5a, C5a-des-Arg and formyl-methionyl-leucylphenylalanine (FMLP). Eosinophils also respond to A23187, C5a and FMLP, and, furthermore, to the eosinophil chemotactic factor of anaphylaxis (ECF-A). PAF is released by endothelial cells in response to A23187, thrombin, vasopressin, angiotensin II, anti-factor VIII and interleukin 1 (IL 1). Even the bacterium E. Coli was recently demonstrated to produce PAF (27).

While it was believed that lymphocytes cannot produce PAF, they can produce 2-lyso-PAF following stimulation with A 23187, and their inability to produce PAF may result from the lack of acetyl transferase (28). Nevertheless, in certain circumstances, appropriate stimulation will cause lymphocytes to synthesize or activate acetyl transferase and thus produce PAF : large granular lymphocytes release PAF after Fc receptor stimulation (29) and a series of human leukemic cell lines of B and T cell origin, as well as Epstein-Barr virus transformed lymphoblastoid cell lines release PAF-like material after stimulation with A23187, acetyl CoA or phytohemagglutinin (PHA) (30).

The cellular origin of PAF produced by various tissues and organs under stress or inflammation is still to be identified.

Many of the cells or tissues producing PAF are themselves targets of PAF-induced bioactions (31-34). Specific receptors or at least binding sites for PAF have been identified in many circumstances (using radiolabelled PAF) (rev. in 35,36). With the discovery and development of PAF receptor-specific antagonists research has been greatly facilitated in this field since they indicate the circumstances in which PAF mediates the observed pathophysiological effects (rev. in 35, 37-38).

REGULATION OF THE IMMUNE RESPONSE : DIRECT EFFECT OF PAF

Without a doubt, PAF participates in the complex system of host defenses as shown by its potent proinflammatory activities. Assisted by the development of PAF receptor antagonists and agonists, recent research efforts have been devoted to the potential of PAF in modulating cellular immune responses.

A concentration-dependent inhibition of lymphocyte proliferation was noted when PAF was added to human peripheral blood lymphocyte cultures (containing 5-10 % monocytes), stimulated with the mitogens PHA or concanavalin A (ConA) (39). Inhibition was observed when PAF was added within the first 3 h of a 72 h culture period, the IC_{50} ranging between 10^{-12} and 10^{-10} M. There was no significant effect on lymphocyte proliferation when PAF was added at a later time. The inhibition was not accompanied by cell toxicity as shown by the trypan blue exclusion method and a similar suppression of lymphocyte proliferation was observed with ethoxy-PAF, a non-hydrolyzable PAF analog (40). The PAF receptor antagonist, BN 52021, prevented PAF-induced this inhibition. Of interest, the cyclooxygenase inhibitor, indomethacin, also reversed suppression, which implies that prostaglandins intervene as second nessengers in the PAF-induced inhibition of lymphocyte proliferation (39). When production of IL 2 by human lymphocytes was measured after mitogen stimulation, a similar effect of PAF and ethoxy-PAF was noted, using the IL 2 dependent CTL-L2 cell line (39). BN 52021 also reversed the suppression of IL 2 production by PAF (39,40). It was also reported recently that PAF abolished CD^4+ T cell proliferation, but this inhibition was observed at higher concentrations of PAF and could be seen when PAF was added late during the culture and was not linked with inhibition of IL 2 production (Dulioust et al.) (41).

Seeing that inhibition of lymphocyte proliferation and IL 2 production may be due to the activation of suppressor cells, lymphocytes were preincubated with PAF for 3-18 H, washed and then added to fresh autologous lymphocytes stimulated with mitogens. This coculture assay showed that PAF could activate suppressor cells which, subsequently, inhibited lymphocytes cultures in the absence of any further contact with PAF (42). An increase in the number of CD8+ T cells and a slight decrease in CD4 + T cells after 18-48 h accompanied the induction of suppressor cells.

Of interest, the PAF antagonists _per se_ generated some suppressor cell activity, although to a lower degree than PAF.

A remarkable spectrum of activities emerged when specific leukocyte subsets were isolated and purified before preincubation with PAF (43) : PAF-preincubated monocytes exerted suppression by means of an indomethacin-sensitive mechanism, while PAF-preincubated CD8+ T cells had a suppressive effect _via_ an indomethacin-resistant mechanism. It is interesting to note that PAF-preincubated CD4+ T cells enhanced lymphocyte proliferation in a pronounced manner and this effect was not blocked by BN 52021.

In a different system, PAF and two non-hydrolyzable PAF analogs raised the proliferation of IL 2-stimulated human lymphoblasts, while some antagonists (CV-3988 and L-652,731, but not WEB 2086 and BN 52021) inhibited such proliferation (44,45). Even when the drugs were added 48 h after the beginning of a 72 h culture period this effect of the antagonists was observed, which implies that the drugs interfered with a later event in T cell activation. Furthermore, inhibition of IL 2-induced proliferation of T-lymphoblasts by the PAF synthesis inhibitor, L-648,611 suggested the possibility that endogenously produced PAF could be implicated in some step(s) of the inhibition (44,45).

An important corroboration of some of the _in vitro_ observations outlined above was provided by the _in vivo_ instillation of PAF into rats _via_ an implanted mini-pump (46). After seven days of instillation, PAF-treated rat splenocytes showed enhanced IL 2 production in response to ConA.

T-cells are known to regulate IgE production (47-49) and are therefore believed to play an important role in the pathogenesis of atopic diseases. Peripheral blood T cells from atopic individuals show decreased suppressor cell activity and numbers (50-53). In normal individuals, mediators such as histamine (54), leukotriene B4 (55-57) and PAF (42,43) can activate suppressor cell function. Thus, it would appear plausible that these mediators could contribute to dampen immune responses and prevent or abort the development of allergic reactivity. While this is supported by studies with histamine (58) further work is necessary concerning the precise effect of lipid mediators in the modulation of the various steps involved in the regulation of the IgE response and related events.

Since accessory cells are essential for many T cell functions, it was interesting to study the effect of PAF on monocyte-macrophage IL 1 production. A biphasic response was observed when increasing concentrations of PAF were added to lipopolysaccharide - or muramyl dipeptide-stimulated human monocytes or mouse peritoneal macrophages (59,60). Low concentrations of PAF (10^{-12} - 10^{-10} M) significantly enhanced IL 1 production while high concentrations (10^{-8} - 10^{-6} M) were markedly inhibitory. These effects were more evident when platelets were added to the monocyte cultures (59). All these effects of PAF on the IL 1 production were blocked by the PAF receptor antagonist BN 52021. Under certain experimental conditions (R)PAF or the PAF analogs PR 1501 and PR 1502 by themselves but not (S) PAF were able to induce IL 1 production, although to a lower extent than with LPS (61).

The PAF/IL-1 interaction may also play a major role in the immune injury of the arterial wall observed in shock, acute respiratory distress syndrome or sepsis : indeed PAF is a potent amplifier of platelet and leukocyte responses : at very low concentrations ($10^{-16} \rightarrow 10^{-11}$ M), it dramatically potentiates not only the release of IL 1 by monocytes/macrophages (see above) but also the production of leukotrienes (62,63) and free radicals [$O_2^{\cdot}$, $OH^{\cdot}$ (64)] from polymorphonuclear cells induced by various stimuli. Similarly, the ether lipid activates platelets to form thrombin, ATP... which in turn, as IL 1, act on endothelial cells to produce more PAF (65). This amplification process causes the formation of a large, dense platelet thrombus, invaded and surrounded by neutrophils, then by eosinophils and macrophages, and spreading over the adjoining vacuolized endothelium (66,67). Indeed, PAF alter the molecular organization of cytoskeletal proteins which control endothelial permeability (68) : human endothelial cells stimulated by PAF retract and lose reciprocal contact while stress fibers disappear or become less regular ; such impairments induce bleb formation.

Another area in which lymphocytes and various phagocytes exert a potent effector function is cytotoxicity against tumor, virus-infected and allogeneic target cells and parasites. Since host defenses against these foreign or transformed cells involve various degrees of inflammation, and since other lipid mediators have been shown to modulate cytotoxicity in either a positive [e.g. leukotrienes (69), TxA_2 (70)] or negative [e.g. prostaglandins (70,71)] manner, recent studies have looked at the effect of PAF on cytotoxic effector cell functions. Natural Killer (NK) cell-mediated lysis of the erythroleukemia target cell line K562 was markedly enhanced by picomolar concentrations of PAF (72). Preincubation of NK cells with PAF followed by washing before their culture with K562 target cells, or addition of PAF as late as 60 minutes after the initiation of the 4 h cytotoxicity culture also produced enhanced NK activity. BN 52021 could not block this PAF-induced effect. Natural cytotoxic cells and macrophages mediate their cytotoxic activities, at least in part, through synthesis and release of tumor necrosis factor (TNF), a protein with many other biological activities similar to IL 1. When preincubated, or co-incubated with PAF, LPS-treated human monocytes and mouse peritoneal macrophages elaborated significantly higher quantities of TNF (73). In addition large granular lymphocytes, which comprise the effector natural killer (NK) cells responsible for lysis of K562 target cells, have been shown to produce PAF under certain conditions (29). Therefore it appears that endogenous PAF may play a role in the regulation of NK function. Interestingly, phospholipid analogs of PAF, 1-0-akyl-lyso-phospholipids (ALP) may provide a new approach to cancer chemotherapy (59). These ether lipids possess an unusually broad range of biological activities-macrophage activation, malignant cell differentiation, and direct cytotoxicity, all thought to be membrane mediated. Unlike most antitumoral agents, these analogs do not appear to have a direct effect on DNA synthesis or function and are non-mutagenic(60). The antitumour activity of ALPs may partly be mediated by the generation of highly tumouricidal immuno-competent cells from the monocyte-macrophage lineage(61,62). The process of macrophage activation is still unclear(62), but an activation of interleukin-1 (IL 1) by ALPs is possible (B. Pignol and P. Braquet, unpublished).

Whether similar mechanisms underlie graft rejection is not yet defined, but PAF is generated by isolated, perfused kidneys (78), and released from renal allografts during hyperacute rejection (79). The PAF receptor antagonist, BN 52021 was shown to increase cardiac allograft survival in rats, with a synergistic effect with azathioprine and cyclosporin A (80). Whether the latter findings are due to the antagonism of PAF-enhanced cytotoxic cell activity, the generation of suppressor cells or the action of the antagonist on other cellular or fluid-phase protagonists of graft rejection remains to be elucidated.

The exact role of PAF in the primary and secondary mixed lymphocyte reaction and in the generation of cytotoxic lymphocytes in-vitro is not well assessed. However, Gebhardt et al. (81) have recently shown that BN 52021 potentiates alloantigen recognition in primary and secondary mixed lymphocyte culture and enhances the generation of cytotoxic lymphocytes in vitro. In this study the platelet-activating factor antagonist was added to and removed from primary and secondary mixed lymphocyte cultures at various times after culture initiation. Similarly, BN 52021 was added to bulk cultures in which cytotoxic lymphocytes were being generated and the effect on the level of cytotoxicity as determined by the release of ^{51}chromium in the cell mediated cytotoxicity assay were determined. The presence of BN 52021 throughout the duration of primary and secondary mixed lymphocyte cultures had the greatest enhancing effect on the proliferative capacity of the responding cells measured by [^{3}H] thymidine incorporation. Addition of BN 52021 24 hours or later after the initiation of cultures had less enhancing effect on lymphocyte proliferation. The removal of BN 52021 from mixed lymphocyte cultures up to 48 hours after their initiation also was found to eliminate the potentiating effect of this antagonist. Similar observations were made in the mixed cultures employed to generate cytotoxic T lymphocytes. The continuous presence on BN 52021 during the entire 72 hour culture period produced an enhanced level of cell mediated cytotoxicity, whereas the removal of the antagonist up to 24 hours after culture initiation eliminated the enhancing effect. Since the secondary mixed lymphocyte cultures and the bulk cultures used to generate cytotoxic T lymphocytes contained exogenous IL 2 in amounts adequate to support the growth of the cells, it is concluded that the potentiating effect of BN 52021 on alloantigen recognition is not due to the enhanced production of IL 2 in antagonist-treated cultures.

Cytotoxic/suppressor T-lymphoblasts accumulate in Langherans cells and may participate in diabetes development (82) ; in vitro splenic lymphocytes induce cytotoxicity against rat Langherans islets (83) which is dose-dependently inhibited by the PAF antagonist BN 52021. This result suggests that PAF may be involved in the cytotoxic effect of splenic lymphocytes and provides a rationale for using PAF antagonists for the prevention of autoimmune diabetes (84).

Preliminary findings also suggest that keratinocyte killing of Candida was blocked by the antagonist BN 52021. Here again, these results indicated that PAF enhanced cell cytotoxic activity.

As reviewed in this section, PAF exerts numerous effects on the various cellular components of the immune system. Although some of these effects may appear contradictory, they may be due to interactions with several cell subsets or different receptors. Many ongoing and future studies will be needed to unravel the complete picture.

REGULATION OF THE IMMUNE RESPONSE : THE INDIRECT EFFECT OF PAF

Apart from its direct effect on the immune response, PAF can also cause the release and/or generation of various mediators which in their turn will influence the regulation of lymphocyte functions. This would also apply for cationic proteins and other suppressive factors, released from activated eosinophils and for leukotrienes, prostaglandins and neuropeptides.

Activation of eosinophils

Both in vivo and in vitro (rev. in 14), it has been evidenced that PAF is a powerful modulator of eosinophil functions :

(i) For human eosinophils, PAF is one of the most powerful chemotactic agents. As the PAF antagonist BN 52021 dose-dependently inhibits eosinophils locomotion (86), the effect is receptor mediated.

(ii) Hemoconcentration and increased total leukocyte counts are induced by systemic PAF injections ; the underlying eosinophil and monocyte subsets decreases by up to 33 % and this is suppressed by PAF antagonists.

(iii) In comparison to those from normal individuals, eosinophils from patients with eosinophilia, including asthmatics, have an increased capacity to release PAF (87). A pronounced eosinophilic infiltration was noticed solely in allergic patients who had been injected PAF intracutaneously, a large quantity of the cells being degranulated (88). Moreover, BN 52021 (89) inhibited two phenomena : PAF induced LTC_4 and $O_2^{\cdot-}$ formation by human eosinophils.

(iv) In vitro, PAF induces the release of cationic proteins, including Major Basic Protein (MBP) (P. Braquet, unpublished). This may explain why in airway hyperreactivity, the PAF antagonist BN 52021 affords protection.

(v) PAF-acether in the presence of Schistosoma mansoni previously coated with C_3b and specific antibodies (90) induces an increase in eosinophil cytotoxicity which is dose-dependently blocked by PAF antagonists. IgE-induced cytotoxicity of rat and human eosinophils is also inhibited by BN 52021 and related Ginkgolides (M. Capron, pers. Com.).

(vi) Finally, electronic microscopy studies have evidenced that the two PAF antagonists : BN 52021 and WEB 2086, antagonize eosinophil recruitment in the lung elicited by both PAF and antigen in the guinea-pig (A. Lellouch Tubiana, in preparation). In PAF-induced thrombosis of the mesenteric artery a similar result is obtained, the artery being infiltrated by eosinophils 10 minutes after challenge. It is interesting to notice that eosinophils were surrounded by degranulated platelets in both lung and mesenteric tissues (R.H. Bourgain, in preparation). In bronchoalveolar lavages from sensitized rabbits challenged with ragweed allergen, the PAF antagonist BN 52021 also antagonizes eosinophil accumulation (91).

PAF is thus a major mediator of eosinophil recruitment and degranulation and indirectly, it may therefore also regulate the immune response : indeed, eosinophil granules contain two highly basic proteins (eosinophil cationic protein (ECP) and eosinophil protein X (EPX)). At appropriate in vivo concentrations (10^{-10} M), these proteins induce suppressive effects on peripheral blood mononuclear cells from the normal donor, which implies that eosinophils have a regulatory role in immunological reactions (92). Indeed, lymphocyte function was reported to be inhibited by non dialyzable and heat sensitive factor produced by zymosan-stimulated eosinophils (93). A similar mechanism could be implicated in familial reticuloendotheliosis, a disease characterized by severe combined immune deficiency with hypereosinophilia (94).

It will become possible to evaluate the role of PAF in eosinophil activation and the putative consequent disturbance in the lymphocyte response by undertaking systematic study of diseases involving eosinophil infiltration with impairment of lymphocyte response.

Prostaglandin and leukotriene release

PAF induces prostaglandin and leukotriene generation in lungs and various cell systems which, in turn, are potent modulators of lymphocyte functions (rev. in 57, 95).

Neuropeptide release

The development and function of the immune system is directly influenced by endocrine factors from the pituitary and hypothalamus (rev. in 96) : indeed, both pituitary and hypothalamic hormones interfere with lymphocyte proliferation and function and it has been shown recently that human peripheral mononuclear cells possess specific receptors for these hormones, and other immuno-regulatory peptides.

It seems that PAF plays an important role in regulating CNS functions : (i) after [^{3}H] PAF injection in vivo in gerbils, the binding of PAF is 250 times higher in the brain than in other tissues ; the PAF antagonist, BN 52021, inhibits this phenomenon (P. Braquet and B. Spinnewyn, unpublished) ; it has recently been shown that there are high affinity binding sites for PAF in the gerbil brain (E. Chabrier and P. Braquet, in preparation) ; (ii) several behavioural parameters in mice have been modified by PAF antagonists of several chemical series (97) ; (iii) in gerbils, the PAF antagonists BN 52021, kadsurenone and brotizolam inhibit the post ischemic phase of cerebral ischemia (98) ; (iv) PAF modulates neuropeptide release from various cerebral tissues (99 ; F. Dray, in preparation ; P. Braquet unpublished). In addition, the hormonal response is changed, following PAF injection in animals (100). Consequently, PAF may indirectly alter the immune response via neuro-peptide release.

Prolactin and growth hormone. PAF causes prolactin (PRL) and growth hormone (GH) to be released from the rat anterior pituitary tissue (99) which effect is blocked by PAF antagonists (P. Braquet, unpublished). It seems that PRL and GH play an important role in regulating lymphocyte functions since hypophysectomized rats lose their immune response (101). The immunological responsiveness of these animals was restored when treated with PRL or GH, but not with corticotropin or other pituitary hormones (102). In addition, on human lymphocytes, specific binding sites for PRL have been evidenced. Of interest, ciclosporin A displaces PRL from these sites (103,104) and on the contrary, stimulation of PRL secretion reverses the immunosuppression induced by ciclosporine (104). These data imply that PRL release may take part in the exacerbation of the immune response afforded by ALPs (see above) and conversely may account for the delay in graft rejection obtained by a combination of ciclosporine and PAF antagonists (80).

Substance P. Substance P (SP) is an 11 amino acid peptide, located in nerve endings of several species, including man (rev. in 105). SP constricts pulmonary airways leading to bronchoconstriction, contracts smooth muscles, stimulates epithelial cell secretion in the lungs and the gut and increases the permeability of microvasculature (rev. in 105). The release of mediators from mast cells such as leukotrienes or histamine is also stimulated by SP in an indirect manner ; in vitro, SP stimulates human monocyte chemotaxis with an EC_{50} as low as 10^{-13} M. SP generates TxA_2, $O_2^{\cdot}$ and H_2O_2 by C. parvum-activated macrophages, activates lysosomal enzyme release and stimulates phagocytosis of yeast cells (rev. in 105).

PAF induces SP release from guinea-pig lung preparations, as determined by radioimmunoassay (106). PAF-induced release of SP is mainly observed in large bronchi and trachea. On a molar basis, PAF seems to be as active as capsaicin in inducing SP release (106). However, from chopped lung parenchyma, no net release of SP elicited by PAF or capsaicin is observed, although the amount of neuropeptide extracted at acidic pH from lung tissue is markedly increased. This observation implies than SP is released from nerve endings upon stimulation with PAF and capsaicin but that it does not spread within the tissue (106). The inhibition of airway hyperreactivity exerted by the PAF antagonists BN 52021 and WEB 2086 may partly be explained by the PAF-induced increase of SP release.

The prior body of evidence indicates that PAF may be a potent mediator of SP release and may therefore also indirectly regulate the immune response : indeed, the uptake of [³H] thymidine and [³H] leucine by purified human peripheral blood T lymphocytes is significantly improved by SP, in vitro at low concentrations (107) ; the same degree of proliferative response to SP has been demonstrated by similar experiments with mouse lymphocytes from mesenteric lymph nodes and Peyer's patches (107). Moreover, the in vitro production of IgA in lymphocytes from spleen and Peyer's patches is significantly enhanced by SP by up to three times (108).

Consequently, in this manner, PAF may once again indirectly influence the immune response.

CONCLUSION

PAF has numerous effects on many organ and tissue targets. The action of the mediator on various cell types can account for several of these effects. PAF has been related with a great majority of inflammatory and immune responses, right from discovery of its production in allergic reactions to the recent and current studies of its effects on cellular immune functions. The direct involvemement of the mediator in these phenomena can now be studied since specific PAF agonists, antagonists and inhibitors, have been developed.

REFERENCES

1. Barbaro J.J. and Zvailer N.J.(1966) Soc. exp. Biol. Med. 122, 1245-1247 .
2. Siraganian R.P. and Osler A.G. (1970) J. Immunol. 104, 1340 .
3. Siraganian R.P. and Osler A.G. (1971) J. Immunol. 106, 1244 .
4. Henson P.M. (1969) Fed. Proc. 28, 1721-1728.
5. Henson P.M. (1979) J. exp. Med. 131, 287-306.
6. Benveniste J., Henson P.M. and Cochrane C.G. (1972) J. exp. Med. 136, 1356-1377.
7. Benveniste J. (1974) Nature 249, 581-582.

8. Benveniste J., Le Couedic J.P., Polensky J. and Tencé M. (1977) *Nature* 269, 170-171.
9. Benveniste J., Chignard M., Le Couedic J.P. and Vargaftig B.B. (1982) *Thromb. Res.* 25, 375-385.
10. Benveniste J., Tencé M., Varenne P., Bidault J. , Boullet C. and Polonsky, J. (1979) *C.R.C. Acad. Sci.* 289D, 1037-1340.
11. Demopoulos C.A., Pinckard R.N. and Hanahan D.J. (1979) *J. biol. Chem.* 254, 9355-9358.
12. Ludwig J.C. and Pinckard R.N. in : New Horizons in platelet activating factor research (wislow CM and Lee ML eds), pp. in press, John Wiley, New York (1986).
13. Satouchi J., Pinckard R.N., McManus L.M. and Hanahan D.J. (1981) *J. biol. Chem.* 256, 4425-4432.
14. Braquet P., Touqui L., Shen T.Y. and Vargaftig B.B.(1987) *Pharmacol. Review*, 39(2), in Press.
15. Braquet P. and Vargaftig B.B. (1986) *Transplant. Proc.* 5 (suppl. 4), 10-19. 16. Braquet P., Paubert-Braquet M. and Vargaftig B.B. (1987) in : Advances in Prostaglandin, Thromboxane and Leukotriene Research, (B. Samuelsson et al. eds.) Raven Press (New York) 17, 818-823 .
17. Vargaftig B.B. and Braquet P. (1987) *Brit. Med. Bull.* 13, 312-335.
18. Snyder F. (1985) *Med. Res. Reviews* 5, 107-140.
19. Chap H., Mauco G., Simon M.F., Benveniste J. and Douste-Blazy, L. (1981) *Nature* 289, 312-314.
20. Camussi G., Aglietta M., Coda R., Bussolino F., Piacibello W. and Tetta C. (1980) *Nature* 42, 191-199.
21. Betz S.J. and Henson P.M. (1980) *J. Immunol.* 125, 2756-2763.
22. Mueller H.W., O'Flaherty J.T. and Wykle R.L. (1983) *J. biol. Chem.* 258, 6213-6218.
23. Lee T.C., Lenihan D.J., Malone B., Roddy L.L. and Wasserman, S.I. (1984) *J. biol. Chem.* 259, 5526-5530.
24. Mencia-Huerta J.M., Roublin R., Morgat J.L. and Benveniste, J (1982) *J. Immunol.* 129, 804-808.
25. Camussi G., Aglietta A., Malavasi F., Tetta C., Piacibello W. Sanavio F. and Bussolino F. (1983) *J. Immunol.* 131, 2397-2403.
26. Prescott S.M., Zimmerman G.A. and McIntyre T.M. (1984) *Proc. Natl. Acad. Sc. USA* 81, 3534-3538.
27. Thomas Y., Denizot Y., Dassa E., Boullet C. and Benveniste (1986) *C.R. Acad. Sc. Paris* 303, 699-702.
28. Jouvin-Marche E., Nino E., Beauvain G., Tence M., Niaudet, P. and Benveniste J. (1984) *J. Immunol.* 133, 892-898.
29. Malavasi F., Tetta C., Funaro A., Bellone G., Ferraro, E., Colli-Franzone A., Dellabona P., Rusci R., Matera L., Camussi G. and Caligaris-Cappis F. (1986) *Proc. Natl. Acad. Sc. USA* 83, 2443-2447.
30. Bussolino F., Fou R., Malavasi F., Ferrando M.L. and Camussi G. (1984) *Exp. Haematol.* 12, 688-694.
31. Siraganian R.P. and Osler A.G. (1981) *J. Immunol.* 106, 1244-1251.
32. O'Flaherty J.T., Surles J.R., Redman J. ,Jacobson D., Piantadosi C. and Wykle R.L. (1986) *J. Clin. Invest.* 78, 381-388.
33. Handley D.A., Arbeeny C.M., Lee M.L., VanValen R.G. and Saunders R.N. (1984) *Immunopharmacology* 8, 137-142.
34. Hartung H.P. (1983) *FEBS Letters* 160, 209-212.
35. Braquet P. and Godfroid J.J. (1987) In : Platelet Activating Factor (F. Snyder E.) Plenum Press, New York, 191-236.
36. Godfroid J.J. and Braquet P. (1986) *Trends Pharmacol. Sci.* 7, 368-373.
37. Braquet P. and Godfroid J.J. (1986) *Trends Pharmacol. Sci.* 397-403.
38. Braquet P, Chabrier C and Mencia-Huerta J.M. In : Advances in Inflammation Research Vol. XX (in press).
39. Rola-Pleszczynski M., Pignol B., Pouliot C. and Braquet P.

(1987) *Biochem. Biophys. Res. Com.* 142, 754-760.

40. Pignol B., Henane S., Mencia-Huerta J.M., Braquet P. and Rola-Plesczynski M. (1987) Int. Congress of Pharmacology Sydney. (Abst).

41. Dulioust A., Vivier E., Salem P., Deryckz S., Benveniste J. and Thomas Y. (1987) *Fed. Proc.* 46, 784 (Abst.).

42. Rola-Pleszcyznski M., Pouliot C., Pignol B. and Braquet P. (1987) *Fed. Proc.* 46, 743.

43. Rola-Pleszczynski M., Pouliot C., Turcotte S., Pignol B., Braquet, P. and Bouvrette L. (1987) (Submitted).

44. Ward S.G., Lewis G.P. and Westwick J. (1987) In : Lipid mediators in immunology of burn and sepsis, (Paubert-Braquet, M. Ed.) Plenum Press (in press).

45. Barrett M.L., Lewis G.P., Ward S., Westwick J. (1986) *Br. J. Pharmacol.* 89, 505P.

46. Pignol B., Henane, S., Sorlin B., Rola-Pleszczynski, Mencia-Huerta, J.M. *Biochem. Biophys. Res. Commun.* In press.

47. Buckley R.H. and Becker W.G. (1978) *Immunol. Rev.* 41, 288-314.

48. Strannegard O. and Strannegard I.L. (1978) *Immunol. Rev.* 41, 149-170.

49. Rachelefsky G.S., Opelz G., Mickey M.R., Kiushi M., Terasaki P.I., Siegel S.C. and Stiehm E.R. (1976) *J. Allergy Clin. Immunol.* 57, 569-576.

50. Canonica G.W., Mingari M.C., Melioli G., Colombatti M. and Moretta L. (1979) *J. Immunol.* 123, 2669-2672.

51. Rola-Pleszczynski M. and Blanchard R. (1981) *Int. Arch. Allergy Appl. Immunol.* 64, 361-370.

52. Rola-Pleszczynski M. and Blanchard R. (1981) *J. Invest. Dermatol.* 76, 279-283.

53. Engel P., Huguet J., Sanosa J., Sierra P., Cols N. and Garcia-Calderon P.A. (1984) *Ann. Allergy* 53, 337-340.

54. Rocklin R.E. and Haberek-Davidson A. (1981) *J. Clin. Immunol.* 1, 73-79.

55. Rola-Pleszczynski M., Borgeat P. and Sirois P. (1981) *Biochem. Biophys. Res. Com.* 108, 1531-1537.

56. Rola-Pleszczynski M. (1985) *J. Immunol.* 135, 1357-1360.

57. Rola-Pleszczynski M. (1985) *Immunol. Today* 6, 302-307.

58. Beer D.J., Osband M.E., McCaffey R.P., Soter N.A. and Rocklin R.E. (1982) *New England J. Med.* 306, 455-458.

59. Rola-Pleszczynski M., Pignol B., Braquet P., Bouvrette L. and Gingras D. (1987) (in preparation).

60. Pignol B, Hénane S, Mencia-Huerta J.M., Rola-Plezczynski M. and Braquet P. (1987) <u>Prostaglandins</u> (In press).

61. Barrett M.L., Lewis G.P., Ward S. and Westwick J. (1987) *Br. J. Pharmacol.* 90, 113.

62. Chilton F.H., O'Flaherty J.T., Walsh C.E., Thomas R.L., Wykle R.L., DeChatelet L.R. and Waite B.M. (1982). J. Biol. Chem. 257, 5402-5407.

63. Lavaud Ph., Braquet M. and Braquet P. In preparation.

64. Vercelloti G.M., Huh P.W., Yin H.Q., Nelson R.D. and Jacob H.S. (1986) 28th annual Meeting of the American Society of Hematology, Abst.

65. Bussolino F., Brevario F., Tetta C., Aglietta M., Mantovani A. and Dejana E. (1986) *J. Clin. Invest.* 77, 2027-2033.

66. Bourgain R.H., Maes L., Braquet P., Andries R. Touqui L. and Braquet M. (1985) *Prostaglandins* 30, 185-197.

67. Bourgain R.H., Andries R., Braquet P. and Vargaftig B.B. *Nature*, submitted.

68. Bussolino F., Camussi G., Aglietta M., Braquet P., Bosia A., Pescarmona G., Sanavio F., d'Urso N., Marchisio P.C. (1987) *J. Immunol.* In press.

69. Rola-Pleszczynski M., Gagnon L. and Sirois P. (1983) *Biochem. Biophys. Res. Comm.* 113, 531-537.
70. Rola-Pleszczynski M., Gagnon L., Bolduc D. and LeBreton G. (1985) *J. Immunol.* 135, 4114-4119.
71. Koren H.S., Anderson S.J. and Fischer D.G. (1981) *J. Immunol.* 127, 2007-2013.
72. Rola-Pleszczynski M. and Turcotte S. (1987) *Immunobiology* (in press).
73. Rola-Pleszczynski M., Bissonnette E., Dubois C. and Gingras D. (1987) (Submitted).
74. Proceedings of the first international symposium on ether lipid in oncology (1986) (Göttingen/FRG, Dec. 5-7).
75. Berdel W.E., Andreesen R. and Mender P.G. (1985) In : Phospholipids and cellular regulation (Juo, J.K. ed.) CRC Press Inc. Baco Raton, Florida, pp. 41-73.
76. Munder P.G. (1986) In : Second International Conference on Platelet-activating Factor and structurally related alkyl ether lipids. (Gatlinburg TN USA), p. 51 (Abst).
77. Andreesen R., Giese V. and Löhr G.W. (1986) In : Proceedings of the first International symposium on ether lipid in oncology. Göttingen/FRG, Dec. 5-7.67.
78. Pirotzky E., Ninio E. and Bidault J. (1984) Lab. Invest. 51, 567-572.
79. Ito S., Camussi G., Tetta C., Milgrom F. and Andres G. (1984) *Lab. Invest.* 51, 148-161.
80. Foegh M.L., Khirabadi B.S., Rowles J.R., Braquet P. and Ramwell P.W. (1986) *Transplantation* 42, 86-88.
81. Gebhardt P.B., Braquet P., Bazan H.E.P., and Bazan N.G. J. Cell. Immunol. In Press.
82. Quiniou-Debrie M.C., Debray-Sach M., Dardenne M., Czernichow P., Assan, R. and Bach, J.F. (1985) *Diabetes* 34, 373-379.
83. Farkas G., Pusztai R., Mandi Y. and Beladi I. (1983) *Transplantation* 36, 583-584.
84. Farkas G., Mandi Y., Koltai M. and Braquet P. (1987) In : Ginkgolides : Chemistry, Biology, Pharmacology and Clinical Sciences (edited by P. Braquet) Methods and Findings, in press.
85. Dobozy A., Hunyadi J. and Szaboo Denderessy, A. (1987) In : the Ginkgolides : Chemistry, Biology, Pharmacology and Clinical Sciences (Ed. by P. Braquet) Methods and Findings.
86. Wardlaw A.J. and Kay A.B. (1986) *J. Allergy Clin. Immunol.* 77, 236 (Abst.).
87. Lee T.C., Lenihan J., Malone B., Roddy L.L. and Wasserman S.I. (1984) *J. Biol. Chem.* 259, 5526-5530.
88. Henocq E. and Vargaftig B.B. (1986) *Lancet*, i, 1378-1379.
89. Bruijnzeel P.L.M., Koenderman L., Kok P.T.M., Hamelin M.L., Verhagne J.L. (1986) *Pharmacol. Res. Commun.* 18, 61-69.
90. McDonald A.J., Moqbel R., Wardlaw A.J. and Kay A.B. (1986) *J. Allergy Clin. Immunol.* 77, 227 (Abst).
91. Coyle A., Sjoerdsam K., Page C., Brown L., Touvay C. and Metzger W.J. *J. All. Clin. Immunol.* (in press).
92. Peterson C.G., Skoog V. and Venge P. (1986) *Immunobiology* 171, 1-13.
93. Ramesh K.S., Pincus S.H. and Rocklin R.E. (1985) *Cell. Immunol.* 92, 366-375.
94. Le Deist F., Fischer A., Durandy A., Arnaud-Battandier F., Nezelof C., Hamet M., de Prost Y. and Griscelli C. (1985) *Arch. Fr. Pediatr.* 42, 11-16.
95. Goodwin J.S. and Ceuppens J. (1983) *J. Clin. Immunol.* 3; 295-315.
96. Payan D.G., Mc Gillis J.P. and Goetzl, E.J. (1986) In : Advances in Immunology, vol. 39, Academic Press, pp. 299-323.

97. Clostre F, Millerin M., Betin C and Braquet P. *Eur. J. Pharmacol.* Submitted.

98. Braquet P, Spinnewyn B., Blavet N. and Clostre F. (1987) XTh IUPHAR - SYDNEY. Abst.

99. Grandison L. and Camoratto A.M. In : Second International conference on platelet- activating factor and structurally related alkyl ether lipids. (Gatlimburg TN USA), p. 71 (Abst.).

100. Bessin P. (1986) Pharm. Res. Commun. 18, 139-150·

101. Nagy E., Berczi I., Wren G.E, Asa S.L. and Kovacs K. (1983) *Immunopharmacology* 6, 231-243.

102. Nagy E., Berczi I. and Friesen H.G. (1983) *Acta Endocrinol.* *(Copenhagen)* 102, 351-357.

103. Russel D.H., Kibler R., Matrisian L., Larson D.F., Poulos B. and Magum B.E. (1985) *J. Immunol.* 134, 3027-3033.

104. Hiestand P.C., Mekler P., Nordmann R., Grieder A. and Permmongkol C. (1986) *Proc. Natl. Acad. Sci.* USA 83, 2599-2603.

105. Barnes P.J., (1987) *J. Allergy Clin. Immunol.* 79, 285-295.

106 Mencia-Huerta J.M., Rodrigue F. and Braquet P. Submitted.

107. Payan D.G., Brewster D.R. and Goetzl E.J. (1983) *J. Immunol.*, 131, 1613-1620.

108. Stanisz A.M., Befus D. and Beinenstock J. (1986) *J. Immunol* 136, 152-158.

LEUKOTRIENES IN LYMPHOCYTE ACTIVATION : LEUKOTRIENE B$_4$ ENHANCES INTERLEUKIN 2 AND INTERFERON-GAMMA PRODUCTION BY MITOGEN AND ALLOANTIGEN-STIMULATED HUMAN T CELLS

M. Rola-Pleszczynski, C. Pouliot, L. Bouvrette and P-A. Chavaillaz
Laboratoire d'Immunologie, Faculté de Médecine, Université de Sherbrooke, Sherbrooke, QC, Canada J1H 5N4

INTRODUCTION

LTB$_4$ and, to some extent, certain other hydroxyeicosatetraenoic acids (HETE) can regulate several immune cell functions (reviewed in 1). We have initially shown that LTB$_4$ can induce the appearance of suppressor T cells which in turn inhibit the proliferative response of human lymphocytes to mitogens (2, 3). Others have also shown LTB$_4$ to suppress LIF production (4) and Ig synthesis (5). The expression of LTB$_4$-induced suppression appears to require the presence of monocytes in the responder population in most circumstances (3, 6). It also requires the integrity of the cyclooxygenase pathway (6): use of indomethacin in the responder cell culture not only prevents the expression of suppression, but actually results in LTB$_4$-induced enhancement of proliferation. This latter effect may be the consequence of induction by LTB$_4$ of "positive" signals in the forms of IL-1, IL-2 or interferon-gamma when suppression is prevented. We have recently shown that LTB$_4$ can augment the production of IL-1 by human monocytes (7). In this paper, we present initial evidence for the modulation of production of the lymphokines IL-2 and IFN by LTB$_4$.

MATERIALS AND METHODS

Leukotriene B$_4$ was a generous gift from Drs A.W. Ford-Hutchinson and J. Rokash, Merck-Frosst Canada. It was dissolved in ethanol and further diluted in RPMI-1640 medium, final ethanol concentration being less than 0.5%. Recombinant IL-1 (rIL-1) was obtained from Cistron and rIL-2 from Amersham. Nordihydroguaiaretic acid (NDGA), caffeic acid (Caf),

indomethacin (Indo) were obtained from Sigma. BW755c was from
Burroughs-Wellcome, U-60257 was a gift from Dr M. Bach (Upjohn) and
cyclosporin A (CyA) was a gift from Dr G. Holme (Sandoz).

Human peripheral blood mononuclear leukocytes (PBML) were obtained
by sedimenting venous blood from healthy, medication-free volunteers on a
Ficoll-Hypaque gradient. Monocytes were removed by adherence to BHK-
precoated Petri dishes. Lymphocyte subsets were enriched in $T4^+$ or
$T8^+$ cells by treatment with OKT8 or OKT4 (Ortho Diagnostics) plus complement, respectively. This resulted in greater than 90% enrichment of
either subset. Alternatively, $T4^+$ or $T8^+$ cell populations were
selected by panning labeled cells on anti-mouse Ig precoated culture
flasks.

PBML, monocyte-depleted lymphocytes or enriched T cell subsets were
cultured in RPMI 1640 medium, 5% FCS, in the presence of either concanavalin A (ConA, 5 ug/ml; Pharmacia), phytohemagglutinin (PHA, 0.1%; Difco
Laboratories), poly I:C (30 ug/ml; Sigma), or allogeneic mitomycin
C-pretreated (50 ug/ml, 30 min) PBML and varying concentrations of LTB_4
for 24 h at 37°C. Cell-free supernatants were then collected and tested
for their IL-2 and IFN-gamma contents. To assay for lymphocyte proliferation, the cultures were incubated for 72 h (mitogens) or 120 h
(allogeneic PBML) at 37°C and ^{3}H-TdR was added for the last 18 h of
incubation. Cells were then harvested and proliferation was assessed by
measuring the uptake of ^{3}H-TdR.

IL-2 was measured using the IL-2 dependent mouse CTLL line.
Interferon-gamma was measured using a commercial RIA assay for human
interferon-gamma (Centocor).

RESULTS

When human PBML were incubated with PHA for 24 h, the addition of
LTB_4 in conjunction with indomethacin markedly enhanced IL-2 production
by these cells. No significant effect was observed in the absence of indomethacin. A significant effect was seen at an LTB_4 concentration as low
as 1 pM (10^{-12}M) (Figure 1).

Because of the possibility that this enhancement in IL-2 production
was secondary to an enhancement of IL-1 production via a direct LTB_4-
monocyte interaction, monocyte-depletion was combined with the addition of

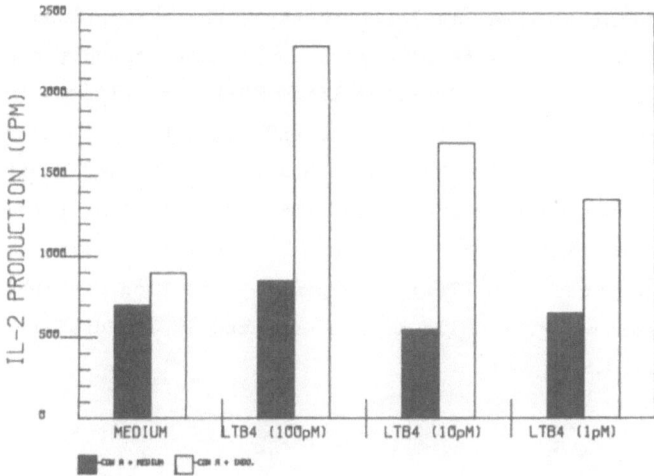

Figure 1. Effect of LTB$_4$ on production of IL-2 by human PBML cultures
stimulated with ConA in the absence or presence of indomethacin
(10^{-6}M). n = 6.

rIL-1 (100 pg/ml) to T lymphocyte cultures stimulated with ConA. This
allowed for a similar increase in IL-2 production in the presence of
LTB$_4$ (Table I), indicating that this mediator was capable of affecting
directly IL-2 production, in the absence of any effect on accessory cells
and IL-1 synthesis.

We had previously shown that T cell subsets could respond diffe-
rently to LTB$_4$ in their immunoregulatory functions on lymphocyte proliferation. A differential effect of LTB$_4$ on the production of IL-2 by T

TABLE I   Effect of LTB$_4$ (10^{-10}M) on IL-2 production by monocyte-
depleted human T lymphocytes (LC) in the presence of rIL-1 and
ConA.

	IL-2 (cpm)	
Lc + rIL-1	567 $\pm$ 69	
Lc + rIL-1 + ConA	1514 $\pm$ 181	
Lc + rIL-1 + ConA + LTB$_4$	2351 $\pm$ 215	p < 0.01

cell subsets could explain, at least in part, these functions. In fact, removal of T8[+] lymphocytes by OKT8 plus complement not only enhanced baseline IL-2 production by PHA-stimulated PBML, but allowed for further augmented production of IL-2 in the presence of LTB$_4$ (Fig. 2). Conversely, depletion of T4[+] cells had the opposite effect, and addition of LTB$_4$ further depressed IL-2 production (Fig. 2). Similar findings were observed when IFN-gamma production was measured in poly I:C-stimulated cultures supplemented with LTB$_4$: IFN-gamma production was inhibited by LTB$_4$ in T4[+]-depleted PBML while it was enhanced in T8[+]-depleted cultures (Fig. 3).

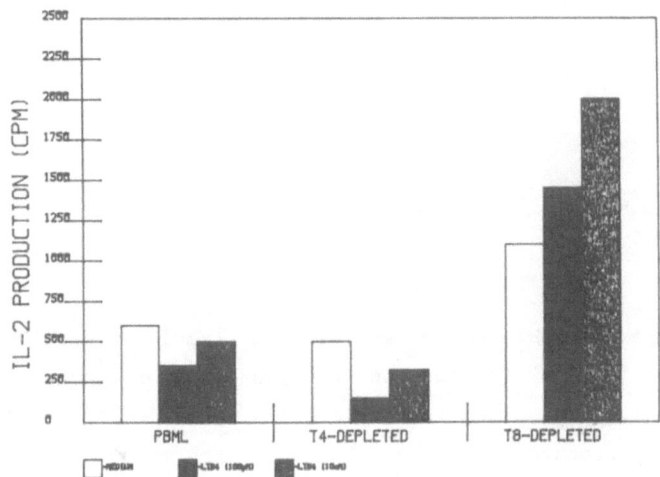

Figure 2. Effect of LTB$_4$ on the production of IL-2 by PHA-stimulated PBML depleted of either T4[+] or T8[+] cells by OKT4 + C and OKT8 + C pretreatment, respectively. n = 4.

Because of its potential to augment lymphocyte responses, LTB$_4$ was tested for its ability to reverse the inhibition of lymphocyte proliferation and IL-2 production by hydrocortisone (H-Cort). H-Cort at 10^{-5} M inhibited proliferative responses to ConA, PHA and allogeneic PBML by 60-80% and this inhibition was not reversed by LTB$_4$ when added simultaneously with H-Cort. (Table II). The smaller inhibition by lower concentrations (10^{-6}, 10^{-7}M) of H-Cort could however be reversed by LTB$_4$ concentrations of 10^{-10}-10^{-8}M. A similar pattern of reactivity was observed when T cells were preincubated with LTB_4 for 18 h before being added to PBML cultures containing ConA and H-Cort (data not shown).

LTB_4 was also capable of reversing, partially or completely, the inhibition of IL-2 synthesis produced by the leukotriene synthesis inhibitor U-60257 and the lipoxygenase inhibitors caffeic acid, NDGA and BW755c. It was also able to restore IL-2 production in the presence of low concentrations of CyA, a powerful inhibitor of IL-2 synthesis (Table III).

TABLE II   Effect of LTB_4 on reversal of inhibition of lymphocyte proliferation by hydrocortisone

Addition to culture		Proliferation (percent of medium controls)		
H-Cort(M)	LTB_4(M)	ConA	PHA	PBML
-	10^{-8}	111	122	109
-	10^{-10}	100	135	100
10^{-5}	-	41	40	20
10^{-5}	10^{-8}	41	43	22
10^{-5}	10^{-10}	38	25	11
10^{-6}	-	60	61	30
10^{-6}	10^{-8}	101*	49	43
10^{-6}	10^{-10}	92*	58	54*
10^{-7}	-	82	80	62
10^{-7}	10^{-8}	114*	106	101*
10^{-7}	10^{-10}	123*	117*	124*

*   Asterisks denote a significant effect of LTB_4 ($p < 0.05$; n = 6).

DISCUSSION

The present studies indicate that LTB_4 can modulate lymphokine production by human lymphocytes. In addition to interacting directly with monocytes and possibly enhancing IL-2 synthesis by increasing IL-1 production (7), LTB_4 appears to be able to affect directly IL-2 producing cells cultured in the presence of IL-1. Depletion of the $T8^+$ suppressor lymphocyte subset further enhances both baseline IL-2 production and LTB_4-induced augmentation of IL-2 synthesis. It thus appears that LTB_4 can generate a number of "positive" signals which regulate the immune response when "negative" signals leading to enhanced suppressor cell function are blocked (Fig. 4).

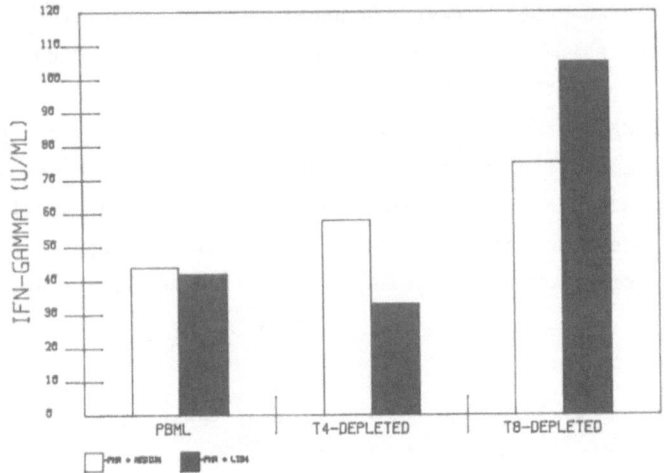

Figure 3.   Effect of LTB$_4$ on the production of IFN-gamma by poly I:C-stimulated PBML depleted of either T4$^+$ or T8$^+$ cells by OKT4 + C or OKT8 + C pretreatment, respectively.   n = 3.

TABLE III   Effect of LTB$_4$ on IL-2 production by human ConA-stimulated T cells in the presence or absence of hydrocortisone (H-Cort). NDGA, caffeic acid (Caf) and cyclosporine A (CyA).

Addition to culture		IL-2 (mean cpm $\pm$ S.E.M.)
-	-	1618 $\pm$ 202
-	LTB$_4$ 10^{-7}M	1715 $\pm$ 150
-	LTB$_4$ 10^{-9}M	1422 $\pm$ 162
U-60257 (10^{-4}M)	-	1140 $\pm$ 98
U-60257	LTB$_4$ 10^{-7}M	1878 $\pm$ 225*
U-60257	LTB$_4$ 10^{-9}M	826 $\pm$ 89
Caf (10^{-4}M)	-	782 $\pm$ 66
Caf	LTB$_4$ 10^{-7}M	1328 $\pm$ 150*
Caf	LTB$_4$ 10^{-9}M	750 $\pm$ 82
NDGA (25 ug/ml)	-	310 $\pm$ 40
NDGA	LTB$_4$ 10^{-7}M	604 $\pm$ 75*
NDGA	LTB$_4$ 10^{-9}M	334 $\pm$ 38
BW755c (50 ug/ml)	-	850 $\pm$ 95
BW755c	LTB$_4$ 10^{-7}M	1525 $\pm$ 182*
BW755c	LTB$_4$ 10^{-9}M	2308 $\pm$ 211*
CyA (10 ng/ml)	-	744 $\pm$ 84
CyA	LTB$_4$ 10^{-7}M	1525 $\pm$ 201*
CyA	LTB$_4$ 10^{-9}M	1342 $\pm$ 160*

* denotes significant effect of LTB$_4$

Indirect evidence for the participation of lipoxygenase activity in lymphocyte function has been brought forth by several investigators (8-13), including ourselves (14). In addition to showing that 5-lipoxygenase inhibitors suppressed natural cytotoxic cell activity, we could restore normal cytotoxic activity with exogenous LTB$_4$ (14). The present studies demonstrate a similar phenomenon using lymphocyte proliferation and production of IL-2. Analogous results were recently reported by Goodwin et al (15).

Studies by Johnson and Torres (16) and Farrar and Humes (11) also showed that LTB$_4$ can replace IL-2 in helper cell-depleted mouse spleen cells or in a IL-2-dependent murine cell line, respectively, and allow for IFN-gamma production. Our studies reported herein indicate that IFN-gamma production by normal human peripheral blood T cells can be markedly enhanced by LTB$_4$. For both IL-2 and IFN-gamma, however, the cell subpopulations involved in the interaction with LTB$_4$ are of crucial importance, with amplification produced in the absence of T8[+] cells and suppression in the absence of T4[+] cells. Whether modulation by LTB$_4$ of the production of these lymphokines may also be responsible for the observed augmentation of natural cytotoxic and NK cell function by LTB$_4$ (14, 17-20) is presently under investigation.

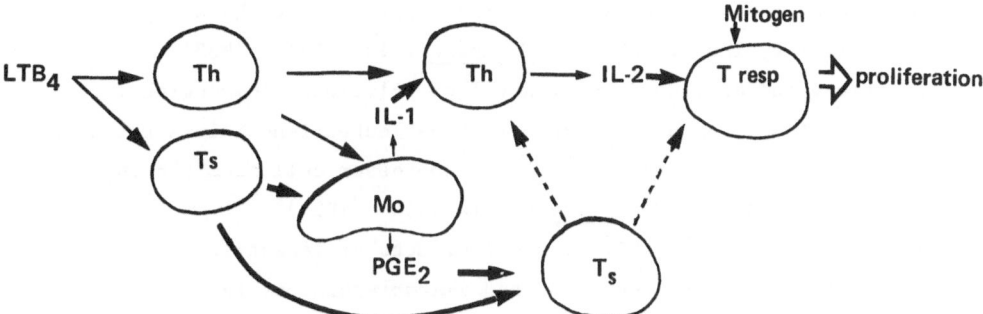

Figure 4. Schematic illustration of the various pathways through which LTB$_4$ interacts with cells of the immune system and modulates positively or negatively the immune response

All these studies would tend to suggest that lymphocytes possess lipoxygenase activity and that this activity is essential to several of their functions. To this date, however, direct and convincing proof of this is lacking, possibly due to the limitations of sensitivity and specificity of the available methodologies. Optimal concentrations of LTB_4, as assessed by its exogenous use in vitro, are in the picomolar range, thus at the lower limit of reliable detectability by high performance liquid chromatography or radioimmunoassays. With the forthcoming elucidation of the structure of 5-lipoxygenase and the subsequent development of appropriate molecular probes for it, it may become feasible to identify its presence in lymphocyte preparations and thus resolve part of the dilemma.

In conclusion, inflammatory mediators such as LTB_4 may contribute to the overall immune reaction by affecting several cell types and the production of various cytokines. The resulting net effect should depend on local conditions such as the presence of monocytes, the proportion of specific T cell subsets or the use of anti-inflammatory drugs.

REFERENCES

1. M. Rola-Pleszczynski, Immunoregulation by leukotrienes and other lipoxygenase metabolites, Immunol. Today 6: 302 (1985).

2. M. Rola-Pleszczynski, P. Borgeat and P. Sirois, Leukotriene B_4 induces human suppressor lymphocytes, Biochem. Biophys. Res. Commun. 108: 1531 (1982).

3. M. Rola-Pleszczynski and P. Sirois, Leukotrienes as immunoregulators in: Leukotrienes and other Lipoxygenase Products, P.J. Piper, ed., Wiley & Sons, London, p. 234 (1983).

4. D.G. Payan and E.J. Goetzl, Specific suppression of human lymphocyte function by leukotriene B_4, J.Immunol. 131: 551 (1983).

5. D. Atluru and J.S. Goodwin, Control of polyclonal immunoglobulin production from human lymphocytes by leukotrienes: leukotriene B_4 induces an OKT8[+], radiosensitive suppressor cell from resting human OKT8[-] T cells, J. Clin. Invest. 74:1444 (1984).

6. M. Rola-Pleszczynski, Differential effects of leukotriene B_4 on T4[+] and T8[+] lymphocyte phenotype and immunoregulatory functions, J. Immunol. 135: 1357 (1985).

7. M. Rola-Pleszczynski and I. Lemaire, Leukotrienes augment interleukin 1 production by human monocytes, J. Immunol. 135:3958 (1985).

8. E.J. Goetzl, Selective feed-back inhibition of the 5-lipoxygenation of arachidonic acid in human T-lymphocytes. Biochem. Biophys. Res. Commun. 101: 344 (1981).

9. J.M. Bailey, R.W. Bryant, C.E. Low, M.B. Pupillo and J.Y. Vanderhoek, Regulation of T-lymphocyte mitogenesis by the leukocyte product 15-hydroxy-eicosatetraenoic acid (15-HETE), Cell. Immunol. 67:112 (1982).

10 K.H. Leung, M.J. Ehrke and E. Minich, Modulation of the development of cell-mediated immunity: possible role of the products of the cyclooxygenase and the lipoxygenase pathways of arachidonic acid metabolism, Int. J. Immunopharmacol. 4: 195 (1982).

11 W.L. Farrar and J.L. Humes, The role of arachidonic acid metabolism in the activities of interleukin 1 and 2, J. Immunol. 135: 1153 (1985).

12 M. Gerber, D. Ball, F. Michel and A. Crastes de Paulet, Mechanism of enhancing effect of irradiation on production of IL-2, Immunol. Lett. 9: 279 (1985).

13 W.E. Seaman and J. Woodcock, Human and murine natural killer cell activity may require lipoxygenation of arachidonic acid, J. Allergy Clin. Immunol. 74: 407 (1984).

14 M. Rola-Pleszczynski, L. Gagnon and P. Sirois, Natural cytotoxic cell activity enhanced by leukotriene B_4: modulation by cyclooxygenase and lipoxygenase inhibitors, in: Icosanoids and Cancer, H. Thaler-Dao, ed., Raven Press, New York, p.235 (1984).

15 J.S. Goodwin, D. Atluru, S. Sierakowski and E.A. Lianos, Mechanism of action of glucocorticoids. Inhibition of T cell proliferation and interleukin 2 production by hydrocortisone is reversed by leukotriene B_4, J. Clin. Invest. 77: 1244 (1986).

16 H.M. Johnson and B.A. Torres, Leukotrienes are positive signals for regulation of gamma-interferon production, J. Immunol. 132: 413 (1984).

17 M. Rola-Pleszczynski, L. Gagnon and P. Sirois, Leukotriene B_4 augments human natural cytotoxic cell activity, Biochem. Biophys.Res. Commun. 113:531 (1983).

18 M. Rola-Pleszczynski, L. Gagnon, M. Rudzinska, P. Borgeat and P. Sirois, Human natural cytotoxic activity: enhancement by leukotrienes (LT) A_4 and B_4 but not by stereoisomers of LTB_4 or HETEs, Prostagl. Leukotr. Med. 13:113 (1984).

19 P. Rossi, J.A. Lindgren, C. Kullman and M. Jondal, Products of the lipoxygenase pathway in human natural killer cell cytotoxicity, Cell. Immunol. 93: 1 (1985).

20 R.A. Bray and Z. Brahmi, Role of lipoxygenation in human natural killer cell activity, J. Immunol. 136: 1783 (1986).

REGULATION OF T-CELL ACTIVATION BY LEUKOTRIENE B4

J. S. Goodwin
Department of Medicine, Medical College of Wisconsin
Milwaukee, Wisconsin

INTRODUCTION

The evidence that endogenous arachidonic acid metabolites are important in normal T cell activation and proliferation is suggested by several experimental findings. Perhaps the most direct evidence was provided by Kelly and his coworkers (1), who showed that drugs which blocked all arachidonic acid metabolism also inhibit mitogen-induced T cell proliferation. Drugs such as nordihydroguaretic acid (NDGA) and BW755c inhibit all lipoxygenase enzymes as well as cyclooxygenase (which is also a lipoxygenase enzyme). There are also specific cyclooxygenase inhibitors with no activity on the lipoxygenases, such as indomethacin and the other non-steroidal anti-inflammatory agents (NSAIA's). More recently, drugs which selectively inhibit 5-lipoxygenase, with little or no activity on 12 or 15-lipoxygenase or on cyclooxygenase, have been described. These inlcude caffeic acid (2), AA861 (3), and Rev 5901 (4). In addition, specific 12-lipoxygenase inhibitors and specific 15-lipoxygenase inhibitors are being developed by some pharmaceutical companies but are not yet available for study.

One can use the effect of specific and non-specific inhibitors of arachidonic acid metabolism on various assays of T cell activation proliferation to gather clues about the possible role of endogenously produced arachidonic acid metabolites in T cell function. For example, the fact that indomethacin and other cyclooxygenase inhibitors cause increased T cell proliferation in mitogen or antigen stimulated cultures was part of the evidence used to demonstrate that PGE is a normal endogenous feedback inhibitor of T cell activation and proliferation (reviewed in 5). Use of non-specific lipoxygenase inhibitors (that is, drugs which inhibit all lipoxygenase enzymes, including cyclooxygenase) suggest that some arachidonic acid metabolite(s) is (are) critical to normal T cell function. The fact that these non-specific lipoxygenase inhibitors stop mitogen induced T cell proliferation was already mentioned above (1). In addition, Johnson and Torros have shown that a non-specific lipoxygenase inhibitor suppressed mitogen-induced interferon production by murine splenocytes (6). This suppression could be reversed by the addition of LTB4, LTC4 or LTD4 to the cultures. LTB4, LTC4, or LTD4 were capable of replacing the helper cell or interleukin-2 (IL-2) requirement for interferon production by Lyt 1- 2+ cells. Similar results have been published by Farrar and Humes (7). The authors concluded that leukotrienes may play an essential role in the IL-2 stimulation of interferon production. This conclusion would have been strengthened had either group of investigators actually demonstrated leukotriene production in their cultures.

Similar data has been presented for natural killer (NK) function. Both Seaman (8) and Rola-Pleszcynski et al (9) have shown that non-specific lipoxygenase inhibitors suppress NK activity, and exogenous LTB4 will overcome that suppression (9), once again suggesting a role for endogenous leukotriene production in lymphocyte function.

In spite of the experimental results described above, most investigators interested in arachidonic acid metabolism or immunology have tended to stay away from experiments looking at a critical role for leukotrienes in T cell activities. There are at least two reasons for this. First, several laboratories reported little or no effect of leukotrienes when they were added to various in vitro assays of T cell function. Second, it is generally accepted that T cells do not metabolize arachidonic acid to leukotrienes. Indeed, the best published studies suggest that T cells do not metabolize arachidonic acid at all. I will discuss both of these points in some detail.

Several laboratories report modest(10-12) or no (13) inhibition of mitogen-induced lymphocyte proliferation by leukotrienes. Webb et al (14) reported that LTD4 or LTE4 in concentrations as low as 10 pM caused some inhibition of 3H-thymidine incorporation in mitogen-stimulated mouse thymocytes. Payan and Goetzl (11) reported that LTB4 but not LTC4 or LTD4 caused a slight inhibition of mitogen-induced 3H-thymidine incorporation and lymphokine production by human lymphocytes. Rola-Pleszcynski et al (10) also found a modest inhibition by LTB4 of human lymphocyte mitogenesis. On the other hand, Guttery et al (13) found no effect of LTB4 over a wide dose range in the same assay system. In sum, the effects of exogenous leukotrienes on some assays of T cell function are not impressive.

Equally unimpressive is the evidence that T cells even produce leukotrienes. Two groups reported that lymphocytes metabolize arachidonic acid to 5-HETE, and LTB4 (15, 16) but others have disputed this claim, arguing that the LTB4 produced comes from contaminating monocytes in the T cell preparation (17-19). Goldyne et al (17) in an article published in 1984 in Journal of Biological Chemistry found essentially no lipoxygenase metabolites produced by human T cells when monocytes were effectively removed prior to culture. These results, as well as similar unpublished results of several other laboratories have discouraged most investigators from looking for an intrinsic role for LTB4 or other 5-lipoxygenase products in T cell function. Any evidence for such a role, such as the inhibition of T cell proliferation by lipoxygenase inhibitors, has been ascribed to an effect on monocyte helper function. As we discuss later, we feel that the failure of most investigators to find leukotriene production by T cells is secondary to difficulties in methodology and not to a real deficiency in 5-lipoxygenase activity in T cells.

One other effect of lipoxygenase metabolites that should be mentioned is the ability of some lipoxygenase metabolites to regulate the synthesis of other lipoxygenase metabolites. The results differ depending on the tissue studied. For example, 15-HETE in relatively high concentrations (10 μM) inhibits 5-lipoxygenase activity in some cell types (20, 21) and actually stimulates leukotriene synthesis several fold in other cell types (22). In addition, inhibition of cyclooxygenase leads to a "shunting" of arachidonic acid into the lipoxygenase pathway, resulting in increased leukotriene production in several systems.

470

MECHANISM OF ACTION OF CORTICOSTEROIDS

The studies described in this paper will deal with our investigation of LTB4 in T cell proliferation and also our studies on lipoxygenase metabolites of arachidonic acid produced by T cells. Since much of our work has been guided by a hypothesis regarding the mechanism of action of glucocorticosteroids, I will start by outlining that hypothesis so that my subsequent reasoning might be more understandable. The hypothesis is that many or all of the immunologic effects of glucocorticoids are secondary to their inhibition of arachidonic acid metabolism (23-26). In the mid-1970's, several groups showed that steroids inhibited the release of arachidonic acid from membrane phospholipid (23, 27-29). Danon and Assouline (30) showed that RNA and protein synthesis was needed for steroids to inhibit arachidonic acid release and subsequent metabolism. Flower and his associates in England (24, 31) and Hirata, Axelrod and their coworkers at the NIH (25) identified a phospholipase A2 inhibitory protein, termed macrocortin or lipomodulin, that was synthesized and released by cells upon exposure to corticosteroids. On the other hand, several investigators have failed to find an inhibition by corticosteroids of the endogenous production of arachidonic acid metabolites in vitro (32) and in vivo (33, 34).

If indeed the effects of corticosteroids in a physiologic system are due to inhibition of arachidonic acid metabolism, then this allows us to construct several testable hypotheses. For example, if in a given physiologic system, corticosteroids have an effect not shared by cyclooxygenase inhibitors (NSAIA's) then that implies that the effect of corticosteroids in that system is due to inhibition of the production of a lipoxygenase metabolite of arachidonic acid rather than a cyclooxygenase product. This can then be tested in several ways. First, one should be able to duplicate the effect of corticosteroids with non-specific lipoxygenase inhibitors. Second, one should be able to reverse the effect of corticosteroids by adding back one or more lipoxygenase metabolites of arachidonic acid. Finally, one should be able to demonstrate that specific lipoxygenase metabolites of arachidonic acid are indeed produced in that system, and that corticosteroids inhibit their production with a dose response relationship of inhibition of lipoxygenase metabolite production similar to the dose response relationship for the physiologic effect of corticosteroids in that system.

I realize that the hypothesis about the mechanism of action of corticosteroids outlined above may seem overly simplistic or naive, but it has helped us generate many useful ideas about the possible role of lipoxygenase metabolites of arachidonic acid in various immunologic functions. There is an extensive literature about the effects of corticosteroids on immunologic function in vivo and in vitro. There are also many previous studies of the effects of NSAIA's in those same systems. Thus, we can use the previously described effects of corticosteroids and NSAIA's to suggest areas where the endogenous production of a lipoxygenase metabolite of arachidonic acid might be critical for normal immunologic function.

LTB4 IN SUPPRESSOR T CELL INDUCTION

Our first investigation of this type was a study of the role of leukotrienes in suppressor cell generation. We reasoned that, since corticosteroids inhibit suppressor cell generation in many in vitro and in vivo models (35-37), and since NSAIA's do not inhibit suppressor cell generation (38-40), then a lipoxygenase metabolite of arachidonic acid might be a critical component of suppressor cell generation. Indeed, Rola-

Plezczynski et al, (10) had reported that human periperhal blood mononuclear cells (PBMC) incubated for 18 hours with LTB4 suppressed the proliferation response of fresh antologous PBMC to mitogens. We found that LTB4, but not LTC4 or LTD4 was a potent suppressor of polyclonal Ig production in pokeweed mitogen (PWM) stimulated cultures of human peripheral blood lymphocytes (41). Preincubation of T cells with LTB4 in nanomolar to picomolar concentrations rendered these cells suppressive of Ig production in subsequent PWM-stimulated cultures of fresh autologous B + T cells. This LTB4-induced suppressor cell was radiosensitive, and its generation could be blocked by cyclohexamide but not by mitomycin C. The LTB4-induced suppressor cell was OKT8 (+), while the precursor cell could be OKT8(-). The incubation of OKT8(-) T cells with LTB4 for 18 h resulted in the appearance of the OKT8 antigen on 10-20% of the cells, and this could be blocked by cyclohexamide but not by mitomycin C. Thus, we showed that LTB4 in very low concentrations induces a radiosensitive OKT8(+) suppressor cell. LTB4 is three to six orders of magnitude more powerful than any endogenous hormonal inducer of suppressor cells previously described. The fact that LTB4 can cause an OKT4(+) to OKT8(+) shift in T cell phenotype has now been reproduced in at least one other laboratory (43). More recently, it has been shown that stimulation of OKT4(+) cells with lectins results in the expression of both OKT4 and OKT8 on some of the cells (44). Given our evidence discussed below that LTB4 is a T cell activation signal, it may be that the OKT4 to OKT8 shift caused by LTB4 is actually a stimulation of OKT4 cells to express both antigens.

We have recently found that 15 HPETE induces both phenotypic and functional suppressor T cells in a manner quite similar to LTB4, the major difference being that much higher concentrations of 15 HPETE (10 nM to 1 uM) were required for the effect (42). In addition to causing induction of OKT8(+) suppressor cells, we have also found that both LTB4 and 15 HPETE cause increased proliferation of OKT8(+) T cells stimulated by mitogen, while the proliferation of OKT4(+) cells is inhibited by these compounds (12). A similar effect of LTB4 had been previously published by Payan et al. (45). The balance between stimulation of OKT8(+) cells and inhibition of OKT4(+) cells is presumably why other investigators found only modest or no effects of LTB4 on the mitogenic response of unfractionated T cell preparations (10, 11, 13). We will return to this issue of differential effects of LTB4 on T cell subpopulations when we discuss our findings of an apparent reversal by LTB4 of inhibition of interleukin-2 (IL-2) production by corticosteroids.

ENDOGENOUS LTB4 IS NECESSARY FOR MITOGEN-STIMULATED IL-2 PRODUCTION AND T CELL PROLIFERATION

After the studies described above we turned our attention to the role of LTB4 in normal T cell proliferation. Our recent findings would lead us to believe that our report on LTB4 in suppressor cell generation was focusing on just one aspect of a more general role for LTB4 in T cell activation. Once again, we were guided in this study by our knowledge of previous studies showing that corticosteroids profoundly suppressed T cell proliferation in mitogen or antigen-stimulated cultures (46, 47), while NSAIA's actually enhance T cell proliferation in these same assays (48). We asked, therefore, whether the effect of corticosteroids in inhibiting T cell proliferation might be secondary to inhibiting the production of some lipoxygenase metabolites of arachidonic acid (26). We first compared the effects of corticosteroids, non-specific lipoxygenase inhibitors, cyclooxygenase inhibitors, and specific 5-lipoxygenase inhibitors on mitogen induced T cell proliferation.

Our results can be summarized as follows:

1. Nonspecific lipoxygenase/cyclooxygenase inhibitors such as BW 755c and NDGA inhibited IL-2 production and T cell proliferation in phytohemagglutinin (PHA)-stimulated cultures in a manner similar to hydrocortisone or dexamethasone, while a specific cyclooxygenase inhibitor, indomethacin, actually enhanced IL-2 production and T cell proliferation.

2. Two Specific 5-lipoxygenase inhibitors, AA861 and Rev 5901 which inhibit 5-lipoxygenase at doses which cause little effect on 12 or 15 lipoxygenase or on cyclooxygenase, also inhibit IL-2 production and T cell proliferation. This suggested that the relevant arachidonic acid metabolite necessary for T cell activation was a product of the 5-lipoxygenase pathway.

3. LTB4 in concentrations as low as 10^{-10}M completely reversed the inhibition of IL-2 production and T cell proliferation caused by steroids or by the 5 lipoxygenase inhibitors. LTB4 did not reverse inhibition caused by PGE2, histamine, or irradiation. LTC4, LTD4 and 5 HETE had no effect on inhibition caused by steroids or lipoxygenase inhibitors. Thus, the reversal of inhibition was specific for LTB4 versus the other 5-lipoxygenase metabolites (LTC4, LTD4 and 5 HETE), and the reversal by LTB4 was specific for inhibition caused by steroids or lipoxygenase inhibitors versus other inhibitors (PGE, histamine, irradiation). This suggested that LTB4 was specifically reversing the inhibition of IL-2 production and proliferation caused by steroids and not stimulating IL-2 production via another unrelated mechanism.

4. LTB4 is produced in PHA stimulated cultures of human peripheral blood T cells. After 18 hours, one million T cells in one ml media produce one nM LTB4. There is a good correlation between inhibition of LTB4 production by various concentrations of hydrocortisone and inhibition of IL-2 production.

We concluded from the findings summarized above that corticosteroids inhibit mitogen-induced IL-2 production and T cell proliferation by inhibiting endogenous LTB4 production. A related conclusion from this study is that LTB4 production is a necessary component of mitogen-stimulated T cell proliferation. Our recent findings using T cell subsets have led us to reject these conclusions as overly simplistic, as will be discussed later.

T CELLS PRODUCE LTB4

Our finding of LTB4 production by T cells was somewhat controversial, especially in the light of published reports that purified T cells do not metabolize arachidonic acid (17, 18). One possibility was that the LTB4 produced in our cultures was produced by the contaminating monocytes. It is clear that a small percentage of accessory cells are required for T cell activation by PHA or other mitogens (49, 50), and our T cell preparations contained 1-2% monocytes as identified by peroxidase staining. We first addressed the issue of the source of the LTB4 by comparing LTB4 production in PHA stimulated cultures of T cells (1-2% monocytes), of peripheral blood mononuclear cells (PBMC, 15-20% monocytes), and of T cells to which 20% adherent cells had been added back. Comparable amounts of LTB4 were produced in each situation (51). In addition, T cell preparations vigorously depleted of monocytes (<0.2%) that no longer proliferated in response to PHA still produced LTB4 upon stimulation by the calcium ionophore A23187. Our method of LTB4 quantitation is discussed below and involves HPLC followed by a specific RIA for LTB4 on the relevant fractions.

A more rigorous method of testing the cell origin of LTB4 is to use T cell lines, as is shown in Figure 1. Jurkat cells, a human T cell line, produced LTB4 when stimulated by either PHA or phorbol-myristate-acetate (PMA). We feel this provides very strong evidence that T cells do indeed metabolize arachidonic acid, and a major arachidonic acid metabolite of human T cells is LTB4. The time course for LTB4 accumulation in cultures of Jurkat cells was similar to that of peripheral blood T cells. LTB4 was always undetectable at 2 hours but present in measureable amounts after 4 hours of culture. Because the limits of sensitivity of our assay of LTB4 is 100 pM, and because levels as low as 1 pM could partially reverse inhibition by steroids or lipoxygenase inhibitors, there may be much earlier production of biologically relevant amounts of LTB4, perhaps in the first few minutes of T cell activation, that are below the level of sensitivity of our detection.

ARACHIDONIC ACID INHIBITS LTB4 PRODUCTION

The question then arose as to why we found relatively large amounts of LTB4 production by T cells when Goldyne et al. (18) and other groups (by various personal communications) had found no LTB4 production. We found one answer lay in the differences in the methodologies employed (52). In the studies of Goldyne et al. (18) as in most studies of arachidonic acid metabolism, exogenous arachidonic acid at 10 M or higher was always added to the incubation media. We found that exogenous arachidonic acid profoundly suppressed LTB4 production by PHA-stimulated T cells. Results of 5 experiments are shown in Table 1, giving a dose response to added arachidonic acid.

Exogenous arachidonic acid also suppressed appearance of 12 HETE and 15 HETE, but to a less extent than LTB4 levels. This dramatic inhibition of lipoxygenase activity in human T cells by arachidonic acid is not necessarily generalizable to other cell populations. Several groups have shown that exogenous arachidonic acid changes the pattern of cyclooxygenase metabolites of arachidonic acid produced in tissue or cell cultures (53, 54). Laclos et al (55) reported some inhibition of 5-lipoxygenase activity and an increase in 15 HETE production by human mixed leukocytes by arachidonic acid at 5 uM or higher. On the other hand, Lee et al. (56) reported no influence of similar concentrations of arachidonic acid on purified human neutrophils, and others, (57, 58) have found an increase in LTB4 produced by human neutrophils when arachidonic acid was added to the media. One obvious conclusion from our findings is that previously reported studies of arachidonic metabolism by various tissues, if done entirely in the presence of exogenous arachidonic acid, should be repeated to see if the arachidonic acid altered the pattern of metabolites produced.

The mechanism of inhibition of LTB4 formation by arachidonic acid is unclear. Several possibilities are testable:

1. Exogenous arachidonic acid or a metabolite promotes further metabolism of LTB4, 12 HETE, etc.. Thus, there is no inhibition of lipoxygenase metabolism by exogenous arachidonic acid, just an acceleration in further metabolism of the products that we measure.

2. The arachidonic acid is preferentially converted to a metabolite that inhibits 5, 12 and 15 lipoxygenase. Vanderhoek, Bailey and their colleagues (20-22) have shown that products of arachidonic acid such as 15 HETE inhibit 5 lipoxygenase.

Table 1. Effect of exogenous arachidonic acid on LTB4 production by PHA-stimulated T cells. Experiments 1, 2 and 3 show results of a direct RIA for LTB4 on the supernates, while in experiments 4 and 5 the supernates were fractionated by HPLC prior to RIA. (Taken from reference 52).

| Arachidonic Acid | LTB4 (pg/tube) | | | | |
	Exp.1	Exp.2	Exp.3	Exp.4	Exp.5
0	210	240	120	100	140
10^{-5}	< 12.5	<12.5	<12.5	<12.5	<12.5
10^{-6}	< 12.5	<12.5	<12.5	20	<12.5
10^{-8}	50	48	50	37	34
10^{-10}			104	70	91
10^{-12}				91	122

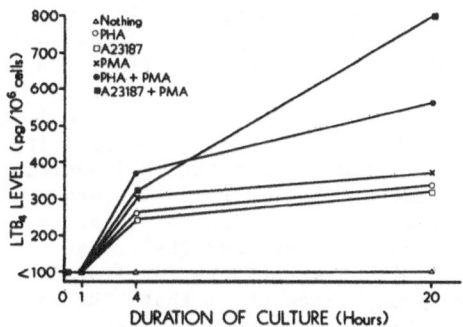

Figure 1. Stimulation of LTB4 production from Jurkat cells by PHA (4.0 ug/ml), A23187 (0.5 ug/ml) and PMA (50ng/ml) alone or in combination (taken from reference 51)

3. Exogenous arachidonic acid might be directly inhibiting lipoxygenase in T cells. There is some indication that lipoxygenases may preferentially act on arachidonic acid while it is still esterified to phospholipid. Free arachidonic acid might then competitively inhibit that lipoxygenation without necessarily being a good substrate.

4. The arachidonic acid could be poisoning the cells through unknown mechanisms, and the decrease in LTB4 production would just be one aspect of a general decrease in cell function. None of our experiments using different sources of arachidonic acid points to such a mechanism, but it has not been completely ruled out.

I should emphasize that the question of whether T cells metabolize arachidonic acid to LTB4 is far from resolved. Borgeat and his colleagues have recently presented data at the International Meeting on Prostaglandins and Related Compounds (Florence, June, 1986) in which they found no LTB4 produced by human lymphocytes, even if no arachidonic acid was added to the system. This brings up an interesting discrepancy in our own results. Our method for measuring LTB4 is by an RIA of the culture supernate. Alternatively, we extract the supernate, deproteinate it, fractionate it by high performance liquid chromatography (HPLC), and then assay by RIA for LTB4 those fractions collected with the retention time of LTB4. Using either method we arrive at similar estimations of LTB4 produced, about 500 pg per/ml of T cells cultured at one million cells per ml with PHA for 18 hours (26, 52). The HPLC runs show a peak with a retention time identical to LTB4 (e.g., see reference 26). However, 500 pg of LTB4 is below the sensitivity of detection of the HPLC. Thus, the HPLC peak seen with the retention time of LTB4 is not all LTB4. On the other hand, the RIA shows that there is immunoreactive LTB4 in the T cell supernates, and this immunoreactive LTB4 has a retention time identical to LTB4 on HPLC. So the peak seen on HPLC with a retention time equal to LTB4 contains LTB4, but the major component may be some other compound.

STIMULATION OF LTB4 PRODUCTION BY IL-2

We have shown that LTB4 is produced by mitogen-activated T cells and that LTB4 is required for IL-2 production (26). We have also found that LTB4 may be a second messenger for IL-2 in IL-2 responsive cells. Exposing T cell blasts or IL-2 dependent murine cell lines, HT-2 or CTLL, to IL-2 resulted in a prompt appearance of LTB4 in the culture media (Figure 2). LTB4 was always measureable in the CTLL and HT-2 cultures by 30 minutes, and low levels (close to the 100 pM limit of sensitivity of our assay) were detectable as early as five minutes after IL-2 addition. This is much earlier than the appearance of LTB4 in PHA stimulated cultures of resting T cells or Jurkat cells (Figure 1). Thus the question arose as to whether LTB4 might act as a second messenger for IL-2. As mentioned previously, there is evidence that LTB4 can substitute for IL-2 in some systems (6, 7, 59). For example, two groups have shown that LTB4 can substitute for IL-2 in the stimulation of interferon production by murine T cells or T cell lines (6, 7). We have shown that LTB4 causes maximal proliferation of IL-2 dependent cells in the presence of suboptimal levels of IL-2 that by themselves only cause minimal proliferation (ref 59, included in appendix).

LTB4 PRODUCTION IS NECESSARY FOR IL-2 RECEPTOR EXPRESSION.

We have recently turned our attention to other actions of LTB4 in T cell activation. One necessary component of IL-2 dependent T cell proliferation is expression of IL-2 receptors. IL-2 receptor expression can take place in the absence of IL-2 production (60), but IL-2 up-regulates IL-2 receptor expression (61, 62). We have found that endogenous LTB4 production is a necessary component for IL-2 receptor expression (63). The experiments were entirely analogous to those described earlier showing that LTB4 was necessary for IL-2 production (26). A representative experiment is shown in Figure 3, showing that LTB4 by itself stimulates the appearance of IL-2 receptors on PHA-stimulated T cells. BW 755c completely inhibits the appearance of IL-2 receptors (as does hydrocortisone or other inhibitors of arachidonic acid) and that inhibition can be overcome by LTB4.

CONFLICTING DATA

The experimental results presented and discussed in the first part of this paper are all consistent with the concept that endogenous production of LTB4 by T cells is a necessary component of T cell activation. Endogenous LTB4 appeared to be necessary for both IL-2 production and IL-2 receptor expression after mitogen stimulation. Results from other laboratories suggest it is also necessary for interferon production. In addition, we have suggested that the inhibition of T cell proliferation by corticosteroids could be explained by inhibition of endogenous LTB4 production. The results of the experiments I will now discuss have thrown some of our previously tidy hypotheses into some disarray. Indeed, it is now impossible to construct a model for LTB4 in T cell activation that accomodates all of our recent experimental results. Because of space limitations, and also because this represents preliminary unpublished work still in progress, I will simply discuss these recent results without presenting the data.

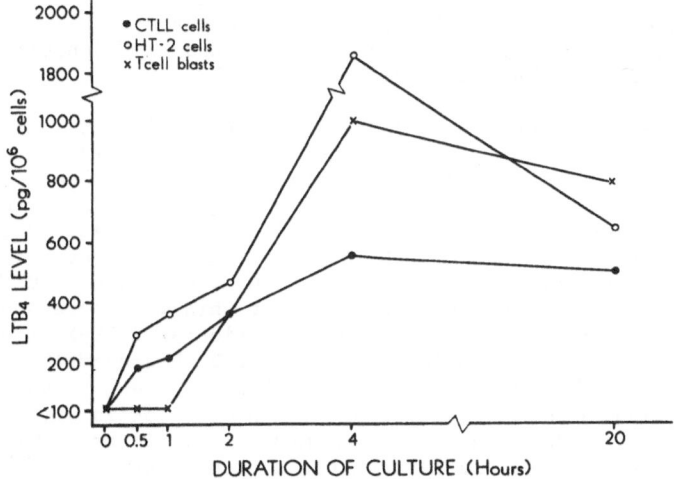

Figure 2. IL-2 stimulation of LTB4 production in IL-2 responsive cells. T cell blasts were made by incubating peripheral blood T cells with PHA for 72 hours. Prior to the experiments the T cell blasts or cell lines were incubated in complete media without IL-2 for 3 hours with hourly change of media in order to reduce the IL-2 bound to the cells. IL-2 at 50 U/ml was then added and the cells were cultured for the indicated times. Control cells containing no IL-2 produced no measureable LTB4 at any time point. (Taken from ref 51).

Our first unsettling finding was that in order for LTB4 to reverse inhibition of T cell mitogenesis by steroids or lipoxygenase inhibitors, adherent cells must be removed during the isolation of T cells. In other words, our routine procedure for isolating T cells involves separation of peripheral blood mononuclear cells (PBMC) on a Ficoll-Hypaque gradient, followed by removal of adherent cells by a 30 min. incubation on glass petri dishes, followed by E-rosetting. By chance, we found that IL-2 production by T cells isolated directly from PBMC without an adherence step was still inhibited by steroids or lipoxygenase inhibitors, but LTB4 had absolutely no effect on this inhibition. On the other hand, LTB4 still completely reversed the inhibition of IL-2 receptor expression in these cells. Thus, we must either remove a population of adherent cells, or activate a population by the adherence step, or both, as a requirement for LTB4 reversal of inhibition of T cell proliferation and IL-2 production by steroids or lipoxygenase inhibitors. One obvious conclusion from this is that not all immunologic effects of steroids can be explained by inhibition of LTB4 production. Another conclusion is that the steroids and lipoxygenase inhibitors are inhibiting T cell proliferation at two or more steps, only one of which is reversible by LTB4.

Another recent finding is that reversal of the inhibition of steroids or lipoxygenase inhibitors by LTB4 or IL-2 is dependent on the type and dose of mitogenic stimulus used. Thus, for PHA and Con A at optimal concentrations, either LTB4 or IL-2 reverses inhibition caused by hydrocortisone or BW755c. For A23187, LTB4 but not IL-2 will reverse the inhibition caused by hydrocortisone or BW755c. This may be related to the fact that A23187 is thought to stimulate T cell proliferation via IL-2 independent pathways (64). Finally, for PHA and Con A at suboptimal doses, IL-2 but not LTB4 reverses the inhibition caused by hydrocortisone or BW755c. These results have several implications. Just as in the previous paragraph, those data indicate that not all effects of steroids can be explained by inhibition of LTB4. On the other hand, we have not been able to separate the effects of steroids from the effects of BW755c or AA861, suggesting that, while the effects of steroids in our systems are not always solely due to inhibition of LTB4 production it is still possible that the effects of steroids can be explained by inhibition of the production of other arachidonic acid metabolites.

Our initial attempts to investigate the cellular basis for the different effects of LTB4 and IL-2 on steroid-induced inhibition of mitogenesis have involved further fractionations of our cell populations into subsets, as well as the addition of various accessory cell factors (IL-1 or supernates from monocyte cultures) to the T cell cultures. We found that corticosteroids and LTB4 had markedly different effects on IL-2 production by different T cell subsets. PHA stimulation of OKT8(+) cells resulted in very little IL-2 production, but this was increased greatly by LTB4. On the other hand, PHA caused considerable IL-2 production by OKT4(+) cells. This was completely inhibited by hydrocortisone, and this inhibition was not reversed by LTB4. Thus the apparent "reversal" by LTB4 of the inhibition by hydrocortisone of IL-2 production is actually a switch in IL-2 production between T cell subsets. This is not true for IL-2 receptors, however. LTB4 did reverse the hydrocortisone inhibition of IL-2 receptors in both OKT4(+) and OKT8(+) cells.

Another recent finding is that phorbol myristate acetate (PMA) can reverse inhibition of IL-2 production by steroids or lipoxygenase inhibitors. Because PMA directly activates protein kinase C, this suggests that endogenous LTB4 is necessary in a step proximal to protein kinase C activation. I should perhaps pause here to state the obvious, that LTB4 currently has no place in the generally accepted paradigm of T cell activation or in the activation of any cell type (reviewed in 65, 66). The currently accepted paradigm involves transduction of cell surface signals (antigen presentation by an accessory cell, in the case of T cells) by hydrolysis of phosphatidylinositol biphosphate, producing diacylglycerol and inositol triphosphate. The former activates protein kinase C and the latter causes increases in intracellular free Ca by releasing endoplasmic reticulum stores (reviewed in ref 65, 66). Phosphatidyl inositol hydrolysis can be bypassed by administering agents that directly activate protein kinase C (such as phorbol esters or diacylglycerols) in combination with calcium ionophores (67-71). Thus there is no "need" in this schema for LTB4 production. It is important to realize, however, that the currently accepted paradigm of T cell activation is insufficient to explain many experimental findings. For example, Nishizuka (72) feels that at least three signals are required for activation of resting T cells, phorbol ester, calcium ionophore and a small amount of PHA. Nishizuka (66) feels that much of the work on T cell activation used calcium ionophore (usually A23187) at doses (0.5 ug/ml) that were much higher than required for increasing intracellular calcium and which had "nonspecific" activating effects on T cells. In addition, findings in T cell lines such as Jurkat, which are already "activated" to some extent (otherwise they would not be continuous cell lines) are important but should not be uncritically accepted as applying to activation of normal human T cells.

Since in humans arachidonic acid is the predominent fatty acid esterified at the two position in phosphatidyl inositol, it is easy to envision how arachadonic acid release and metabolism would follow the production of diacylglycerol from phosphatidyl inositol.

SUMMARY

I have presented what I feel is a confusing array of findings regarding the role of LTB4 in T cell activation. The confusion should not distract us from the strong evidence that LTB4 in picomolar concentrations has powerful effects on T cell activation. When the confusion is finally resolved, I am confident that endogenous LTB4 will be recognized to have a central role in the early events of T cell activation.

REFERENCES

1. Kelly JP, Johnson MC, Parker CW. J Immunol 122:1563 (1979).
2. Koshihara Y, Neichi T, Murota S, Lao A, Fujimota Y, Tatasuno T. Biochem Biophys Acta 792:92 (1984).
3. Yamamota S, Yoshimoto T, Furukawa M, Horie T, Kohno SW. J Allergy Cli Immunol 74:349 (1984).
4. Coults S, Khandwala A, Van Inwegen R, Bruens J, Jariwala N, Dally-Meade V, Ingram R, Chakraborty U, Musser J, Jones, Pruss T, Neiss E, Weinryb I. International Washington Spring Conference:Prostaglandins and Leukotrienes 1984, Washington, D.C. May 8-11 (1984).
5. Goodwin JS, Ceuppens J. J Clin Immunol 3:295 (1983).
6. Johnson HM, Torres BA. J Immunol 132:413 (1984).
7. Farrar WL, Humes JL. J Immunol 135:1153 (1985).
8. Seaman WE. J Immunol 131:2953 (1983).
9. Rola-Plezcznyski M, Gagnon L, Sirosis P. Biochem Biophys Res Commun 113:531 (1983).
10. Rola-Plezczynski M, Boreat P, Sirois P. Biochem Biophys Res Commun 198:1531 (1982).
11. Payan DG, Goetzl EJ. J Immunol 131:551 (1983).
12. Gualde N, Atluru D, Goodwin JS. J Immunol 134:1125 (1985).
13. Guttery JE, Tilden A, Herron DK, Gallagher P, Baker SR, Ades EW. J Clin Lab Immunol 13:151 (1984).

14. Webb DR, Nowowiejsk I, Healy C, Rogers TJ. *Biochem Biophys Res Commun* 104:1617 (1982).
15. Parker CW, Stenson WF, Huber MG, Kelly JP. *J Immunol* 122:1572 (1979).
16. Goetzl EJ. *Biochem Biophys Res Commun* 101:344 (1981).
17. Goldyne ME. *Recent Adv Clin Immunol* 3:9 (1983).
18. Goldyne ME, Burrish GF, Poubelle P, Borgeat P. *J Biol Chem* 259:8815 (1984).
19. Decker DM, Kennedy MS. *Clin Res* 32:39A (1984).
20. Bailey JM, Bryant RW, Low CE, Pupillo MR, Vanderhoek JY. *Cell Immunol* 67:112 (1982).
21. Vanderhoek JY, Bryant RW, Bailey JM. *J Biol Chem* 255:10064 (1980).
22. Vanderhoek JY, Tare NS, Bailey M, Goldstein A, Pluznik D. *J Biol Chem* 257:12191 (1982).
23. Gryglewski RJ, Panczenko B, Korbut R, Grodzinska L, Ocetkiewica A. *Prostaglandins* 10:343 (1975).
24. Blackwell GJ, Carnuccio R, DiRose M, Flower RJ, Parente L, Perrico P. *Nature* 287:147 (1980).
25. Hirata F, Schiffman E, Yenkatasubramanian K, Salomen K, Axelrod J. *Proc Nat Acad Sci, USA* 77:2533 (1980).
26. Goodwin JS, Atluru D, Sierakowski S, Lianos EA. *J Clin Invest* 77:1244 (1986).
27. Kantrowitz F, Robinson DR, McGuire MB, Levine L. *Nature* 258:737 (1975).
28. Floman Y, Zor U. *Prostaglandins* 12:403 (1976).
29. Tashjian A, Yoelkel E, McDonough J, Levine L. *Nature* 258:739–(1975).
30. Danon A, Assouline G. *Nature* 173:552 (1978).
31. Flower RJ, Blackwell GJ. *Nature* 278:456 (1978).
32. Chandrabose KA, Lapentina EG, Schmitges CJ, Siegel MI, Cuatrecas P. *Proc Natl Acad Sci USA* 75:214 (1978).
33. Gold EW, Fox OD, Edgar PR. *J Steroid Biochem* 9:313 (1978).
34. Navay-Fejes-Toth A, Fejes-Toth G, Fischer C, Frolich JC. *J Clin Invest* 74:120 (1984).
35. Waldmann TA, Broder S, Krakauer R, MacDermott RP, Durm M, Goldman C, Meade B. *Fed Proc* 35:2067 (1976).
36. Tosato GI, Magrata I, Koski I, Dooley N, Blaese RM. *N Eng J Med* 301:1133 (1979).
37. Tosato GI, Magrata I, Koski I, Dooley N, Blaese RM. *J Clin Invest* 66:383 (1980).
38. Goodwin JS. *Cell Immunol* 49:421 (1980).
39. Soppi E, Eskola J, Ruuskanen O. *Immunopharmacology* 4:235 (1982).
40. Badger AM, Griswold DE, Walz DT. *Immunopharmacology* 4:149 (1982).
41. Atluru D, Goodwin JS. *J Clin Invest* 74:1444 (1984).
42. Gualde N, Rigaud M, Goodwin JS. *J Immunol* 135:3424 (1985)
43. Rola-Plezczynski M. *J Immunol* 135:1357 (1985).
44. Blue M, Daley JF, Levine H, Schlossman SF. *J Immunol* 134:2281 (1985).
45. Payan D, Missirian-Bastian A, Goetzl E. *Proc Natl Acad Sci USA* 81:3501 (1984).
46. Goodwin JS, Messner RP, Williams RC. *Cell Immunol* 45:303 (1979).
47. Robertson AJ, Gibbs JH, Potts RC, Brown RA, Browning M, Beck JS. *Intl J Immunopharmacol* 3:21 (1981).
48. Goodwin JS, Bankhurst AD, Messner RP. *J Exp Med* 146:1719 (1977).
49. Rosenstreich DL, Farrar JJ, Dougherty S. *J Immunol* 116:131 (1976).
50. Williams JM, Ransil BJ, Shapiro HM, Strom TB. *J Immunol* 133:2986 (1984).
51. Atluru D, Lianos E, Goodwin JS. Submitted (1986).
52. Atluru D, Lianos EA, Goodwin JS. *Biochem Biophys Res Commun* 135:670 (1986).
53. Dimov V, Christensen NJ, Green K. *Biochem Biophys Acta* 754:38 (1983).
54. Coene MC, Howe CV, Claeys M, Herman AG. *Biochem Biophys Acta* 710:437 (1982).
55. Laclos FB, Braquet P, Borgeat P. *Prostaglandins Leukotrienes Med* 13:47 (1984).

56. Lee TM, Mencia-Heurta JM, Shih C, Corey EJ, Lewis RA, Austin KF. J Clin Invest 74:1922 (1984).
57. Borgeat P, Samuelsson B. Proc Natl Acad Sci USA. 76:2148 (1979).
58. Prescott SM. J. Biol Chem 259:7615 (1984).
59. Atluru D, Goodwin JS. Cell Immunol 99:444 (1986).
60. Mills GB, Cheung RK, Grinstein S, Gelfand EW. J Immunol 134:1640 (1985).
61. Reem GH, Yeh NH. Science 225:429 (1984).
62. Reem GH, Yeh NH. J Immunol 134:953 (1985).
63. Atluru D, Koethe S, Goodwin JS. Submitted (1980).
64. Koretzky GA, Daniele RP, Greene WC, Nowell PC. Proc Natl Acad Sci USA 80:3444 (1983).
65. Rasmussen H. New Eng J Med 314:1091 and 1164 (1986).
66. Nishizuka Y. Nature 308:693 (1984).
67. Weiss A, Imboden J, Shoback D, Stobo J. Proc Natl Acad Sci USA 81:4169 (1984).
68. O'Flynn K, Zanders ED, Lamb JR, Beverly PC, Wallace DL, Tatham P, Tax WJ, Linch D. Eur J Immunol 15:7 (1985).
69. Imboden JB, Stobo JD. J Exp Med 161:446 (1985).
70. Weiss A, Wiskocil RL, Stobo JD. J Immunol 133:123 (1984).
71. Niedel JE, Kuhn LJ, Vandenback GR. Proc Natl Acad Sci USA 80:36 (1983).
72. Kaiberchi K, Takai Y, Nishizuka Y. J Biol Chem 250:1366 (1985).

A ROLE FOR PLATELET-ACTIVATING FACTOR (PAF)
IN HUMAN T-LYMPHOCYTE PROLIFERATION

S.G. Ward, G.P. Lewis, J. Westwick
Department of Pharmacology, Hunterian Institute, Royal College of
Surgeons Lincolns Inn Fields, London WC2A 3PN, England

INTRODUCTION

Interleukin-2 (IL-2) is a soluble 15,000 dalton sialoglycoprotein, originally termed T-cell growth factor. Human IL-2 consists of a 133 amino acid polypeptide chain containing a single intramolecular disulfide bridge[1]. Interleukin-2 is released by and activates T-lymphocytes after stimulation of the T-cell antigen receptor complex termed T_3-T_1[2]. Upon interaction with its specific membrane receptors[3], IL-2 generates signals which determine T-lymphocyte proliferation. Interleukin-2 receptor expression on T-lymphocytes is transient and occurs only after appropriate immune stimulation by antigens, T-cell specific monoclonal antibodies, mitogenic lectins such as phytohaemagglutinin (PHA) and phorbol esters, all of which stimulate components of the T-cell surface antigen receptor complex[2,4,5]. Upon receptor binding, IL-2 mediates a switch in T-cells from G_1 into the proliferative S, G_2 and M stages of the cell cycle[6,7,8]. It seems that high and low affinity receptors for IL-2 exist[9] and the biological effects of IL-2 are due to occupancy of the high-affinity receptor[3,9]. Interleukin-2 bound to the high-affinity receptor is internalised[10,11] but the role of the low affinity receptor has not yet been determined. The events which follow IL-2 binding to receptors are not clear but there are conflicting reports as to whether IL-2 induces (poly)-phosphatidylinositol (PPI) metabolism or calcium mobilization[12,13,14].

Platelet activating factor (PAF) is produced by and activates a range of inflammatory cells, including platelets, polymorphonuclear leucocytes and macrophages[15,16]. However, to date, its effects on lymphocyte function, in particular IL-2-induced T-lymphoblast proliferation, have not been characterized. We have examined the role of PAF in lymphocyte mitogenesis by determining the effect of (a) PAF and two non-hydrolysable PAF agonists, PR1501[17] and PR1502[18] (b) three compounds which have been described as potent and selective PAF receptor antagonists, at least with respect to platelet activation, namely L-652,731[19], BN52021[20] and CV3988[21] and (c) a selective PAF synthesis inhibitor L-648,611[22] on IL-2-induced proliferation of human T-lymphoblasts.

We report that IL-2 induced proliferation of human T-lymphoblasts as measured by ^{3}H-thymidine (^{3}H-TdR) incorporation can be dose-dependently antagonized by two structurally distinct PAF antagonists, namely L-652,731 and CV3988, as well as a PAF synthesis inhibitor, L-648,611. Together with results showing enhancement of IL-2-induced ^{3}H-TdR incorporation by PAF and two synthetic analogues of PAF in human T-lymphoblasts, these results suggest a novel role for PAF as a modulator of T-lymphocyte proliferation.

MATERIALS AND METHODS

PAF (1-O-octadecyl-2-acetyl-sn-glycero-3-phosphorylcholine) was obtained from Bachem,Switzerland. PR1501 and PR1502 are non-hydrolysable PAF agonists in which the $O-CO-CH_3$ of PAF is replaced by $-NH-CO-CH_3$ or by $-CH_2-CO-CH_3$ respectively and were generously donated by Dr P. Rasmussen of Leo Pharmaceuticals, Denmark. The PAF receptor antagonist L-652,731 and the PAF synthesis inhibitor L-648,611 were generously donated by Drs T.Y Shen, and J Chabala of the Merck Institute, Rahway, N.J. USA. CV3988 was generously donated by Dr Nishikawa of Takeda Chemical Industries, Osaka, Japan. BN52021 was a gift from Dr P.Braquet of Institut Henri Beaufour, Paris, France. Phytohaemagglutinin (PHA) was purchased from Wellcome. Human recombinant IL-2 was a gift from Dr D.A. Cantrell of the Imperial Cancer Research Fund, London.

Cell cultures

Human T-lymphoblasts were produced essentially according to the method of Smith and Cantrell[23]. Human peripheral blood mononuclear cells (PBMC) were isolated by Ficoll-Paque (Pharmacia) gradient centrifugation of fresh human blood. The PBMC's were cultured at 10^6 viable cells/ml in RPMI 1640 medium (GIBCO) supplemented with 10% heat-inactivated foetal calf serum (HIFCS), 100 u/ml penicillin/streptomycin (GIBCO), in the presence of PHA 1 ug/ml (Wellcome) for 72 hours. After 72 hours, contaminating monocytes were removed by adherence and the remaining non-adherent cells were washed 3 times in serum-free RPMI 1640 before being re-suspended in 10% HIFCS containing RPMI 1640 containing 20 ng/ml recombinant IL-2. The cells were washed, re-suspended and supplemented

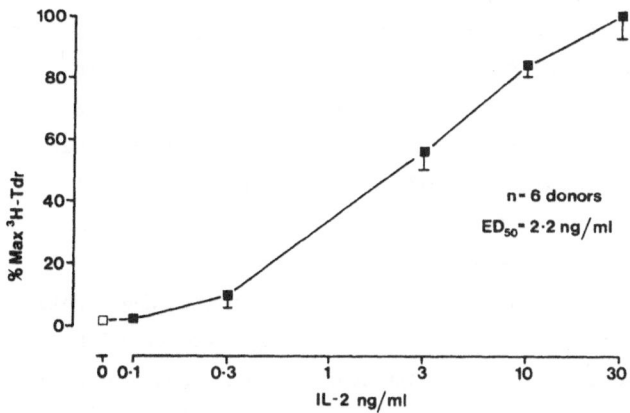

Fig. 1. Dose response curve of IL-2 induced ^{3}H-TdR incorporation by human
T-lymphoblasts. Each point represents the mean ± s.e. mean of
percentage maximum ^{3}H-TdR incorporation using 10 to 14-day-old
lymphoblasts from 6 donors.

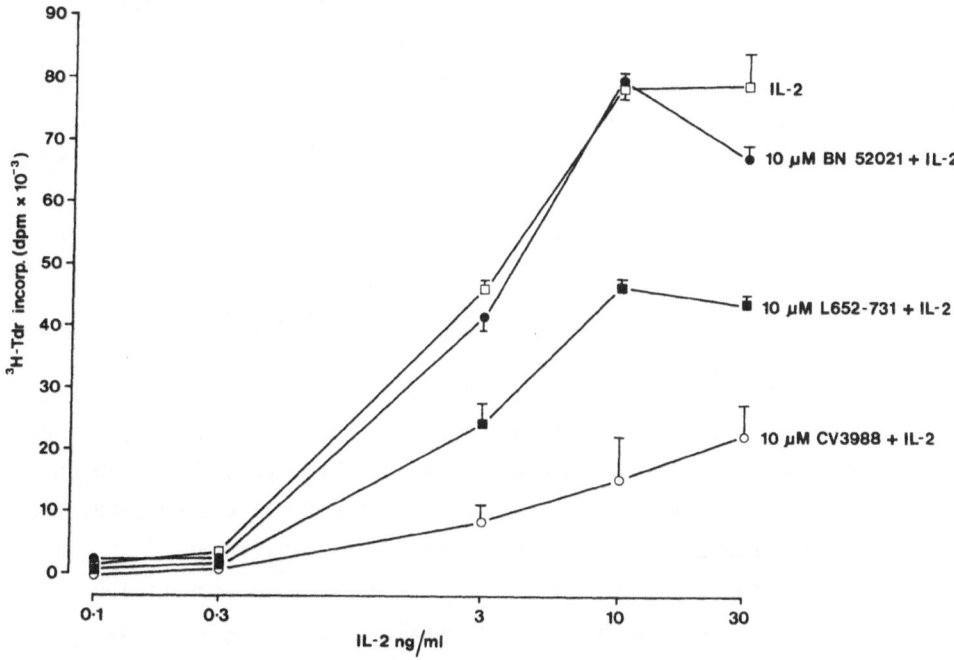

Fig. 2. The effect of three PAF antagonists (BN52021, L-652,731 and
CV3988) on the dose response curve of IL-2-induced human
T-lymphoblast ^{3}H-TdR incorporation. Each point represents the
mean ± s.e. mean of ^{3}H-TdR incorporation induced by IL-2 (0.1 –
30 ng/ml) in 14-day-old lymphoblasts pre-treated for 10 min with
10 μM BN52021 (●), L-652,731 (■), CV3988 (○) or vehicle (□).
The above data is representative of results obtained with
lymphoblasts from 3 donors.

with fresh IL-2 (20ng/ml) every 2 days for 10-14 days. Two days before the cells were to be used for experimentation, they were washed in serum-free media again, re-suspended but not supplemented with IL-2. Six hours before use, the cells were again washed and re-suspended, this time in media supplemented with 1% HIFCS. This treatment ensured that the lymphoblasts accumulated in the G_0-G_1 phase of the cell cycle. After 6 hours in 1% serum- containing media, the cells were washed and re-suspended in media containing 10% HIFCS.

Aliquots (160 ul) of cell suspension containing 1.6×10^5 viable IL-2 receptor positive T-cells (T-lymphoblasts) were added to 96-well, flat-bottomed microtitre plates. Quintuplicate aliquots were then treated with 20 μl drug vehicle, PAF agonists (0.03-300nM), PAF receptor antagonists or the synthesis inhibitor (1-30 μM) followed 10 minutes later (unless otherwise stated) by the addition of 20 μl recombinant IL-2 (0.1-30ng/ml). The cells were then incubated for 72h. For the final 18h of incubation the cells were pulsed with ^{3}H-TdR (1 uCi/ml, 2Ci/mMol) and harvested on to Whatman GF/A filter paper using a multiple automated cell harvester. Incorporation of ^{3}H-TdR was measured using standard liquid scintillation counting techniques and expressed as dpm. To examine the possible cytotoxic effects of these compounds, the PAF antagonists and vehicles were incubated with T-lymphoblasts at the concentration stated above, then trypan blue dye exclusion was examined 24, 48 and 72 h later.

RESULTS

1. Effect of PAF receptor antagonists on IL-2 induced proliferation of human T-lymphoblasts

Interleukin-2 induced a dose-related ^{3}H-TdR incorporation (Fig.1) with an ED_{50} of 2.2+0.18 ng/ml, n=6 donors. When the dose response curve of IL-2-induced proliferation was performed in the presence of 10 μM of each of the three PAF receptor antagonists, L-652,731 and CV 3988, but not BN52021, produced significant inhibition of proliferation (Fig 2). Furthermore L-652,731 and CV3988 produced dose-related inhibition of sub-optimal IL-2-induced ^{3}H-TdR incorporation with IC_{50}'s of 11+2.5 and 13.4+4 μM respectively (n=3 donors). The other PAF receptor antagonist, BN52021, at the highest concentration examined (30 μM), was ineffective at modulating IL-2-induced ^{3}H-TdR incorporation (n=3 donors, fig 3). The PAF synthesis inhibitor L-648,611 was also an effective inhibitor of IL-2-induced proliferation (IC_{50} = 15 μM, fig 3).

Trypan blue exclusion assays performed at 24, 48 and 72 h demonstrated

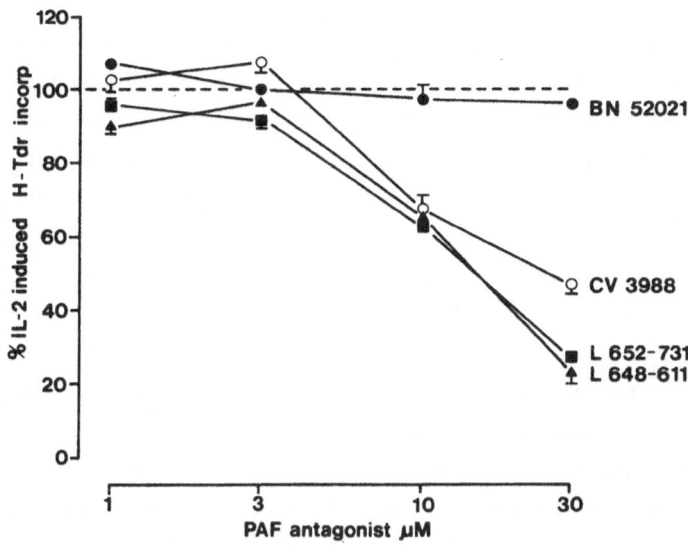

Fig. 3. Mean ± s.e. mean of % sub-optimal (3 ng/ml) IL-2 induced ^{3}H-TdR
incorporation in T-lymphoblasts in the presence of vehicle (- -)
or 1 to 30 µM BN52021 (●), CV3988 (○) L-652,731 (■) or
L-648,611 (▲). The data is representative of results obtained
with lymphoblasts from 3 donors.

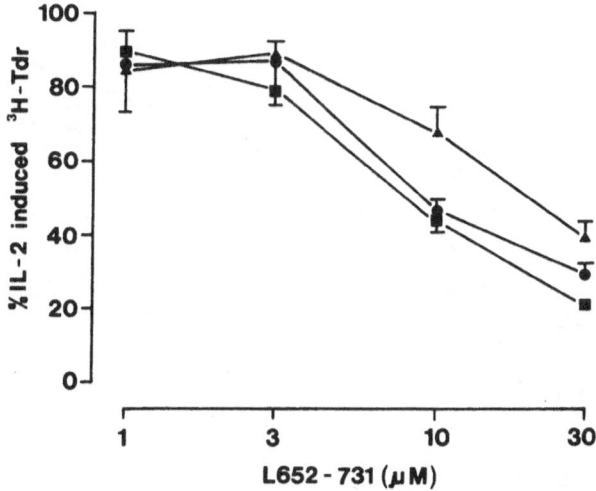

Fig. 4. Mean ± s.e. mean of % sub-optimal (0.3 ng/ml) IL-2 induced ^{3}H-TdR
incorporation in T-lymphoblasts in the presence of 1 to 30 µM
L-652,731 added 10 min before (■) 24 h after (●) or 48 h after
(▲) the addition of IL-2. In the latter two treatments (●,▲)
the lymphoblasts were washed to remove IL-2 then resuspended
prior to the addition of L-652,731. All treatments were pulsed
with ^{3}H-TdR (see methods) at 54 h and harvested at 72 h.

that the PAF receptor antagonists (1 - 30 µM) were not cytotoxic (data not shown).

Time course experiments were carried out on 10-day-old blast cells. PAF antagonists were added at 24 hrs and 48 hrs after the initial IL-2 stimulation of the cells and after the IL-2 had been washed off. Fig.4 shows that, even under these conditions, the PAF antagonist L-652,731 was still effective in reducing proliferation. It can be seen from Fig. 4. that L-652,731 was less effective against IL-2 at 48 hrs than it was at 24 hrs post IL-2 or 10 min prior to IL-2.

2.   Effect of PAF agonists on sub-optimal IL-2-induced proliferation

PAF and two stable analogues of PAF (PR1501 and PR1502) produced a marked and dose-dependent potentiation of sub-optimal IL-2-induced ^{3}H-TdR incorporation (Fig.5). Furthermore, PAF or PR1501 could reduce the inhibitory effects of 10 µM L-652,731 if added 10 min prior to the addition of L652,731 (Fig.6). However, the inhibitory effect of 10 µM L-648,611 could not be reversed by similar treatment (data not shown).

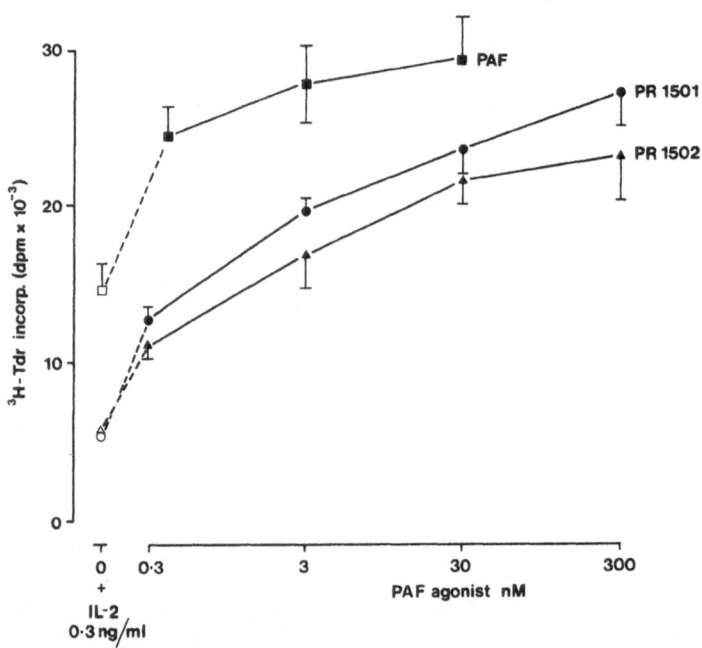

Fig. 5. Mean ± s.e. mean of sub-optimal (0.3 ng/ml) IL-2 induced ^{3}H-TdR incorporation in human T-lymphoblasts in the presence of vehicle (open symbols) or 0.3 to 300 nM PAF (■), PR1501 (●) or PR1502 (▲).

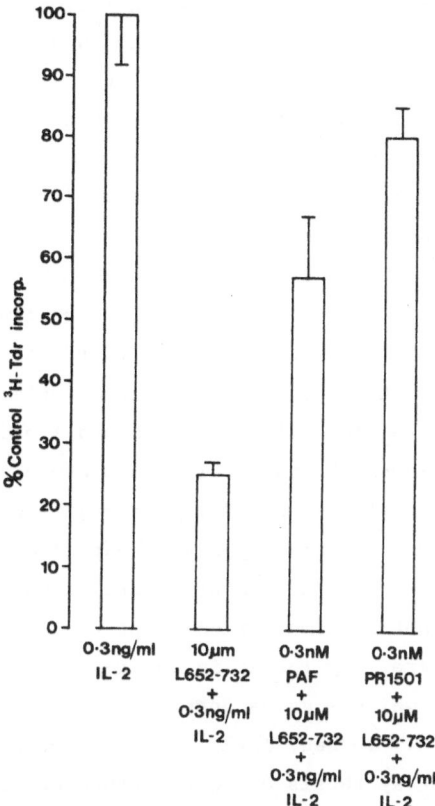

Fig.6. Reduction of the inhibitory effect of L-652,731 by PAF and PR1501, on sub-optimal IL-2-induced proliferation of human T-lymphoblasts. Each histogram is the mean $\pm$ s.e. mean of % control (0.3ng IL-2/ml) ^{3}H-TdR incorporation.

DISCUSSION

These results suggest that PAF can exert a modulatory role in human T-lymphoblast proliferation. Two structurally unrelated compounds which have been described as potent and selective PAF receptor antagonists, namely L-652,731[19] and CV3988[21], as well as an inhibitor of PAF synthesis, L-648,611,[22] are effective inhibitors of IL-2-induced proliferation, as measured by ^{3}H-TdR incorporation. The results presented here also suggest that the putative PAF receptor on lymphocytes may well have different structural specificities to that present on platelets. This was suggested by the finding that only L-652,731 and CV3988 inhibited IL-2 actions, whilst BN52021, which is at least as active as both these compounds against PAF on platelets, failed to affect IL-2-induced proliferation. To examine this proposal, we are currently examining other compounds on human T-lymphoblasts which are platelet PAF receptor antagonists.

Further evidence of a role for PAF in human T-lymphoblast proliferation was provided by the PAF agonists. PAF and two non-hydrolysable PAF agonists enhanced sub-optimal IL-2-induced proliferation of human T- lymphoblasts.

Until recently, it was thought that, unlike other leukocytes, lymphocytes were not capable of producing PAF.[24,25] However, Jouvin-Marche et al demonstrated that lymphocytes are capable of producing 2-lyso-PAF when stimulated with the Ca^{++} ionophore, A23187, and their failure to produce PAF is thought to be due to the lack of acetyl transferase activity rather than the absence of phospholipase A_2 activation. Furthermore, Malavesi et al[26] have found PAF to be released from large granular lymphocytes after F_c receptor stimulation.

A series of human leukaemic cell lines of B and T origin and of Epstein-Barr virus (EBV) lymphoblastoid cell lines are capable of releasing PAF-like material after appropriate stimulation with A23187, acetyl CoA and/or PHA[27]. The authors suggested that circulating human lymphocytes may also release PAF after appropriate stimulation. IL-2 may be the appropriate stimulant and the method of preparation of T-lymphoblasts, as used here, may result in the T-lymphoblasts becoming capable of producing PAF, which - as demonstrated here - plays a role in IL-2-induced proliferation. Indeed, IL-2 itself may be the stimulus required for activation or synthesis of acetyl transferase, the vital enzyme required for PAF generation from 2-lyso-PAF. Since IL-2 has been shown to stimulate phosphorylation of proteins[28], it may achieve such enzyme activation by way of complex protein phosphorylations. The leukaemic cell lines reported by Bussolino et al[27] may be so activated.

A role for the elevation of intracellular free calcium ($[Ca^{++}]_i$) has been shown in the production of IL-2.[29,30] Perhaps the elevation of increased $[Ca^{++}]_i$ following T_3-T_1 stimulation by either antigens, monoclonal antibodies or mitogenic lectins can result in the formation of 2-lyso-PAF by activation of the appropriate calcium-dependent phospholipase A_2. In support of this, Bussolino et al reported that stimulation with A23187 could result in 2-lyso-PAF formation. At present, we are attempting to detect PAF production by human T-lymphoblasts at various stages of the cell cycle.

Furthermore, the results of time course experiments shown in Fig. 4 suggest that the PAF receptor antagonist L-652,731 can be added at any time during the incubation period and still prove an effective inhibitor of IL-2-induced proliferation. The possibility of the PAF antagonists and the synthesis inhibitors sequestering or inactivating the IL-2 molecule itself, as reported by Robb[31] for ganglioside inhibition of IL-2 dependent proliferation, is unlikely because compounds were still effective following removal of the IL-2 stimulus.

Since L-652,731 inhibited proliferation to some extent, no matter what time it was added, it might be possible that cytoplasmic events are affected rather than initial membrane receptor operated events. Gutowski and Coles[32] reported activated lymphocyte cytoplasmic extracts contained a substance of greater than 100,000 Mr, which triggered DNA synthesis in isolated nuclei but not intact cells. Interleukin-2 may regulate the production of such 'second messenger'-type protein and the action of this messenger may well be the site of action of PAF.

ACKNOWLEDGEMENTS

We are grateful to Ciba-Geigy USA for financial assistance. S G Ward is an MRC scholar.

REFERENCES

1. R. J. Robb, Interleukin-2 - the molecule and its function, Immunol. Today, 5(7):203 (1984).
2. S. C. Meuer, R. E. Hussey, D.A. Cantrell, J. C. Hodgdon, S.F. Schlossman, K. A. Smith and E.L. Reinherz, Triggering of the T_3-T_i antigen receptor complex results in clonal T-cell proliferation through an Interleukin-2 dependent autocrine pathway, PNAS, 81:1509 (1986).
3. R. J. Robb, A. Munck and K. A. Smith, T-cell growth factor receptors - quantitation, specificity, biological relevance, J. Exp. Med. 154:1455 (1981).
4. D. A. Cantrell and K. A. Smith, Transient expression of IL-2 receptors - consequences for T-cell growth, J. Exp. Med., 158:1895 (1983).
5. T. A. Waldmann, Structure, function and expression of IL-2 receptors on normal and malignant lymphocytes, Science, 232:727 (1986).
6. K. A. Smith, Interleukin-2, Ann. Rev. Immunol., 2:319 (1984).
7. G. G. B. Klaus and C. M. Hawrylowicz, Cell-cycle control in lymphocyte stimulation, Immunol. Today, 5(1):15 (1984).
8. D. A. Cantrell and K. A. Smith, The Interleukin-2 T-cell system - a new cell growth model, Science, 224:1312 (1984).
9. R. J. Robb, W. C. Greene, C. M. Rusk, Low and high affinity cellular receptors for interleukin 2 - implications for the level of Tac Antigen, J. Exp. Med., 160:1126 (1984).
10. A. M. Weissman, J. B. Harford, P.B. Svetlik, W. L. Leonard, J. M. Depper, T. A. Waldmann, W. C. Greene and R. D. Klausner, Only high-affinity receptors for Interleukin-2 mediate internalization of ligand, PNAS, 83:1463 (1986).
11. M. Fujii, K. Sugamura, K. Sano, M. Nakai, K. Sugita, Y. Hinuma, High-affinity receptor-mediated internalization and degradation of Interleukin-2 in human T-cells, J. Exp. Med., 163:550 (1986).
12. G. B. Mills, D. J. Stewart, A. Mellors and E.W. Gelfand, Interleukin-2 does not induce phosphatidylinositol hydrolysis in activated T-cells. J. Immunol. 136(8):3019 (1986).
13. G. B. Mills, R. K. Cheung, S. Grinstein and E.W. Gelfand, Interleukin-2-induced lymphocyte proliferation is independent of increases in cytoplasmic-free calcium concentration, J. Immunol. 134(4):2431 (1985).
14. A. J. H. Gearing, M. Wadhwa and A. D. Perris, Interleukin-2 stimulates T-cell proliferation using a calcium flux. Immunol. Letts. 10:297 (1985).

15. R. Roubin and J. Benveniste, Release of lipid mediators from macrophages and its pharmacological modulation, in: "Reticuloendothelial system, Comprehensive Treatise," 8:73 (1983)

16. M. Chignard, E. Coeffler and J. Benveniste. Role of PAF-acether and related ether-lipid metabolism in platelets, in "Adv. in Exp. Medicine and Biology," Vol.192, "Mechanisms of Stimulus-Response Coupling in Platelets," ed. J. Westwick et al, Plenum, New York. (1985).

17. S. B. Hadvary, J-M. Cassal, G. Hirth, R. Barner and H. R. Baumgartner, Structural requirement for the activation of blood platelets by analogues of platelet-activating factor (PAF-acether), in: "Platetet Activating Factor, INSERM Symposium No.23", ed. J. Benveniste and B. Arnoux, Elsevier, Amsterdam (1983).

18. M. L. Lee, A. Frei, C. Winslow and D. A. Handley. Isosteric analogs of PAF-acether: synthesis and biological activity of 1-0-octadecyl-2-deoxy-2-(2'-oxopropyl)-glycero-3-phosphorylcholine, in: "Platetet Activating Factor, INSERM Symposium No.23", ed. J. Benveniste and B. Arnoux, Elsevier, Amsterdam (1983).

19. S. B. Hwang, M-H. Lam, T. Biftu, T. R. Beattie and T-Y. Shen, Trans,2,5-bis-(3,4,5-trimethoxyphenyl)tetrahydrofuran - an orally active specific and competititve receptor antagonist of PAF. J. Biol. Chem., 260(29):15639 (1985).

20. P. Braquet, B. Spinnewyn, M. Braquet, R. H. Bourgain, J. E. Taylor, A. Etienne and K. Drieu, BN52021 and related compounds: a new series of highly specific PAF-acether receptor antagonists isolated from Ginkgo bilda, Blood Vessels, 16:558 (1985).

21. Z. Terashita, S. Tsushima, Y. Yoshioka, H. Nomura, Y. Inada and K. Nishikawa, CV3988 - a specific antagonist of platelet-activating factor (PAF), Life Sci., 32(17):1975 (1983).

22. J. C. Robbins, H. Machoy, M-H. Lam, M.M. Ponpipom, K. M. Rupprecht and T. Y. Shen, A synthetic phospholipid inhibitor of PAF biosynthesis, Fed.Proc. 44:1269 (1985).

23. D. A. Cantrell and K. A. Smith, Interleukin-2 regulates its own receptors, PNAS, 82:864 (1985).

24. G. Camussi, M. Aglietta, R. Coda, F. Bussolino, W. Piacibello and C. Tetta, Release of platelet-activating factor (PAF) and histamine, II. The cellular origin of human PAF: monocytes, polymorphonuclear neutrophils and basophils. Immunology, 42:191 (1981).

25. E. Jouvin-Marche, E. Ninio, G. Beauvain, M. Tence, P. Niaudet and J. Benveniste, Biosynthesis of PAF-acether, VII. Precursors of PAF-acether and acetyl-transferase activity in human leukocytes, J.Immunol., 133(2):892 (1984)

26. F. Malavasi, C. Tetta, A. Funaro, G. Bellone, E. Ferrero, A. Colli Franzone, P. Dellabona, R. Rusci, L. Matera, G. Camussi and F. Caligaris-Cappio, F_C receptor triggering induces expression of surface activation antigens and release of platelet-activating factor in large granular lymphocytes, PNAS, 83:2443 (1986).

27. F. Bussolino, R. Foa, F. Malavasi, M. L. Ferrando and G. Camussi, Release of platelet-activating factor (PAF)-like material from human lymphoid cell lines, Exp. Haematol., 12:688 (1984).

28. A. R. Mire, G. Wickremasinghe, R. Michalevicz and A. V. Hoffbrand. Interleukin-2 induces rapid phosphorylation of an 85-kilodalton protein in permeabilized lymphocytes. Biochim.Biophys.Acta, 847:159 (1985).

29. J. B. Imboden, A. Weiss and J. D. Stobo, Transmembrane signalling by the T_3-antigen receptor complex. Immunol. Today, 6(11)328 (1985).

30. G. B. Mills, J. W. W. Lee, R. K. Cheung and E.W. Gelfand, Characterization of the requirements for human T-cell mitogenesis by using suboptimal concentrations of phytohaemagglutinin, J. Immunol. 135(5):3087 (1985).

31. R. J. Robb, Suppressive effect of gangliosides upon IL-2-dependent proliferation as a function of inhibition of IL-2 receptor association. J. Immunol., 136(3):971 (1986).

32. J.K. Gutowski and S. Cohen, Induction of DNA synthesis in isolated nuclei by cytoplasmic factors from spontaneously proliferating and mitogen-activated lymphoid cells. Cell Immunol., 75:300 (1983).

Chapter 6

New drugs
in shock

LACK OF EFFECT OF CYCLOSPORINE A ON ENDOGENOUS ARACHIDONIC ACID METABOLISM IN HUMAN NEUTROPHILS

O. H. Nielsen, J. Elmgreen and I. Ahnfelt-Ronne*
Department of Medical Gastroenterology C, Herlev Hospital
University of Copenhagen, and *Department of Pharmacology
Leo Pharmaceutical Products, Ballerup, Denmark

INTRODUCTION

Arachidonic acid (AA) may undergo metabolism via either the cyclo-oxygenase pathway to prostaglandins (PGs) and thromboxanes (TXs) or via the lipoxygenase pathways to hydroxyeicosatetraenoic acids (HETEs) and leukotrienes (LTs) (1). Different cell types metabolize AA according to their relative activity of the AA metabolizing enzymes.

An important modulator of the inflammatory reaction, cyclosporine A (CS-A), was studied as a potential modifying agent of AA release and metabolism in human neutrophils (PMNs) by a newly developed technique for immunopharmacological investigations.

MATERIALS AND METHODS

Blood from 10 healthy volunteers was drawn in EDTA (10 mM) and the leukocytes were isolated by a modification of Böyum's method (2). Suspensions of PMNs (5×10^6 cells/ml) were incubated with $1-{}^{14}C$-AA (1 µCi/ml, 58 mCi/mmol) (Amersham International, U.K.) in autologeous serum (5%) at $37°C$ under a stream of 5% carbon dioxide until steady state conditions for incorporation of AA into the phospholipids were achieved (5h). Excess $1-{}^{14}C$-AA was removed by washing. After preincubation (1-30 min) with CS-A (100-10.000 ng/ml) (${}^3H$-CS-A, Sandoz, Switzerland served as internal standard) the cells were challenged with calcium ionophore A23187 (Calbiochem, CA, U.S.A.) (10 µM, 15 min). Activation of the cells was terminated by centrifugation of the neutrophils through dibuthyl phthalate:dinonyl phthalate (3:1). The eicosanoids were isolated from the extracellular fluid by extraction and thin-layer chromatography, and quantitated by autoradiography and laser densitometry of the autoradiographs (3). Identification of the radioactive spots was performed by co-chromatography with pure standards, and further identification as well as determination of the specific activity were performed by quantitative high pressure liquid chromatography. The test system was validated with well-known inhibitors of AA metabolism, indomethacin and nordihydroguaiaretic acid (NDGA).

RESULTS

Following the 5h incubation period, 35% of the added $1-{}^{14}C$-AA was in-

Table 1. Total release of radioactivity (nCi/5x10^6 cells), and relative contribution of the eicosanoids, LTB$_4$, and 5-HETE (in per cent of radioactivity) as revealed by laser densitometry of autoradiographs during preincubation with serial dilutions of CS-A. Medians with ranges in parentheses (n=10).

| | Concentration of CS-A (ng/ml) | | | |
	0	100	1000	10000
5-HETE (%)	13.0 (8.4–15.2)	12.1 (8.4–15.9)	12.3 (8.5–14.9)	12.1 (8.0–14.6)
LTB$_4$ (%)	5.0 (2.3–7.9)	4.7 (2.4–8.0)	5.1 (2.1–8.0)	5.2 (2.5–7.9)
Radioactivity (nCi/5x10^6 cells)	18 (12–23)	18 (13–24)	18 (12–23)	18 (11–22)

corporated into the PMNs - 30% into the phospholipids, and 70% into the triglycerides. Radioactivity released from the PMNs consisted mainly of free AA, 58% (44–71), LTB$_4$, 5% (2–8%), 5-HETE, 13% (8–15%), and less than 3% of a cyclooxygenase product identified as 12-hydroxy-5,8,10-heptadecatrienoic acid (HHT). The main fraction of the residual radioactivity released was more lipophilic than AA, indicating esterification to cholesterol or glycerol.

Three weeks exposure was found to be optimal for autoradiography (Fig. 1). The relation between radioactivity and the laser densitometric response is shown in Fig. 2. A positive correlation was found up to approximately 7500 dpm (3.4 nCi) per spot (p<0.01). No significant changes in formation or release of AA or its metabolites were detected using CS-A within the range of pharmacological concentrations (Table 1).

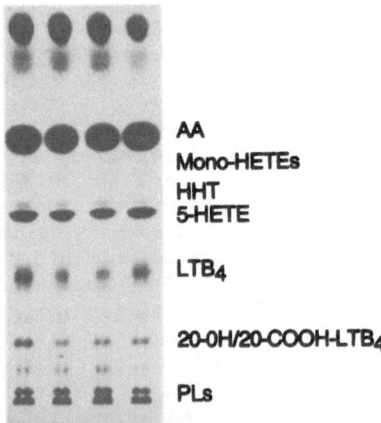

AA
Mono-HETEs
HHT
5-HETE
LTB$_4$
20-OH/20-COOH-LTB$_4$
PLs

0 100 1000 10.000 ng/ml

Fig. 1. Autoradiographic presentation of TLC separated eicosanoids released from peripheral human neutrophils prelabelled with 1-^{14}C-AA and challenged with A23187 after exposure to CS-A. Lack of effect of cyclosporine A is shown.

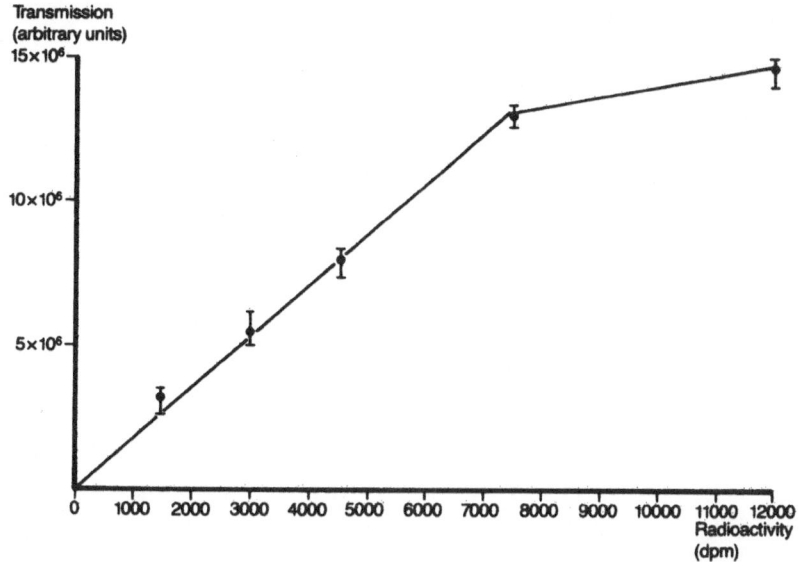

Fig. 2.   Correlation between radioactivity and laser densitometric
response (3 weeks of exposure time) (n=5). Medians with
ranges as bars.

Quantitative HPLC revealed specific activities of LTB_4 and 5-HETE in the
range of 13-20 mCi/mmol, corresponding to a 3 fold dilution of radiolabelled
AA. The lipoxygenase inhibitor NDGA (10^{-5}M) nearly abolished the LTB_4 and
5-HETE production, and indomethacin (10^{-6}M) inhibited HHT production by 70%.
The coefficient of variation within assay's was less than 0.15 for assessment
of released AA-metabolites.

DISCUSSION

CS-A is a lipophilic polypeptide which has been shown to suppress humoral and cell mediated immunity, an action which has been postulated to involve interference with the AA metabolism (4,5). However, therapeutic concentrations of CS-A (100-500 ng/ml) (6) as tested in this assay did not affect
the AA release or the lipoxygenase activity, thus suggesting that other factors may be important for the action of CS-A.
The present method based on incorporation of $1-^{14}C$-AA into phospholipids of cell membranes is of physiological relevance both for studies of
various cell types in relation to chronic inflammation and for immunopharmacological investigations.

REFERENCES

1.   Samuelsson B, Hammarström S, Borgeat P. Adv Inflam Res, 1:405 (1979)
2.   Böyum A. Scand J Clin Lab Invest, 21 (suppl. 97): 9 (1968)
3.   Nielsen OH, Bukhave K, Ahnfelt-Rønne I, Elmgreen J. Int J Immunopharmac,
     in press (1986)
4.   Lindsey JA, Morisaki N, Stitts JM, Zager RA, Cornwell DG. Lipids, 18:
     566 (1983)
5.   Whisler RL, Lindsey JA, Proctor KVW, Morisaki N, Cornwell DG. Trans-
     plantation, 38: 377 (1984)
6.   Keown PA, Stiller CR, Ulan RA, Sinclair NR, Wall WJ, Carruthers G,
     Howson W. Lancet, 1: 686 (1981)

SULOTROBAN (BM 13.177) AND BM 13.505, TWO TX-RECEPTOR BLOCKERS, ARE EFFECTIVE IN EXPERIMENTAL MODELS OF ISCHAEMIA AND SHOCK.

K. Stegmeier and J. Pill
Research Laboratories Boehringer Mannheim GmbH
Sandhoferstraße 116, D-6800 Mannheim 31, Germany

INTRODUCTION

There is much evidence that arachidonic acid metabolites may participate in the pathophysiology of ischaemia and shock. One of them, thromboxane A_2 (TXA$_2$), is a potent platelet aggregator and vasoconstrictor and its concentration in plasma is elevated in experimental models of ischaemia and circulatory shock. This suggests that it is an important mediator in the pathogenesis of circulatory shock, myocardial ischaemia and sudden death[1]. The use of specific TXA$_2$ receptor blocking drugs offer the possibility to further characterize the role of TXA$_2$ in these processes leading to severe functional and structural changes in some organs, especially in the heart, the lung and the kidneys. We have previously reported that Sulotroban (BM 13.177) is a selective and competitive thromboxane receptor blocking drug[2,3]. More potent analogues of BM 13.177 have since been identified and the results of pharmacological investigations with one of them, BM 13.505, have been recently presented[4].

In this paper we present some data from our own investigations together with some findings reported by other investigators which assess the effects of these drugs in animal models of acute renal failure, myocardial ischaemia, sudden death and shock and demonstrate the role of TXA$_2$ in the pathogenesis of these disorders.

RESULTS AND DISCUSSION

Prevention of glycerol-induced acute renal failure in rabbits

Acute renal failure (ARF) in humans remains a poorly understood response of the kidney to hypoperfusion, presumably mediated by an acute ischaemic insult[5]. There is no experimental model whose pathology resembles that of the common hypotensive form, in which vasoconstriction is believed to play a considerable role. One of the well established models used to study the pathogenic mechanism in ARF is glycerol-induced ARF in rabbits. A decreased renal blood flow and vasoconstriction of large- and medium-sized renal vessels have been demonstrated[6]. Furthermore, a direct correlation between the level of renal function as determined by serum creatinine and the metabolic potential of microsomal preparations from kidney for TXA$_2$ formation have been reported[7].

This suggests that the potent vasoconstrictor TXA$_2$ is generated by the kidney following glycerol administration and may mediate the ARF in this model.

We estimated renal function by determination of the serum creatinine concentration, renal TX formation by calculating the 24 h urinary excretion rate of TXB$_2$ determined by radioimmunoassay and examined kidney morphology after autopsy 72 h after injection of glycerol. ARF was induced by subcutaneous injection of 14 ml glycerol per kg (50 % in saline). The thromboxane receptor blocker BM 13.505 (BM, 5 mg/kg) or the thromboxane synthase inhibitor dazmegrel (D, 40 mg/kg) or the cyclooxygenase inhibitor indomethacin (I, 8 mg/kg) were injected intraveneously 30 min before glycerol administration. Two and six hours after glycerol injection the same doses of the drugs were injected subcutaneously.

Table 1. Change of serum creatinine and TXB$_2$ excretion rate within 24 h after glycerol injection

	Serum creatinine (μmol/l)		Thromboxane B$_2$ (ng/24 h)	
	before	24 h after	before	after
G	105(73–125)	177 (93–417)*	401(142–645)	1256(399–2057)**
G + BM	94(65–119)	111 (82–179)	374(253–630)	1184(287–2555)*
G + D	94(79– 97)	121(107–198)	313(14–588)	205(54– 318)
G + I	109(99–228)	348(162–621)***	282(182–548)	728(307–2345)***

Median and range, n = 6 - 8
*P = 0.05, **P < 0.05, ***P < 0.01

Twenty-four hours after glycerol injection the most prominent change in serum parameters was an increase in serum creatinine concentration in G and G+I groups compared with the values before treatment (Tab. 1). In contrast, a moderate increase of serum creatinine concentration was observed in only one animal of the G+BM group and two animals of the G+D group whereas in all other animals no change occurred.

The data in Tab. 1 also show that subcutaneous injection of G stimulated renal excretion of TXB$_2$. In the G+D group TXB$_2$ excretion rate was lower than in the pretreatment period. In the G+I group the rise in TXB$_2$ excretion rate was reduced to about 2/3 of the G group. Whereas BM 13.505 did not influence TXB$_2$ excretion rate stimulated by co-administration of G. The morphological changes in individual animals correlated very well with the serum creatinine concentrations. Pathological findings with G alone and G+I were tubular dilatation, casts, necrosis and interstitial oedema or fibrosis. Two animals of the group treated with G+D showed moderate changes in morphology and serum creatinine concentration. The other animals of that group exhibited mild to moderate morphological changes without impairment of renal function. In contrasts, only minimal changes were seen in a few animals treated with G+BM.

These results indicate that TXA$_2$ plays an important pathogenic role in G-induced ARF. I even enhanced G-induced ARF although it reduced TXB$_2$ excretion. This is consistent with previous observations of Torres et al.[8] and may result from simultaneous inhibition of renal biosynthesis of prostaglandin I$_2$ and prostaglandin E$_2$ which play an important compensatory role in the protection of renal blood flow and glomerular filtration rate. The excellent effect of BM 13.505 is in accordance with its ability to block the receptors of TXA_2 and prostaglandin endoperoxides without inhibiting prostaglandin snythesis.

U 46619-induced sudden death in mice

Injection of 800 ug/kg of the thromboxane mimetic U 46619 into male, anaesthetized mice causes lung emboli and sudden death within a few minutes. The mortality induced by this thromboxane mimetic may be due to a combination of vasoconstriction and thrombosis[9].

If the drugs (0.01 - 1 mg/kg) were injected into the tail vein 5 min before the U 46619 injection, the survival rate characterizes the antagonistic potency. BM 13.505 protected mice from death induced by U 46619 with an ID_{50} value of 0.025 mg/kg and BM 13.177 with an ID_{50} of 0.06 mg/kg.

In order to investigate the time course of the protective action, the drugs were administered orally. U 46619 was injected into the tail vein 2 to 48 hours after pretreatment with the antagonist. A dose of 1 mg/kg BM 13.505 protected animals at least for 6 h against the lethal effects of U 46619 injection (Tabl. 2). In contrast BM 13.177 was of minor efficacy, a dose of 50 mg/kg protected only 50 % of the animals from death. At 24 hrs after administration of 5 or 10 mg/kg BM 13.505 about 50 % of the animals survived U 46619 injection.

Table 2. Time course of the protective effect of BM 13.505 and BM 13.177 against U 46619-induced sudden death in male mice

Substance	Dose mg/kg p.o.	Number of surviving mice / number of tested mice at the given hours after drug administration*					
		2 h	4 h	6 h	24 h	30 h	48 h
BM 13.505	1	-	-	5/5	1/15	-	-
	5	5/5	5/5	9/9	7/10	4/5	2/5
	10	10/10	10/10	10/10	6/10	4/10	0/5
BM 13.177	50	15/20	8/15	4/13	2/10	0/4	0/5

*sudden death was were provoked by the injection of 800 ug/kg U 46619 at the time given.

Treatment of acute myocardial ischaemia

1. Cats

Increased local TXA_2 formation in ischaemic myocardium has been considered to be a major pathophysiological event contributing to tissue damage in the course of myocardial ischaemia[10]. Cats were subjected to 5 h of permanent occlusion of the left anterior descending coronary artery (LAD) and 30 min later infusion with BM 13.177 (5 mg/kg x h, i.v.) started[11]. In comparison with vehicle-treated LAD-occluded cats, BM 13.177 significantly attenuated the loss of creatinine phosphokinase activity from the ischaemic myocardium and antagonized the ischaemia induced rise in the ST-segment of the electrocardiogram. Under these conditions BM 13.177 did not reduce plasma thromboxane levels or ischaemia-induced platelet aggregate formation but considerably antagonized TX-dependent platelet secretion ex vivo.

Brezinski et al.[12] reported similar findings on changes of creatinine phosphokinase activity and ST-segment, however they infused higher doses of BM 13.177 (bolus of 20 mg/kg, followed by 20 mg/kg x h).

2.   Dogs

BM 13.177 was also studied in an experimental model of ischaemia and reperfusion-induced arrhythmias[13]. BM 13.177 was given to greyhounds in a bolus of 5 mg/kg 15 min prior to occlusion of the left anterior descending coronary artery, followed by an infusion of 9 mg/kgxh. The TX receptor blocker markedly reduced the severity and incidence of arrhythmias resulting form both ischaemia and reperfusion.

Models of experimental shock

Arachidonic acid-induced vasoconstriction in isolated lungs from rabbits

The pulmonary vascular bed responds with acute vasoconstriction to increased availability of free arachidonic acid (AA), wether exogenously applied or released from endogenous sources upon stimulation. The influence of BM 13.177 on the arachidonic acid-induced vascular effects in isolated lungs form rabbits was studied[14].

BM 13.177 dose-dependently inhibited the pressor responses evoked by repetitive direct application of AA or by repetitive stimulation of endogenous AA-release with the calcium-ionophore A 23187.

Endotoxin-induced shock in mice

The protective effect of BM 13.177 and BM 13.505 on endotoxin-induced mortality was studied in NMRI mice[15]. A significant decrease in and a prolonged onset of lethality was observed after pretreatment. Both drugs were also capable of reducing the lethality rate in BCG-sensitized animals.

Endotoxin-induced pulmonary vasoconstriction in sheeps

Intravenous injection of E. Coli endotoxin (1 ug/kg) caused a transient increase of pulmonary artery- and airway pressure paralleled by large increases of TXB_2 concentration in plasma. Pretreatment with BM 13.177 (bolus 5 mg/kg, followed by 0.75 mg/kg x min) abolished the rise of pulmonary artery and airway pressure[16]. Plasma concentrations of TXB_2 in BM-treated animals were similar to endotoxin-treated controls.

CONCLUSIONS

These studies indicate that TXA_2 plays a significant role in the pathogenesis of acute renal failure, sudden death, ischaemia and shock in these animal models, and that TX receptor blocking drugs attenuate or prevent the deleterious effects of TXA_2. These drugs may be useful tools for investigating the importance of TXA_2 in the pathogenesis of other experimental models, especially in that of sepsis and burn. Clinical trials (sulotroban) and studies in healthy volunteers (BM 13.505) showed that both drugs are well tolerated. Ongoing clinical trials will help to clarify the specific role of TXA_2 and the therapeutical benefit of a thromboxan receptor blockade in certain pathological states, among which is septic shock.

REFERENCES

1. Lefer A.
   Fed. Proc. 44 : 275 (1985).

2. Patscheke H., Stegmeier K.
   Thromb. Res. 33 : 277 (1984).

3. Stegmeier K., Pill J, Müller-Beckmann B., Schmidt F., Witte E., Wolff H. and Patscheke H.
   Thromb. Res. 35 : 379 (1984).

4. Stegmeier K., Pill J. and Patscheke H.
   Naunyn-Schmiedeberg's Arch. Pharmacol. Suppl. 332 : R36 (1986).

5. Brezis M., Rosen S., Silva P. and Epstein F.H.
   Kidney Int. 26 : 375 (1984).

6. Solez K., Altman J., Rienhoff H., Riela A., Finer P. and Heptinstall R.
   Kidney, Int. 10 153 : 159 (1976).

7. Benabe J., Klahr S., Hoffman M. and Morrison A.
   Prostaglandins 19 : 333 (1980).

8. Torres V,, Strong C., Romero J. and Wilson D.
   Kidney Int. 7 : 170 (1975).

9. Myers A., Penhos J., Ramey E. and Ramwell P.
   J. Pharmcol. Exp. Ther. 224 : 369 (1983).

10. Parrat J. and Cocker S.
    In: Prostaglandins and other eicosanoids in the cardiovascular system, (editor: K. Schrör) Karger (Basel) p. 172 (1985).

11. Schrör K. and Thiemermann Ch.
    Br. J. Pharmacol. 87 : 631 (1986).

12. Brezinski M., Yanagisawa A., Darius H. and Lefer, A.
    Am. Heart. J. 110 : 1161 (1985).

13. Parrat J. and Wainwright C.
    Br. J. Pharmacol. 88 : 291 Proc. Suppl. (1986).

14. Seeger W., Ernst Ch., Walmrath D., Neuhof H. and Roka L.
    Thromb. Res. 40 : 793 (1985).

15. Urbaschek R., Patscheke H., Stegmeier K. and Urbaschek B.
    Circulatory Shock 16 : 71 (1985).

16. Hüttemeier, P
    Clin. Physiol.: in press

GINKGO BILOBA EXTRACTS INHIBITS OXYGEN SPECIES PRODUCTION GENERATED BY PHORBOL MYRISTATE ACETATE STIMULATED HUMAN LEUKOCYTES

J. Pincemail[1], A. Thirion[2], M. Dupuis[1], P. Braquet[3], K. Drieu[3], P. Hans[4] and C. Deby[1]

[1] Laboratoire de Biochimie et Radiobiologie, Institut de Chimie, B6, Université de Liège, Sart-Tilman, 4000 Liège 1, Belgium
[2] Centre de transfusion Sanguine de Liège, rue Dos Fanchon, 4000 Liège, Belgium [3] IHB Ipsen Institute for Therapeutic Research, 92350 Le Plessis Robinson, France [4] Service d'Anaesthésiologie, Hôpital de Bavière, Université de Liège, Boulevard de la Constitution, 66, 4020 Liège, Belgium

INTRODUCTION

Recent studies (1) have shown that phorbol myristate acetate (PMA) induced superoxide anion ($O_2^{\bullet-}$) release in polymorphonuclear (PMNs) cells from whole body gamma irradiated rabbits was significantly reduced in animals treated preventively with a Ginkgo biloba extract (Gbe, IPS 200 Institut Henry Beaufour, France). In an effort to define further the mechanism of action of Gbe, experiments were performed to study its effects on the respiratory burst in whole stimulated PMNs preparations (employing PMA as stimulus) as well as on the NADPH-oxidase activity. Then, the interactions of Gbe on the release of toxic oxygen species ($O_2^{\bullet-}$, H_2O_2, $OH^{\bullet}$) during PMNs stimulation were studied. At least, it was also of interest to examine the effect of Gbe on the activity of myeloperoxidase, enzyme contained in neutrophils and responsible of the generation of strong oxidants such as hypochlorous acid (HOCl) or chloramines.

METHODS

Polymorphonuclear (PMNs) cells were isolated as previously described (2) and suspended in a phosphate buffer (pH 7.4) medium (PBS). After addition of $CaCl_2$-$MgCl_2$ (2 mM - 0.5 mM), the cells were stimulated with phorbol myristate acetate (8 X 10^{-8}M final) during 30 minutes at 37 °C.
a) Measurement of oxygen consumption on 2 X 10^7 stimulated cells/ml of PBS was determined with the Clark oxygen electrode (3). NADPH-oxidase activity was assessed in the 27,000 g particulate preparation from neutrophils stimulated with PMA (4) by the measurement of oxygen consumption. The 27,000 g pellet was resuspended in 0.34 M sucrose-Tris at a concentration of 1.3 mg protein/ml. The oxygen uptake was followed during 10 minutes after addition of 1 mM NADPH in the absence (control) or in the presence of Gbe.
b) Superoxide generating activity was measured by following the superoxide

dismutase inhibitable reduction of ferricytochrome C at 550 nm in a double beam spectrophotometer (5). 1.5×10^6 cells/ml of PBS were pre-incubated with Gbe during 15 minutes. After centrifugation, the yellow supernatant was removed. Cells were resuspended in PBS, stimulated and the release of $O_2^{\bullet}$ was followed.

c) The amount of H_2O_2 generated by 2×10^6 stimulated PMNs/ml of PBS after 30 minutes was determined in the presence of 10^{-4} M azide according to the method using potassium thiocyanate (6).

d) 2.5×10^6 cells/ml of PBS were stimulated in the presence of 1 mM α-keto-γ-methiol-butyric acid (KMB). The hydroxyl radical generated during the PMNs activation reacted with KMB to produce ethylene (7), measured by gas-liquid chromatography, using a Porapak column in Barber-Colman 3000 gas chromatograph.

e) Myeloperoxidase was purified from human neutrophils (8) with an absorbance ratio (A_{430}/A_{280}) of 0.65. Enzyme activity was measured spectrophotometrically at $20°$ C : 0.9 mU (4.4×10^{-8}) of myeloperoxidase was combined with 2.9 ml of 50 mM phosphate buffer, pH 6.0 containing 0.167 mg/ml dianisidine hydrochloride and 1.4×10^{-4} M hydrogen peroxide. The change in absorbance at 460 nm was measured during 1 minute (9).

RESULTS

Table I indicates that Gbe, at 500 and 250 µg/ml, significantly inhibits the oxygen uptake of PMA stimulated cells. A similar effect is observed when the broken cells preparation (27,000 g fraction) containing the NADPH-oxidase activity is used (Table II).

The release of activated oxygen species is also decreased when cells are stimulated in the presence of Gbe (Table III). For example, an inhibitory effect of 29% and 36% is respectively observed in $O_2^{\bullet}$ and H_2O_2 assay with a concentration of 250 µg Gbe/ml.

Table I. Inhibitory effect of Gbe on the oxygen con-
         sumption by PMA stimulated human leukocytes.
         Values are the percent of oxygen present in
         the medium after PMA addition in the absence
         (control) or in the presence of Gbe at 500,
         250 and 125 µg/ml.

Time	Control	Gbe 500 µg	Gbe 250 µg	Gbe 125 µg
0	100	100	100	100
1	100	100	100	100
2	96	100	90	97
3	65	93	72	73
4	25	76	51	41
5	14	58	29	11
6	0	34	11	0
7	0	11	4	0
8	0	0	0	0

Table II. Inhibitory effect of Gbe on the NADPH oxidase activity contained in the 27,000 g particulate preparation. Values are the percent of oxygen present in the medium after NADPH addition in the absence (control) or in the presence of Gbe at 500, 250, 125 µg/ml.

Time	Control	Gbe 500 µg	Gbe 250 µg	Gbe 125 µg
0	100	100	100	100
1	66	84	80	76
2	37	70	61	52
3	21	60	45	33
4	12	51	31	20
5	4.5	40	21	12
6	1.5	33	13.5	6
7	0	29	9	4
8	0	25	7.5	3

Table III. Inhibition effect of Gbe on O_2^-, H_2O_2 and $OH^{\cdot}$ release during PMNs stimulation and on the myeloperoxidase activity. Values are mean $\pm$ S.D.

Gbe (µg/ml)	% inhibiton			
	$O_2^{\cdot}$ (n=3)	H_2O_2 (n=3)	$OH^{\cdot}$ (n=3)	Myeloperoxidase activity (n=3)
Control	0	0	0	0
500	34 $\pm$ 2	54 $\pm$ 3.5	–	100
250	29 $\pm$ 2	36 $\pm$ 2.3	86 $\pm$ 1.7	100
125	15 $\pm$ 1.6	17 $\pm$ 1.3	84 $\pm$ 1.1	100
62.5	0	6 $\pm$ 1.1	75 $\pm$ 2.3	78.6 $\pm$ 2.4
31.2	–	–	58 $\pm$ 5.9	62.9 $\pm$ 0.7
15.6	–	–	42 $\pm$ 4.7	53 $\pm$ 0.5
10	–	–	–	51.3 $\pm$ 1

The ethylene production indicates that Gbe strongly diminishes the generation of OH$^\cdot$. Indeed, it is able to give an inhibition of 42% at a concentration as low as 15.6 µg/ml. At much higher concentrations, the inhibitory effect is more marked and can reach 85% at 250 µg/ml, for example.

At least, in a concentration range of 10-100 µg, Gbe considerably reduces the myeloperoxidase activity. The enzyme activity is completely inhibited in the presence of 125 µg Gbe/ml. (Table III)

DISCUSSION

The inhibition of the respiratory burst in whole PMNs preparation, observed in the presence of Gbe, seems to be exerted at different levels.
a) Gbe inhibits NADPH-oxidase activity, as shown on the broken cells preparation. Tauber et al. (10) have recently demonstrated that quercetin and kaempferol, two liposoluble flavonoids, were able to permeate PMNs and so to inhibit the respiratory burst enzymatic system. Ginkgo biloba extract contains flavonoids such as coumarinic esters of kaempferol and quercetin heterosides which may be so responsible of the extract activity. As expected, the inhibitory effect of Gbe on NADPH-oxidase activity leads to a decrease in the release of superoxide anion and hydrogen peroxide by PMA stimulated human PMNs.
b) Gbe strongly scavenges OH$^\cdot$ generation in these experiments on PMNs. It is efficient at concentration as low as 15.6 µg/ml while no effect is observed on O_2 uptake or O_2^- and H_2O_2 release at this concentration. This can be explained by the free radical scavenging activity of Gbe, previously described in in vivo and in vitro studies (11,12).
c) The myeloperoxidase activity which is responsible of the generation of strong oxidant species (HOCl, chloramines) is also significantly reduced by Gbe, even at low concentration (IC$_{50}$ = 10 µg/ml).

Our results can explain why PMA induced superoxide anion release in polymorphonuclear cells from whole body gamma irradiated rabbits is significantly reduced in animals treated-preventively with Gbe (1). By its regulator action of PMNs functions, Gbe appears to be an interesting therapeutical agent. Beside its radiobiological protection, Gbe has been proposed in the reduction of the post-radiotherapic oedema in larynx cancer treatment where there is an increase of O_2^- release by PMNs (13). On the other hand, Gbe could be experimented in the treatment of disease where free radical production by PMNs is suspected as it is the case in the Adult Respiratory Distress Syndrome (ARDS) (14) or in burn-injured patients (15).

REFERENCES

1. Braquet M., Lavaud P., and Ducousso R. Fourth International Conference on Superoxide Dismutase, Rome, 1-6 September 1985, Abstract book P 152.
2. Borgeat P., and Samuelsson B. Proc. Natl. Acad. Sci. USA 76:2148 (1979).
3. Kvarstein B. Scand. J. Clin. Lab. Invest. 25:337 (1970).
4. Light D.R., Walsk C., O'Callaghan A.M., Goetzl E.J., and Tauber A.I. Biochemistry 20:1468 (1981).
5. McCord J.M., and Fridovich I. J. Biol. Chem. 224:6049 (1969).
6. Bielefeldt H., and Babink L.A. Inflammation 8:251 (9184).
7. Weiss S.J., Rustaji P.K., and Lo Buglio A.F. J. Exp. Med. 147:316 (1978).
8. Bakkenist A.R., Wever R., Vulsma T., Plat H., and Van Gelder B.F. Biochim. Biophys. Acta 524:45 (1978).
9. Krawisz J.E., Sharon P., and Stenson W.F. Gastroenterology 87:1344 (1984).
10. Tauber A.I., Fay J.R., and Marletta M.A. Biochem. Pharmacol. 33:1367 (1984).

11. Etienne A., Chapelat M.Y., Braquet M., Clostre F., Drieu K., Defeudis F.V., and Braquet P. in : Cerebral Ischemia (Eds. A. Bes, P. Braquet, R. Paoletti and M.O.K. Sjesjö) Excerpta Medica (Amsterdam, New-York, Oxford) p 379 (1974).
12. Pincemail J., Deby C., Lion Y., Braquet P., Hans P., Drieu K., and Goutier R. in : Flavonoids and Bioflavonoids (Eds. L. Farkas, M. Gabor and F. Kallay) p 423 (1985).
13. Bally et al. Personal communication.
14. Lamy M., Deby-Dupont G., Deby C., Pincemail J., Duchateau J., Braun M., Damas P., and Roth M. in : Pulmonary Circulation in Acute Respiratory Failure, Barcelone, June 19-22, 1984 (Eds. C. Lemaire, P. Suter and W. Zapol) Churchill Livingstone (1985) (in press).
15. Braquet M., Lavaud P., Dormont D., Garay R., Ducousso R., Guilbaud J., Chignard M., Borgeat P., and Braquet P. Prostaglandins 29:747 (1985).

THE CORONARY, INOTROPIC, CHRONOTROPIC AND ARRHYTHMO-GENIC EFFECTS OF PAF-ACETHER ON ISOLATED GUINEA-PIG HEART AND THEIR SELECTIVE INHIBITION BY BN 52021

I. Viossat, P. E. Chabrier, M. Chapelat and P. Braquet
Institut Henri Beaufour, 72 avenue des Tropiques
91940 Les Ulis, France

INTRODUCTION

A phospholipid mediator identified as 1-0-alkyl-2-acetyl-sn-glyceryl-3-phosphorylcholine (1,2), platelet-activating factor (PAF-acether) has been demonstrated to play an important role in a variety of inflammatory, respiratory and cardiovascular disorders (3). Intravenous injection of PAF-acether induces strong systemic reactions mimicking typical anaphylactoid reactions such as bronchospasm (4), increased vascular permeability (5) and hypotension (6). Such similarities are found in vitro in perfused organs. In isolated perfused guinea-pig heart, injection of PAF-acether produces a coronary constriction and heart failure with a decrease in contractile force (7) which are characteristics of in vitro cardiac anaphylaxis (8). BN 52021, a specific PAF-acether receptor antagonist has been shown to inhibit PAF-acether effects in various models : contraction of guinea-pig lung strips (9), hypotension in rats (10). It also antagonizes the hemodynamic modifications occurring in shock states such as endotoxemia in guinea-pig (11) and anaphylaxis in rat (12). Here we report the effects of PAF-acether and BN 52021 on isolated perfused guinea-pig heart.

MATERIALS AND METHODS

Drugs

- PAF-acether (1-0-hexadecyl-2(R)-acetyl-glycero-3-phosphorylcholine) was purchased from Bachem (Bubendorff, Switzerland) and stored at -80°C in a 0.5 % bovine albumin serum solution.
- BN 52021 [3-(1,1-dimethylethyl)hexahydro-1,4,7b-trihydroxy-8-methyl-9H-1,7α-epoxy methano)1H,αH-cyclopenta[c]furo[3',2':3,4]cyclopenta [1,2-d] furan-5,9,12(4H)-trione].(IHB-IPSEN Institute for Therapeutic Research, Le Plessis-Robinson, France) was dissolved in dimethyl sulfoxide.

Experimental procedure

Male Hartley guinea-pigs (350 - 450 g) were anaesthetized with pentobarbital (30 mg/kg i.p.). One minute after heparin injection (1000 UI/kg i.v.) hearts were excised and rapidly cannulated through the aorta on a non-recirculating Langendorff perfusion apparatus. The heart preparations were perfused with a modified Krebs Henseleit buffer pH 7.4 at 37°C and oxygenated with a 95 % O_2 - 5 % CO_2 gas mixture under a constant hydrostatic pressure of 80 cm of water. Hearts were firstly submitted to a stabilization period of 15 minutes, receiving a normal pefusion medium. They were then perfused for 10 minutes with either BN 52021 at various doses or the control medium containing 2 °/$_{\circ\circ}$ DMSO. PAF-acether was then injected in a bolus of 0.1 ml above the cannula and hearts were submitted to another 15 minutes perfusion period.

Coronary flow was measured by sampling at regular time intervals and expressed as $ml.min^{-1}.g^{-1}._{w.w}$. Heart rate and contractile force were recorded by means of an isotonic transducer and a Gemini II Ugo Basile recorder and respectively expressed as mm and beats per minute (B.P.M.).

Data analysis

Statistical comparisons were determined by Fisher-Snedecor one way analysis of variance on the experimental values. Results were presented as percent change of control or as means ± S.E.M.

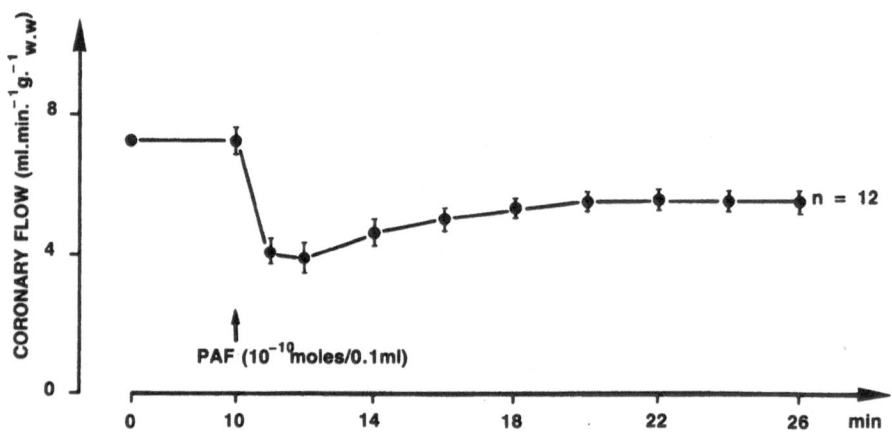

Figure 1 : Time-course of the effect of PAF-acether on coronary flow

Table I
Effect of PAF-acether on coronary flow, heart rate and contractile force

Dose of PAF-acether (moles)	Coronary flow (% decrease) 2 min.after PAF	Heart rate (% decrease) 2 min.after PAF	Contractile force (% decrease) 6 min.after PAF
10^{-11} (n = 7)	0	4 ± 2 n.s.	4 ± 6 n.s.
5.10^{-11} (n = 4)	30 ± 3**	-	-
10^{-10} (n = 6)	55 ± 5***	6 ± 3 n.s.	6 ± 4 n.s.
10^{-9} (n = 6)	65 ± 4***	26 ± 2***	35 ± 4***
10^{-8} (n = 7)	85 ± 1***	34 ± 9**	30 ± 4***
10^{-7} (n = 7)	91 ± 4***	41 ± 7***	31 ± 8***

Results are expressed as % decrease versus initial value
** p < 0.01 - *** p < 0.001 - n : number of animals

RESULTS

Effects of PAF-acether and BN 52021 on isolated perfused guinea-pig heart

PAF-acether induced a dose-dependent reduction of coronary flow (table I). Vasoconstriction appeared with a maximum at 2 min after PAF-acether injection. Coronary flow returned and stabilized after 4 min to a lower level than initial value as shown in figure 1.

With a similar time course, PAF-acether dose-dependently decreased heart rate (table I). In contrast, effect on contractile force was observed later with a maximum at 6 min (table I). It is interesting to note that the concentrations necessary to affect cardiac function (contractile force, heart rate) were higher than those affecting coronary flow. Moreover, we observed that the highest dose (10^{-7} moles) of PAF-acether induced arrhythmias in 40 % hearts.

Whatever the dose used, BN 52021 alone was devoid of any effect on cardiac function. In our control experiments, the vehicle DMSO was also without effect.

Effect of BN 52021 versus PAF-acether

A high dose of PAF-acether (10^{-7} moles) which induced alterations in all the parameters measured was used to demonstrate the action of BN 52021. At this dose, PAF diminished by 91 % the coronary flow. BN 52021 inhibited in a dose-dependent manner the fall in coronary flow (15 % at 10^{-5} M; 63 % at 2.10^{-4} M) (Fig. 2). It inhibited the decrease of heart rate induced by PAF-acether by 82 % at 2.10^{-4} M, although it was ineffective at 10^{-5} M (Fig. 3). In contrast, at both concentrations BN 52021 did not prevent the fall of contractile force caused by PAF-acether (Fig. 3). No arrhythmias were observed in presence of BN 52021.

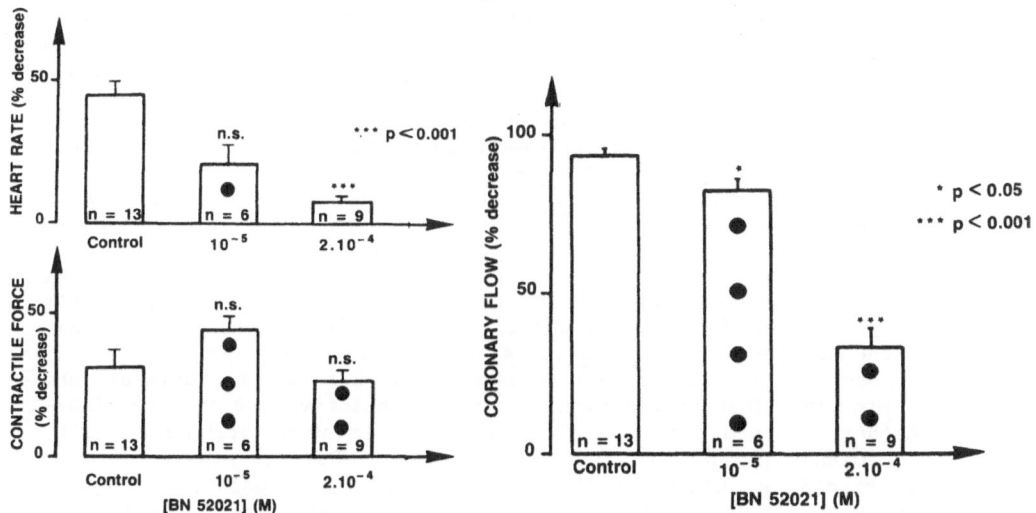

Figure 2 : Inhibition of PAF-induced (10^{-7} moles) coronary spasm by BN 52021

Figure 3 : Effect of BN 52021 on PAF-induced alterations of heart rate and contractile force

DISCUSSION

PAF-acether exerts a direct effect on cardiac function as previously demonstrated (9,10) and confirmed by us in our present experiments. These effects are represented by dramatic changes in coronary flow, cardiac rhythm and contractile force, effects which are also observed in cardiac anaphylaxis (11, 12). In perfused guinea-pig heart, BN 52021 is effective against PAF-induced coronary vasoconstriction and greatly protects hearts against heart rate diminution and the appearance of arrhythmias. However, BN 52021 does not totally inhibit the decline of contractile force induced by a high dose of PAF-acether. Therefore, it is possible that this effect is just a consequence of myocardial ischemia due to the non-totally blocked coronary spasm which cannot be reversed specifically by PAF-acether receptor antagonists. BN 52021 is known to inhibit competitively the effect of PAF-acether on platelet aggregation (13), on hypotension and on brochonspasm and to antagonize many manifestations related to shock states. In our study, BN 52021 has been demonstrated to inhibit also the direct cardiac action of PAF-acether indicating that this drug competes with PAF-acether in cardiac tissue.

REFERENCES

1. C.A. Demopoulos, R.N. Pinckard and D.J. Hanahan Platelet-activating factor : evidence for 1-0-alkyl-2-acetyl-sn-glyceryl-3-phosphoryl-choline as the active component (a new class of lipid chemical mediator), J. Biol. Chem., 254:9355 (1979).

2. J. Benveniste, M. Tencé, P. Varenne, J. Bidault, C. Boullet and J. Polonsky Semi-synthèse et structure proposée du facteur activant les plaquettes (PAF) : PAF-acéther, un alkyl ether analogue de la lysophophatidylcholine, C.R. Acad. Sci. Paris, 289D: 1037 (1979).

3. P. Braquet, L. Touqui, T.Y. Shen and B.B. Vargaftig Perpectives in platelet-activating factor research, J. Med. Chem. (1986)(in press).

4. B.B. Vargaftig, J. Lefort, M. Chignard and J. Benveniste Platelet-activating factor induces a platelet dependent bronchoconstriction unrelated to the formation of prostaglandin derivatives, Eur. J. Pharmacol., 65:185 (1980).

5. P. Braquet, R.F. Vidal, M. Braquet, H. Hamard and B.B. Vargaftig Involvement of leukotriene and PAF-acether in the increased microvascular permeability of rabbit retina, Agents and Actions, 15(1/2):82 (1984a).

6. M.L. Blank, F. Snyder, L.W. Byers, B. Brooks and E.E. Muirhead Antihypertensive polar renomedullary lipid, a semisynthetic vasodilator, Biochem. Biophys. Res. Comm., 90:1194 (1979).

7. J. Benveniste, C. Boullet, C. Brink and C. Labat The actions of PAF-acether (platelet-activating factor) on guinea-pig isolated heart preparations, Br. J. Pharmac., 80:81 (1983).

8. R. Levi, J.A. Burke, Z.G. Guo, Hattori Y., C.M. Hoppens, L.M. McManus, D.J. Hanahan and R.N. Pinckard Acetyl glyceryl ether phosphocholine (AGEPC) : a putative mediator of cardiac anaphylaxis in the guinea-pig, Circ. Res., 54:117 (1984).

9. C. Touvay, B. Vilain, A. Etienne, P. Sirois, P. Borgeat and P. Braquet Characterization of platelet-activating factor (PAF-acether)-induced contractions of arachidonic acid metabolism and by PAF-acether antagonists, Immunopharmacol. (1986) (in press).

10. J. Baranes, A. Hellegouarch, M. Le Hegarat, I. Viossat, M. Auguet, P.E. Chabrier, F. Clostre and P. Braquet The effects of PAF-acether on the cardiovascular system and its inhibition by a new highly specific PAF-acether receptor antagonist : BN 52021, Pharmacol. Res. Commun., (1986) (in press).

11. S. Adnot, J. Lefort, V. Lagente, P. Braquet and B.B. Vargaftig Interference of BN 52021, a PAF-acether antagonist with endotoxin-induced hypotension in the guinea-pig, Eur. J. Pharmacol., (1986) (in press).

12. M. Sanchez-Crespo, S. Fernando-Gallardo, M.L. Nieto, J. Baranes and P. Braquet Inhibition of the vascular actions of immunoglobulin G aggregates by BN 52021, a highly specific antagonist of PAF-acether, Immunopharmacol., 9:45 (1985).

13. P. Braquet Treatment and prevention of PAF-acether-induced sickness by a new series of highly specific inhibitors. GB Patent 8;418,424, July 19 1984, US Patent, July 19 1984.

THE EFFECT OF BW 755C ON WOUND HEALING IN A SCALD BURN INJURY PIG MODEL-MORPHOLOGICAL AND BIOCHEMICAL FINDINGS

J. A. Bauer[1], P. Conzen[2], K. Wurster[3]

[1] Chirurgische Klinik Innenstadt und Chirurgische Poliklinik der Universität München, Nußbaumstr. 20, D-8000 München 2
[2] Institut für chirurg. Forschung der Univ. München, Machioninistr. 15 D-8000 München 70. [3] Pathologisches Institut des Städt. Lehrkrankenhauses München-Schwabing Kölner Platz 1, D-8000 München 40

Aim of study

The significance of the arachidonic acid metabolites as mediators of thermal lesions remains controversial (1,2,3). We have therefore examined the effect of BW 755C - a known inhibitor of cyclooxigenase and lipoxigenase-on wound healing. In this study 2 nd$^{\circ}$ scald burns effecting 20 % of the body surface of anesthetised pigs were induced (3). The effect of systemic administration of BW 755C (10 mg/kg/die)(n=7) five minutes following induction of the scald burn was compared with that of a control group (n=4).

Material and Methods

The scald burn was created by immersing the flank of the animal in hot water (75°C) for 10 seconds. The burned area was regulated via application of an adhesive dressing and represented 20 % in 20 kg animals. The increased skin thickness was regularly controlled within the first three hours with a 10 MHz ultrasonic head transducer (4). Blood samples were taken before burning, 4 h after burning and daily for a week. On the seventh day the animals were killed and biopsies were obtained. In the plasma

519

samples prostanoid levels were measured (TxB_2, 6-ketoPGFlalpha, PGF2alpha, 13, 14-dihydro-15-ketoPGF2alpha).

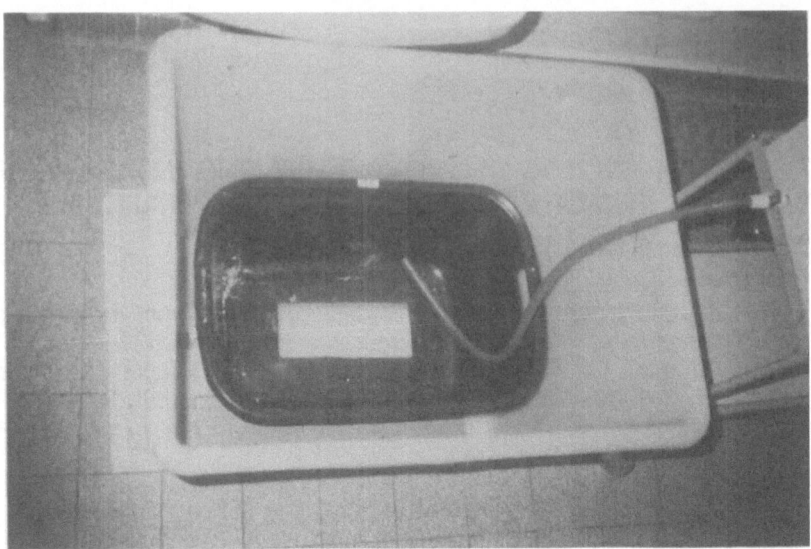

Fig. 1) Scald burn injury model (waterboiler, connection tube, bath, basket with cut out piece)

Results

The animals treated with BW 755C exhibited adequate wound healing on inspection and microscopy. Half of the control animals, but none of the treated animals, developed wound infection. Lower prostanoid levels compared with the control group were found in the group of treated animals. The other parameters remained unchanged (kininogen, prekallikrein, C_1-INA., clotting parameters).

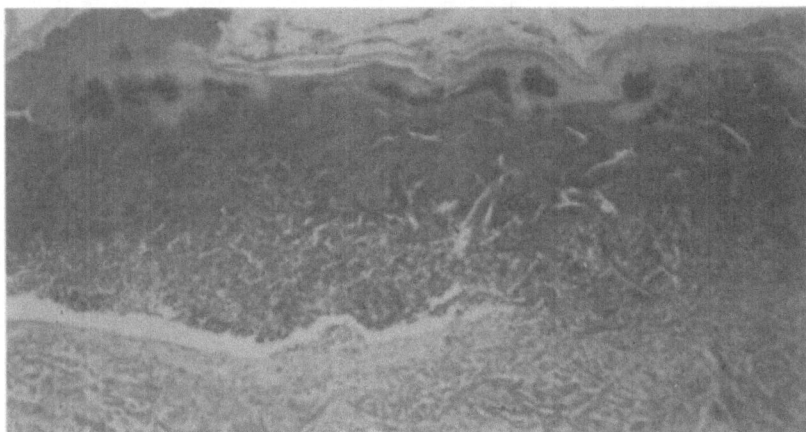

Fig.2) Microscopical findings on the treated animal at the 7th day(healing sandwich: denaturated od epidermis/unified scorf, a tiny bit granulocytes and bacterias / new epidermis in the depth)(HE,100x)

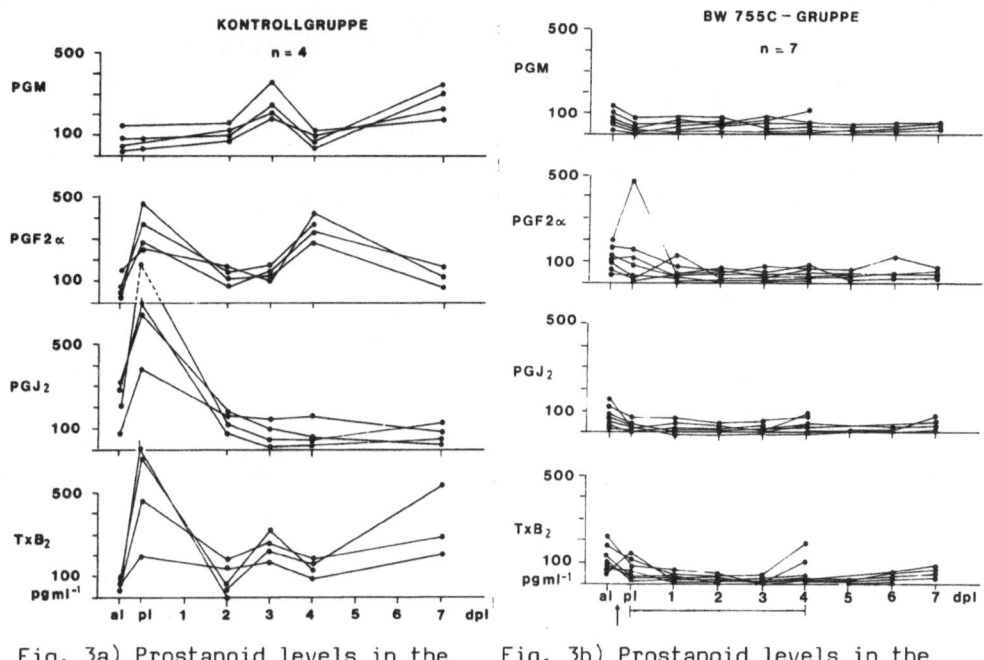

Fig. 3a) Prostanoid levels in the control group

Fig. 3b) Prostanoid levels in the teated group

Conclusion

On the basis of our experimental model we conclude that the administration of arachidonic acid inhibitors can significantly improve wound healing in scald burn injuries.

References

1. Alexander F., Mathieson M., Teoh K.H.T., Huval W.V., Lelcuck S., Valeri C.R., Shepro D., Hechtman H.B.
   Arachidonic acid metabolites mediate early burn edema,
   J. Trauma, 8:709 (1984)
2. Denzlinger C., Rapp S., Hagmann W., Keppler D.
   Leukotrienes as mediators in tissue trauma,
   Science, 230:330 (1985)
3. Bauer J.A., Lehn N., Sauer Th.
   Use of a real-time scanner with a 10-MHz transducer head for quantitative ultrasound assessment of depth and area of skin scald injuries
   in rats,pigs,and humans and of therapeutic effect of etofenamate on
   scald injuries in rats.

Langenbecks Arch. Chir. Suppl. Chir. Forum
(ed.: H.J. Streicher, M. Schwaiger), Springer, (Berlin-Heidelberg-New
York-Tokyo), p. 45 (1986)
4. Bauer J.A., Schiller K., Eitel F.
Die Anwendung der 10 MHz-Ultraschallsonographie zur Bestimmung der Ver-
brennungstiefe am Dorsum von Ratten, in: Jacob, S.W., Herschler, R.J.,
Schmellenkamp, M.: The use of DMSO in medicine,
(Springer, Berlin-Heidelberg-New York-Tokyo), p 148 (1985)

A NEW SONOGRAPHIC MODEL FOR EFFECTIVE QUANTIFICATION OF ANTIEDEMATOUS AND ANTIINFLAMMATORY DRUGS IN BURNS

J. A. Bauer[1], A. Wendelberger[1], W. Felix[2] and

[1] Chirurgische Klinik Innenstadt und Chirurgische Poliklinik der Universität München, Nußbaumstr. 20, D-8000 München 2.

[2] Walther-Straub-Institut für Pharmokologie und Toxikologie der Universität München, Nußbaumstr. 26, D-8000 München 2

Aim of study

An experiment set up to test and quantify the effect of antiedematous and anti-inflammatory drugs in thermic lesions (1) must be simple, quick and safe. Surface area and depth of the lesions must be exactly defined.

Materials and methods

The aim conditions can be fulfilled by using 10 MHz ultrasonics to measure the depth of a scald wound of standardised area on the back of a rat or the flank of a pig (water temperature -75°C; period of exposure: 10 sec.) The area of the scald is kept constant by standardising the body weight and the size of the opening of the apparatus. Only these factors, the temperature, and time of exposure affect the depth of the scald.

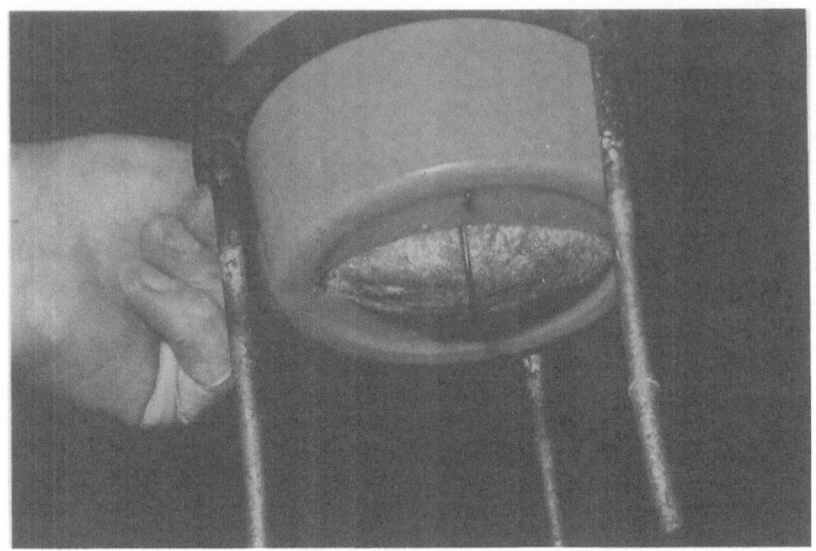

Fig. 1)   Rat dorsum after scalding (15 % of body surface)

10 MHz-ultrasonics is so effective in demonstrating that the lesions is standardised that biopsies are superfluous. Inspection and evaluation according to a simple scale are adequate methods of controlling the progress of woundhealing during the subsequent weeks.

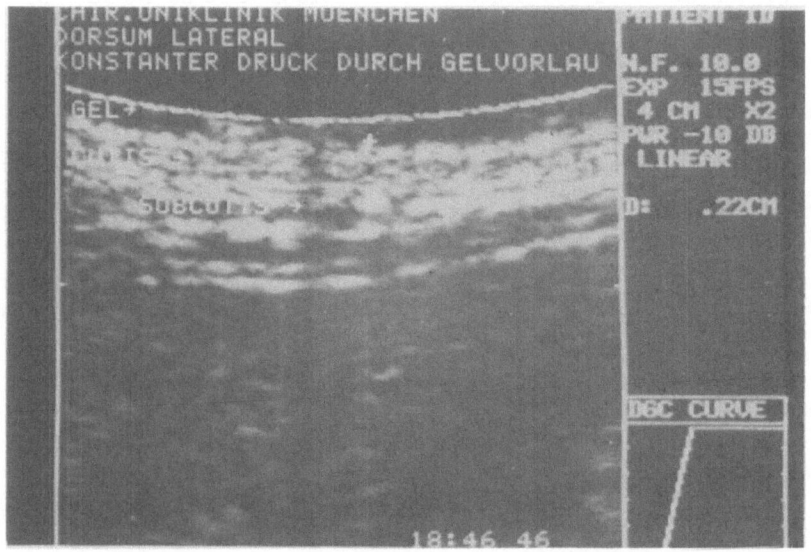

Fig. 2)   Suspension sonogram through the non compressed gel mass.

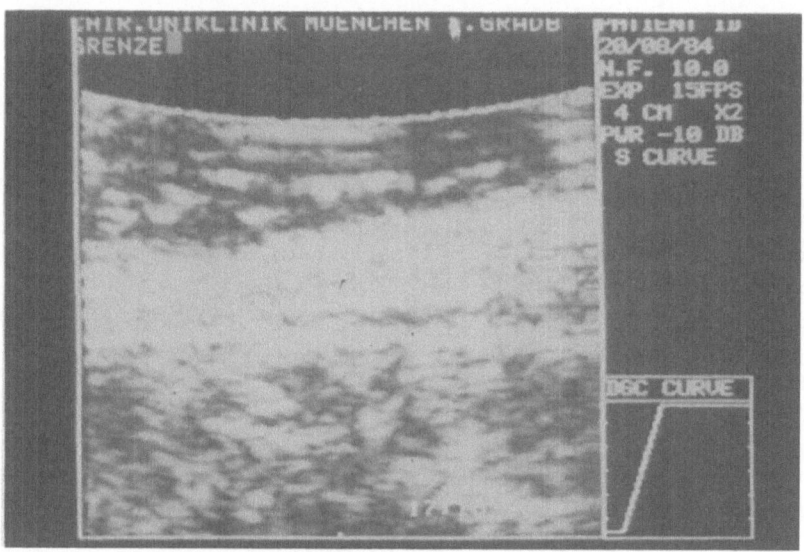

Fig. 3) Transition area non burned tissue/burned tissue 3 h after
scalding (Sonogram)

Results

A trial of various drugs shows that some drugs present scalding edema,
while others are anti-edematous but do not improve wound healing, and a
third group does not prevent, and indeed may even increase the edema,
but improves wound healing. (In all doubtful cases a biopsy should be
carried out on the seventh postoperative day). A forth group shows no
edema and good healing.

Flammacerium, BW 755C, Indomethacin, Aspirin, Diclofenac, Flavichroms
and others compounds were also tested.

The drugs were distributed into the four groups:
Group 1 (presenting scalding edema).
Bufedil, Calalase, SOD, Aspirin, Indomethacin, Diclofenac, Aescin locally
given.
Group 2 (antiedematous, but not improving wound healing)
HR, Aescin i.p., Tolyprin
Group 3 (not antiedematous, but improving wound healing)

Flavichroms, DMSO 60 %

Group 4 (antiedematous and good wound healing)
Etofenamat
BW 755 C

Conclusion

Only the inhibitors of PG and LT formation (dual NSAID's) do work very
well.
As seems possible the arachidonic acid cascade (2,3) is at the onset of
the inflammation burst.

References

1. Arturson G.
   Antiinflammatory drugs and burn edema formation, in:
   Care of the burn wound (ed.: S.R.May, G.Dogo).
   Karger (Basel München Paris) p 49 (1985)

2. Alexander F., Mathieson M., Teoh K.H.T., Huval W.V.,
   Lelcuck S., Valeri C.R., Shepro D., Hechtman H.B.
   Arachidonic acid metabolites mediate early burn edema,
   J. Trauma, 8: 709 (1984)

3. Denzlinger C., Rapp S., Hagmann W., Keppler d.
   Leukotrienes as mediators in tissue trauma, Science,
   230:330 (1985)

INDEX

Lymphocyte proliferation
  in burns, 342
    leukotriene effect on, 459-466,
      469-479
    lipid mediator effect, 384, 390
    platelet-activating factor in,
      448-449, 483-491
    after surgery, 363
Lymphotoxine, 315, 316
Lysosomal enzymes in burn, 88

Macrophage in burn, 7, 11
Mast cell
  degranulation (after burns), 40
Malonaldehyde (following burns), 40,
  89
Mixed lymphocyte reaction and PAF,
  451
Multiple organ failure, 139-149
Myocardial infarction, effect on NK
  cells, 369-371

Natural killer cells
  after acute myocardial infarction,
    369-371
  in burns, 258, 325, 339
  after cortisol infusion, 365-367
  after norepinephrine infusion,
    365-367
  and platelet-activating factor,
    451
  after surgery, 292, 361-364

Neural release mechanisms in injury,
  285-287, 329, 330
Neuropeptides and PAF, 453, 454
Neutrophil activity
  adhesiveness, 418-425
  arachidonic acid metabolism,
    497-499
  in burns, 259
  after cortisol infusion, 366, 367
  after norepinephrine infusion,
    366, 367
  respiratory burst, 507-510
  after surgery, 289, 290, 362

Oxygen free radicals
  definition, 79
  generation, 84
  physiopathological roles, 84
  in shock and trauma, 85
  sources, 81

Pancreatic failure in burns, 233-235
Pentane as a marker of peroxidation,
  97-108
Peroxidation
  in burns, 40, 87-94
  definition, 80

Peroxidation (continued)
  determination (general methodes),
    97, 98
  in endotoxemia, 87-94
  ethane as a marker, 103, 104
  pentane as a marker, 98-108
Phagocytosis impairment
  in burns, 259
  after surgery, 298, 299
Phosphatase (acid) activity
  in burns, 89
  in endotoxin, 92
Plasminogen in burns, 59
Platelet-activating factor
  acute lung injury, 197-203
  anaphylactic shock, 209-215
  and angiotensin II, 132-134
  antagonists, 134, 135, 169-173,
    207, 208, 487-489
  in burn, 14, 15
  cardiac impairment by, 513-516
  cell sources, 447, 448
  circulatory collapse induced by,
    177-185
  generation in sepsis, 167-169
  and immune response, 447-454,
    483-491
  and interleukine-1, 409-414
  with kidney failure, 130-135
  lymphocyte proliferation, 483-491
  and prostaglandin, 132-134
  pulmonary impairments induced by,
    181-183
  in shock, 163-173
  tissue plasminogen activation by,
    244, 245
  transplant rejection, 388, 389, 451
  and vessel wall, 443
Prekallikrein in burn, 59
Prostaglandins
  lymphocyte proliferation, effect,
    388
  production by anaphylactic lung,
    187-195
  in sepsis, 109-121
  and tissue plasminogen activation,
    245
Prostaglandin E
  immunosuppression by, 376, 377
Proteases in burns, 47, 57, 227-235
Pseudonomas exotoxin (in burns), 39
Pyrogens (exogenous), 332

Sepsis
  and biogenic amines, 221-224
  in burns, 31-41
  and eicosanoids, 109-120
  endothelial cells, 429-433
  and fibrinolysis, 246
  immune function in, 305, 306